NEW USES OF
ION ACCELERATORS

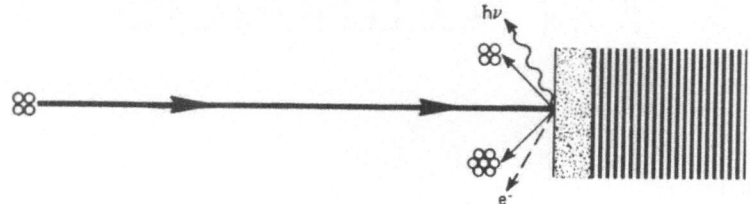

AUTHORS

Thomas A. Cahill
James A. Cairns
Wei-Kan Chu
Billy L. Crowder
Geoffrey Dearnaley
Leonard C. Feldman
Quentin Kessel
James W. Mayer
Otto Meyer
Samuel T. Picraux
Winthrop W. Smith
Eligius A. Wolicki
James F. Ziegler

NEW USES OF ION ACCELERATORS

Edited by
James F. Ziegler
IBM Research
Yorktown Heights, New York, USA

PLENUM PRESS·NEW YORK AND LONDON

Library of Congress Cataloging in Publication Data

Main entry under title:

New uses of ion accelerators.

Includes bibliographical references and index.
1. Ion bombardment. 2. X-rays—Industrial applications. 3. Materials—Analysis. 4. Ion accelerators. I. Ziegler, James F.
QC702.3.N48   539.7'3   75-16315
ISBN-13: 978-1-4684-2171-2   e-ISBN-13: 978-1-4684-2169-9
DOI: 10.1007/978-1-4684-2169-9

PREFACE

The use of ion accelerators for purposes other than nuclear physics research has expanded to the point where 'other uses' are now the most typical. The point has been reached where there are as many ion accelerators in industry, as in universities; and the bulk of new accelerator purchases appears to be for applied purposes.

We mention this as introduction to a tribute to an earlier book: "New Uses of Low Energy Accelerators" (1968). The authors of this book were almost all nuclear physicists. This book addressed itself to new uses other than nuclear research. And in great part because of the widespread seminal influence of this book, many of the new uses discussed became mature fields of research with their own conferences and publications.

We have attempted in this book to both update with topics not included in the first book, and to present in a more tutorial and detailed manner the topics discussed.

This book is in many ways a joint book. All chapters were the result of considerable collaboration between the authors. We hope that, above all, we have written with clarity. We welcome comments and questions from any reader.

James F. Ziegler
IBM-Research

CONTENTS

NEW USES OF
ION ACCELERATORS

ION-EXCITED X-RAY ANALYSIS OF ENVIRONMENTAL SAMPLES

Thomas A. Cahill

Department of Physics, Crocker Nuclear Laboratory, and
Director, Institute of Ecology

University of California, Davis   95616

I. INTRODUCTION

Major advances in the application of accelerators to environmental analysis have occurred in the past four years, largely though not entirely based upon the detection of ion-excited x-rays by energy dispersive Si (Li) and Ge (Li) detectors. While these advances parallel other improvements made in the area of quantitative elemental analysis in the past decades (neutron activation analysis, mass spectroscopy, optical methods), it is important to realize that the information delivered to workers in the biological and physical sciences by these accelerator-based techniques is very different from anything normally available in the past. Ion-excited x-ray analysis (IXA) is capable of delivering to a user quantitative multi-elemental analyses for all elements heavier than about neon to good detection limits, parts per million (ppm), at an extremely low cost per element. For example, recent programs in aerosol analysis have been delivering positive elemental determinations in aerosol samples at $0.20 per element using IXA,[1] while an optical method, atomic absorbtion analysis, costs about $20.00 per element for the same information. This low cost per element is mostly due to the multi-element nature of IXA, in that all elements present above a threshold level are seen, whether they are of interest to the user or not. This fact, common to both IXA and x-ray fluorescent analysis (XRF) when Si (Li) and Ge (Li) detectors are used, is a disadvantage in trace element studies below the 1 ppm level, as will be shown in a later section. However, the broad element coverage is an advantage in many types of studies, and strongly encourages correlation analyses in relative

and absolute elemental abundances, a tool of considerable
importance that can now be economically utilized.

In addition, once the decision has been made to use IXA,
the way is open to doing analyses of very light elements, hydrogen through fluorine, using ion scattering analysis (ISA). Thus,
through IXA and ISA, a target can be "weighed" by adding up the
amounts of all elements present above the threshold of detection.
Other options are available through use of ion beams, including
the ability to effectively analyze extremely small amounts of
sample and to explore the spatial distribution of elements in a
sample, both laterally and in depth.

It will be the major purpose of this chapter to discuss
methods by which IXA can be performed and optimized so as to
match the advantages that it possesses to the needs of actual and
potential users of elemental analysis in the biological and
physical sciences, with particular emphasis on environmental
problems.

II. GENERAL CONSIDERATIONS FOR ION BEAM ANALYSIS OF ENVIRONMENTAL SAMPLES

Any success for accelerator-based analyses of environmental
samples must arise from both the capabilities of accelerators to
perform such analyses and from unmet analytical needs of the
environmental community. Although either subject could be chosen
to initiate the discussion, the former will be chosen, since the
latter is very complicated and demands controversial estimates of
analytical capabilities.

As soon as one wishes to discuss the use of primary ion
beams to perform elemental analyses, a few basic points must be
raised: 1) The primary excitation must reach the element or
isotope under study, and 2) the secondary radiation must exit
from the target and reach a detector.

In these discussions, a target is defined as an environmental sample prepared in some manner so as to be presentable to
a primary ion beam.

The first condition is in reality dominant in any consideration involving ions of only a few MeV, for very little penetration
will occur in a target, even if the beam is allowed to degrade
to zero energy. (Table 1).

It is evident that these limits can be substantially increased by increasing the energy of the primary ion beam or using
the primary ion beam to produce penetrating secondary radiations

Table 1

Penetration of a 4 MeV Proton in a Low Z Matrix[*]

Target Configuration	Thickness
Thin Target (10% energy loss)	5 mg/cm^2
Thick Target (stopped beam)	28 mg/cm^2

*Effective Atomic Number, $Z_{eff} = 9$

such as neutrons or gamma rays.

The second criterion, that the radiation induced by the primary ion beam reaches the detector, also places limits upon the total thickness of the target, which may or may not include a backing material or distributed matrix such as a filter or a binder. (Table 2). It should be stated at this point that only prompt secondary radiation will be considered, although interesting possibilities exist for post-irradiation analytical procedures.

Table 2

Transmission of Secondary Radiation through a Low-Z Sample, 5 mg/cm^2
(approximate)

Secondary Radiation	Energy	Criterion
Photons	3 keV	10% attenuation
Electrons	140 keV	10% energy loss
Protons	4 MeV	10% energy loss
Deuterons	5.5 MeV	10% energy loss
Tritons	6.5 MeV	10% energy loss
^{3}He	14.5 MeV	10% energy loss
^{4}He	16 MeV	10% energy loss
Neutrons	Thermal	10% attenuation, good geometry

In order to illustrate the nature of these limits, if electro-magnetic radiation is to be used as the secondary radiation, characteristic x-rays of elements lighter than potassium will suffer attenuations greater than 10%. Since in this energy region this attenuation will increase roughly as $Z^{-2.5}$, and since the attenuation is highly matrix-dependent, any attempt to quantitatively measure elements such as silicon will demand target thicknesses <u>thinner</u> than that required by limiting primary ion beam energy loss to 10%.

Thus, in summary, the use of low energy ion beams for analytical purposes must squarely face the fact that there will not be much material to analyze due to limited incident ion beam penetration. There are at least two cases in which this limitation becomes advantageous, however:

1. When little material is available for analysis, and
2. when the secondary radiation limits available target thickness.

Thus, if either or both of these two conditions are met, accelerator-based analyses may prove to be highly advantageous.

III. FORMALISM AND OPTIMIZATION

The relationship between the primary ion beam present at a thin target and secondary radiation present at a detector can be written as follows:

$$N_x = N_i \frac{N_o \rho t}{A} \sigma_{\theta, \epsilon} \, d\Omega \, \eta_x \qquad\qquad \text{III. 1}$$

where N_x = the number of x-rays present at a detector,

$\quad N_i$ = the number of primary ions

$\quad N_o$ = Avogadro's number

$\quad \rho$ = target density (gm/cm^3), and

$\quad t$ = target thickness (cm), of the target component emitting the x-rays whose atomic weight is A.

$\sigma_{\theta, \epsilon} = \dfrac{d\sigma}{d\Omega}$ = the cross section in cm^2 per atom for x-ray emission for each transition of element Z, per steradian of solid angle, at angle θ and primary ion energy E.

$\quad \eta_x$ = the detector efficiency at energy E_x

$\quad d\Omega$ = the solid angle in steradians of the detector.

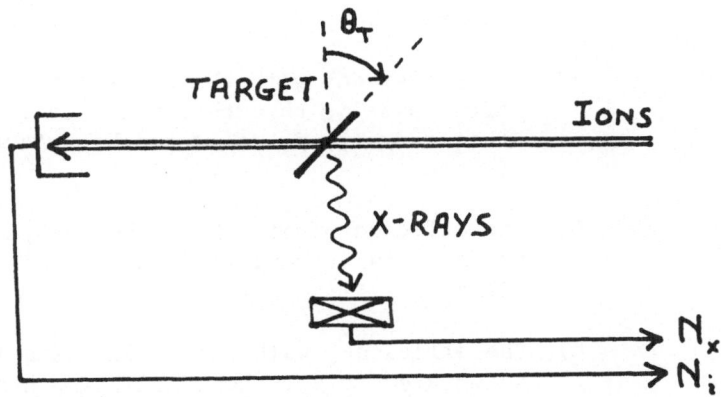

Figure 1.   Schematic of x-ray analysis system.

For an analysis of a target, the desired quantity is $(\rho t)_z$ gms/cm^2, the areal density of element Z.

$$(\rho t)_z = \frac{N_x}{N_i} F_{x,z} \qquad \qquad \text{III. 2}$$

where

$$F_{x,z} = \frac{A_z}{N_o \, \sigma_{\theta,E} \, d\Omega \, \eta_x} \qquad \qquad \text{III. 3}$$

$F_{x,z}$ is a constant for a given element, transition, primary ion beam, detector, and geometry, and can thus be tabulated.

The accuracy with which $(\rho t)_z$ can be found is thus dependent upon the uncertainties in variables such as these:

1. N_x a)   The statistical error in N_x

            b)   x-ray attenuation between target elements and the detector.

            c)   the efficiency with which this event is tallied in some storage device, including dead time effects

        d)   the accuracy with which the peak represent-
ing the characteristic event can be
integrated.

2. N_i a)   the efficiency with which the primary ion,
after passing through the sample (for a
thin target), is collected by a Faraday
device.

        b)   the efficiency with which the charge on the
primary ion is converted into an electrical
signal

        c)   the efficiency with which this signal is
integrated to get the charge, and hence, N_i

3. $F_{x,z}$ a)   the effectiveness of the system in unambiguously determining Z and A

        b)   the accuracy of $\sigma_{\theta,E}$ due to methods used to
fix σ , to finite primary beam energy
degredation, and to geometric effects, etc.

        c)   the accuracy of the solid angle determina-
tion

        d)   the efficiency with which an incoming x-
ray is detected

Equation III.3 turns out to be very useful in everyday
operations, so useful that for our system, which is based on
alpha-induced x-rays, wallet cards have been made up that allow
easy manual calculation of elemental abundances. This is important as if prevents careless mistakes and builds intuitive
strength among operators of the analytical system. An example
of a few such entries is given in Table 3.

For this table, N_i was reduced to the integrated charge, Q,
in microcoulombs, and additional constants were included in F.

Before discussing the details of how the parameters N_x, N_i,
and $F_{x,z}$ are fixed, one should discuss optimization of the
x-ray system. Optimization will mean as many different things as
there are different people who wish to use an analytical system.
In one case, a user may care only for how well one can measure
cobalt in a stainless steel alloy. Another may want a series of
specific elements in organic matrices to high relative precision
but the user may care nothing about absolute errors. Each of
these demands requires a different experimental arrangement.

Table 3

Selected $F_{x,z}$ Factors for the UCD/ARB System

$$(gt) \ \text{ng/cm}^2 = F \ \frac{N_x}{\rho \ (\mu c)}$$

Element	Z	X-ray Energy (K)	F (18 MeV α)
.	.	.	.
.	.	.	.
.	.	.	.
Cr	24	5.411	0.234
Mn	25	5.895	0.290
Fe	26	6.400	0.351
.	.	.	.
.	.	.	.
.	.	.	.

Our system was designed for a primary user, the California Air Resources Board (ARB), that desired quantitative determinations of the widest possible range of elements, hydrogen through uranium, at low cost, in aerosol samples that rarely exceed 300 µg/cm^2 in total areal density. Thus, while the system has been utilized by the two other users described above (and about 30 more), it is not optimized for their problems.

In general, however, one can define two criteria that will be applicable to a wide range of cases:

1. Detection of an areal density, D_x, for element Z, and

2. Detection of a fractional mass, D_f, for an element Z.

These criteria will depend upon the statistical significance of N_x with respect to background events. Although ion-excited x-rays will be chosen to illustrate the optimization procedures, similar techniques such as ion scattering and ion-induced prompt reactions have equivalent formalisms. (See other chapters in this book).

We will assume that the x-rays induced by the incident ion beam are counted in such a way as to provide a spectrum consisting of a given number of x-ray counts for each interval of x-ray energy ΔE_x. In order for the characteristic radiation of an element to be detectable, the number of counts in an interval close to the characteristic energy, the peak counts, must be greater than the number of counts present in the area due to background generating processes in a statistically meaningful sense. That is, one must see the peak

above the background. The standard definition of statistical
significance in such cases, defined by the International Union
of Pure and Applied Physics (IUPAP) is that the number of counts
in the peak must be three times greater than the uncertainty in
the background under the peak taken over the same interval as
the peak. Actually, this definition is not correct for situations
in which a small number of counts are involved. If we define the
number of counts in a peak as N_x, and the number of counts in the
background under the peak as N_{Bx}, then the IUPAP definition is:

$$N_x \gtrsim 3\sqrt{N_{Bx}} \qquad\qquad\qquad\qquad III. 4$$

The correct definition, according to the student test, is:

$$N_x \gtrsim 2 + 2\sqrt{2N_{Bx} + 1} \qquad\qquad\qquad III. 5$$

These definitions are close to each other for background counts
greater than 10.

In order to apply these tests to thin targets, we will define
a situation in which one has a backing, assumed to possess no
elements whose characteristic radiation is seen by the detector,
and a deposit which contains the elements under study (Figure 2).
The areal density of the backing is defined as $(\rho t)_B$, while the
areal density of the deposit under study is $(\rho t)_s$. These materials
will create a background spectrum that is dominated by two contributions:[2]

1. Bremsstrahlung created by the slowing down of free or
lightly bound electrons in the target (backing plus deposit)
scattered by the primary ion beam. Since the maximum velocity of
the knock-on electrons is about twice the velocity of the primary
ion beam in a non-relativistic approximation, the bremsstrahlung
background will cut off at about

$$E_x (keV) \simeq \frac{4E_o (MeV)}{m_I} \qquad (m_I = \text{the mass of the incident ion in amu}) \qquad III. 6$$

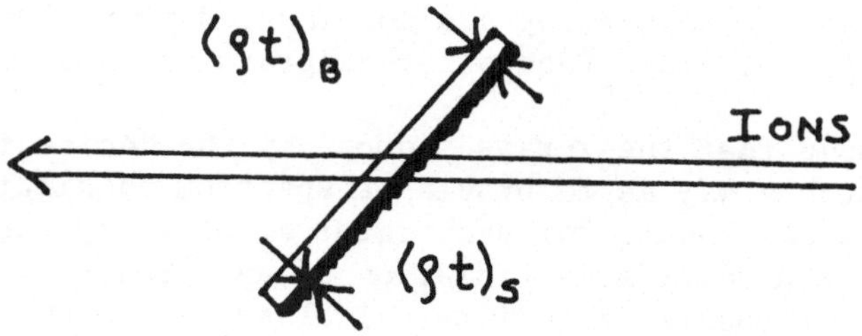

Figure 2.   Schematic of a Backed Target

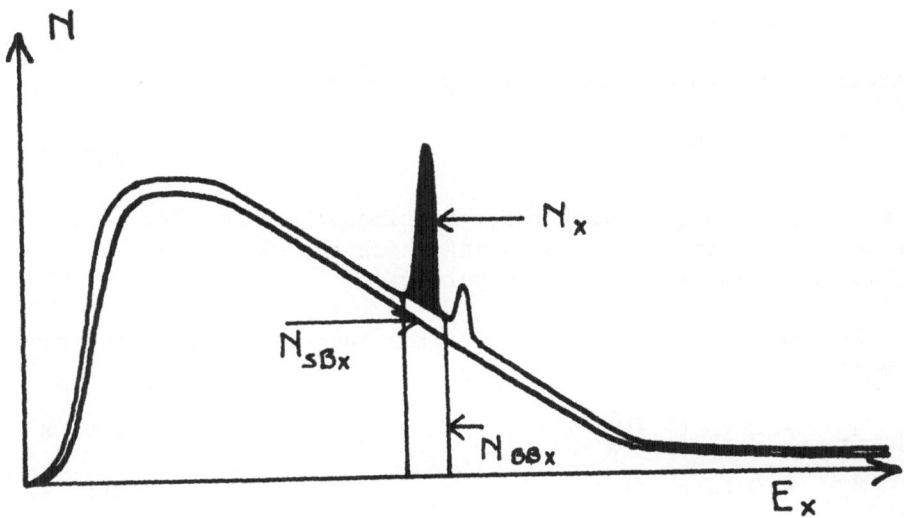

Figure 3. Ion-excited X-Ray Spectrum.

2. Compton events from x-rays and gamma rays, bound electron bremmstrahlung, and other effects. The spectrum one could obtain from such a target system might appear as in Figure 3.

Thus, N_x = number of characteristic x-rays in ΔE_x

N_{sBx} = number of background x-rays in ΔE_x generated by the presence of the sample deposit.

N_{BBx} = number of background x-rays in ΔE_x generated by the presence of the backing

Where ρt_B = areal density of the backing

and ρt_s = areal density of the sample deposit.

Further, we can define

$\sum N_{BBx} = N_B$ = the total number of background x-rays due to the backing

$\sum N_{sBx} = N_{sB}$ = the total number of background x-rays due to the deposit

$\sum N_x$ = the total number of characteristic x-rays due to the deposit

Therefore, the total number of counts in a spectrum is

$$N_T = N_B + N_{sB} + \sum N_x \qquad\qquad \text{III. 7}$$

This is equal to the count rate, R, times the live time, T_L, defined as the total period during which the electronic acquisition system can receive pulses.

We can express all values for the number of x-rays in terms of equation II.2, so that

$$N_x = Q\ (\rho\ t)_x/F_x \qquad\qquad \text{III. 8}$$

$$N_{sBx} = Q\ (\rho\ t)_s/F_{sBx} \qquad\qquad \text{III. 9}$$

$$N_{BBx} = Q\ (\rho\ t)_B/F_{BBx} \qquad\qquad \text{III. 10}$$

These simplified relationships assume spectra that allow resolution of neighboring contributions from various elements, and in realistic samples, these limits may not be possible in all cases. Nevertheless, we can write the criterion for the detectable limit of an element whose characteristic radiation occurs at E_x as:

$$N_x \geqslant 3\sqrt{N_{BBx} + N_{sBx}} \qquad\qquad \text{III. 11}$$

then,

$$\text{III. 12}$$

$$D_{x,min.} \equiv (\rho t)_{x,min.} \geqslant \frac{3F_x}{Q}\sqrt{N_{BBx} + N_{sBx}}$$

This can be written as

$$\text{III. 13}$$

$$D_{x,min.} \geqslant \frac{3F_x}{Q}\sqrt{Q\ (\rho t)_B/F_{BBx} + Q\ (\rho t)_s/F_{sBx}}$$

The continuous background from most samples is similar in shape and magnitude to that produced by low-Z backings, since the gross compositions and the background causes are similar. Thus,

$$F_{BBx} \simeq F_{sBx} \equiv F_{Bx} \qquad\qquad \text{III. 14}$$

Under this assumption,

$$D_{x,\ min.} \geqslant \frac{3}{\sqrt{Q}}\sqrt{\frac{F_x^2}{F_{Bx}}}\ \left\{(\rho t)_B + (\rho t)_s\right\}^{1/2} \qquad \text{III. 15}$$

This form illustrates many of the strategies that can be used to improve (i.e. lower) $D_{x, min}$.

One can obviously improve D_x by increasing Q. This can be done by running at higher beam currents, but limitations are usually set either by target damage or by the maximum count rate that can be achieved. Target damage can be decreased, but not avoided, by increasing the target irradiation area and by arranging to have the beam cover this area uniformly. Damage can also be decreased by increasing the solid angle $d\Omega$ so that, for a fixed count rate R, the beam current is minimized. Since the count rate is proportional to the beam current, one improves D_x as the count rate capabilities of the system improve.

While there are both obvious and subtle ways to increase R electronically, basic limitations are set by the pulse integration time which, for a high resolution, broad range system, limit performance to about 20,000 counts per second, time average rate. Much can be accomplished, however, through use of filters that modify the x-ray spectrum before it reaches the detector. These prove to be especially useful for ion-excited x-rays, since most of the count rate lies at low energies and can be easily removed by a simple low-Z filter, if the element under study has a reasonably energetic x-ray. This is not the case in x-ray fluorescence, where most of the count rate is at the high end of the spectrum.

One can also increase Q by increasing the irradiation time, T. Limitations on T are often economic in nature, and since the improvement in D_x goes as $T^{-1/2}$ while the cost of an accelerator is usually linear in time, this route for improvement has limited applicability.

One can effectively manipulate the ratio (F_x^2 / F_{Bx}) by adjusting the primary ion velocity and, to a much lesser extent, its charge[2,3]. This ratio is essentially:

$$(\sigma_{Bx} / \sigma_x^2 d\Omega).$$ III. 16

The importance of maximizing $d\Omega$ is shown by this equation, although limits are set by the area of the detector for an adequate energy resolution, geometrical considerations involving finite target size, and the limitation upon the count rate, R, that can be accommodated.

The optimization of σ_{Bx} / σ_x^2 depends critically upon the element chosen, as σ_x varies strongly with the atomic number of the element, Z, and the velocity and (to a much smaller extent) the charge of the primary ion beam. For a given Z, the optimum value for the velocity of the primary ion will be constrained at both the limit of low velocities and high velocities. At very low velocities, σ_x falls much more rapidly with decreasing energy than σ_{Bx}, and the detectable limit climbs. At very high velocities, the rapid rise of σ_x with energy slows and finally stops, while σ_{Bx} rises steadily. Thus, again, overall performance degrades. A consensus appears to be forming that the optimum velocity for the primary ion beam lies near 0.1 c, or somewhere between one and 5 MeV/amu, for a broad range of elements, if one uses L transitions for heavy elements.

One can also manipulate the ratio F_x^2/F_{Bx} by reducing the effective integration width E in which N_x is determined. A reduction of ΔE_x by energy dispersive $Si(Li)$ or $Ge(Li)$ detectors to wavelength dispersive systems, results in an improvement of $\sqrt{20}$ for equal counts in N_x. However, collection efficiency may drop by a factor of 100, so that it may be difficult to obtain equivalent N_x.

Finally, although F_x is essentially isotropic, F_{Bx} is not, being forward peaked. Thus, angles forward of about 90° may not be desireable, and detailed investigation of the angular dependance of F_{Bx} may bring some improvement.

One can manipulate $[(\rho t)_B + (\rho t)_s]^{1/2}$ in several ways, but the maximum improvement occurs for $(\rho t)_B = 0$. Further reduction in $(\rho t)_s$ is in some ways illusory, for, if we define the fractional detection limit

$$D_f \equiv \frac{(\rho t)_{x,\ min.}}{(\rho t)_s} \propto \frac{[(\rho t)_B + (\rho t)_s]^{1/2}}{(\rho t)_s} \qquad \qquad III.\ 17$$

Then this $\rightarrow \infty$ as $(\rho t)_s \rightarrow 0$. Another way of saying this is that a measurement of $(\rho t)_{x,min.}$ at 10^{-12} gms/cm^2 in a total sample, $(\rho t)_s$, of 10^{-6} gms/cm^2 is only one part per million (ppm).

The theoretical basis for optimization can obviously be expanded in greater detail, but since the premises upon which each individual system must be measured are highly variable, not much can be achieved. The next section will discuss optimization of one particular system--that developed at Davis primarily for analysis of atmospheric contaminants.

Table 4

Variation of Minimum Detactable Limit, $(\rho t)_{x, \, min}$

A. Detector   Variation

    1. Solid Angle $(d\Omega)$ $(d\Omega)^{-1/2}$

    2. Energy Resolution (ΔE) $(\Delta E)^{1/2}$

    3. Count Rate (R) $(R)^{-1/2}$

B. Target

    1. Backing Thickness $(\rho t)_B$ $[(\rho t)_B + (\rho t)_s]^{1/2}$

    2. Characteristic Cross
       Section (σ_x) $(\sigma_x)^{-1}$

    3. Backing Cross Section (σ_{Bx}) $(\sigma_{Bx})^{1/2}$

C. Ion Beam

    1. Ion Velocity   Optimum at ~ 0.1 c

    2. Ion Charge (Z_i) $(Z_i)^{-1/2}$

    3. Ion Integrated Current (Q) $(Q)^{-1/2}$

D. General

    1. Time, T $(T)^{-1/2}$

IV. THE UCD/ARB AEROSOL ANALYSIS SYSTEM

As was shown in the first section, the design of an analytical system will depend critically upon the needs that it is meant to fulfill. Thus, description of a "general" system serves little purpose, since few if any systems will be built along these lines.

The specific system designed and built at the Crocker Nuclear Laboratory of the University of California, Davis, (UCD), was in response to the needs of the California Air Resources Board (ARB) for analysis of atmospheric aerosols.[4] Analytical needs were not being adequately met due to severe constraints imposed by the nature of the problem and by existing air sampling instrumentation. Some of the most important were:

1. Information on atmospheric particle size was vital for
air quality studies. However, existing particle sizing devices
generally delivered very small amounts of material for subsequent
chemical analysis. Therefore, ability to effectively analyze
small amounts of material was required. In addition, the ability
to perform these analyses using common filter media or impactor
substrates was highly desireable.

2. Information was desired on as many elements as possible,
but certainly those elements that
         a) dominated the total aerosol mass, or
         b) were known to be toxic.
As is shown in Table 5, this demands an ability to analyze the
very light elements, hydrogen through fluorine, that constitute
about 65% of the mass of an average urban aerosol.[5]

3. Due to variations in meteorological parameters, a large
number of analyses were often required in order to obtain a
statistically meaningful result from a study. Another way to
put it is that, for a fixed amount of funds, one often preferred
100 analyses at a medium sensitivity to 10 analyses at a sensitivity an order of magnitude better. Thus, costs were a significant,
perhaps dominant, factor.

4. A final criterion of paramount importance was that the
system had to be quantitative and accurate.

Many other objectives were able to be achieved to a greater
or lesser degree in the program, but without the four primary
criteria, they would have been of little value. As the system
developed, it seemed that these criteria were all achieved at
the expense of the sensitivity or minimum detectable limit that
could be routinely attained. The nature of these choices will be
the topic of this section. The basic requirements of the system
will be discussed in the next section.

The system that will now be described was not arrived upon
without much trial and error. It is in a continuous process of
upgrading, and the status as described is as of about January 1,
1975. Its development has been a group effort, with major contributions from numerous individuals, faculty, staff, and student
(see Acknowledgements and Bibliography).

Table 5

Prevalence of Elements in an Average Urban Aerosol [5]

Element Constituents	Prevalence	
	$\mu g/m^3$	%
Very Light Elements (H→F)	68	65%
Light Elements (Na→Cl)	22.6	22%
Medium Elements (K→Ba)	12.7	12%
Rare Earths (La→Lu)	0.1	0.1%
Heavy Elements (Hf →U)	1.5	1.4%
	105	100%

A.   The Primary Ion Beam

The accelerator upon which the ion beam is performed is a 76" (190 cm) isochronous cyclotron possessing the following characteristics (Table 6).

Table 6

CNL Cyclotron Beams

a) Primary Ions and Energies

protons	2.5 MeV to 65 MeV
deuterons	12 MeV to 45 MeV
^{3}He	30 MeV to 90 MeV
^{4}He (alphas)	16 MeV to 85 MeV
heavy ions	(largely undevelopped)

b) Beam Currents   variable to 40 µA (external)

c) Microstructure of the beam in time

For ^{4}He, 18 MeV   2 nsec on, 170 nsec off

d) Macrostructure of the beam in time   None

e) Cost of Beams $1.50/minute

Based upon early studies and work in other laboratories, an ion energy of 18 MeV for alpha particles was chosen (v ≈ 0.1 c). This choice was made for several reasons:[6]

a) It was routinely obtainable from the accelerator.

b) It allowed in principal for simultaneous or sequential analysis of samples for very light elements (H➔F) by elastic alpha forward scattering as well as light elements (Na➔Cl), medium elements (K➔Ba), Rare earths (La➔Hf), and heavy elements (Ta➔U) by ion excited x-rays.

c) Alphas appeared to be equivalent to protons at equal velocity for IXA. (Figures 4 and 5, and Table 7).

d) It appeared to lie in an acceptable velocity for optimal sensitivity based upon work done at many laboratories. (See Figure 6 for a comparison between 16 MeV alphas and 30 MeV alphas.)

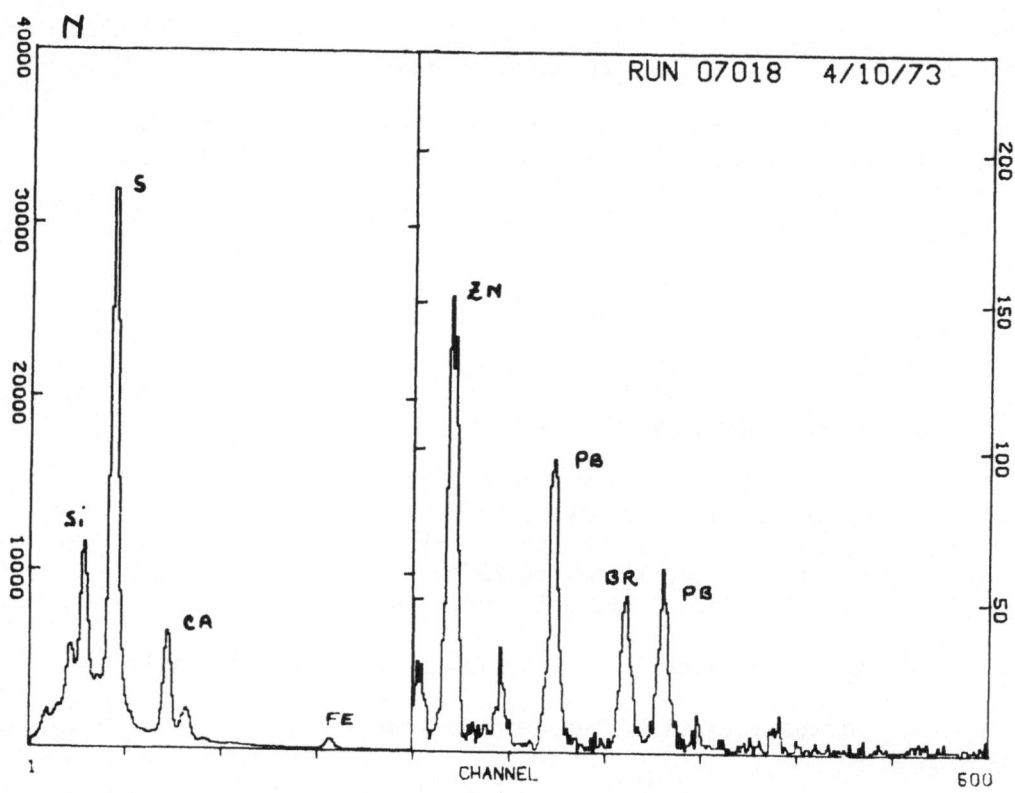

Figure 4.   Spectrum of 10 μc of 4 MeV Protons on an Average
             Aerosol Sample

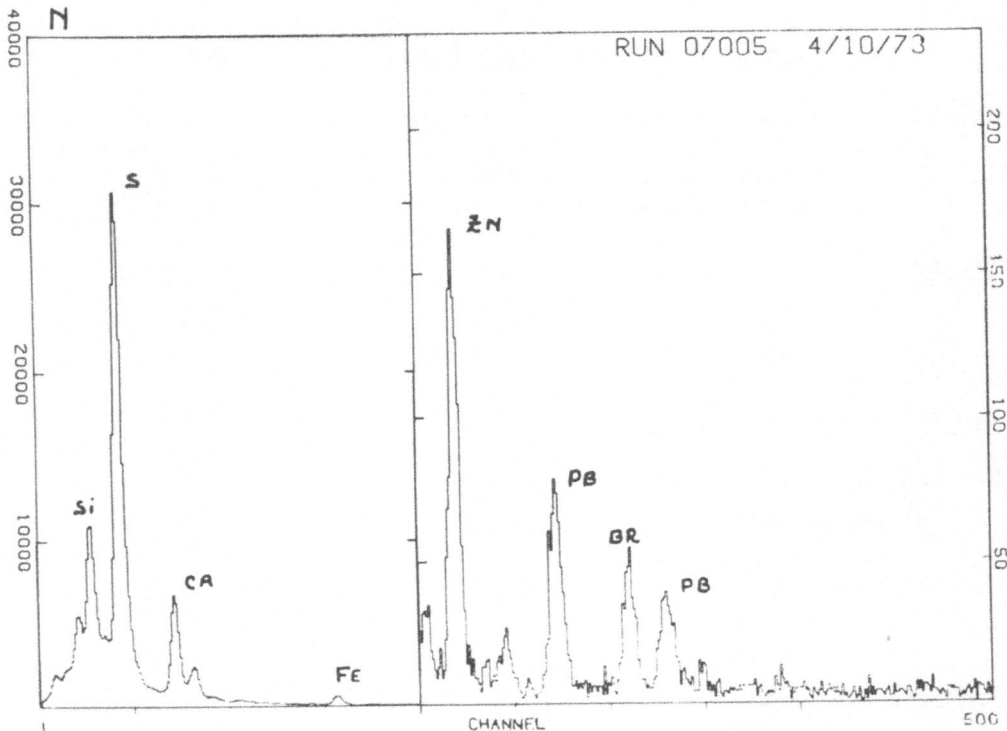

Figure 5.   Spectrum of 5 μc of 16 MeV Alpha Particles on an
             Average Urban Aerosol Sample.

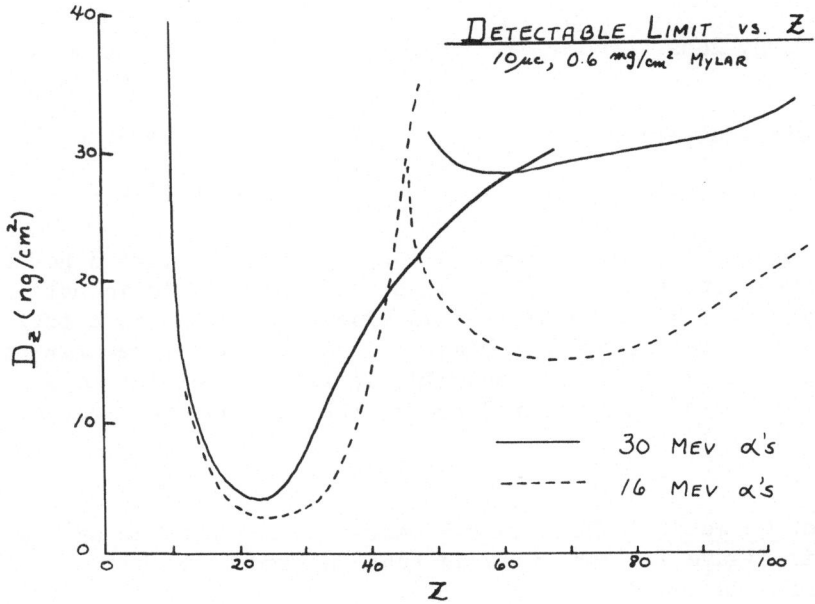

Figure 6.   Comparison of Minimum Detectable Limits, 30 MeV [4]He
             and 16 MeV [4]He beams.

Table 7

Comparison of Proton and Alpha Beams (Approximate)

Assume: Each beam has 4 MeV/amu, and that one uses 10 μc of protons and 5 μc of alphas. Values are relative to proton values, Z_i being the incident charge and M_i being the incident mass.

Factor	Protons	Alphas	Alphas
			numbers for 0.00015" mylar
1. X-ray Generation, N_x	1	1	–
2. Background, N_b			
a) Bremsstrahlung	1	1	–
b) Compton	1	somewhat greater than 1	–
3. Detectability	1	1	–
4. Target Damage			
a) Energy loss rate (per particle)	1	$4 \ (Z_i^2)$	200 keV
b) Energy loss	1	1	20 mw at 0.1 μA
5. Multiple Scattering	1	$0.5 \ (Z_i/M_i)$	0.3°
6. Kinematic Loss-Scattering	1	$4 \ (M_i)$	–

Summary: 10 μc of 4 MeV protons is roughly equivalent to 5 μc of 16 MeV alpha particles. The only major differences lie in multiple scattering (less for alpha beams) and kinematical loss in elastic scattering (greater for alpha beams, and hence, easier separation of light elements). Similar reasoning applies to heavier ions for integrated beam currents reduced by $1/Z_i$ to that for protons.

Target Geometry. The target geomatry chosen for the system is shown in Figure 7. The criteria that entered into this configuration included:

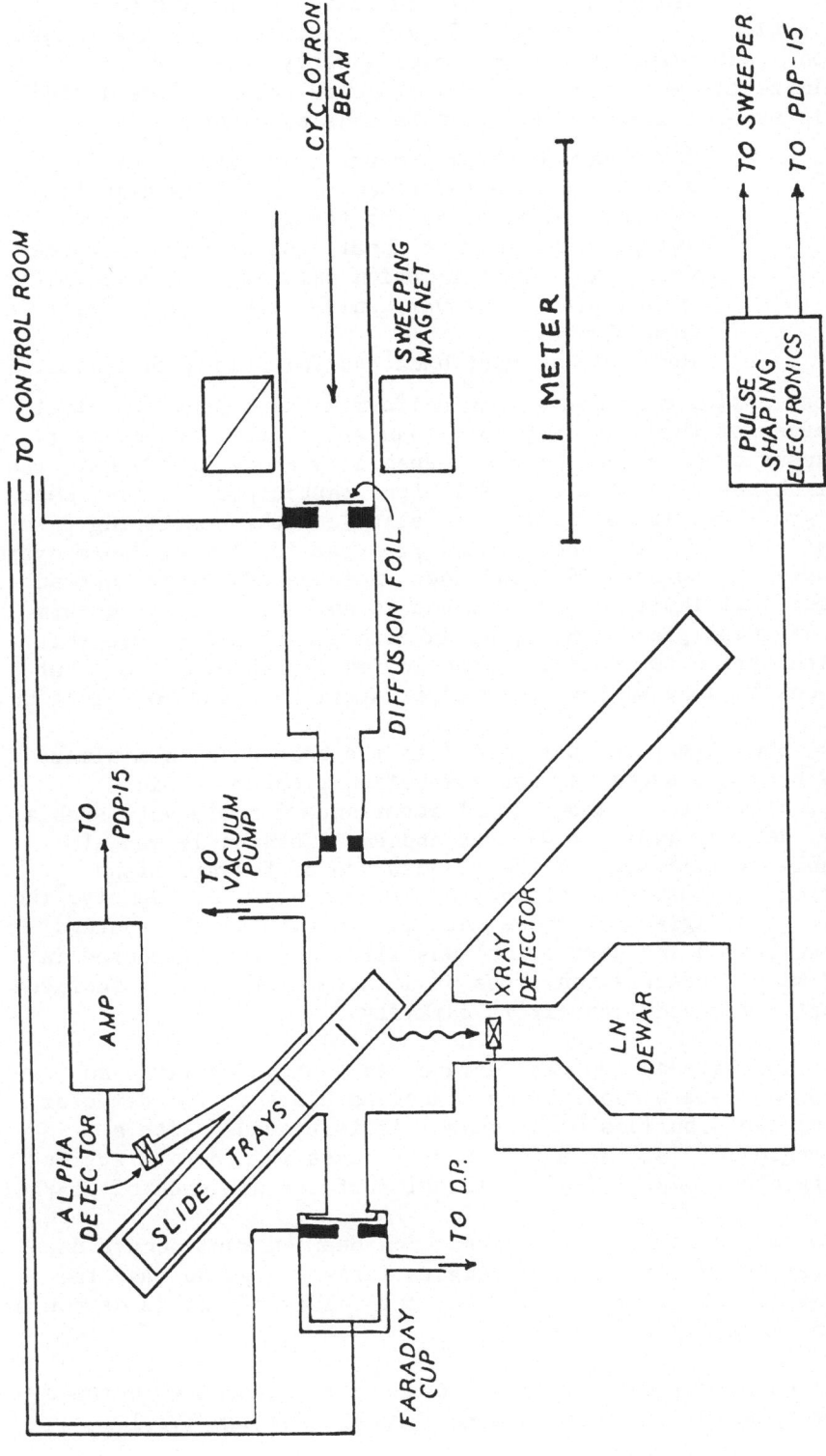

Figure 7.   UCD-ARB Aerosol Analysis System.

1. A uniform beam (to $\pm$ 10%) be delivered to any target
configuration up to one inch (2.5 cm) in diameter at the target
location. The value of 2.5 cm was originally chosen in order
to minimize target damage effects, but soon became closely tied
with air sampler design. This can be done by either

   a) diffusing the beam through a few mg/cm^2 of Al
       well before the target and then collimating to
       the desired shape at the target, or
   b) Sweeping the beam at about 1 hz across the target
       face. This latter option only insured uniformity
       in one plane, however, unless the beam is swept in
       both planes.
   c) Moving the target back and forth in both planes.

The first option provides a beam uniform in both planes, but at
the expense of throwing away more than 2/3 of the beam current.
This causes unwanted background, especially above E_x = 9 keV, the
bremsstrahlung cut-off for 18 MeV alpha particles. It also degraded energy resolution so badly that elastic alpha scattering was
not possible.   The second option produces far better "beam hygiene"
with losses at perhaps 20%, and does not degrade energy spread.
It is more difficult to tune up, and is not necessarily uniform
in the un-swept plane. Sweeping in both planes could cure this.
The third option is routinely done in ion implantation work but
it can not be done without extensive target handling modification.

2. The target was set at 45° to the beam axis to minimize
primary beam and secondary ion absorbtion problems. Since
these problems are of about equal importance for elements such as
calcium, an exit angle of 45° was chosen. This angle was also
reasonably compatible with the planned use of forward alpha
scattering for detection of very light elements. Incidently, the
$\sqrt{2}$ that this angle introduces into all results of the system
(both excitation and absorption) was all too often neglected in
hurried manual calculation. Use of a table of $F_{x,z}$ that included
this factor reduced such errors markedly.

3. The Faraday cup was designed to subtend as large an
angle as possible without being placed so close to the detector
as to provide a serious background. It is provided with a
static magnetic field of a few hundred gauss in order to supress
secondary electrons, and all internal surfaces are made of graphite.

However, some loss of incident ion beam current occurs due
to scattering in the target at angles large enough so that the
beam ions do not strike the Faraday cup. This effect is of course
vital to ISA analysis.

The calculation of particle loss due to outscatter between
the target and Faraday cup is made through Equation IV. 1.[7]

$$y_{rms} \simeq 7.5 \frac{Z_i L_F}{E_0} \left\{ \frac{(\rho t)_B + (\rho t)_s}{L_{RAD}} \right\}^{1/2} (1 + E)$$

where y_{rms} is the rsm displacement of a multiply-scattered ion beam of charge Z_i and energy E_0 seen at a distance L_F downstream of a target of areal density $[(\rho t)_B + (\rho t)_s]$. L_{RAD} for a target of $Z_{T, eff} = 6$ is 42.4 gms/cm^2, while $(1 + E)$ is less than or approximately equal to 2.

The thickest target that falls within the system specifications of the UCD/ARB system has $\Delta E \simeq 20\%$, which corresponds to $(\rho t)_B + (\rho t)_s$ of about 10 mg/cm^2. Whatman 41 filter paper which is normally about 9 mg/cm^2 is an example of such a target. In this situation, $y_{rms} \simeq 2.5$ cm for $L_F = 100$ cm. Thus, the 10 cm diameter Faraday cup is subject to corrections of a few percent. These can be checked by noting beam reduction when a thick target replaces an open hole at the target location.

4. The vacuum at the Faraday cup was separated from the vacuum in the target changer by a Havar foil. It is essential to maintain an excellent vacuum in the Faraday cup, and a baffled oil diffusion pump was used for this purpose. Even with the presence of a several hundred gauss magnetic field, an error can be introduced of approximately 1% per micron pressure at the Faraday cup. (This bad vacuum is explained below.)

5. Care was taken to insure that the primary ion beam impinged only upon graphite surfaces in order to avoid as much as possible fluorescence problems caused by energetic characteristic x-rays from metals.

The choice of a target handling mechanism was constrained by the need to mount, handle, and store large numbers of samples for the aerosol analysis programs of the California Air Resources Board. When calculations of target damage and beam uniformity indicated optimum irradiation areas of a few centimeters squared, 35 mm slide frames were chosen as the target mounts. This allowed maximum use of existing slide-handling technology. It also allowed "customers" to mount their own slides. A commercial slide changer was used to insert the slide mounts into the beam. Linear trays with 36 positions are loaded three at a time so that, allowing for blanks and standards, up to 102 targets can be handled at a time. The device is maintained at a roughing pump vacuum. Vacuums much better than a few microns are hard to achieve when one has loaded 102 filthy pieces of blotter paper (Whatman 41) into the target changer, regardless of pumping speed. However, the situation is greatly ameliorated when plastic or metal slide frames are used and surface deposition filters or impaction foils are the targets. The device operates under computer control, and considerable logic is included to guarantee that the target is properly positioned as indicated by light emitting diodes.

B. Detection of X-Rays

The Detector. The constraints upon the choice of the x-ray detector allowed two alternatives for energy dispersive systems-- Si (Li) and Ge (Li). The relative efficiencies of each detector are shown in Figure 8.[8] The choice made for the UCD/ARB system was based upon in-beam evaluation tests that compared various options. It was found that the only major difference between Si (Li) and Ge (Li) systems involved the additional sensitivity of the Ge (Li) systems at x-ray energies above about 40 keV. This sensitivity, when coupled with the presence of low energy gamma rays associated with the energetic ion beam, provided a useless count rate, since the cross section for characteristic K x-rays from rare earths and heavy elements dropped so low as to be virtually useless for analytical purposes.

The L x-rays of these heavy elements were used in the analysis as the cross sections are larger by several orders of magnitude than the K x-rays. The question as to the energy at which the system uses K x-rays, for elements Na to I (1.04 keV $\leq$ $E_x \leq$ 28.5 keV), while L x-rays are used for the elements Cs and heavier (4.3 keV $\leq E_x \leq$ 13.6 keV), the choice of the transition energy, was set by the presence of calcium as a major constituent of many typical samples. Thus, the L x-rays of heavier elements had to lie safely above the K_β transition of calcium, which lies at about 4 keV. Also, the L transitions had to be resolvable in order to allow unambiguous determinations of infrequently observed rare earths and heavier elements.

This cross-over choice has a major disadvantage, however. Sensitivities in the region between Mo (K_α = 17.44) and I (K_α = 28.51 keV) are poor, as much as 10 to 100 times poorer than most lighter elements. Unfortunately the dozen elements in this region include some rather important elements, such as cadmium, tin, iodine, and barium. Attempts to rectify this problem will be discussed in the section on Filtering.

Detector Placement. The detector chosen for the system was a 10 mm^2 Si (Li) detector 3 mm thick, cooled to liquid nitrogen temperatures and directly coupled to a pulsed optical charge sensitive preamplifier.   The window is 0.0005" Be (1.27 X 10^{-3} cm), which allows reasonable efficiencies for x-rays as low as 1 keV (sodium). The detector subtends a solid angle of 7 X 10^{-4} sr, a value that is surely smaller than is optimum. This will be decreased by about a factor of 10 in the next scheduled rebuilding of the system. The detector possessed a resolution of 165 ev

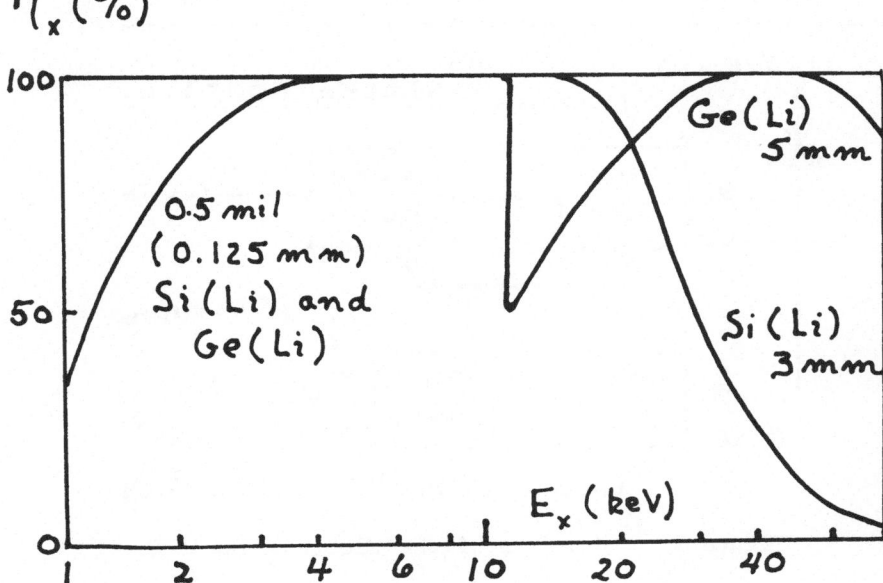

$\eta_x (\%)$

Figure 8.   Schematic Efficiencies for Si (Li) and Ge (Li)
            Detectors.

at 6.4 keV in beam-off, low count rate conditions.

     Electronics. The electronics originally supplied with the
system were the standard pulsed optical system supplied by the
manufacturer, including pulse pileup and deadtime correction
circuitry.

     Originally, only the detector and preamplifier were located
at the site of analysis, 250 feet away from the counting room.
Noise pick-up on the shielded lines both degraded resolution and
limited the detection of x-rays from Na, Mg, and Al. This was
easily corrected by placing all the electronics up to but not
including the ADC's in the cave area next to the sample and
detector. A remote coarse gain changer was supplied by the
manufacturer, which aided operations. One problem, however, was
that the electronics could not be tuned up in a realistic beam-
on condition, since the area is slightly radioactive when the beam
is on. Sources are not adequate, since it is hard to get high
count rate sources in the critical Na → Cl region and the background
noise level is worse in a beam-on condition. Experience has allowed us to develop procedures for setting up the electronics before a run. Checking their operating characteristics during a

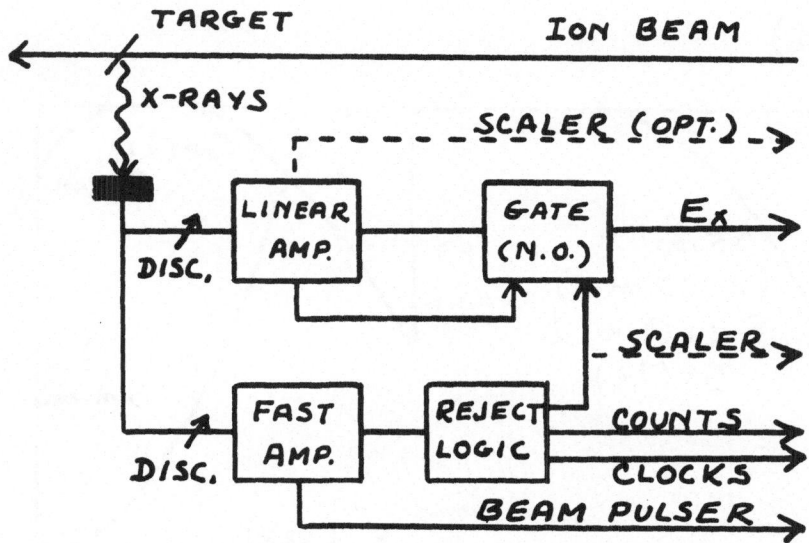

Figure 9.   Schematic of the Electronics.

run is accomplished by sending almost every available logic pulse
to the counting room, where they are scaled. Thus, one can easily
verify proper operation of the system. Considerable effort was
expended in order to obtain these logic signals from the commercial
system, since no provision had been made to obtain many of them.
In this configuration, an effective lower sensitivity limit of
about 700 ev. could be achieved in operating conditions, and resolution was about 185 ev. at E_x = 6.4 keV and count rates below about
2000 c/sec.

   Count Rate and Dead Time Correction. The system as it existed
still had a major fault, one that proved very difficult to correct.
The signal arriving in the electronics from the preamplifier was
processed by two parallel chains of electronics, the slow linear
amplifier and the fast logic circuits (used in pile-up rejection).
Since the bandpass of the fast amplifier had to be greater (since
it was "fast") than that of the slower linear amplifier, the noise
level was higher in it than in the linear amplifier. A discriminator

could be set at the fast amplifier just above its noise level. There was an internal discriminator on the linear amplifier that rides on top of the D.C. baseline offset. In normal operation, it would be set at a noise level considerably lower than that of the fast amplifier. Since all logic operations were based upon the fast amplifier, there was a considerable region of energy E_x, perhaps 2 keV wide in which a pulse could enter the linear system without triggering the logic operations.

To be clear, we used two electronic paths—one for linear analysis to determine the x-ray energy, and one logic signal analysis to determine pile-up. It is very difficult to set both of these at the <u>same identical threshold level</u> above the background noise. This had several effects:

1. The count rate meter was erroneous, since it was based upon the output of the fast leg of the system. It could be in error, for certain samples, by a factor of 5 to 10. This could be detected by running for a fixed interval, say, 100 seconds, at a count rate of 1000 c/sec., and then looking at the total number of counts accepted by the computer. It could be 500,000 counts or more. This error was important, since normally the system is run at an optimum count rate, and to push beyond that point is to incur penalties in resolution and dead time correction.

2. The dead time correction circuits were useless. Errors of an order of magnitude could arise easily, although a factor of 3 was more common on realistic samples.

Please note that, since it is difficult to obtain intense sources in the Na→Cl region, evaluating such effects on the bench is difficult. Also, the efficiency of IXA in this region, and its characteristic background, also peaked in this region, make the problem much worse for IXA than for XRF.

First, all significant logic pulses were obtained from the system and sent to the counting room where they were counted. The problem was then immediately evident in detail, and corrective measures could be taken. A pulser input was added to the system in the cave after the preamplifier to allow the two discriminators to be set at the same effective E_x. This cost us Na, Mg, and Al, since these were hidden in the noise of the logic circuit. A new, lower noise fast amplifier with a longer clip time constant was provided by the manufacturer, which allowed the system to perform logic operations on Na, Mg, and Al. However, due to the longer clipping time constant in the fast amp, pile-up was detected which made necessary the addition of a base line restorer to the fast amplifier which the manufacturer now supplies as standard. This program has resulted in a system that can now be trusted, providing

that normal procedures for set-up and operation are followed.
however, the system can not operate at as high a count rate as
before since the fast amp was compromised in its pulse pair
resolutions versus noise merit in order to extend dead time
correction and reject capability into the low energy range. Dead
times tend to be higher at the same effective count rates than if
a filter is used to eliminate low energy x-rays or the sample is
such as to have most x-rays well above 3 keV or so.

The whole problem was further improved by the application of
on-demand beam pulsing.[9] The problem of dead time correction can
be minimized or eliminated if, whenever the detector accepts a
valid x-ray, the exciting source is turned off until the system
is ready to accept another pulse. The schematic of the system is
shown in Figure 10. A set of parallel plates is placed on each
side of the ion beam well before the sample. One of these plates
noramlly is grounded, while the other carries +4000 volts. The
bend introduced by the field is cancelled by a bending magnet, and
the beam normally then passes through a set of slits before
striking the sample. Once a pulse is detected, the electronics
sends a logic pulse to the sweeper, which crow bars (shorts to
ground) the high voltage, reducing the voltage to about 400 volts

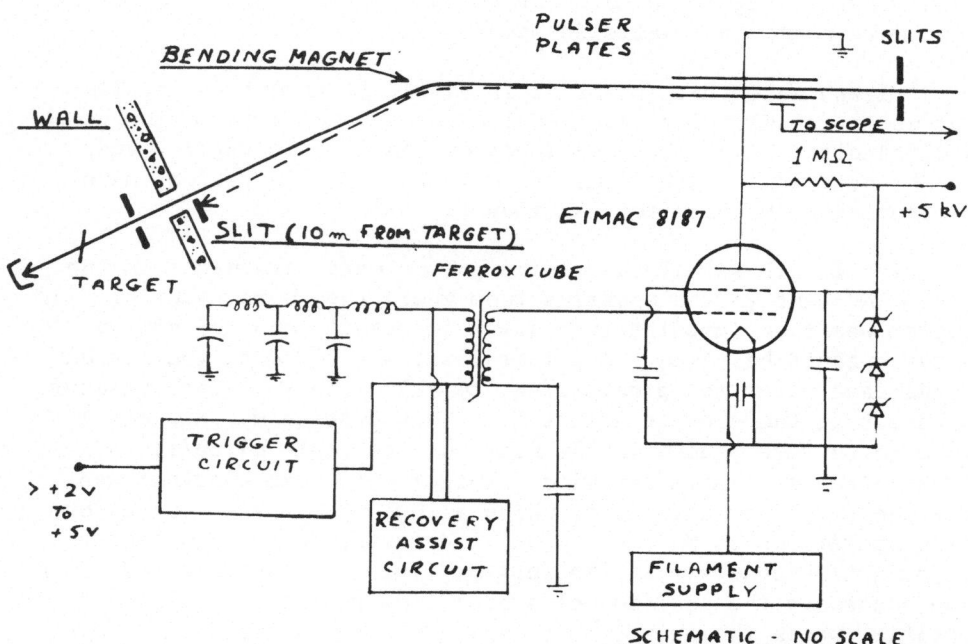

Figure 10.   Schematic of the On-Demand Beam Pulsing System.

in about 150 nsec. The beam now is bent so as to strike the set of slits, and the beam current at the sample is zero. The voltage is held off for about 40 μsec, by which time the system has processed the pulse, and then the beam is re-established. An example of the operation of the system is shown in Figure 11.

The usefulness of on-demand beam pulsing was not fully evident at first, although the ability to run at 10,000 counts/second with dead time correction better than 5% was much appreciated. The first point to note is that the microstructure of the 18 MeV alpha beam (2 nsec on, 170 nsec off) and the area geometry allows only one additional pulse to strike the detector after the system is triggered off by a previous pulse. If the count rate is 10,000 cts/sec, the probability of an x-ray of any type, background or characteristic, triggering the detector during a single 2 nsec beam pulse is only about 0.2%. Thus, this second pulse has a small probability of creating a pile-up event, and the existing dead time correction and pulse pile-up rejection circuits have little to do. The second point to note is that the Faraday cup is now an accurate measure of beam on target with the electronics in a receptive mode. Thus, one can use preset charge to control the run duration, making it fairly simple to control a set detectable limit.

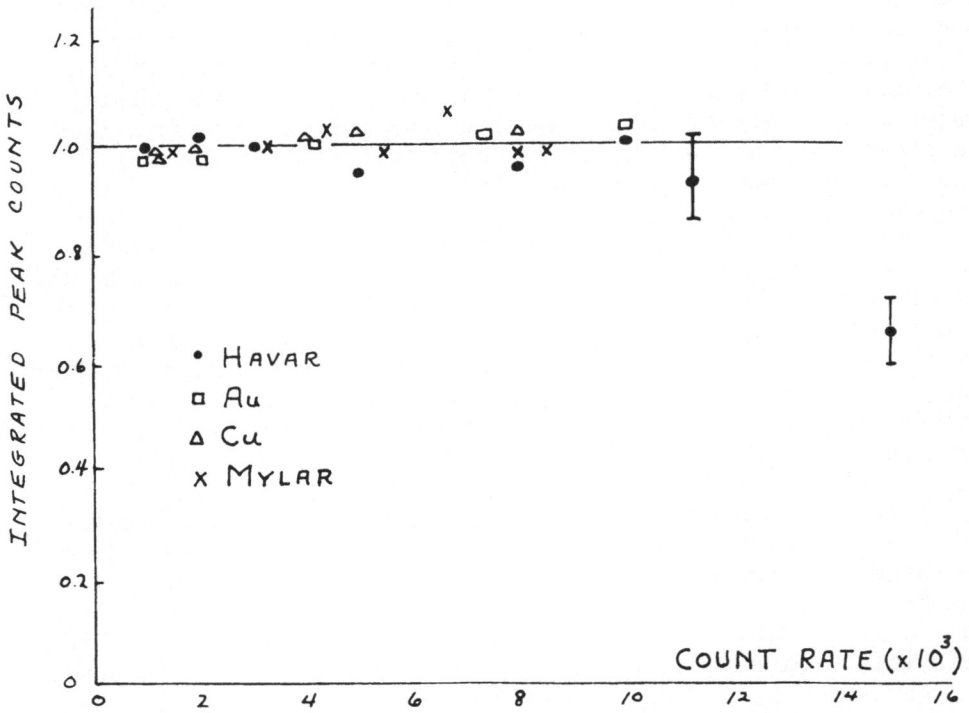

Figure 11. Integrated Counts vs Count Rate with Beam Pulsing.

A third point of some importance is that the instantaneous
beam current that generates the count rate is now much higher than
the average beam current that governs target damage and activation
of beam line components. If the clock time for a run is T, and the
total charge integrated is Q, than the average beam current, I_{AVE},
is equal to

$$I_{AVE} = Q/T \qquad\qquad\qquad\qquad IV.\ 2$$

If the beam is turned off for a recovery time T_R, slightly longer
than the electronic reset time, and if the count rate is R, then
the instantaneous beam current I_{INST} is related to I_{Ave} by

$$I_{AVE} = I_{INST}\ (\ 1\ -\ RT_R) \qquad\qquad\qquad IV.\ 3$$

If this system were not used, I_{AVE} would of course equal I_{INST}.
Thus, for $R = 10^4$ counts/second and $T_R = 50 \times 10^{-6}$ sec., $I_{AVE} =$
0.5 I_{INST} and target damage is halved per system live time event,
assuming that damage is linear with integrated charge at fixed
current value, I_{INST}.

The final point, and one that has proved very useful, is that
the on-demand beam pulsing system tends to provide a "feedback
loop" that insures that every target is run close to an optimum
count rate without the attention of the system or accelerator
operator. If the system is tuned up at 5,000 counts/second, and
a heavily loaded sample arrives during the run, the beam pulsing
system effectively lowers the average beam current to insure that
the electronics are not driven too hard (above 10,000 counts/
second). The same effect occurs when there are fluctuations in
accelerator beam current. It has proved useful to have a count
rate meter at the accelerator operator's console, as it relieves
him of anxiety when his beam current suddenly drops due to a heavily
loaded target. It also allows for more efficient operation.
In the air quality work, many samples are often quite lightly
loaded and thus, beam currents can be raised to produce an optimum
count rate. This feedback loop on count rate is planned to be
included in our accelerator beam current control system in the
near future.

Although the system is self-regulating, it is important to
have a method to make sure that the beam is being swept away from
the target when an event arrives at the x-ray detector. This is
done by placing a small pick-up capacitor near the deflector plates
and feeding it back to an oscilloscope in the counting area. When
the system is working properly, a pulse is seen, followed by about
50 μsec interlude, and then additional pulses are seen at quasi-
random intervals centered around multiples of 50 μsec. If the
system is not sweeping effectively, events will continue to arrive

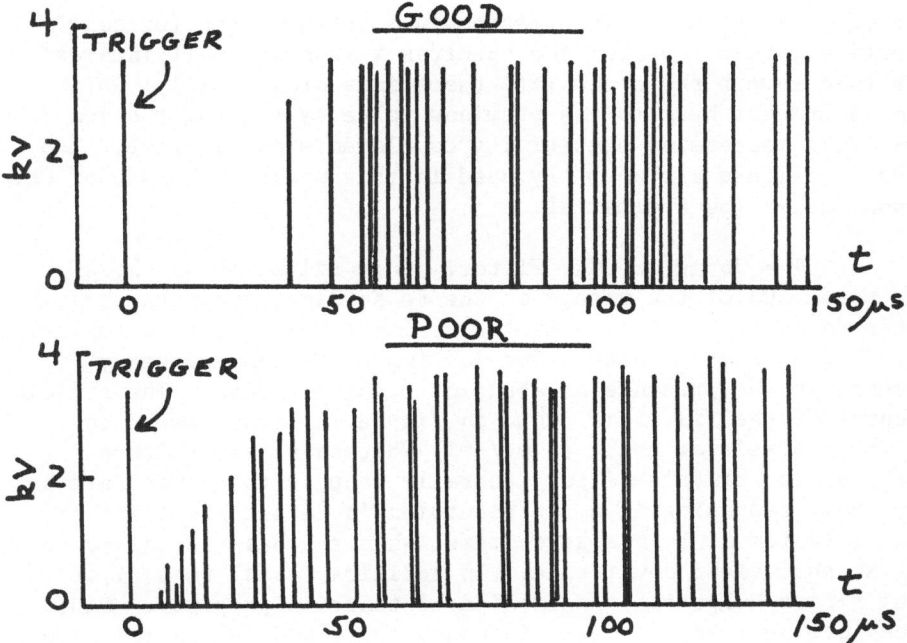

Figure 12.   Beam Sweeper Operation.

from the logic circuit of the detector system that will try to fire the plates. Figure 12 shows schematically what is seen on the oscilloscope when the system is operating properly, and operating poorly or not at all. Normally, this latter situation occurs when the beam optics are not set so that the swept beam strikes the limiting collimator.

With the changes made above, the electronic system has proven stable and reliable. Normally, the system is turned on and checked with a source before operation, without any adjustments being made. A set of diagnostic samples are run at the beginning of each analysis run, designed to verify proper operation of the system. A most important point is that all members of the group can quickly identify any failing of the system through a number of simple checks.

Filtering. Between the sample and the detector, it is possible to interpose a mechanical filter. The rationale for doing this may not always be obvious, but filtering can greatly aid IXA analysis in some cases. The key points are that the IXA-based system is always count rate limited, that the background mostly occurs at energies below that of the iron K line, that the sensitivity is relatively uniform for all elements, light (Na → Ca) as well as medium and heavy, and that most samples are abundant in these light elements. The question then becomes

that of maximizing the information in a certain time (or cost) by
selectively removing from the spectrum x-rays of small interest.
This then lowers the count rate (sometimes dramatically) which
then is brought back to its previous value by raising the ion beam
flux, thus increasing sensitivity to elements of interest. Two
types of filters are normally used in this work, both made of CH_2
or some other low-Z material.

    1. The 76 mg/cm^2 CH_2 Filter. This filter is so chosen
that about 50% of the x-rays of the Fe K line penetrate through
to the detector. Sensitivity for elements below Ca is essentially
nil. This filter is useful for looking at medium and heavy
elements in the presence of abundant light elements. The efficiency curve of the filter is shown in Figure 13. One common use is
for the analysis of thick (9 mg/cm^2) Whatman 41 air filters.
Interposition of this filter generally reduces the count rate by
about 90%, resulting in a net reduction in detectable limit by
about a factor of 3 in a given time, when the beam is raised to
give an equivalent count rate. In addition, total removal of
light x-rays helps the electronics to function properly. Other
cases involve trace elements in bone, soil, and in general almost
any biological material. Limits are often set by target damage
in these conditions.

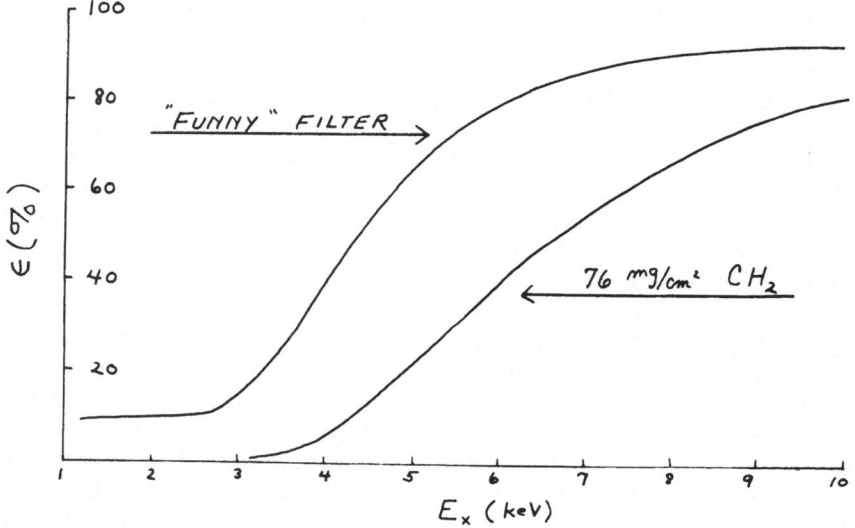

Figure 13.   Efficiency of Commonly Used Filters.

2. The "Funny" Filter. During the course of the air
pollution studies, the system faced the following dilemma. Light
elements, Na→Cl, are dominant in most air samples and must be
detected. Yet, medium and heavy elements are biologically
important, and since they occur in small amounts, the lowest
detectable limit possible was needed. The early answer was to
run each sample twice, one short run (30 sec.) without any filter,
and one longer run (2 min.) at higher beam currents using the 76
mg/cm^2 filter. The net result that the detectable limit for
light elements was increased by about a factor of 2, while the
detectable limit for medium and heavy elements was decreased
by a factor of 3. Thus, the sensitivity curve was "tipped"
towards the heavy element end by a net factor of 6, but each
sample had to be run twice, increasing cost and paperwork.
The development of the "Funny" Filter solved this problem in a
simple fashion. A thin filter of some preselected thickness (es)
was pierced by a hole (s) whose dimension (s) were chosen to
allow a selected fraction of light element x-rays to penetrate
to the detector. Thus the relative sensitivity of the system
can be tailored to match the parameters of the samples under
investigation. An example of the efficiency curve of these
filters is given in Figure 13. Calibration of the device is
trivial. A complicated sample containing many elements is run
first with no filter present and then with the "Funny" Filter
present. The ratios of the peaks provide a ready filter correction factor that can be applied by the analysis codes to further
samples run in this fashion. The gain and loss in sensitivity can
be calculated as follows. Suppose that the normal filter (without a hole) decreases the count rate to 10% of its previous value
without any filter. Suppose the area of the hole is 10% of
the active area of the detector. The new count rate will be about
19% of its previous value. By raising the beam back to a level
sufficient to attain the previous count rate, one gets the
kind of detectable limits shown in Table 8.

The "Funny" Filter comes as close to "something for nothing"
as any part of the entire UCD/ARB system. For the expense of a
few mils of Kapton, or any other low Z foil, with a 1 mm^2 hole
drilled in the center, one gains as much sensitivity for the
medium and heavy elements as if one had run more than 5 times as
long (i.e. about 5 times the cost!) In addition, since the light
elements Na→Cl, impinge only upon the center of the detector,
the energy resolution of the detector is increased for these
elements. This is again very useful, for effective separation of
these light elements is critically dependent upon the best energy
resolution possible. All samples are normally run with the
"Funny" Filter in place. Occasionally, the 76 mg/cm^2 CH$_2$ filter
will also be interposed if one is not interested in light elements
at all. However, such samples are normally run on the x-ray
fluourescence system at CNL.

Table 8

Use of Filters to Modify the Detectable Limit

Assume: (1)   Total run time t is used in all cases

(2)   Average detectable limit for light elements
      (Na→Ca) is about 10 ng/cm^2, while for medium
      and heavy elements, it is about 40 ng/cm^2.

(3)   90% of the count rate is due to $E_x < 4$ keV (Sc),
      10% from x-rays $E_x > 4$ keV.

(4)   Count rate is fixed at R_{max} for all runs.

| | Detectable Limit | |
Measurement Strategy	Light Elements	Medium and Heavy Elements
No Filter	10 ng/cm^2	40 ng/cm^2
t/3 No Filter, 2t/3 76 mg/cm^2 Filter	16 ng/cm^2	16 ng/cm^2
Funny Filter	13.5 ng/cm^2	17 ng/cm^2

C. Data Acquisition and Reduction

Data acquisition will depend critically upon facilities available at each laboratory. The availability of a PDP 15/40 computer at the Crocker Nuclear Laboratory allowed on-line acquisition and simultaneous reduction of data. However, this is certainly not essential to any program, although some interactive capacity should exist in every system in order to verify proper system operation, especially during set-up.

A schematic of the automatic data acquisition system is shown in Figure 14. The energy spectra are accumulated in the computer through the ADC interface. The operator issues commands to the acquisition program through a switch panel via the CAMAC interface. The program controls the data accumulation apparatus through the system controller via the CAMAC interface, under a software control/ data acquisition package called ACE.

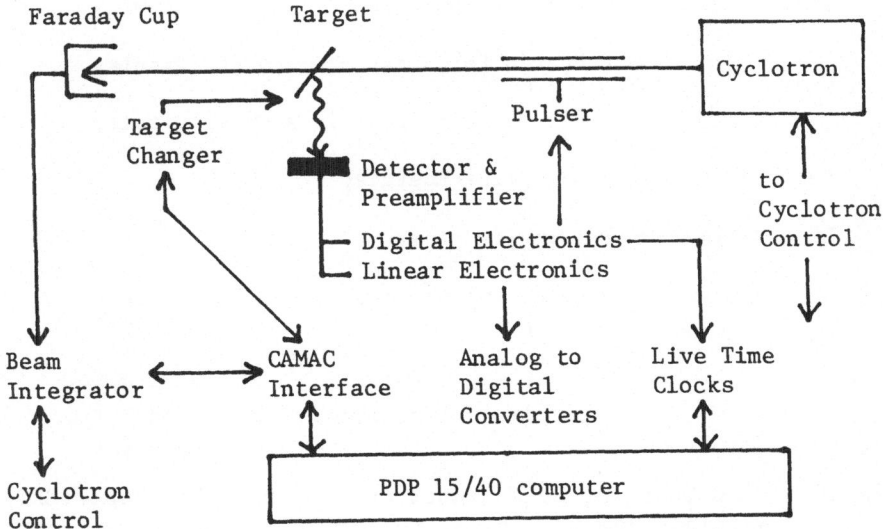

Figure 14.   Data Acquisition System.

The sequence of acquisition steps is:
1. Insert slide.
2. Turn on current integrator, enable ADC, and turn on beam.
3. Count until preset charge has accumulated.
4. Disable ADC, turn off beam, and record data.
5. Reset current integrator and remove slide.
6. Advance tray to next sample.

The energy spectrum accumulated during step 3 is stored in 500 channels, and various acquisition parameters are stored in the following 12 channels. These include live time, real time, charge, tray number, slide number, and the sample's conversion factor from square cm to cubic meters. In step 4 these data are recorded on magnetic tape for permanent storage and on magnetic disk for the reduction routine.

Data Reduction. While data acquisition is a highly individualized procedure in each laboratory, data reduction can largely be done off line (if necessary). Thus, the data reduction code, RACE, is available in Fortran IV, although key elements are in machine language at Davis to speed up the analysis.[10]

The data reduction is performed while the spectra for following samples are being acquired, using the data files written on the magnetic disk. A flow chart for the reduction is shown in

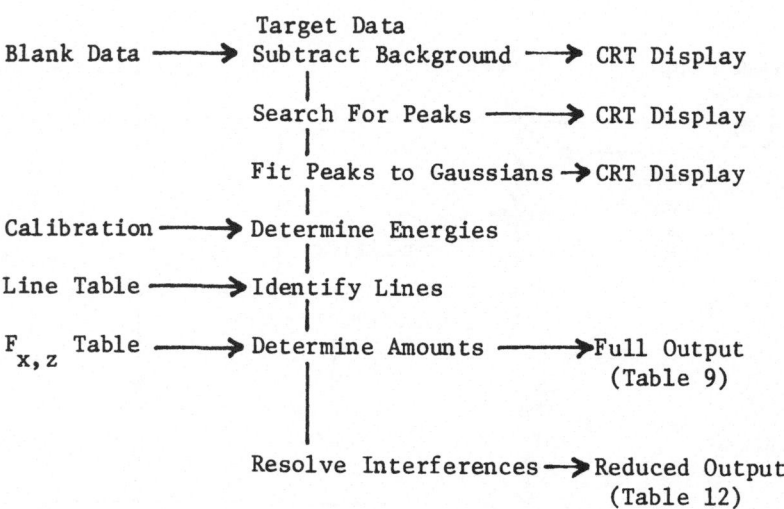

Figure 15.   Flow Chart of Data Reduction.   Cathode Ray Tube (CRT)
             displays can be copied on-line for a hard copy record.
             Original spectra, run data, full and reduced output
             are recorded on magnetic tape.

Figure 15.   Up to 20 seconds are needed to reduce the data from
each sample.   The various segments are chained so that the total
program, including the acquisition segment but excluding the
monitor and input/output handlers, fits into 14 K words of computer
memory   (18 bit words).

     In addition to the sample spectrum, the disk contains the
spectrum from a blank (same substrate but without loading), a
linear energy calibration, and a table of energies and 'cross
sections' for all reasonable x-ray lines.   (The 'cross section'
includes detector efficiency and solid angle.

     At the Crocker Lab, the data are entered into the computer via
an ND 2200 ADC under software control.   All relevant parameters
are entered into the open file, and when the run is over, written
on magnetic tape.   Acquisition is concurrent with analysis of
previous spectra.   Acquisition parameters are controlled through
a   switch module in the CAMAC system, with changes being entered on
the teletype unit.

     A major effort was centered on the development of the data
reduction codes.   The philosophy that governed development of the
codes was set by the desire to utilize fully the broad elemental
range of IXA, since this aspect is most important to its utiliza-
tion.   This demanded that any element present had to be detected
and identified, even if it occurred only once in 1000 samples.

The ability to have reasonable sensitivity for all elements, Na→U, is practically unique to IXA, and should be exploited. This is accomplished in the following manner. The spectrum, once recorded on magnetic tape for permanent storage, is then transmitted into the 14 K background of the computer for stripping and data reduction, along with all relevant parameters. The process then proceeds--

1. A blank spectrum, previously recorded, is compared with the spectrum to be analyzed. It is "floated" (normalized) until it matches the minima in the spectrum, and then subtracted. The "float" value is recorded and later printed out. This process can be displayed on the screen and, if desired, photocopied. (Figure 16, A,B. The "float" value was 1.04.)

2. The residual background is then fit by a background parameterization, whose values can be modified by the operator, if desired. This process also is displayed and can be photocopied.

3. The program now identifies peaks by a correlation function method, using a gaussian of a preselected width. The result is a series of peak locations and "cut-points" at either side of the peak. This is then displayed and often photocopied, since it is the only hard copy that generally exists of what the spectrum looked like and how well the program detected the peaks. A square root display is used in order to see the background, and because in some ways it is, statistically speaking, the most appropriate display. (Figure 16, C.)

4. The program now integrates the peaks, from cut point to cut point, and records the values. These are checked for statistical significance (Equation III. 4), and, if significant, printed in the output under the label, "Linear Sum."

5. The program now examines peaks to see if peaks lie close enough to each other to constitute a multiplet. The order of the multiplet is then determined, and the peaks are gaussian fitted. The data and fits can be displayed and copied (Figure 17). The peak location found by the gaussian fit is used to obtain the peak location accurately. This location, the integrated counts and uncertainty, the total background under the peak, its width, the chisquare value of the fit, and the order of the multiplet are printed out.

6. The program then uses a previously entered energy calibration to set the energy values of each peak, along with expected errors.

7. The program then identifies all transitions of various elements within an energy window set around each peak location.

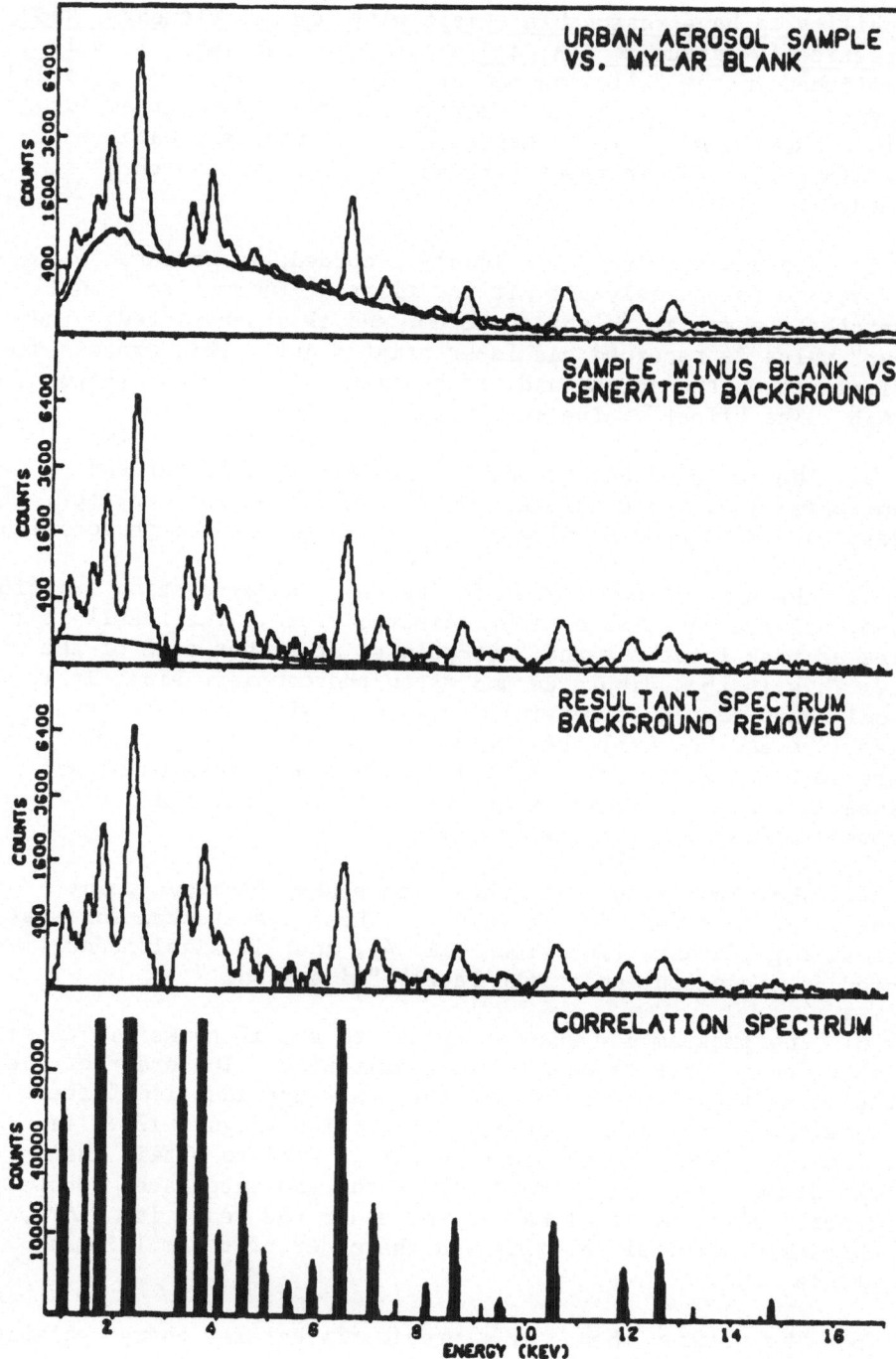

Figure 16.   Urban aerosol Sample on Mylar Substrate at Various
Stages of Data Reduction.   Square Root of Counts per
Channel vs. Energy.

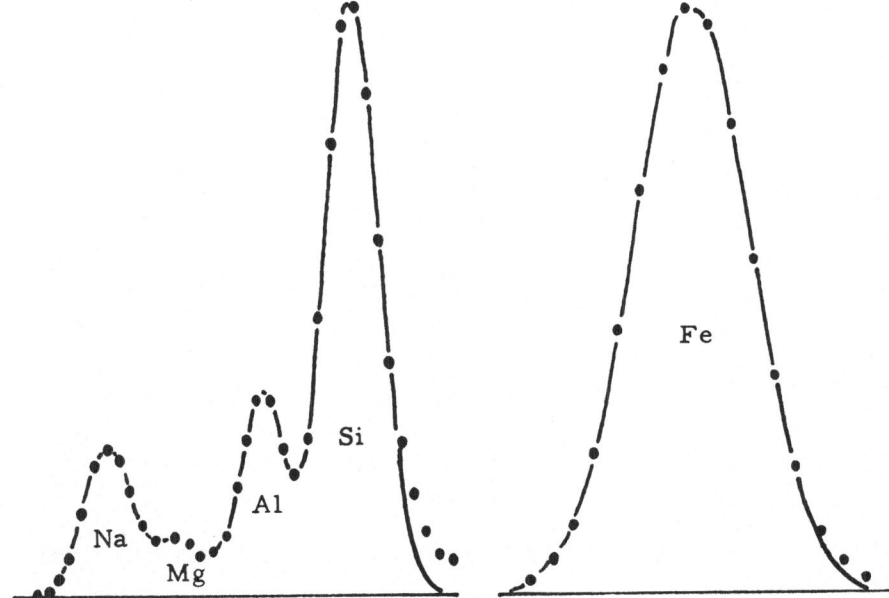

Figure 17. Gaussian Fits for Quadruplet and Singlet. Counts
per Channel vs. Channel.

8. The program then calculates the amount present in ng/cm^2,
using a more explicit form of Equation III. 2. Errors are propogated into the same format. If the conversion factor is 1.000,
the results are in ng/cm^2. If not, the program will write the
result in ng/m^3 (of air, naturally) or ppm, depending on one's
interest. Previously calculated conversion factors are resident
upon the disk, and are entered for each sample.

9. Minimal detectable limits are calculated for unobserved
elements.

10. The result is printed out in the form shown in Table 9.

Interferences. The output as shown in Table 9 represents all
that can be derived from the spectrum in terms of the criterion
of significance. There are a number of factors, however, that can
make these results erroneous, misleading, or ambiguous, even if the
system runs to perfection. The major cause of this type of
problem is interferences.

There are a number of forms of interference, all based
upon the fact that present energy dispersive x-ray detectors have
energy resolutions that are not good enough to separate transitions
from different elements in all cases. A few examples are--

Table 9

Full Output of Race (Partial)
(See Table 12 for Reduced Output of Same Target)

UC DAVIS XRAY ANALYSIS 00606 12/13/74
TRAY 99111 SLIDE 28

CONVERSION FACTOR = 0.180
LIVE TIME = 65.0/81.7 = 0.796
BLANK FLOAT (300 CHANS) = 1.280
ENERGY = 0.6676 + 0.0326*CHAN (R)
COUNTS = 185414. REJECTS = 32826.

CHARGE = 5.000 MICROCOULOMBS
FWHM = 7 CHANNELS
BKGND AVG PARAMETERS = 33 4 14
TABLE FFT18 * 0.870
RATE = 2671.2 COUNTS/SEC

PEAK	LOOPS (ITERATIONS)	CHANNEL	ENERGY (ERROR TO ALL LINES)	ELEMENT	COUNTS (GAUSSIAN)	NANOGRAMS PER M**3	LOC	CUTA CUTB	LINEAR SUM	BACKGROUND	WIDTH	CHIS χ^2	MULT
1	8	12.0	1.058	NA K	5897.	579.	11	3	6013.	7443.	8.61	49.4	3
		+/- 0.1	0.017		+/- 168.	+/- 60.		19			+/-0.16		
2	8	24.8	1.474	AL K	2791.	108.		20	3328.	8324.	4.36	7.8	3
		+/- 0.1	-0.013		+/- 157.	+/- 12.	25	28			+/-0.13		
3	8	33.0	1.741	SI K	15869.	522.		29	15559.	14684.	5.68	5.3	3
		+/- 0.0	0.002		+/- 256.	+/- 53.	33	42			+/-0.06		
4	5	50.7	2.319	S K	3462.	110.		43	4244.	10094.	5.29	1.5	2
		+/- 0.1	0.012		+/- 176.	+/- 12.	51	55			+/-0.15		
5	5	60.2	2.626	CL K	24265.	836.		56	25246.	9798.	6.23	8.1	2
		+/- 0.0	0.004		+/- 278.	+/- 84.	60	73			+/-0.06		
6	5	81.5	3.320	K K	4198.	103.		74	4193.	6227.	5.76	6.0	3
		+/- 0.0	0.008		+/- 147.	+/- 11.	81	86			+/-0.11		
7	5	92.8	3.687	CA KA	13269.	243.		87	13150.	6075.	6.44	1.6	3
		+/- 0.0	-0.003		+/- 212.	+/- 25.	93	98			+/-0.08		
8	5	102.5	4.004	CA KB	1951.	219.		98	1962.	5808.	7.23	0.3	3
		+/- 0.2	-0.008		+/- 149.	+/- 28.	103	111			+/-0.42		
9	0	118.0	4.509	BA LA	1561.	105.		112	1479.	4968.	4.67	10.7	2
		+/- 0.1	0.050		+/- 114.	+/- 13.	118	124			+/-0.12		
				OR TI KA		32.							
			0.001			+/- 4.							
10	0	131.0	4.932	TI KB	307.	40.		125	210.	4820.	4.67	12.6	2
		+/- 0.1	0.001		+/- 101.	+/- 14.	131	140			+/-0.25		
				OR V KA		7.							
			-0.017			+/- 2.							

(remainder of table omitted)

1. The K lines of light elements are so close together that they can not be fully resolved from neighboring elements if energy resolution is poor. (Example: Na--Mg).

2. The K_α line of element Z falls close to the K_β line of element Z-1 or Z-2. (Example: Br-Rb).

3. An L transition falls close to a K transition. (Example: As K_α and Pb L_α).

4. An M line of a heavy element falls close to a K or L transition. (Example: Pb M, Mo L_α, S K_α).

5. A spurious peak associated with the electronics or the detector, and not with a characteristic x-ray transition energy, may lie close to a valid transition. (Example: the S K sum peak, due to pile-up, falls close to Ti; the Fe K single escape peak falls close to V K_α).

The problem is, in essence, unsolvable, if one allows all possible target compositions, assumes a 200 ev resolution, and pushes the detectable limit to very low values. Many of the peaks normally seen at this resolution are unresolved or partially resolved multiplets, and the number of resolvable transitions is, in itself, very large in heavy elements. For example, there are 18 resolvable transitions associated with the L lines of U. We think that this point must be kept in mind when one attempts to do trace element analysis using x-rays (regardless of excitation modes). The same broad range capability that makes x-ray analysis so useful as a means of elemental analysis causes problems when one tries to do trace work and see a small amount of one element in the presence of large amounts of neighboring elements.

The problems associated with interferences vary according to the energy of the detected characteristic x-rays, and each region has its own types of problems. Examples will be taken from the aerosol analysis work unless otherwise noted.

0.8 keV $<$ E$_x$ $<$ 3.0 keV

This region, corresponding to characteristic K x-rays from Ne to Ar, is characterized by unresolved or partially resolved K_α-K_β transitions that do not seriously interfere with adjacent elements at normal resolution. (Figure 17). Interferences are due to unresolved L transitions of elements from Co to Pd, and unresolved M transitions from Ho through U.

The problems associated with this region are not major, since IXA has good sensitivity for the K transitions associated with Co through Mo and the L transitions of Ho through U. Thus, through detection of the higher lying line, the counts associated with the lower lying transition can be subtracted.

For aerosol samples, the absence of the noble gasses Ne and Ar (assuming the system is run in vacuum), the low values generally found for lanthanides, actinides (except possibly Th and U), Tc, Ru, Rh, and in fact, most heavy metals except Pb and possibly Hg, and the abundance of elements Na through Cl, all tend to make corrections minor and quite accurate. It was necessary to obtain cross sections for all of these lower transitions, however, which required use of standards. The Zn-Na, Br-Al, Sr-Si, and Pb-Mo-S interferences are the most significant.

$E_x > 8$ keV

In this region, interferences are an annoyance rather than a tragedy. With a few exceptions, the detector resolution is able to resolve most interferences. Strong secondary lines are usually available. Beyond 13 keV, the interfering heavy elements are actinides, which, hopefully, will not be abundant in ambient air analysis. The Pb-As, As-Br, and Br-Rb interferences are the most significant ones in this region.

3 keV $< E_x < 4$ keV

This region is dominated generally by strong K-Ca peaks. Since these elements are abundant, and since Ar is absent, the K K_β and Ca K_α interferences can be generally solved. Only U provides an M interference (at 3.17 keV), and the U value is known from the L transitions and can be subtracted.

L interferences occur for the elements Ag through I. Ag and Cd are resolved from K, and in fact, these lines generally provide better detectable limits for these elements than their K lines. However, In, Sn, Sb, and Te are heavily involved in the K-Ca lines. Sn can be seen between K and Ca, if present in amounts equal to or better than about 1/4 the Ca value. The energies are well separated and this causes no problem. Generally, corrections for K and Ca are small or non-existent due to the small amounts of In, Sb, and Te in aerosols, but independent information on the amount (or upper limit) of elements In through I is desireable. This problem is serious enough to cause the UCD/ARB system to be modified through the addition of a large area (300 mm^2), low resolution (300-300 keV) Si(Li) detector in a close in position but placed behind a 300 mg/cm^2 CH$_2$ filter. This detector will observe only events from about 15 keV to 35 keV, and will hopefully improve sensitivity by about an order of magnitude.

Table 10

Transitions in the Ti-V-Ba Region (except Lanthanides and Ce)

Energy of Transition (keV)

	K_α	K_β	L_α	$L_{\beta 1}$	$L_{\beta 2}$	L_γ
Relative Intensity	1.00	0.14	1.00	0.62	0.19	0.07
Element						
Sc		4.459				
Ba			4.467			
Ti	4.508					
Ba				4.828		
Ti		4.931				
V	4.949					
Ba					5.156	
Cr	5.411					
V		5.427				
Ba						5.531
Mn	5.895					
Cr		5.947				

$4 \text{ keV} < E_x < 8 \text{ keV}$

This region is dominated by K_α-K_β interferences for almost every element. The L transitions of Xe through Er interfere in this region. The generally low values of Sc, Ce, and the lanthanides in air samples helps an otherwise grim situation. The most serious interference is the Ti-V-Ba system, since generally a string of K_α-K_β interferences alone can be handled. As an example of a problem that needs more work, consider the energy region from about 4 keV to 6 keV. Table 10 gives the energy of the major transitions and their intensity relative to the K_α line (for K) or the L_α line (for L transitions) in this region.

The analysis code sets relatively wide windows around each peak, ± 50 ev, and thus, ignoring Sc, Ce and the lanthanides (which would be seen and read out if present, but would further confuse the issue), one could get the results shown in Table 11 if all these elements were present.

There are a number of ways to attack the problem.

First, the mean deviation of a detected peak energy from the calibration value is about 12 eV. A code can calculate the

Table 11

Possible Interferences

	E_x (keV)	N_x (# cts. at E_x)
Ti K$_\alpha$ or Ba L$_\alpha$	4.500 ± 0.009	$N_1 \pm \Delta N_1$
V K$_\alpha$ or Ti K$_\beta$ or Ba L$_{\beta 1}$	4.900 ± 0.009	$N_2 \pm \Delta N_2$
Ba L$_{\beta 2}$	5.200 ± 0.009	$N_3 \pm \Delta N_3$
Cr K$_\alpha$ or V K$_\beta$	5.420 ± 0.009	$N_4 \pm \Delta N_4$
Ba L$_\gamma$	5.520 ± 0.009	$N_5 \pm \Delta N_5$

probabilities that a peak location belongs to each element. The
product of these probabilities for each transition, properly
normalized, results in an elemental probability.

Second, some lines are non-interfering under normal
resolution, such as the Ba L$_{\beta 2}$ amd Ba L$_\gamma$ transitions. These
are relatively weak, however, and thus there is a region of
uncertainty in which the L$_{\alpha 1}$ and L$_{\beta 1}$ lines could be detected but
not the (non-interfering) L$_{\beta 2}$ and L$_\gamma$ lines. This would raise
the "certifiable" detectable limit by about a factor of 7 to 10.

Third, the transition ratios are well known to high
precision ($\pm$ 2%), and a range of possible elemental mixtures can
be eliminated by this technique.

Fourth, the Ba value can be independently determined
from the K transitions. However, the detectable limit is considerably worse for IXA work, and thus one again must accept a region of uncertainty in which Ba might (or might not) be present.

The code RACE performs these calculations and calculates
such probabilities. However, it is sometimes difficult for the

user to accept the statement that there is a 2% chance of Ba being present!   (Table 12).

Thus, there are procedures that can minimize problems associated with interferences and allow one to get correct answers under most conditions.

1. Make the energy resolution of the detector as good as possible. Make sure the system is stable against gain shifts and minimize spurious electronic peaks.

2. Detect K lines for all elements whose L line interference causes problems, and likewise detect L lines for all elements whose M line interference causes problems.

3. Use the information contained in the intensity ratios of secondary peaks to unravel interferences.

4. Make no assumptions about what elements are present in a sample in major amounts. Detect all peaks, since one of them may provide the clue to solve an interference. At the very minimum, the code could "flag" an analysis as one presenting special problems.

5. Know your sample type as well as you can. In the last resort, it may be possible to make value judgements that give probable results. Be careful how such data are presented, however.

6. Be very cautious about using energy dispersive x=ray methods on samples where interferences are really serious. Wavelength dispersive methods or other techniques may be superior in these cases.

In summary, the data acquisition and reduction codes hold the key to unlocking the potential of x-ray analysis. If one desires to analyze large numbers of samples at low cost, automatic reduction of the x-ray spectra to the relevant values for element content is essential. If one uses automatic analysis, then a high degree of sophistication must be built into the acquisition-reduction code both in terms of peak location and stripping and in terms of interference minimization. It would be wrong, however, to despair of ever obtaining accurate values from realistic samples because of interferences. In most cases, environmental and medical samples are not all that complicated, and the task can be done routinely if reasonable care is exercised. For those cases in which things do get complicated, a sub-routine that recognizes that a problem exists and alerts the investigator to it is essential.

Table 12
Reduced Output of Race

```
          UC DAVIS XRAY ANALYSIS  00606      12/13/74
             TRAY 99111      SLIDE 28
        CONVERSION = 0.180 CM**2/M**3

              ELEMENT              NANOGRAMS
                                   PER M**3

          1     NA K          579. +/-        60.

          2     AL K   -        92. +/-        14.

          3     SI K           522. +/-        53.

          4     S  K   -        62. +/-        20.

          5     CL K           836. +/-        84.

          6     K  K           103. +/-        11.

          7     CA KA          243. +/-        25.
                   KB          219. +/-        28.

          8     TI KA           32. +/-         4.
                   KB           40. +/-        14.

          9     MN KA  *         5. +/-         2.

         10     FE KA          413. +/-        42.
                   KB          389. +/-        43.

         11     CU KA            8. +/-         2.

         12     ZN KA           23. +/-         4.
                   KB           30. +/-         9.

         13     PB LA          151. +/-        21.
                   LB          163. +/-        38.

         14     BR KA           41. +/-         6.

          - COMPETING LINE SUBTRACTED
          * BELOW MINIMUM SENSITIVITY

          PROBABILITY FOR BA = 0.02

        UPPER LIMITS OF ELEMENTS NOT FOUND
          1  MO    21.            8  AU    15.
          2  BA    22.            9  HG    16.
```

D. System Calibration

Referring back to Section III,

$$(\rho t) \; gms/cm^2 = (N_x/N_i) \; F_{x,z} \hspace{3cm} III.2$$

we have discussed factors affecting the accuracy with which we
can determine N_x, the number of x-rays, and N_i, the integrated
beam current. We still have not determined $F_{x,z}$, and methods
for doing this will be referred to as "calibrating the system."
As before,

$$F_{x,z} = \frac{A_z}{N_0 \, \sigma_{\theta,\epsilon} \, d\Omega \, \eta_x} \hspace{3cm} III.3$$

where A_z is the mean atomic weight of element Z, N_0 is
Avogadro's number, $\sigma_{\theta,\epsilon}$ the cross section for emitting an
x-ray in angle θ per unit solid angle, and $d\Omega$ is the solid angle
in steradians subtended by the x-ray detector, and η_x the detector
efficiency at E_x. $F_{x,z}$ refers to a given x-ray transition
occuring after element Z has been excited, and many such transitions exist for any element and any chemical state of any element.
For analytical purposes, the strongest transitions are normally
used, providing that
   a)   the transition can be observed by the x-ray detection
        system used,
   b)   That the transition provides for an unambiguous deter-
        mination of the element Z under study, and
   c)   That in the specific sample under study, this transition
        is not masked or confused by interferences.
For a normal Si (Li) detector with thin (0.5 mil = 0.013 mm)
window, condition a) limits one (for 10% efficiency) to the K
lines of elements Na the rare earths, L lines of Ni to U,
and M lines of rare earths and heavier elements. Condition b)
and condition c) are linked together, but generally result
in the fact that strongly excited M lines are not useful with
Si (Li) or Ge (Li) detectors, due to inability to separate
neighboring heavy elements and competition from abundant light
elements. For example, the M lines of lead lie on top of the
L lines of molybdenum which in turn are superimposed on the K
lines of sulfur. Thus, we normally treat M lines only as an
annoyance, an interference to be handled, and rely on K and L
transitions for analytical purposes.

   Within the resolution of most Si (Li) and Ge (Li) detectors,
one then has the following transitions as possible analytical
probes:
   1.   Unresolved K lines, Na$\rightarrow$Cl
   2. K_α and K_β transitions K$\rightarrow$Mo or so
   3. K_{α_1}, K_{α_2}, K_{β_1}, K_{β_2} transitions Ru$\rightarrow$Rare Earths.
While heavier elements can be observed via K transitions, cross

sections become very small at $\sim$ 0.1 c compared to L transitions
(about 1% at tin, decreasing at higher Z)

4. Unresolved L_α transitions, Ni $\rightarrow$ Y
5. L_α , L_β , and L_γ transitions, Zr $\rightarrow$ rare earths
6. Numerous (12 or more) resolvable L transitions, rare
    earths to U. (Figure 18)^{II}

For lead, the UCD/ARB system normally uses the L_η , L_α , L_β ,
L_{γ_1}, and L_{γ_2} transitions in its reduction code RACE, with a separate $F_{x,z}$ for each transition. In situations involving large
amounts of lead, caution must be exercised and the code should consider more transitions. The lead M transition is also calculated
from the L transitions in order to correct for the S interference.

At this point, two options can be taken. One involves
establishing the presence of an element by using all strong transitions, transition intensity ratios, and energies, and then using
an effective F_z for that element. The other, used at Davis, treats
every statistically significant x-ray peak separately, using numerous
$F_{x,z}$'s for each element. These are then displayed, and only in the
last step does the program give a final value for the amount
present. This is extremely useful in resolving interferences, and
we strongly encourage this approach.

The first step in establishing $F_{x,z}$ is to establish A_z through
determining Z. This involves a calibration of peak amplitude recorded by the pulse height analyzer in terms of x-ray energy. The easiest way is often to have either a special target made up with convenient elements scanning the range of E_x or to use a typical analytical target that you are familiar with such as soil dust, an urban
aerosol, etc. This calibration should be done carefully, with a
linear or polynomial fit made to several transitions, aiming for
a precision of $+$ 1 ev at around 6 keV. It may not be obvious why
such precision is needed, since the detector will normally operate
at about 170 to 200 ev at 6 keV in operational conditions and
normal count rates. However, when a strong x-ray peak is gaussian
fitted, the centroid can be often determined to $\pm$ 5 ev, greatly
reducing the number of interferences encountered. In practice,
the standard automatic code used at Davis achieves $\pm$ 12 ev (1σ)
during a 1000 sample run, using simple gaussian shapes without
tails or skewness.

Once one has an energy calibration that allows an unambiguous
determination of the element (and hence A_z), one is still faced
with the crucial task of determining the amount of element Z
present. There are a number of strategies, some of which are:

1. Referring to Equation III.3, one can obtain $\sigma_{\theta,\varepsilon}$
from cross section tables, and $d\Omega$ from the detector solid angle
to the target, and the detector efficiency versus energy from
calibrations or manufacturer's tables. $F_{x,z}$ can then be calculated.

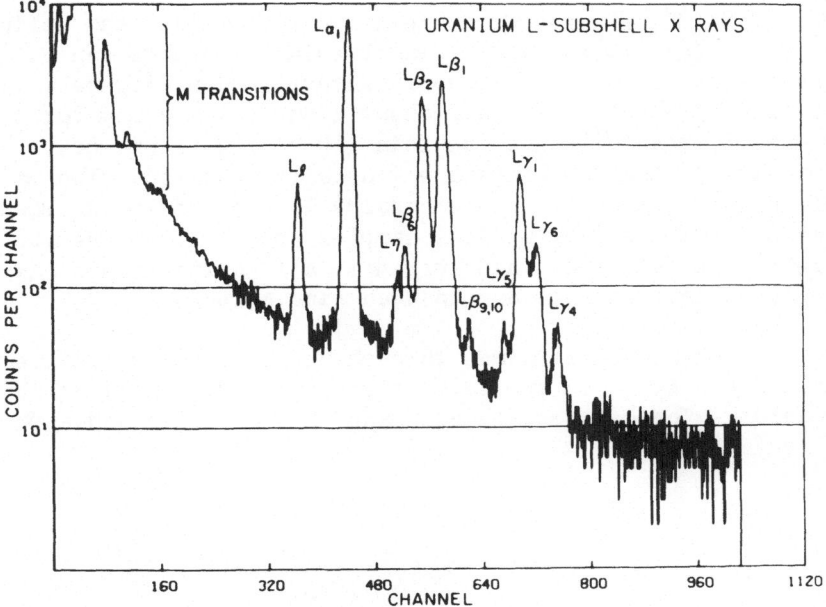

Figure 18. Uranium L-Subshell X-Rays. (Reproduced by
courtesy of The Physical Review.[11])

Although this approach is probably the fastest way to calibrate the system, it suffers through dependence on published
cross section values that may not cover the elements of
interest or all transitions at your energy or ion. It also
assumes you know the effective detector area accurately, which
is not always the case, especially with small area detectors.
Although standard sources are available for x-ray energies
above 6 keV, care must be taken with absorbtion in the source
holders, airpaths, etc. in establishing the detector efficiency.
Finally, standard x-ray sources are entirely absent or largely
unavailable below E_x = 3 keV, a region in which detector
efficiency starts to change rapidly with energy.

    2. $F_{x,z}$ can be determined through use of standard, gravemetric
non-radioactive sources. These sources are known amounts (gms/cm^2)
of elements or compounds generally deposited upon substrates such
as mylar. Such a foil is placed in the ion beam, N_x and N_i are
recorded, and $F_{x,z}$ is directly determined for all transitions of
interest.

    A major advantage of this technique is that these foils are
commercially available at moderate prices from several sources,
accurate to $\pm$ 5%. Such foils can be made from virtually any
material and thus calibration in the crucial E_x = 1 to 3 keV
energy region is easy. Since all parameters vary smoothly with

E_x and Z (σ, A_z and η_x), relatively few foils can calibrate
a system for all energies. Intermediate elements can be
quantitatively obtained through interpolation (Figure 19).
In order to keep x-ray self-absorbtion to a minimum for light
elements, the foils must be thin (10 to 100 µg/cm² of
deposit) so that backings are generally required. The major
disadvantage is that such standards look nothing like realistic environmental or medical samples, so that, unless your
electronics are sound and tested at all energies, you may
get erroneous results due to dead time effects.

3. $F_{x,z}$ can be determined through use of standard reference
materials, such as those available through the U.S. National Bureau
of Standards. These materials are made into a target and exposed

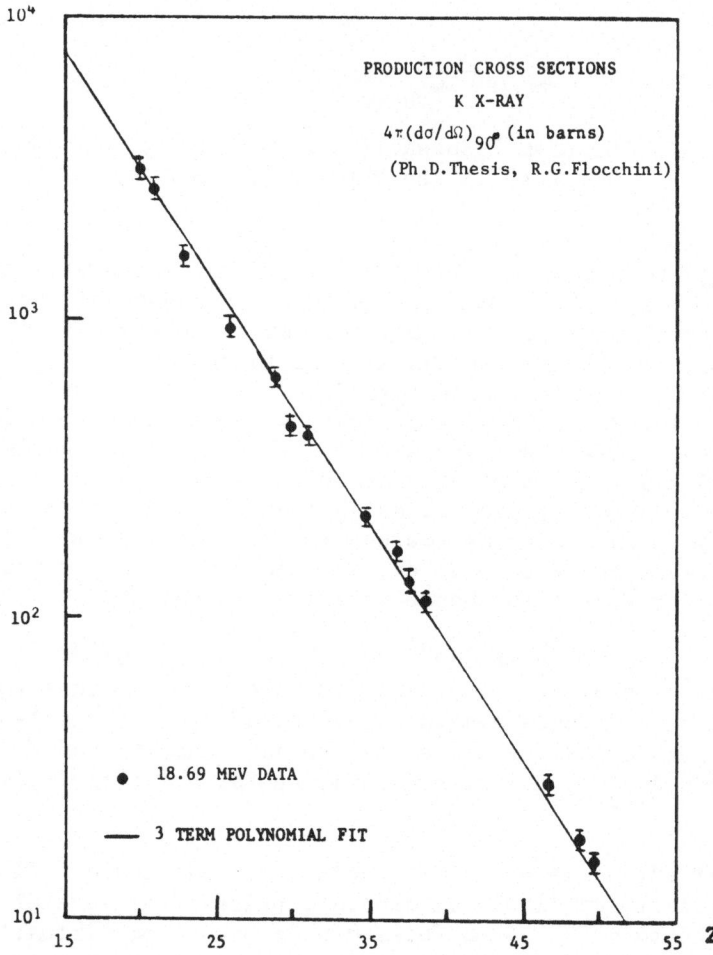

Figure 19. Interpolation between Standard Foils by Polynomial Fit.

to the ion beam. Several elements are present in known amounts. The results should give $F_{x,z}$ for these elements, which can then be interpolated for other elements.

This method has the advantage that the samples are now very similar to the samples one plans to analyze. However, target matrix effects are inherent in this approach, and to separate this effect from σ, $d\Omega$, and η_x is not easy, especially for elements below about $E_x = 3$ keV. The standards have particle sizes that cause such problems even when loadings on the target are made very thin. Nevertheless, such an attempt should be made, and for medium and heavy elements, the results should be accurate. If one can talk a calibrated laboratory out of a few analyzed targets, this can be a most useful technique, indeed, as corrections will be generally known.

4. One can give up an attempt to fix $F_{x,z}$ analytically, but instead use non-radioactive chemicals as an internal calibration. One adds a known amount of an element known not to be present in the sample in significant amounts. The result that one obtains for a similar element of interest is then obtained by taking the ratio of the known to the unknown.

This method often is applicable when other methods fail, such as for thick targets that stop the beam when N_i may not be available. It also allows one to perform complicated sample to target reductions, as long as one is sure that one is not differentially eliminating the tracer element or the element of interest.

In summary, the first three methods have been extensively used at Davis, with method 2) being the primary method. It is in the agreement between different approaches that confidence can be gained that the system is truly calibrated, and an enormous amount of effort must be continually expended in this regard.

E. Target Preparation and Matrix Effects

Reduction of an environmental or medical sample into a target form suitable for ion beam analysis, is a problem of infinite complexity and manifold variations. There is no way in which a short section in this chapter can adequately treat this problem, since so much depends on what the initial material is (gas, liquid, solid, biological or physical, etc.), what the analysis system does (ion beam configuration, energy, vacuum, heating problems, etc.), but most importantly, what information the user wants (all elements, a few, one, to high sensitivity, or high

precision, etc.) However, there are a number of criteria that can
be applied to a large number of cases, and these will be discussed as they have some generality.

The key to optimum target preparation lies in Tables 1
and 2. The primary ion radiation must reach the element under
study, and the secondary radiation must reach the detector.
There are uncertainties in σ introduced by beam energy degredation through the target which become larger as E increases,
and there are uncertainties in N_x introduced by matrix corrections
of the secondary radiation which become larger as the attenuation
increases. Sensitivity increases (i.e. lower minimum detectable
limits exist for fixed N_i) as the target thickness $(\rho t)_s$ increases
for fixed backing thickness $(\rho t)_B$, (Equation III.15), but so
do these corrections in σ and N_x. At some point, which is
determined by the lowest energy x-ray that must be detected and
the precision desired in the quantitative analysis, $(\rho t)_s$ reaches
its maximum thickness, fixing the minimum detectable limit for
all elements. An example of how one does such an optimization
procedure is outlined below.

Assume that a user
1. Wishes to measure elements Z_1, Z_2,...Z_n
2. To a experimental error Δ (%)
3. With the best sensitivity obtainable at a fixed cost,
    which usually fixes a maximum N_i.

One then determines the minimum energy x-ray that must be
used for one of the elements, $E_{x, min}$. The question then becomes
how well one thinks one can correct for matrix effects (absorbtion,
refluo rescence, etc.) due to the sample thickness $(\rho t)_s$. These
corrections of course depend upon the composition of the target,
which is initially unknown, although with experience, educated
guesses can be made for the effective mass attenuation coefficient
due to all elements present in the target. If one can assume
that variations in composition, uniformity, etc. limit the reliability of the corrections to $\pm 30\%$, then in order to perform a $\pm$
10% measurement (Δ) set by the user, one can not tolerate a
correction of more than about 30%. Knowing $E_{x, min}$ and guessing
μ_{eff} (generally a $Z_{eff} \simeq 9$ works well),one can calculate $(\rho t)_{s, max}$.
Some examples are given in Table 13.

For x-ray energies greater than 4 keV, incident ion penetration corrections soon become more important than x-ray attenuation
effects.

Unfortunately, other effects are often much more important
sources of uncertainty than are the relatively simple self
absorbtion effects. One involves penetration of material into

Table 13

Maximum Sample Thickness for a 30% Self Attenuation Correction in a Z_{eff} = 9 Target Cocked at 45° to the Detector (Approximate)

Element	$E_{x,min.}$ (keV)	$(\rho t)_{s,max.}$	Correction at $(\rho t)_s = 300 \mu g/cm^2$
Sodium	1.041	150 $\mu g/cm^2$	1.70
Magnesium	1.255	230 "	1.39
Aluminum	1.487	390 "	1.23
Silicon	1.739	640 "	1.14
Phosphorus	2.014	1 mg/cm^2	1.09
Sulfur	2.307	1.5 "	1.06
Chlorine	2.622	2.2 "	1.04
Potassium	3.312	4.3 "	1.02

a filter, if that method is used to prepare a target. The filter material itself then causes absorbtion of x-rays, independent of loading but strongly dependent upon particle size. Another involves the "granularity" of any sample, in that even for lightly loaded samples deposited on an impervious surface, particle size effects and uncertainties in particle morphology [13] (Is sodium wrapped around sulfur, sulfur wrapped around sodium, or are they uniformly dispersed in each particle?) can be large sources of error. In Table 14, examples are given for these effects for some common materials.

The preceding discussion indicates the necessity for treating correction factors associated with target matrix effects seriously. The time to think of these problems is when the sample is being reduced to a target prior to analysis. Nothing is as discouraging as obtaining a fine looking analysis of an interesting sample only to find out that the matrix corrections are impossible to perform because the target was too thick.

Relating these questions to the UCD/ARB aerosol analysis system, it was evident that the elements sodium, aluminum, and silicon had to be quantitatively determined in atmospheric aerosols, since these three elements comprise by themselves 15% of the mass of an average aerosol, or 42% of the mass of all elements sodium and heavier in the aerosol (Table 5). Thus, loadings of no more than about 300 $\mu g/cm^2$ could be tolerated, and, for even silicon, particle sizes much greater than 20μm could not be effectively analyzed. Since the thinest material capable of taking the stresses of impaction and sample handling in routine air monitoring operations was 600 $\mu g/cm^2$ (0.00015")

Table 14

Some Typical Corrections for
Filter Matrix Absorption and Particle Size

True Value = Correction x Observed Value

| Element | E_x (keV) | Filters | | Particle Size (Diameter) | | |
		Whatman 41 (9 mg/cm^2)	GelmanGA-1 (3.6 mg/cm^2)	30μm Rock	10μm Aerosol	1μm Aerosol
Sodium	1.041	60	2.70[13]	4.7	2.5	1.32
Magnesium	1.255	25	1.80	4.2	2.0	1.22
Aluminum	1.487	12	1.57	2.6	1.7	1.16
Silicon	1.739	8	1.41	2.3	1.5	1.12
Phosphorus	2.014	5	1.29	2.1	1.37	1.09
Sulfur	2.307	3.3	1.19	1.9	1.26	1.07
Chlorine	2.622	2.5	1.13	1.8	1.20	1.05
Potassium	3.312	1.8	1.07	1.5	1.12	1.025
Calcium	3.690	1.5	1.05	1.4	1.09	1.02
Titanium	4.508	1.4	1.03	1.3	1.07	1.017
Iron	6.400	1.15	1.01	1.1	1.03	1.01

mylar[14] type S, $C_{10}H_8O_4$, it was evident that a loss in minimum detectable limit had to be incurred. Filter penetration effects, even into a membrane type Gelman GA-1[15] filter, were likewise very serious, so that a surface deposition filter (Nuclepore[16] ~1 mg/cm^2) had to be used. These constraints upon target thickness, set basically by x-ray attenuation, proved fortunate in two ways, however. The collection of adequate sample onto the filter and film surfaces was easy, since very little mass was needed, and the thin targets proved suitable for analysis of very light elements, hydrogen through fluourine, by elastic ion scattering. (See Appendix.)

Once one has determined the optimum $(\rho t)_s$ for a given analytical problem, it still remains to reduce an environmental sample to a suitable target. Gaseous and liquid targets will not be discussed, although many of the same criteria apply as for solid targets. Gas cells have been used at Davis since 1970 with good results for gaseous targets, especially when care is taken so that the entrance and exit windows for the ion beam are screened from the x-ray detector by proper low-Z collimators.

For solid targets, the key point is that the formalism provides a result, $(\rho t)_z$, that is proportional to the total ion flux, N_i, and does not depend upon the area of irradiation.

The result $(\rho t)_z$ is also generally of little interest to most
users, who want a result in parts per million, micrograms per
cubic meter of air, or other such form.

In order to make this calculation, there are two straightforward approaches:
1. Construct a target that extends beyond the area of ion
irradiation in all directions and has a uniform thickness, $(\rho t)_s$.
Then, $D_f = (\rho t)_z / (\rho t)_s$, irregardless of the exact ion beam
irradiation area or beam spatial nonuniformities.
2. Construct a target that is fully contained within a
known ion beam irradiation area. The beam must then be spatially
uniform. If the amount of sample within this area is S gms,
and the area of the beam is A cm^2, then $(\rho t)_s = S/A$, and

$$D_f = \frac{(\rho t)_z}{(\rho t)_s} = \frac{(\rho t)_z}{S/A}$$

Although these numerical values are equivalent, there are other
important considerations. One involves the beam configuration,
in that a spatially non-uniform ion beam demands uniform targets.
Self-absorption corrections are much easier to make if the target
is uniform. If only a very small amount of material is available,
the freedom to modify the irradiation area can be most useful.
For example, if one has a normal minimum detectable limit of 10
ng/cm^2 for a certain element and flux N_i, and uses an irradiation
area of 1 cm^2, the mass detection limit for that element is 10 ng.
By reducing the irradiation area to 0.1 cm^2, the mass detection
limit drops to 1 ng. This method for increasing sensitivity is
often limited by beam damage. At Davis, the beam irradiation
area is normally around 1 cm^2. Occasionally, the beam is made
uniform in only one plane by beam sweeping, rather than in both
planes by beam diffusion. The reason that the beam can be uniform
in only one plane is that most of the air samples analyzed as
part of the ongoing UCD/ARB programs are also uniform in at
least one plane. Therefore, uniformity has to be guaranteed in
only the non-uniform direction. The reason that this approach
was used is that the non-uniformity is in elemental concentration
as a function of time of day (a result of the sample collection
device). So, although the beam is normally swept back and forth
at 1 hertz (cycle/second), giving uniform coverage from midnight
to midnight, it can also be moved stepwise, giving results in
3 hour increments over that period with no loss in sensitivity.
This is an example of the flexibility that can arise through
use of an ion beam.

By this point, the parameters of the target have been
established in some detail. Target holders are available that
are compatible with the ion beam irradiation area. The sample
is also presumably well known, and the problem now becomes how

to reduce the sample, which is usually in an awkward form, into a usable target. This problem is so strongly dependent on the specific inquiry that one is pursuing that little can be said that is of general applicability. All that can be done is to point out some considerations, and illustrate methods that have been found useful in our work at Davis.

1. Is the target going to be gaseous, liquid, or solid?

Gaseous and liquid targets require confinement in order to be compatible. The only key point is that the windows should not be visible to the detector through proper low-Z (graphite, plastics) collimation.

2. If the target is solid, is the target going to be thin (passing the beam), or thick (stopping the beam)?

In the latter case, procedures are quite different and will not be described here.[17]

3. Assuming the target is solid, is a backing required?

Some samples, such as leaf cuttings, hair strands, and such items, do not require any backing. A leaf segment, cut with scissors to a suitable area, clipped into a target holder, and dried, is a perfectly suitable target for elements potassium and heavier. Generally, one is not so fortunate, and a backing is usually required.

4. What type of backing should be used?

From earlier discussions, it should be as thin as possible for best sensitivity and contain no elements heavier than fluorine. The limit will be set by how carefully one is willing to handle the target, and how easy it is to put the sample onto the backing. Three thickness regimes are common:
   a) thin, 10 - 50 $\mu g/cm^2$ foils of graphite, Zaponite, etc.
   b) medium, 600-1000 $\mu g/cm^2$ foils and filters of commercially available plastics such as Mylar, Kapton,[14] Teflon, cellophane, and Nuclepore filters in polycarbonate substrates.
   c) thick, 3000-10,000 $\mu g/cm^2$, generally filters such as made by Gelman, Millipore, and papers such as Whatman 41.
The Davis system normally operates in the medium thickness regime, but occasionally it handles thick backings.

5. How can the sample be placed on the backing?

There are both physical (freezing, drying, ashing, filtering, etc.) and chemical (chemical reductions such as chelation, coagulation, etc.) methods for taking a sample (gas, liquid, or solid) and making it into a solid target. For gaseous samples, it is easy to remove particulate admixtures through filtering and these filters can often be used directly as a target. In this case, reduction to a value demanded by most users ($\mu g/m^3$) requires a knowledge of the number of cubic meters of air that passed through each cm^2 of the filter, so that (a ng/cm^2)$\times$(b cm^2/m^3)=(c $\mu g/m^3$). Particulate components of gaseous samples can also be removed[18] through impaction devices onto smooth, impervious substrates, but the substrate must be coated with some sticky substance to allow the particulates to adhere.[19] A 3% parafin in toluene dip (5 seconds) puts on about 50 $\mu g/cm^2$ of parafin, and this appears adequate. Apiezon grease disolved in acetone has good properties, too, in this regard.

Solid samples generally must be finely ground, or particle size corrections may become a major source of uncertainty. (Table 14). These powders can be put onto filters by sucking air through the filter while stirring up the dust in a closed container, although checks should be made that no chemical differentiation has taken place in the process. Non-uniform targets can be made by placing a spot of material in the center of the irradiation area. Adhesion can be aided by spraying with clear plastic spray. This procedure also aids the stability of any target under ion beam heating and vacuum drying during analysis, and it is widely used. However, be sure to:
  a) Analyze each spray can before you use it.
  b) Include a sprayed blank in each target set.
In some cases, especially for biological materials, it may be advantageous to concentrate a solid sample by ashing (wet or dry) or otherwise reducing the very light element component. Great caution should be exercised, as many important elements can be lost through such processes. Some laboratories have had good results through such procedures, but only after very extensive checks. Even if the procedure works for one type of sample and several different elements, there is no guarantee that it will work in other cases.

For liquid samples, any procedure that results in a solid residue can be handled as a solid (ground, dusted onto a filter, or placed on a substrate). This process can result in changes, as some components may be preferentially removed during the reduction process. In drying, the residue is chemically non-uniform, with some elements (those near saturation) coming out of solution before others. One way this can be handled is through atomizer sprays that allow a myriad of small drops to strike the backing, and then dry, before more liquid is sprayed.[24]

Although each drop is non-uniform, the ensemble represents a
very uniform target of controllable thickness. Obtaining heavy
deposits by this method can be onerous, but for heavy deposits,
one can often dry and scrape up a powder residue. Freeze drying
a liquid sample can sometimes help avoid nonuniform targets.

Calculation of the minimum detectable limit of the analysis in
terms meaningful to a user may be handled in the following manner.
Suppose one knows $(\rho t)_{s,MAX}$ set by the lightest element one
wishes to quantify. This sets the minimum detectable limit, D_z,
for some element, in ng/cm^2, for a fixed ion flux. One must know
the relationship between the original liquid sample and the dried
residue to reduce the result and the sensitivity to a unit such
as parts per million. This is obtained from the total solid
residue as a fraction of the mass of the original liquid. If
the liquid has a residue amounting to r of the liquid mass, then

$$D_f(ppm) = \frac{D_z(ng/cm^2) \; r \; (gms \; residue/gms \; liquid)}{(\rho t)_{s,max}(mg/cm^2)}$$

This illustrates how closely the fractional detectable limit is
tied to total residue, so that very low fractional limits can be
achieved from clean liquid samples because one must evaporate
large volumes of liquid to get a target of optimal thickness.

Some samples can best be handled by liquid filtration, with
the dried filter providing an excellent target. These samples involve suspended particulates in fluids, whether natural in origin
or prepared by homogenizing a physical or biological sample in
a grinder or blender with a carrier fluid. In the case of natural
fluids suspected of containing particulate matter, best results
seem to be achieved when the sample is filtered first, and then
dired, thus providing two separate sample reductions. Chemically,
the results are usually vastly different.

In all these techniques, good analytical practices should be
followed. The cleanliness of the working area, the volume or mass
of the samples, the thickness of the backing, and the reduction
factor from sample to target should be carefully established and
verified. A good mass balance is a must. Humidity stablization of
filter blanks is very important, as considerable fluctuation of
filter mass can occur as a function of humidity.

Once the target has been prepared, there is no assurance that
it will not change during storage before analysis, or during
analysis. While storage problems do not appear serious in most
cases, target degradation under vacuum and ion beam irradiation
has been observed. Extreme examples involve the drilling of neat,
rectangular holes in a membrane filter (Gelman GA-1, $3.6 \; mg/cm^2$)
with 10 μcoulombs of 16 MeV alphas run at a power level of about
500 $mwatts/cm^2$. Entrapped water in the filter appeared to be the

reason for this effect, not the destruction of the substrate through heating.[20] Many substrates will show a darkening of the material after 10 μcoulombs irradiations at 300 mwatt/cm^2 rates. However, <u>for the case of atmospheric aerosols</u>, negligible change in analytical results occur for these damage levels. Specifically,[21]

1. No change ($\pm$ 5%) was seen after a sample was exposed to vacuum for 24 hours (1 micron).
2. Irradiation to 640 μc (64 times the normal irradiation) showed a possible 20% loss of bromine from a PbBrCl aerosol target on mylar. This then predicts an 0.4% loss during normal irradiation, if the effect is linear with dose. Dose rates were normal.
3. Random re-analysis of Nuclepore targets after one year's storage (room temperature) showed no changes in the value of any element silicon or heavier, to $\pm$ 3%.
4. Analyses done first by XRF in air, then IXA in vacuum, and then XRF in air, showed no significant variation ($\pm$ 5%) in any element heavier than calcium (lighter elements were not available from our XRF at that time).

To say that such changes do not occur in other targets would be very unwise. However, covering the target with a thin coating of plastic spray appears to reduce such effects, based on nuclear physics experience with mercury-containing targets.

In summary, target preparation is the area of IXA that requires the greatest care and caution. Most people with access to energetic ion beams are physicists, who generally have little trouble with vacuum problems, electronic data collection, computer reduction of results, and such areas in an IXA program. Sample reduction to targets requires different skills and approaches, and it is very useful to have a collaborating analytical chemist on any such program.

F. Estimation of Analytical Costs

Realistic estimates of the cost incurred in performing an analysis of a sample must be developed if charges are to be made to external users. As was mentioned in the previous section, most practitioners of IXA at this time are physicists, not economists, and the errors introduced in estimating analytical costs can be many times larger than would ever be acceptable in quoting a result. This error, if propogated into the analysis of a large number of samples at a predetermined cost, can cause severe financial stress to any laboratory.

One method that can be followed in costing out analyses is to predetermine a target cost figure ($/sample) and then to adjust N_i so that the fixed cost (i.e. fixed analysis time in many cases) can be met. This means that the detectable limit will fluctuate from analysis to analysis, when the samples vary a lot.

The other method, running at a fixed detection limit and N_i but allowing the irradiation time to vary (generally because a sample produces a target so heavily loaded that the count rate capability of the system is reached, thus requiring longer irradiation times for fixed N_i), can produce rather impressive cost overruns.

The components that enter into the cost of an analysis at the UCD/ARB system are shown in Table 15. The assumption is that the cost of the accelerator beam on target is charged in a non-subsidized basis. This means that, except to the capital cost of the accelerator, all costs with operating the device, maintaining it, upgrading it, providing space, supplies, etc., are charged to every user on a beam charge basis. Such costs, when they have been calculated, vary widely from accelerator to accelerator. Large sector-focussed accelerators such as the one at Davis generally charge between $75. and $200./hour of beam. Smaller accelerators cost less, but when such calculations are made fairly, costs are often found to range from $15. to $40./hour, even in quite small machines tucked in the basment of a Physics Department. I would not advise anybody to raise the curiosity of a buisiness manager about this matter unless it is absolutely essential.

<u>Table 15</u>

<u>Typical Cost Calculations Done for the UCD/ARB Aerosol
Analysis System</u>

Assumptions: There are 630 targets on hand. Each is to be run
115 seconds, clock time (about 10 μc for aerosols).
It takes 10 minutes to load 100 targets into the
target changer, and 5 seconds to go from target to
target. Thus, 22 hours are required to run the set.
Validations and checks take two hours of beam time.

Charge Catagory	Calculation	Cost	Cost/target
1. Accelerator beam ($90./hr)	a) 2 hours checks b) 22 hours run	$180. $1980.	$0.29 $3.14
2. Labor ($ 5./hr)	a) Mounting. listing.. − b) 26 hours during run $130. c) Post-run wrap-up − d) Inspection of results −		$0.20 $0.21 $0.10 $0.50
3. Supplies	a) Detector replacement − b) Liquid nitrogen,etc. − c) Spectrum copy,mag.tapes		$0.20 $0.05 $0.25
4. Indirect costs	34.5% of all direct costs		$0.60

TOTAL COSTS PER ANALYSIS $5.54

G. Validation of System Operations

The quality of the information delivered by IXA systems and
the very low cost per elemental identification guarantee a bright
future for the technique. Nevertheless, an IXA system is considerable more complicated than most alternative analytical methods, and
there are many possibilities for system failure. Thus, even after
a system has been completely calibrated and checked, there are few
assurances that the very next target analyzed is not in error.
Attempting to set reasonable constraints on what is an essentially
unprovable premis (the system never makes mistakes) is an unend-
ing task, yet one that is vital to building confidence in proper
system operation among actual and potential users. This problem is
compounded by the fact that an IXA system delivers so many results
per analysis due to its multielement capabilities, and each of these
many elements is supposed to be correct and proven to be accurate.
There are two aspects to such a validation program – internal, within the laboratory, and external, involving interlaboratory and
intermethod comparisons.

Internal Validations. The first component of an internal
validation program involves guaranteeing proper operation of all
system components during each analysis run. A detailed check list
must be developed and faithfully followed before any analysis run.
At Davis, this list consists of some 80 or so items, including
vacuum system, target handler, beam integration system, detector,
filters, and in-cave electronics, counting room electronics, computer hardware, beam pulsing system, beam sweeping system (at 1 hz),
and computer software. The ion beam characteristics must be set
(ion, energy, beam spatial configuration). During the run, the
following parameters are monitored by the system operator:
  1. Intermittently – during each tray change
        Faraday cup vacuum   – by gauge in cave
        Target changer vacuum   – by gauge in cave
        Tray number   – 3 trays per loading
  2. Continuously – during the run
        Beam current   – meter on beam integrator
        Count rate   – count rate meter
        Beam pulsing system   – Oscilloscope (Figure 12)
        Slide number   – T.V. camera
        Beam sweeper   – T.V. camera (mylar glows)
        Spectrum; RACE operations   – Display tube; hard copy
The computer records all run parameters, including conversion
factors, and prints out the results on-line about 20 seconds after
the data have been acquired. It is the operator's responsibility to
periodically check these results to insure that dead time is
small, calibrations are still good, etc.

Once the beam parameters have been established at the beginning of the run (they are then monitored by the accelerator operator during the run), the first task is to run a standards tray in order to obtain an energy and amount calibration. Once the energy calibration is set, about 12 gravimetric standards are run using the predetermined table of $F_{x,z}$ values stored in the computer. These range from NaCl to Pb, and the values are compared to the gravimetric values through:

$$\text{Ratio} = \frac{(\rho t)_z \text{ gravimetric}}{(\rho t)_z \text{ system}}$$

This ratio is almost always close to unity. If the ratio differs from unity by less than $\pm$ 3%, no changes are made,and the system is ready to move to the next step. If the ratio lies between 3% and about 10% away from unity, checks are made to see if any source of error can be detected. If not, the assumption is made that the accelerator has delivered a beam energy slightly at variance with the calibration energy, and a correction is entered into the code as a normalization designed to bring the ratio to unity. Any larger error stops the run until the problem is found.

The next task is to run realistic targets, occasionally from previous runs, and the full computer print-out of the results is the final check on the system. The run is then started. Gravimetric standards are occasionally placed at the beginning of a tray (the first slot is reserved for standards, as the second is for blanks). Samples run early in the run are often reanalyzed late in the run to check for drift. Any variation during the run results in the entire run being scrapped, so that there is real incentive to do regular checks.

It should be evident that a significant amount of time is involved in these checks, so that at Davis, the system is generally only run once every two weeks. This allows samples to be stored until enough are available for a run of at least 18 hours. Normal rates of analysis are between 600 and 1000 targets per day.

The re-analysis of targets previously analyzed as part of the normal work load is a very useful validation of the system, as it bears upon the routine nature of the operation, and not a special test run. This program, labelled "random re-analysis", generally involves about 10 targets at a time, drawn from analyses done in the previous year. Such tests done at Davis have always resulted in mean ratios within $\pm$ 5% of unity for all elements silicon and heavier.

Once all these checks and tests have been done, the system may be considered as "internally validated". A table of such tests and validations, including normal correction factors, should be given each year to all users. An example, prepared as part of the

Table 16

Internal Validations of the UCD/ARB System [1]

Summary of Analytical Corrections and Uncertainties

1. Statistical uncertainty in peak counts   –   variable
2. Uncertainty in gravimetric standards   –   ± 5%
3. Integration of Ion beam   –   ± 2%
4. Ion beam attenuation   –   < 5%
5. Electronic corrections, accuracy   –   ± 5%
6. Peak integration and background
      uncertainties   –   ± 7%, (variable)
      Thus, reproducibility   –   ± 5%
             system uncertainty   –   ±10%
7. Loss of volatile elements (beam +
      vacuum)   –   < 5%
8. Particle Size Corrections

	Stage #1	Stage #2	Filter Stage
Na	2.49	1.32	< 10%
Mg	1.96	1.22	"
Al	1.68	1.16	"
Si	1.49	1.12	"
P	1.37	1.09	"
S	1.26	1.07	"
Cl	1.31	<10%	"
K	1.12	<10%	"
others	<10%	<10%	"

9. Loading Corrections (for 120 $\mu g/m^3$ aerosol)

	Stage #1	Stage #2	Filter Stage
Na	<10%	1.70	1.22
Mg	"	1.39	1.11
Al	"	1.23	1.07
Si	"	1.14	1.04
others	"	<10%	<10%

Uncertainty in particle size and loading corrections are estimated
at ± 20% for most ambient aerosols, although, especially for the
particle size corrections, worse cases are common. These corrections thus increase the total error for the four lightest
elements.

ARB work, is included as Table 16. In this table, "Stage 1" aerosols
include particles between 3.6 and 20 microns, "Stage 2" between
0.65 and 3.6 microns, and "Filter" between about 0.1 and 0.65 microns.
The filter was Nuclepore, and thus had negligible emmeshment.

External Validations. At this point, the IXA system must still pass the acid test of external, interlaboratory comparisons before the results will be widely accepted. It is one of the most remarkable achievements of laboratories doing IXA work that their results have been shown to be consistently accurate in such tests. IXA, XRF, and NAA have all demonstrated the ability to generate results accurate to a few percent for a wide variety of elements in a wide variety of matrices, all doing better in this regard than some other widely used methods. [22]

Before participating in such tests, however, it is wise to compare results on an informal basis with other laboratories and other methods. The existence of a validated XRF system at CNL in Davis is a resource in such checks for both methods. Normally, the XRF system is used for all samples thicker than 10 mg/cm^2 and the IXA for all samples thinner than 1 mg/cm^2, with some area of over-lap, and target handling and data reduction are fully compatible.

If all these checks appear satisfactory, participation in a formal interlaboratory comparison is the next step in validation. Such comparisons have recently been run by a consortium of Columbia Scientific Industries, Lawrence Livermore Laboratory, and Battelle Northwest[22], and the National Bureau of Standards[23]. The first comparison involved pre pared targets of solutions standards, rock dusts, and aerosols, while the second involved samples of gasoline, fuel oil, coal, and fly ash. It is fervently hoped that support be given to such work so that these comparisons can be continued on a regular basis.

Obviously, if pre-prepared targets are being used, great care should be taken to verify that the target is suitable for the IXA system. Often, they will be too thick, and large correction factors may be involved. Treat the whole procedure with great conservatism, as a faulty result can get spread far and wide.

No laboratory using IXA cooperated in the N.B.S. comparison, but seven laboratories were involved in the CSI/LLL/BNW comparison which was reported at the annual Denver x-ray meeting. These laboratories did well as a group, especially in their ability to handle diverse targets and a large number of elements. These laboratories were:

Department of Chemistry   Department of Physics
Brigham Young University   Duke University
Provo, Utah 84602   Durham, North Carolina
        U.S.A.   U.S.A. 27706

Departments of Physics and
      Oceanography
Florida State University
Tallahassee, Florida,U.S.A.
              32306

Illinois Institute of
   Technology Research
   Center
Chicago, Illinois,U.S.A.

Lund Institute of Technology
Solvegatan 14, S-223 62
Lund, SWEDEN

Niels Bohr Institute
University of Copenhagen
Copenhagen, DENMARK

Department of Physics and the
      Crocker Nuclear Laboratory
University of California
Davis, California, U.S.A.
              95616

V. ION-EXCITED X-RAY ANALYSIS PROGRAMS

The flexibility of ion-excited x-ray analysis to meet the
needs of a variety of users is bound to result in operational IXA
systems optimized in many different ways. While the UCD/ARB system
has been described in detail in order to illustrate the nature of
the choices that have to be made, it is most unlikely that any
other system will ever look precisely the same. It appears highly
desireable that many systems should have some elements in common,
including the ability to use 35 mm slide frames as a standard
mounting in order to aid intercomparisons and share workloads.
In any case, a growing number of laboratories could write an
equivalent but highly variant description of their optimization
programs, as demanded by the purposes to which the system is
being directed. As publications from these laboratories continue
to appear, the nature of these choices is becoming clearer. The
bibliography that accompanies this chapter is an attempt to
make this literature more accessible, although gross errors of
omission have doubtless been committed. After all, there are over
100 accelerators in the United States alone that are well suited
for IXA programs. Since the variety of potential users in the
biological, Physical, environmental, medical, material, and forensic
sciences and applications is enormous, one can be assured that an
increasing number of IXA programs will be established, each contributing ideas and insights into this vigorous area of research.

APPENDIX

Forward Scattering

Although this report is on IXA analysis, it is very much in keeping with the previous treatment to include a few words on ion scattering analysis (ISA) for very light elements, H → F, that can not be handled by IXA due to matrix corrections.

If one limits target thickness (backing and reduced sample) to less than about 1 mg/cm^2, elastically scattered alpha particles at angles greater than about 45° and protons at backward angles can be easily used to quantitatively detect elements fluorine and lighter.[24] With thinner targets and higher energies, elements heavier than fluorine have been analyzed. A major reason that the UCD/ARB uses 18 MeV alpha beams is that this ion/energy combination allows good IXA analyses to be done simultaneously with ISA analysis, thus weighing the target with an ion beam. These data are useful in air quality work, since about 2/3 of all airborne particulate matter consists of very light elements. Analysis of clean waters and some problems in organic chemistry may also benefit from such an analysis. The major new problems that arise involve the presence of elements of interest in the blanks. Fluctuations in blank values thus set the minimum detectable amount of some elements, rather than peak counts or smooth backgrounds. Otherwise, much of the work described in this chapter is relevant to this new method of analysis. Minimum detectable limits are in the order of micrograms per cm^2, rather than nanograms per cm^2, for a 10 μc analysis of a 600 g/cm^2 mylar target. Figure 20 shows a typical spectrum for NaCl on mylar for 10 μc of 18 MeV alph particles at 55° lab. Note the hydrogen peak from the ^{1}H(α,p)^{4}He scattering.

Acknowledgements

Much of the description of the UCD/ARB system is drawn from documents developed by the Air Quality Group in a cooperative effort, with the major contributors and their areas of effort indicated on publications of the group. However, special recognition is due for the efforts of Bill Cline, Bob Eldred, Pat Feeney, Bob Flocchini, Jim Harrison, and Dan Shadoan for the IXA system and Gordon Wolfe for the ISA work. This work was supported by the University of California's "project Clean Air", The California Air Resources Board, and the TraceContaminants Division of the National Science Foundation, Research Applied to National Needs. The staff of the Crocker Nuclear Laboratory has supported this effort enthusiastically from the beginning. I would also like to thank Jim Ziegler for his advice and encouragement in this work, as well as for alerting me to the implications of the Student test, equation III.5.

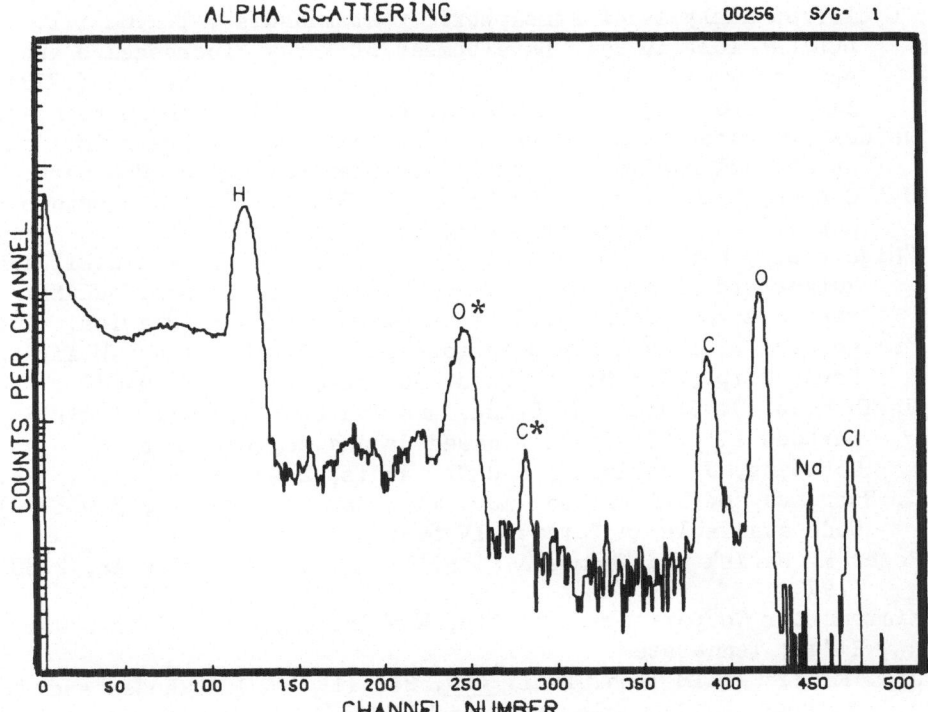

Figure 20. Spectrum of 10_2 µc of 18 MeV alphas on NaCl layered on a 600 µg/cm² mylar backing. Detector angle, 55°.

REFERENCES

1. "Monitoring of Smog Aerosols with Elemental Analysis by Accelerator Beams" , T.A.Cahill, R.G.Flocchini, R.A.Eldred, P.J. Feeney, S. Lange, D. Shadoan, and G. Wolfe, Proc. 7th Materials Research Symposium, National Bureau of Standards (1974); "Regional Monitoring of Smog Aerosols", T.A.Cahill and R.G.Flocchini, report to the California Air Resources Board, UCD/CNL Report 184 (1974)

2. F. Folkmann, C, Gaarde, T. Huus, K. Kemp, Nucl. Instr. and Methods 116, 487 (1974)

3. R.G.Flocchini, P.J.Feeney, R.L.Sommerville, and T.A.Cahill, Nucl. Instr.& Methods 100, 397 (1972); T.A.Cahill, R.G. Flocchini, P.J.Feeney, and D.J.Shadoan, Nucl. Instr. & Methods 120,193 (1974)

4. "Cyclotron Analysis of Atmospheric Contaminants", T.A.Cahill, UCD/CNL 162 (1972); "Development of X-ray Fluorescence and Applications", F.P.Brady and T.A.Cahill, UCD/CNL 166 (1973), final report to the RANN division, National Science Foundation

5. Values presented by J. Cooper, 4/73, Battelle Northwest mmeting on aerosol analysis (Seattle). Oxygen value from UCD work.

6. J.A.Cooper, Nucl. Instr.& Methods $\underline{106}$, 365 (1973) and numerous unpublished results from many laboratories.

7. "High Energy Nuclear Data Handbook", W. Galbraith, W.S.C.Williams, Rutherford Laboratory, Chilton, Didcot, Berkshire,ENGLAND, reprinted on pocket cards by Reviews of Modern Physics.

8. "Large Area Silicon X-Ray Detectors", R.S.Frankel and D.W.Aitken, Kevex Corp., 898 Mahler Road, Burlingame, Calif. 94010

9. H. Thibeau, J. Stadel, W. Cline, and T.A.Cahill, Nucl. Instr.& Methods $\underline{111}$, 615 (1973); based upon a suggestion by F. Goulding, J. Jaklevic, and D. Landis, Lawrence Berkeley Lab.

10. J. Harrison and R.A.Eldred, Adv. in X-Ray Analysis $\underline{17}$, 560(1973) Code available in Fortran IV from UCD.

11. Roger K. Wyrick and Thomas A. Cahill, P hysical Review A8, 2288 (1973)

12. Micromatter Corporation, Seattle, Washington; Ortec, Inc., Oak Ridge, Tennessee

13. J.R. Rhodes, ASTM Tech. Publ. $\underline{485}$, 243 (1971); J.R.Rhodes and C. B. Hunter,X-Ray Spectrometry $\underline{1}$, 113 (1972);"Self Absorption Corrections",T.G.Dzubay and R.O.Nelson (1974) Denver X-ray.

14. Trademark, Dupont Corporation, Mylar $(C_{10}H_8O_4)_n$; Kapton, $(C_{22}H_{10}N_2O_4)_n$

15. Trademark, Gelman Corporation

16. Trademark, Nuclepore Corporation, Vallecitos, California

17. R.L.Walter, R.D.Willis, W.F.Gutknecht, and J.M.Joyce, Anal.Chem. $\underline{46}$, 843 (1974)

18. D.A.Lundgren, J.Air Pollution Control Assoc. $\underline{17}$, 4 (1967)

19. "Contribution of Freeway Traffic to Airborne Particulate Matter", T.A.Cahill and P.J.Feeney, UCD/CNL 169 (1973). Report to the California Air Resources Board. (NTIS)

20. Unpublished observation, T.A.Cahill

21. Ref.4, plus unpublished study for the Calif. Dept. of Public Health, Air and Industrial Laboratory, Berkeley

22. D.C.Camp, J.A.Cooper, and J.R.Rhodes, X-Ray Spectrometry $\underline{III}$, #1 (1974); D.Camp, A.N.S. Transactions $\underline{17}$, 106 (1973); D. Camp, Proc. 23rd Annual Conference on Applications of X-Ray Analysis, Denver (7/1974).

23. Proc. 7th Material Research Symposium, National Bureau of Standards, (10/1974).

24. R.K. Jolly and H.B. White, Nucl. Instr. and Methods $\underline{97}$, 103 (1971).

Bibliography

Ion-Excited X-Ray Analysis

This partial bibliography includes some materials that can be described as "fugitive", in that they consist of final reports, progress reports, etc., that will never appear in the literature. Yet, these reports are often the only detailed documentation that exist on a system. As such, they have been included when their existence was known to the author. Information on other literature that has been omitted will be gratefully received.

1970

Coockson, J. and M. Poole, 1970, New Scientist, p. 404.

Duggan, J. L. (editor), March 1970, The Second Conference on the Use of Small Accelerators in Teaching and Research, AEC Conf.-700322.

Duggan, J. L., 1970, "The Use of Small Accelerators in Teaching", Symposium on the Use of Small Accelerators (New York: American Nuclear Society.).

Johansson, T. B., and R. Akselsson, S. A. E. Johansson, 1970, "X-ray analysis: Elemental trace analysis at the 10^{-12} g level", Nuclear Instruments and Methods 84, 141-143.

Needham, P. B. and J. R. Sartwell, B. D. Sartwell, 1970, Adv. X-ray Anal. 14, 184.

Spaulding, J., 1970, "Charged Particle Induced X-Ray Experiments", The Second Conference on the Use of Small Accelerators, AEC Conf.-700322.

Jarvis, O. N., and C. Whitehead, M. Shah, 1970, AERE Report No. R 6612, (unpub.).

1971

Akselsson, B and T. B. Johansson, 1971, "A beam mapping method", Nuclear Instruments and Methods 91, 663-664.

Cahill, T. A. and R. G. Flocchini, R. Sommerville, P. K. Mueller,
    A. Alcocer, March 1971, "Chemical Analysis of Atmospheric
    Aerosols: Alpha-Excited X-Ray Emission", Amer. Chem. Soc.,
    Kendall Symposium, Los Angeles (March, 1971).

Cahill, T. A. and R. Sommerville, R. Flocchini, August 1971,
    "Elemental Analysis of Smog by Charged Particle Beams:
    Elastic Scattering and X-Ray Fluorescence", Nuclear Methods
    in Environmental Research, A.N.S. Symposium, Columbia,
    Missouri, Pg. 6.

Campbell, J. L. and P. O'Brien, L. A. McNelles, 1971, Nuclear
    Instruments and Methods 92, 269.

Duggan, J. L. and W. M. Beck, August 1971, "X-Ray Trace Analysis
    Experiments," The Conference on the Use of Nuclear Tech-
    niques in the Solution of Environmental Problems held in
    Columbia, Mo.

Johansson, T. B. and R. Akselsson, S. A. E. Johansson, 1971, Adv.
    X-Ray Analysis 15, 373.

Jolly, R. K. and M. B. White, 1971, Nuclear Instruments and
    Methods 97, 103.

Watson, R. L. and J. R. Sjurseth, R. W. Howard, 1971, "An investi-
    gation of the analytical capabilities of X-ray emission
    induced by high energy alpha particles", Nuclear Instruments
    and Methods 93, 69-76.

1972

Cahill, T. A., October, 1972, Report to the California Air
    Resources Board and Project Clean Air, University of Cali-
    fornia, Davis, Rep. No. UCD-CNL 162.

Cahill, T. A., 1972, Bull. Am. Phys. Soc. 17, 505.

Cahill, T. A., 1972, "Accelerators in Env. Analysis", Transactions
    A.N.S. 15, 178.

Cahill, T. A. and R. Sommerville, 1972, "Automatic Analysis of
    Atmospheric Aerosol for Research and Monitoring Purposes",
    Transactions A.N.S. 15, 70.

Cookson, J. A. and A. T. G. Ferguson, F. D. Piling, 1972, "Proton Microbeams, their production and Use", Chemical Analysis by Charged Particle Beams, 39 (1972); Elsevier Sequois S. A., Lausanne; J. Radioanal. Chem. 12, 39.

Deconninck, G., 1972, "Quantitative Analysis by (p,X) and (p, gamma) reactions at low energies", Chemical Analysis by Charged Particle Bombardment, 157, Elsevier Sequoia S. A., Lausanne.

Duggan, J. L. and W. L. Beck, L. Albrecht, L. Munz, J. D. Spaulding, 1972, Adv. X-Ray Analysis, 15, 407.

Flocchini, R. G. and P. J. Feeney, R. J. Sommerville, T. A. Cahill, 1972, Nuclear Instruments and Methods 100, 397.

Gordon, B. M. and H. W. Kraner, 1972, Trace substances in environ- mental health-V (ed. D. D. Hemphill; University of Missouri Press, Columbia, Missouri) p. 231.

Gordon, B. M. and H. W. Kranerm, 1972, "On the development of a system for trace element analysis in the environment by charged particle x-ray fluorescence", Chemical Analysis by Charged Particle Beams, 181; Elsevier Sequois S. A. Lausanne, J. Radioanal. Chem. 12, 39.

Lochmuller, C. H. and I. Galbraith, R. Walter, J. Joyce, 1972, Anal. Lett. 5, 943.

Perry, S. and F. P. Brady, T. A. Cahill, R. Flocchini, N. S. P. King, R. Wyrick, 1972, Bull. Am. Phys. Soc. 17, 120.

Purser, K. H., 1972, The University of Rochester, reprint UR-NSRL-72.

Rudolph, H. and J. K. Kliwer, J. J. Kraushaar, R. A. Restinen, W. R. Smythe, May 1972, Proc. 18th Ann. ISA Analysis Instru- mentation Symp. (San Francisco) p. 151.

Rudolph, H. and R. A. Ristinen, W. R. Smythe, A. Alfrey, S. R. Contiguglia, 1972, Bull. Am. Phys. Soc. 17, 548.

Stupin, D. M. and P. Fintz, A. Gallmann, H. E. Gove, G. Guillaume, A. Pape, J. C. Sens, October 1972, Rapport interne, CRB-LPNIN 7202, Centre de Recherches Nucleaires et Universite Louis Pasteur, Strasburg, France.

Verba, J.W., J.W.Sunier, B.T.Wright, I.Slaus, A.B.Holman, J. Kulleck
    "Chemical Analysis with Charged Particle Beams", 1972, 171
    Elsevier Sequois S.A., Lausanne, and J. Radioanal. Chem. 12.

1973

Barnes, B.K., L.E. Beghian, G.H.R. Kegel, S.C. Mathur, P.W. Quinn,
    Bull. Am. Phys. Soc. No. EO 13, Wash., D.C., 4/73

Bond, C.D., L.S. August, P. Shapiro, C.M. Davisson, W.I. McGarry,
    Bull. Am. Phys. Soc. No. EO 12, Wash., D.C., 4/73

Brady, F.P., T.A. Cahill, "Development of X-ray Environmental Instr-
    umentation", U.Calif., Davis, report CNL-172 (1973)

Cahill, T.A., P.J. Feeney, "Contribution of Freeway Traffic to Air-
    borne Particulate Matter", U.Calif., Davis report CNL-169, to
    the California Air Resources Board, June, 1973 (N.T.I.S.)

Cahill, T.A., R.G.Flocchini, P.J.Feeney, "Alpha Particle Excitation
    using Cyclotrons", Transactions American Nuclear Society
    17, 102 (1973)

Cooper, J.A., Advan. X-ray Anal. 16, (1973);

Cooper, J.A., Nucl. Instr. and Methods 106, 525 (1973)

Hansen, J.S., J.C. McGeorge, D. Nix, W.D. Schmidt-Ott, I. Unus,
    R.W.Fink, Nucl. Instr. and Methods 106, 365 (1973)

Herman, A.W., L.A. McNelles, J.L. Campbell, J. Applied Radiation
    and Isotopes 24, 677 (1973)

Herman, A.W., L.A.McNelles, J.L. Campbell, Nucl. Instr. and Methods
    110, 429 (1973)

Nelson, J.W., I. Williams, T.B. Johansson, R.E. Van Grieken, K.R.
    Chapman, J.W.Winchester, IEEE Nuclear Science, November, 1973

Perry, S.K., F.P. Brady, Nucl. Instr. and Methods 108, 389 (1973)

Thibeau, H., J. Stadel, W. Cline, T.A. Cahill, Nucl. Instr. and
    Methods 111, 615 (1973)

Umbarger, C. J. and R. C. Bearse, D. A. Close, J. J. Malanify,
    1973, Advan. X-Ray Anal. 16, 102-110.

Valkovic, V. and D. Miljanic, R. M. Wheeler, R. B. Leibert, T.
    Zabel, G. C. Phillips, 1973, _Nature_, 243, 543.

Zabel, T., 1973, M. Sc. Thesis, Rice University.

                                1974

Akselsson, R. and R. B. Johansson, 1974, _Z. Physik 266_, 245.

Azevedo, J. and R. T. Flocchini, T. A. Cahill, P. R. Stout, 1974,
    J. Environ. Quality 3, 171.

Cahill, T. A. , R. G. Flocchini, P. J. Feeney and D. J. Shadoan,
    Nucl. Instr. and Methods 120, 193 (1974).

Camp, D. C. and J. A. Cooper, J. R. Rhodes, 1974, _X-ray Spectrom-
    etry 3_, 47.

Campbell, J. L. and A. W. Herman, L. A. McNelles, R. A.
    Willoughby, 1974, "Advances in X-ray analysis", Pergamon
    Press, Vol. 17.

Folkmann, F. and C. Gaarde, T. Huus, K. Kemp, 1974, "Proton
    Induced X-ray Emission as a Tool for Trace Element Analysis",
    Nuclear Instruments and Methods 116, 487.

Harrisson, J. and R. Eldred, 1974, "Advances in X-ray analysis",
    Pergamon Press, Vol 17.

Mandler, J. W., and R. A. Semmler, "Trace Element Analysis Using
    Charged Particles", _Nuclear Methods in Environmental Research_
    (ANS), Columbia, Missouri (1974).

Robinson, D. C., N. E. Whitehead and G. E. Coote, "Some Analysis
    of Air Filters by Proton-Induced X-ray Fluorescence", Inst.
    of Nuclear Science, R-131, Lower Hutt, New Zealand.

Thomas, J. P. and L. Porte, J. Engerran, J. C. Viala, J. Tousseel,
    1974, _Nucl. Instr. & Methods 117_, 579.

Valkovic, V. and R. B. Liebert, T. Zabel, H. T. Larson, D. Miljanic,
    R. M. Wheeler, G. C. Phillips, 1974, _N.I.M. 114_, 573.

Walter, R. L. & R. D. Willis, W. F. Gutknecht, J. M. Joyce, 1974,
    Anal. Chem. 46, 843.

CHAPTER 2

MATERIAL ANALYSIS BY NUCLEAR BACKSCATTERING

INTRODUCTION: James F. Ziegler

APPLICATIONS: James W. Mayer

FORMALISM: Wei-Kan Chu

MATERIAL ANALYSIS BY NUCLEAR BACKSCATTERING

Introduction

James F. Ziegler

IBM - Research

Yorktown Heights, New York 10598

Historically, the first use of ion backscattering for material analysis was by Geiger and Marsden[1] (1909), whose effects were explained by the Rutherford[2] atomic model (1911). This was followed immediately by suggestions of ion channeling in single crystals by Stark and Wendt[3] (1912). Recently, the most widely publicized example of nuclear backscattering was the first atomic compositional analysis of the moon by Surveyor 5 (1967). This first soft-landing package contained a radioactive source of 5MeV α-particles which were backscattered from the moon's surface and provided information on the elemental composition of the moon's surface.[4,5]

Nuclear backscattering is useful in obtaining the concentration vs. depth profiles of elements in the upper micron of smooth solids. It is a powerful analytic tool because it is simple, quantitative, reliable, and non-destructive. Its primary limitation is that it provides only limited information on more than half the samples one might be interested in. In simplest terms: if a sample can be clearly analyzed by nuclear backscattering it will provide fast and quantitative profiles, nondestructively and reliably.

In this chapter we shall explain the basic concepts of nuclear backscattering by considering in detail one experiment. This experiment will show the primary physical processes and the limitations of the technique. There are other experimental arrangements which involve other ion beams and energies. These will be illustrated in the next chapter.

The basic experimental arrangement is shown in Figure 1. The ion beam is usually H or He because reliable and accurate detectors are available (silicon surface-barrier detectors) to detect these backscattering ions and measure their final energy. The energy chosen for the beam is usually kept below the threshold for nuclear reactions, typically H^+ at 200-400 keV, or He^+ at 1-3 MeV. These energies allow analysis of the upper micron of the surface and are below the nuclear reaction threshold for most elements with atomic number above 8.

As an example we shall consider a thin film of Hf on a silicon substrate.[6,7] Most of the ions will penetrate to depths of $\sim$ 10 microns and produce no measurable damage. About 10^{-5} of the projectiles will approach a target nucleus close enough to undergo a large-angle scattering. Those which are backscattered will come back out of the target and can be detected by a silicon detector, where their final energy can be determined.

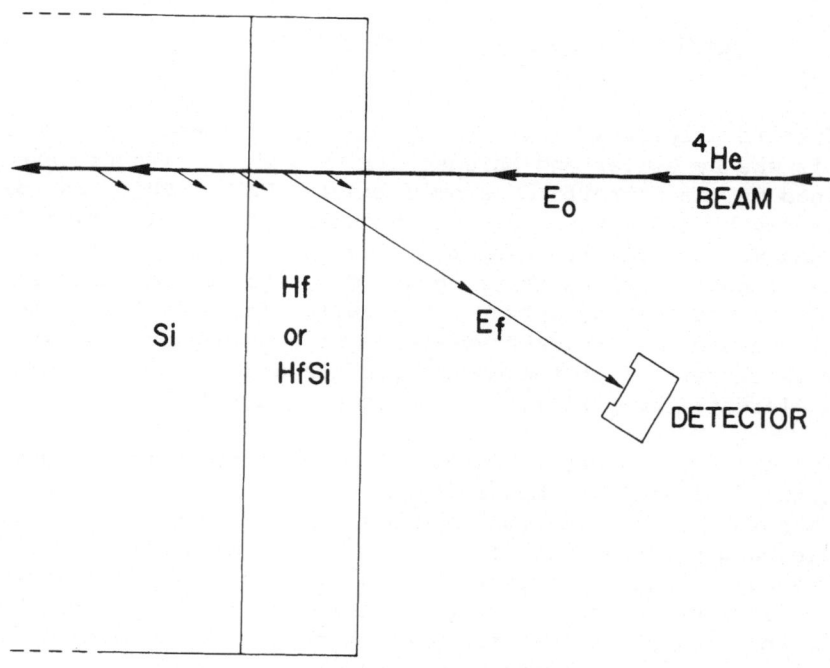

Figure 1) The basic experimental arrangement for nuclear backscattering is shown. The incident beam is usually $^1H^+$ (200-400 keV) or $^4He^+$ (1-3 MeV). Most of the beam penetrates deep into the sample. A few ions undergo a large-angle collision with target nuclei and come back out of the sample. They are detected by a surface-barrier detector, which also can determine each backscattered ion's energy to better than 1%. The target shown is a thin metal layer on a substrate of silicon. This target will be used for an example throughout this chapter.

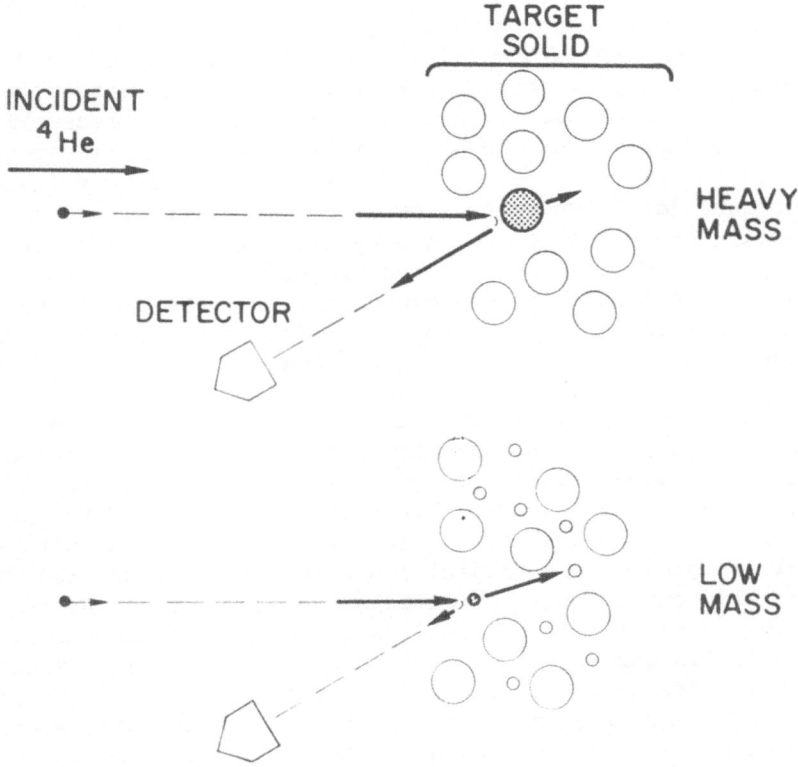

Figure 2)   The primary effect in nuclear backscattering is conservation of momentum.   This effect provides a correlation between the
final backscattered ion energy and the mass of the target nucleus
it hit.   The upper drawing illustrates that if the ion hits a very
heavy target nucleus, this nucleus will absorb little energy, and
the backscattered ion will retain most of its original energy.   In
the lowest figure it is shown that if the ion backscatters from a
light target nucleus, most of its energy will be transferred to
that nucleus (to conserve momentum), and the backscattered ion will
retain only a small part of its original energy.

     The physics of the experiment is illustrated in Figure 2. The
incident ion beam is chosen so as to allow deep penetration, but
with an energy kept below where nuclear resonance effects would
complicate the scattering. The scattering is simple Coulombic
repulsion, and is accurately described by the Rutherford cross-
section (see chapter on Formalism). The primary effect of the
scattering is to change significantly the energy of the backscattered projectile (^{4}He in the illustration) because of conservation
of momentum. If the projectile hits a very heavy target nucleus
as shown in the upper part of Figure 2, the target nucleus will
absorb little energy, and the backscattered ^{4}He ion will retain
most of its energy. If, however, it scatters from a very light
target nucleus (as shown in the lower part of Figure 2), the target
nucleus will absorb most of the energy leaving the projectile at a
low energy. Since the backscattered projectile's energy can be
measured to better than 1%, one can deduce from the final projectile
energy the mass of the target atom. Simply, we use the projectile
as a probe to mass analyse the target elements.

     The above concept is summarized in Figure 3. The upper right
inset shows an experimental diagram, with the target nucleus absorb-
ing some of the projectile's energy. The middle inset graphs the
amount of final ^{4}He energy as a function of the mass of the target
nucleus from which it scatters. The lower diagram shows with arrows
the final energies of ^{4}He projectiles which have encountered various
elements. Two other important effects are also shown. As the projectile penetrates the thin film it will lose small amounts of
energy to collisions with the electron sea of the solid. These
perturbations are almost continuous and spread out the final energies, giving information as to the total path length of the projectile in the solid. In fact, we have a concentration vs. depth profile (to first order) of each element with the high energy side
(right hand side) of each peak corresponding to atoms lying on the
surface. As we follow each individual peak to lower energies we
are looking at concentrations of that element deeper in the target.

     We therefore have two basic effects: Large changes in the
projectile energy can identify atomic masses in the target; small
changes in energy can be correlated to depth in the target.

     A final effect must be mentioned: the cross-section for scat-
tering is proportional to $(Z_{TGT})^2$, and more scattering occurs from
high atomic number target atoms, Z_{TGT}, and less from lower ones.
The technique is therefore more sensitive to small amounts of
heavier elements than to lighter ones.

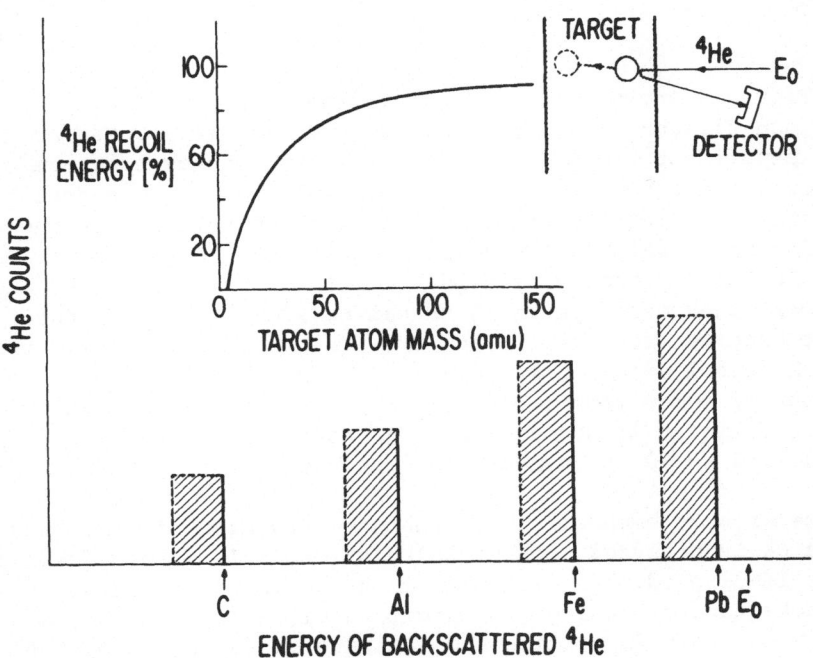

Figure 3) The experimental arrangement of nuclear backscattering
is shown in the upper right of the figure. Because of conservation
of momentum the backscattered ion (the projectile ion is assumed to
be $^4He^+$) will lose energy to the target nucleus it hits. The
amount of energy retained by the backscattered ion varies with
target nuclear mass as shown in the upper left inset. The spectrum
of final ion energies is shown in the bottom figure for various
target materials (E_o is the original ion energy). As the ion penetrates the film it loses small amounts of energy to the electron
sea of the target. These small energy losses act to spread out
each film spectrum so that ions scattering from deeper in each film
appear as counts at slightly lower energy. Thus each peak directly
gives a profile of target element concentration versus depth, with
the surface being on the high energy side of each peak, and lower
energy counts coming from below the surface.

We have simplified the physical basis of nuclear backscattering by not discussing second order effects, and variations which occur when one uses ions other than H$^+$ ($\sim$ 200 keV) and He$^+$ ($\sim$ 2 MeV). These details will be discussed at the end of this chapter. Actual numerical calculations used in nuclear backscattering are shown in the Appendix to this chapter.

Figure 4 shows actual raw data from a target of Hf ($\sim$ .15 µm thick) on silicon. The projectiles were ^{4}He$^+$ at 2.5 MeV and the total experiment time was 10 minutes. This energy spectrum of the backscattered projectiles shows how simple is the interpretation of the backscattering spectrum. At the high energy side of the spectrum is the Hf which shows a finite thickness because of its flat top. At the low energy side is the silicon substrate. We also see two small unexpected peaks which can be identified from their surface (right hand) energies as Ar and Zr. The Zr impurity comes from the basic Hf metal since normal high-purity Hf contains $\sim$ 3% Zr. The Ar impurity comes from incorporation in the metal film during sputter deposition in an Ar gas ambient. We note that the Ar is not flat topped, and shows an increase in concentration with depth.

We wish to emphasize here that the raw data of this target can be basically analyzed without calculations. It is necessary to look up the leading edge of each peak in a convenient table to convert ^{4}He backscattering energy to target recoil mass. Once this is done, one may identify impurities and estimate gross variations of concentration with depth. Actual relative concentrations can be obtained by simple slide-rule calculations (see "Formalism").

As one final point we note that the differences in cross section make the Hf peak some 20X the silicon peak.

It is instructive to carry further the analysis of the film of Hf on Si after it has been heated. We will not learn very much more about nuclear backscattering, but we will make clear that considerable insight can be obtained into physical processes in the target because of the versatility of nuclear backscattering.

Figure 5 indicates changes which would occur in the backscattering spectrum if the Hf should react with the Si and form a layer of HfSi. We note that the front edge of the Hf in HfSi remains stationary, but the peak height drops as the Hf is diluted with Si. The drop is not 50% because of changes of the energy loss of ^{4}He ions in HfSi relative to Hf (pure). The Hf peak in HfSi is wider than the original peak - and to first order the two areas

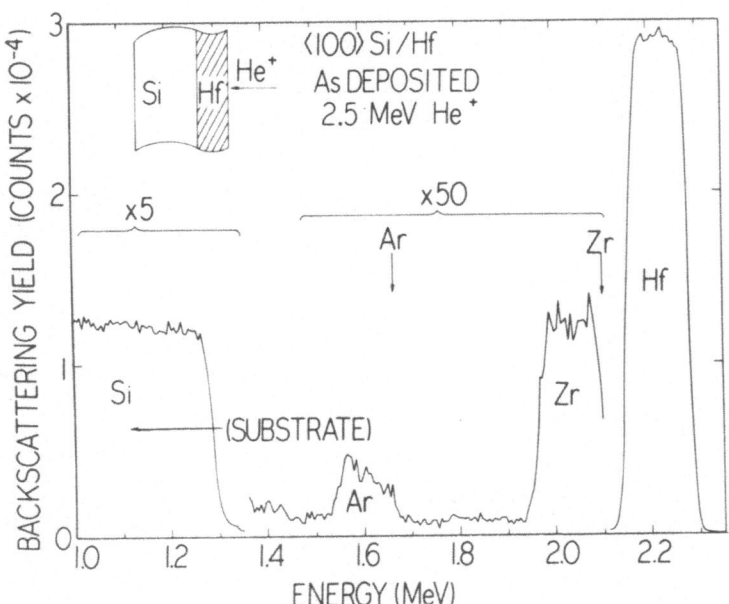

Figure 4)   Raw data is shown for [4]He ion backscattering from a target of a metal film (Hf) on a Si substrate. The Hf peak occurs at
high energy since [4]He ions backscattering from Hf retains 91% of
their original energy. This peak width indicates a Hf layer width
of .16 μm. The peak is flat topped indicating a uniform layer. At
lower energies there are two small peaks of Zr and Ar. The Zr is
naturally present in Hf to about 3% since it can not be extracted
chemically. The Ar is present because the Hf layer was sputtered
and Ar was incorporated from the sputtering ambient. The tilt to
the Ar peak indicates it is not uniform in the film. At the lowest
energy in the figure is the Si substrate peak. This spectrum was
accummulated in about 10 minutes.

will be the same since there are the same number of Hf atoms in the
target.   In the silicon peak we see a foot projecting out to the
arrow pointing to the energy of [4]He ions scattering from Si on the
target surface.   The Si in the original spectrum was displaced
below this arrow about the energy width of the Hf peak.   In the
original sample there was no Si on the surface, and [4]He scattering
from the Si substrate had to penetrate the Hf layer both going in
and out.

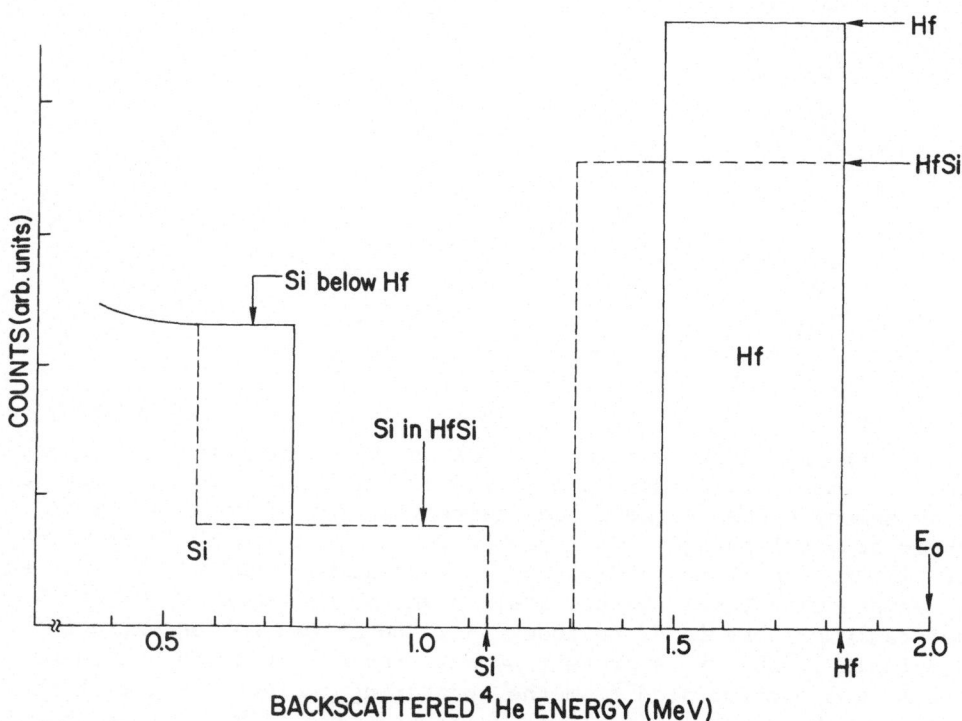

Figure 5) Simplified spectra are shown for Hf on Si (solid lines).
and for HfSi on Si (dashed lines).   The Hf peak changes with the
formation of HfSi as the Hf is reduced in density by the Si, and
the Hf layer becomes thicker.   The Si substrate peak originally is
displaced below its surface arrow because [4]He ions must penetrate
through the Hf film to reach the Si.   The Si peak is offset to
lower energy by about the width of the Hf peak.   With the formation
of HfSi, there is Si up to the Si surface arrow.

Figure 6 shows raw data from ^{4}He backscattering from three samples: One without heat treatment; one which has had a reaction between the Hf and Si forming HfSi; and a final superimposed spectrum from a sample where further heat treatment has formed HfSi$_2$. Looking at the spectrum labeled II (cross-hatched areas) we see changes in the spectrum similar to those indicated in our schematic representation in Figure 5.

In all the spectrum peaks we notice a second order effect: all the peak tops tilt slightly with their lower energy side having more counts. This is because the scattering cross section changes as $(1/E^2)$, and as the ^{4}He projectiles penetrate they have increased cross sections for scattering.

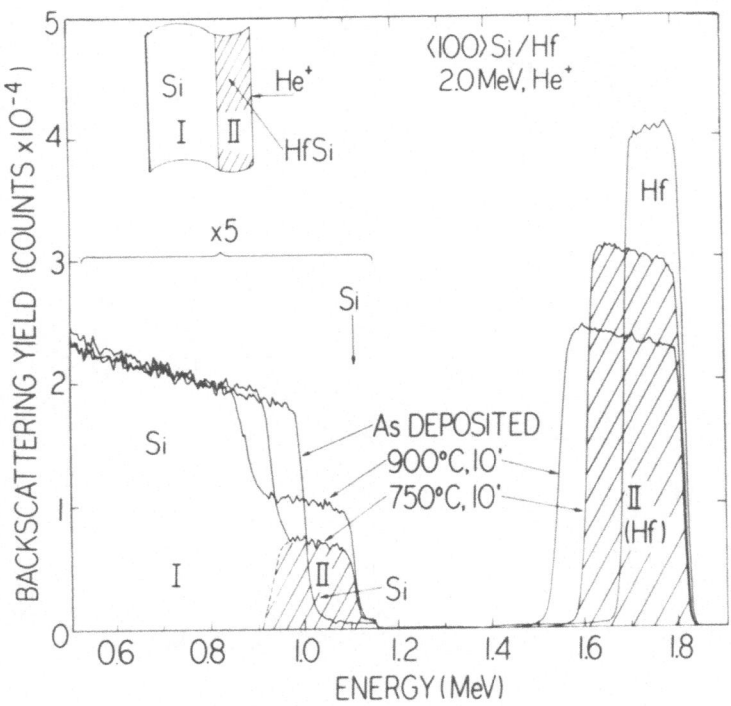

Figure 6) Raw data is shown for three superimposed spectra: Hf on Si; HfSi on Si; and HfSi$_2$ on Si. The Hf peak gets lower and wider as expected from the discussion of Figure 5. The original Si peak is displaced from the Si surface; but with the two silicides it is found at the surface. In each case, the experimental running time was 10 minutes.

We may go deeper into the analysis of the solid state reaction
between Hf and Si. In Figure 7 we schematically represent what
might occur if we quenched the interaction when it was only partially
complete. Here we hypothesize that the reaction might occur as a
film which grows from the Hf/Si interface. We speculate that either
(or both) the Hf or Si must diffuse through the HfSi which has
already formed, and when it reaches some unreacted material, some
more HfSi is formed. This type of reaction is called "diffusion
limited". Another similar type of interaction will be described
later.

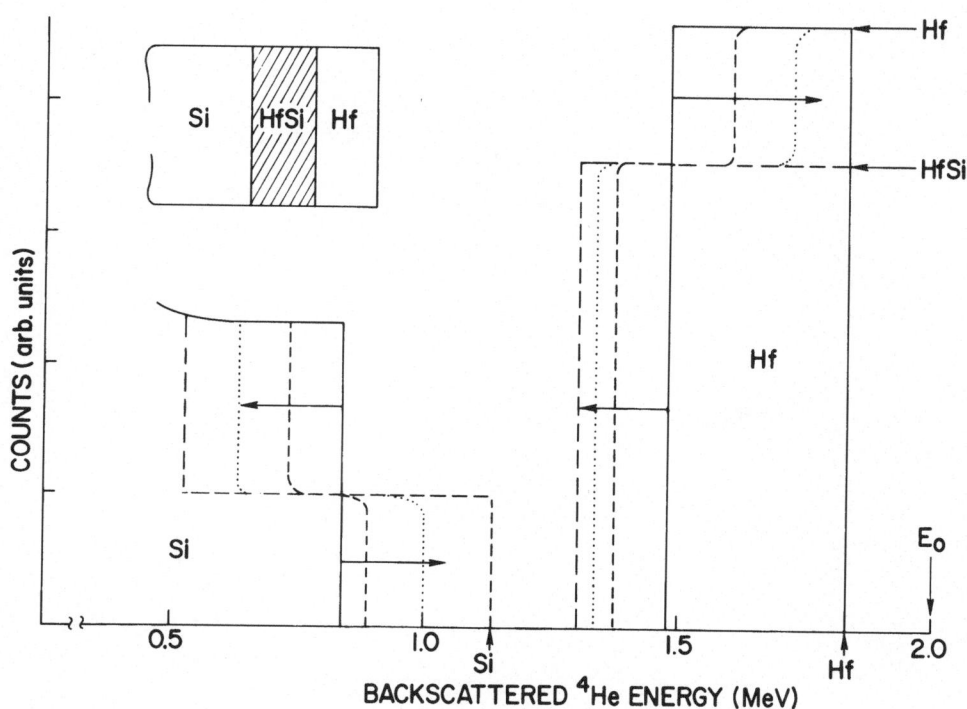

Figure 7) Simplified representation of spectra showing one formation process for HfSi. The upper left inset shows the assumption
of HfSi formation at the interface of the Hf and Si materials. The
HfSi layer expands as the formation continues, until the Hf is
entirely converted. The Hf peak remains at full height near its
surface energy, but at some depth it drops abruptly to the diluted
level of Hf in HfSi. The Si substrate is originally lower than the
Si surface energy because it is covered with Hf. As the HfSi formation process continues the Si extends out towards the Si surface.

This diffusion limited reaction, if stopped before it goes to completion, would show a spectrum (see Fig. 7) with pure Hf at the Hf surface channel, and at some depth the Hf peak would drop to that of Hf in HfSi. As we let the chemical reaction continue, the depth of pure Hf on the surface would decrease, and more of the Hf would be in the HfSi. As also shown in Fig. 7, the Si peak would show a foot pushing out towards the Si surface arrow as the reaction progresses. In the schematic we note another second order effect in nuclear backscattering spectra. In the layer of HfSi, the Hf atoms are diluted by 50% with Si atoms. However, we find the Hf peak height does not drop 50%, but only about 30%. This is because the energy loss of the ^{4}He ions is different in HfSi than in Hf (pure). The difference is due primarily to the simple change in electron density. The changes in peak height of a pure substance with dilution by another element can be directly calculated (see "Formalism"). Or stated another way, by determining the drop in the spectral peak height, one can determine the amount of dilution by a second element (usually to $\pm$ 4%).

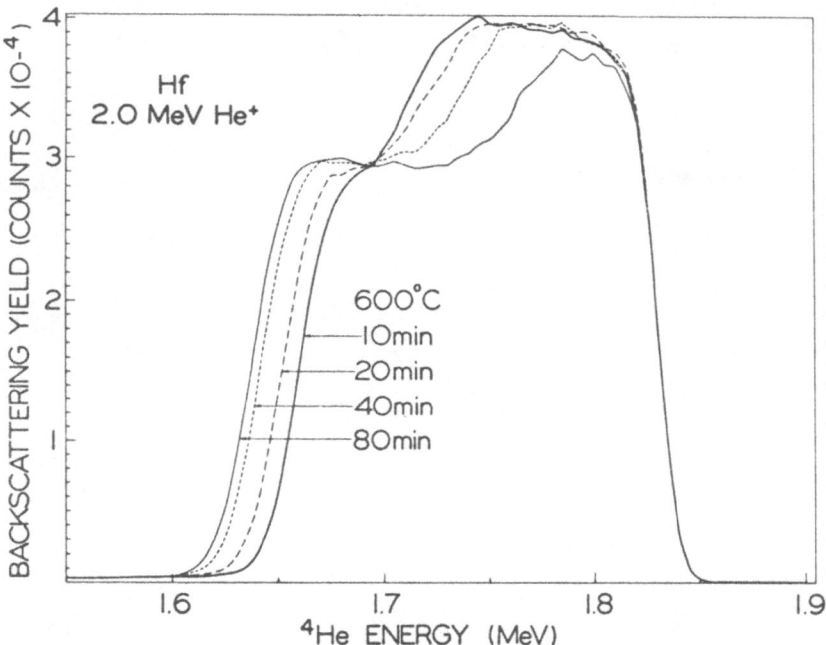

Figure 8) The raw data from the Hf portion of a series of spectra are shown. HfSi is first forming at the deepest part of the Hf layer, and this conversion of Hf to HfSi continues as the sample is heated for longer times. The energy width, ΔW, indicates the amount of HfSi formed.

In Figure 8 we see again the raw data of several spectra superimposed. We have expanded the figure to show the details of the Hf peak only. We see that successive heat treatment of Hf on Si forms a dilution of the Hf peak which starts at the Hf/Si inter-face. In the spectra, the Hf remains pure at the Hf surface (right side of the peak), and the dilution consumes more and more of the Hf as the heat treatments are made longer. We can measure the width of the new diluted layer as shown in Figure 8 by the arrow marked ΔW. This width is determined as an energy (about 0.1 MeV in Figure 8), but this may be converted to a physical depth as shown in the chapter on Formalism of Nuclear Backscattering.

This width of the new layer of HfSi can be plotted as a function of $t^{1/2}$ as shown in Figure 9. If the interaction occurs by simple diffusion, the result is a straight line, and one can obtain directly the diffusion constant of the reaction. One should note that no chemical information is contained in nuclear backscattering. The

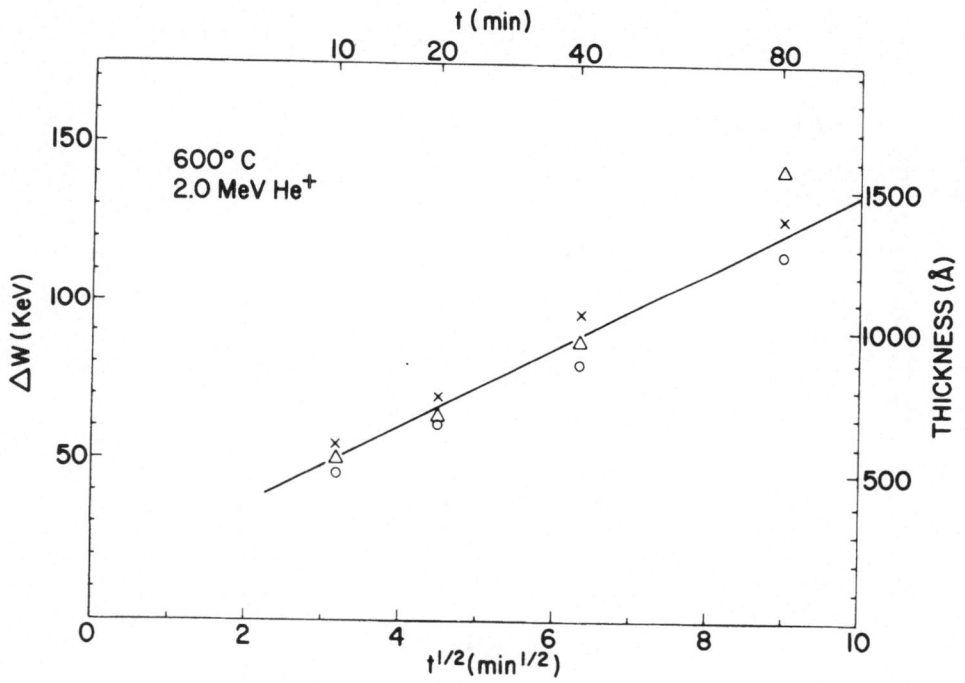

Figure 9) The value of ΔW from Figure 8 is plotted against the square root of the heat treatment time, t, for several targets. From this curve we can calculate the diffusion constant of the reaction.

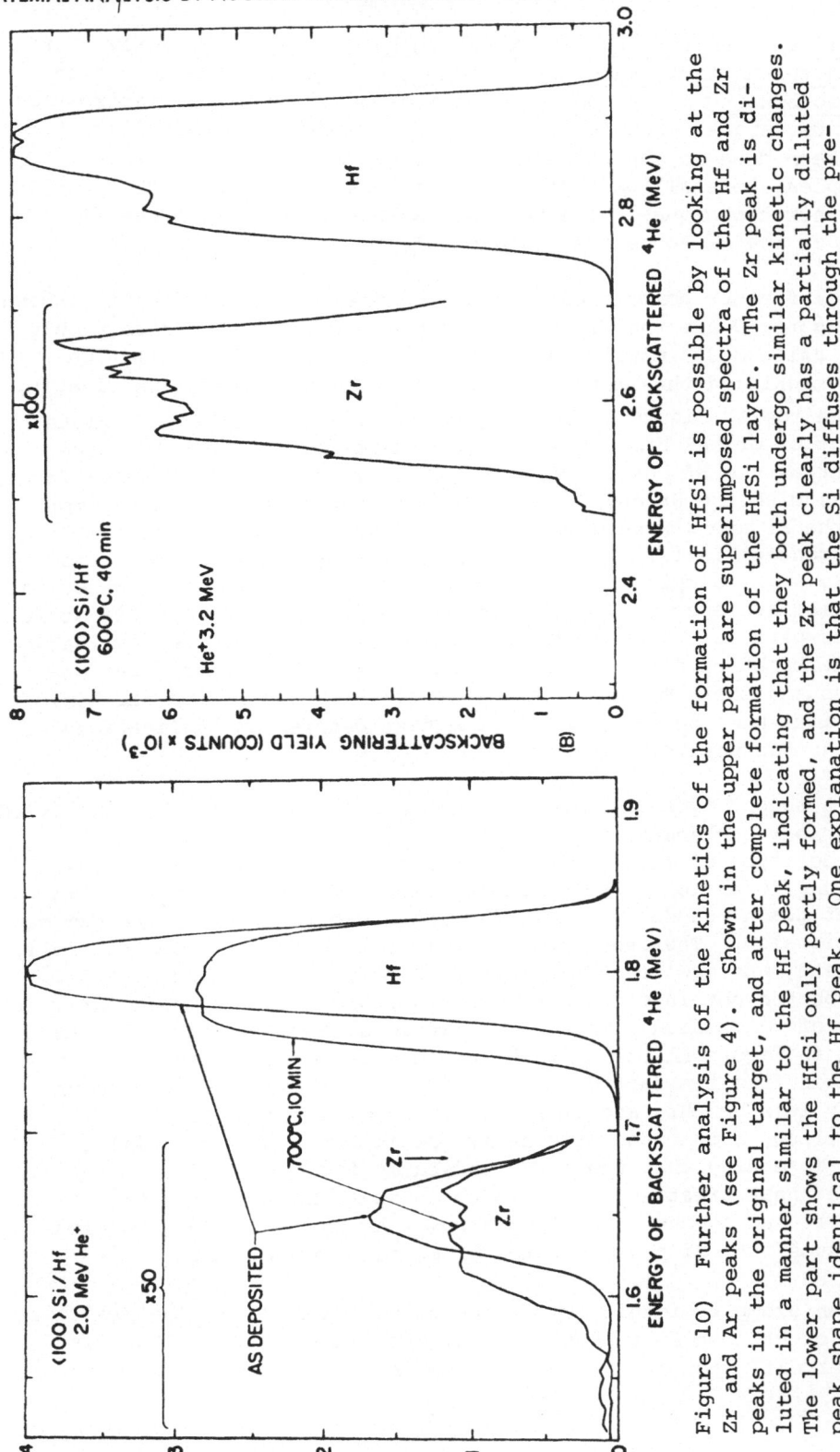

Figure 10) Further analysis of the kinetics of the formation of HfSi is possible by looking at the Zr and Ar peaks (see Figure 4). Shown in the upper part are superimposed spectra of the Hf and Zr peaks in the original target, and after complete formation of the HfSi layer. The Zr peak is diluted in a manner similar to the Hf peak, indicating that they both undergo similar kinetic changes. The lower part shows the HfSi only partly formed, and the Zr peak clearly has a partially diluted peak shape identical to the Hf peak. One explanation is that the Si diffuses through the previously formed HfSi and dilutes the pure Zr and Hf in similar manner in the formation of new HfSi.

simplicity and reliability of the technique can be directly attri-
buted to the fact that the ⁴He ions are insensitive to the electronic bonding of the solid. One must obtain chemical information
from techniques such as electron or x-ray diffraction. Since this
has been done in the present example of HfSi, and for later
examples, we shall omit mention of further analysis of the sample
by other techniques. The **He ion** backscattering shows the final
form of the Hf to be 50% Hf, and 50% Si.

     A further comment can be made on the formation of HfSi. Shown
in Figure 10 are the Hf and the Zr peaks (the Zr was present only
as a uniform 3% impurity as originally shown in Figure 4). We
have magnified the Zr peak in Figure 10 to show its shape clearly.
In the top of Figure 10 the Zr peak is shown (after the complete
formation of the HfSi layer) to have expanded in width the same
amount as the Hf peak. One might guess that this could be explained
by assuming the Hf and Zr were stationary during the interaction,
and the Si atoms were diffusing up and diluting them both similarly.
This concept is further strengthened by stopping the interaction
before it is complete. In the lower part of Figure 10 is shown the
intermediate spectrum, and it shows the Zr partially diluted at its
deeper part in a manner similar to the Hf peak. Further evaluation
of the kinetics of the chemical reaction can be obtained by ion-
implanting inert markers such as Ar or Xe in the Hf and Si. This
technique will be described in the next chapter on Applications of
Nuclear Backscattering.

     A second type of thin film compound formation can be identified
by nuclear backscattering. In Figure 11 we consider a "reaction-
limited" type of interaction. The upper left inset illustrates a
compound formation in which the silicon diffuses very fast along
the Hf grain boundaries, but the Si/Hf reaction forming the compound
HfSi is slow. The inset shows the film with the reaction partially
complete, and isolated grains of pure Hf are surrounded by HfSi.
The spectra obtained from such interactions differ significantly from
those of "diffusion limited" interactions shown in Figure 7. As
shown in Figure 11 the Hf peak should show immediate and uniform
dilution from the fast diffusing Si. The Hf peak should show uniformly less height, and it should broaden as the layer thickens.
The silicon front edge originally (solid lines in Figure 11) is
lower in energy than the arrow indicates for surface Si. After the
initial heat treatment Si is found in small amounts from the Si
substrate up to the Si surface arrow. As the compound formation
continues, the Si peak near the Si surface grows higher.

     An example of this type of reaction can occur if the previous
layer of HfSi on Si is heated to higher temperatures. There is
further diffusion of Si from the substrate and the new compound

$HfSi_2$ forms. Figure 12 shows raw data from various samples of
$HfSi_2$ on Si heated to higher temperatures. The Hf peaks show the
uniform decrease we described in the "reaction-limited" type of
interaction. In this example the interpretation is not completely
accurate, as local stress in the film also is important and affects
the degree of compound formation.

Since the [4]He nuclear backscattering is essentially non-destructive, it is possible to perform other analytic measurements on
the same sample which was profiled by backscattering.

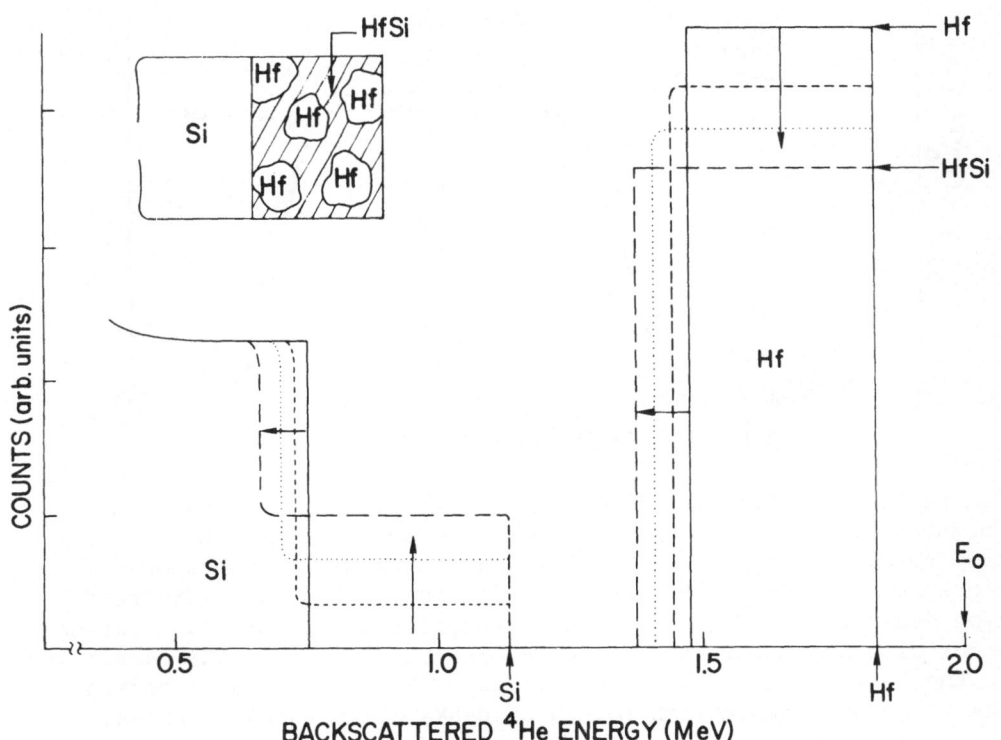

BACKSCATTERED [4]He ENERGY (MeV)

Figure 11) A second possible manner of compound formation is
"reaction-limited" (this should be compared to "diffusion-limited"
shown in Figure 7). We assume here that the Si diffuses very rapidly
along the grain boundaries, but reacts slowly with the Hf grains
forming HfSi. The intermediate spectra are different from those of
Figure 7, identifying the differences in compound formation kinetics.
Note that Si is found all the way at the surface after the shortest
heat treatment. Also the Hf peak is immediately diluted in a manner
uniform with depth.

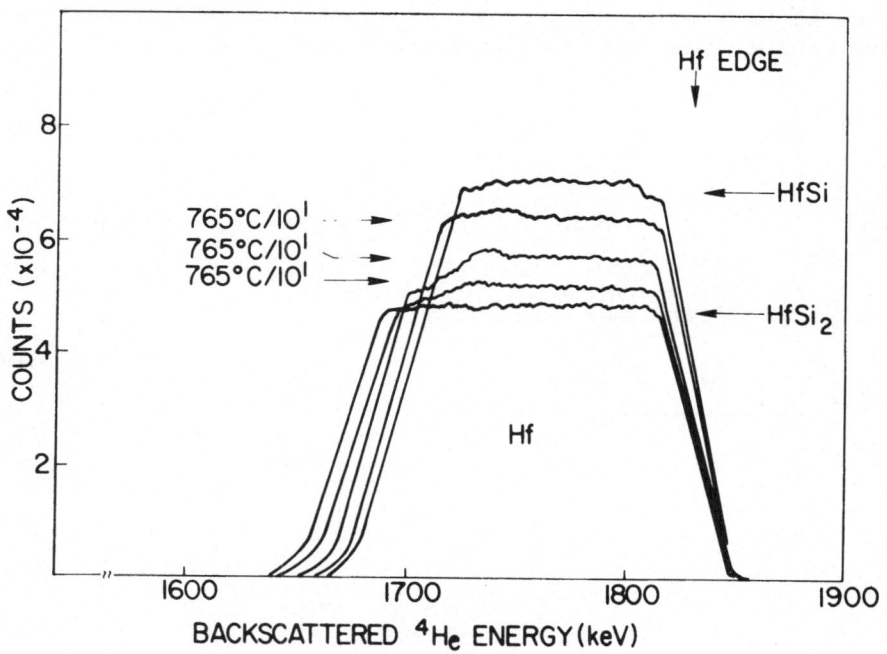

Figure 12) Further heat treatment of the HfSi layer causes the
formation of $HfSi_2$. Shown are superimposed spectra of the Hf peak
from several samples showing that the intermediate stages of the
conversion of HfSi to $HfSi_2$ is similar to the "reaction limited"
type described in Figure 11. The fact that all intermediate levels
occur with identical heat treatment reflects possibly the sensitivity
of the reaction to local stress in the HfSi film. The Si portion of
these spectra are similar to the Si peaks illustrated in Figure 11.

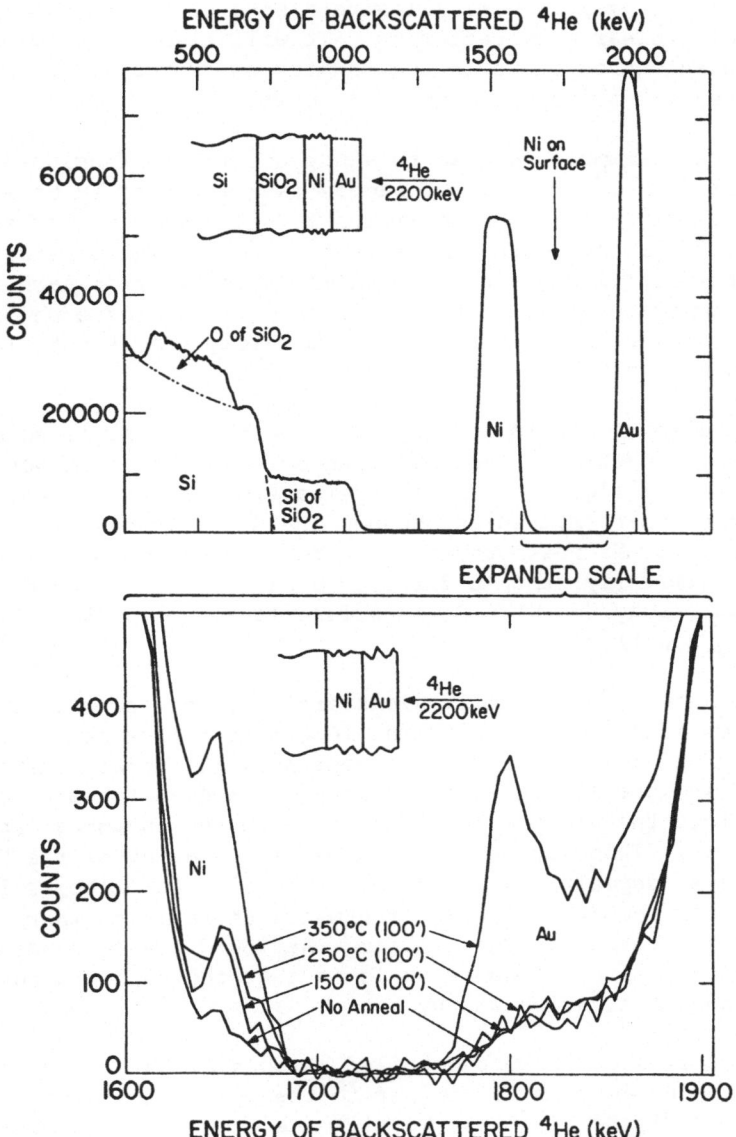

Figure 13) The interdiffusion of two thin metal films is shown for
Au and Ni on SiO$_2$. The upper spectrum is from the original target
showing the relationship of the various target peaks. Note the
Expanded Scale notation (between the Ni and Au peaks) which is shown
between the upper and lower drawings. The lower drawing expands
this region, and shows superimposed spectra for various heat treated
targets. The Ni up-diffusion and the Au down-diffusion are clearly
shown.

One final example will illustrate some of the second order
effects of nuclear backscattering. It will also serve to show how
the calculations outlined in the chapter on "Formalism of Nuclear
Backscattering" are applied to real spectra.

For this example we wish to study the interdiffusion of two
thin metal films.[8] The target sample is a layer of Au on a layer
of Ni as shown in an inset to Figure 13. The complicated substrate
of SiO_2 on Si is an inert substrate for the two interacting films.
The upper spectrum shows the two films separated from each other
and the substrate because of the different masses of each of the
elements. The Au film is about .06 μm thick, and is not flat on
top because of detector resolution.

In the lower portion of Figure 13 we show a magnified expansion
of the upper spectrum (note the expanded scale notation between the
upper and lower figures). In this expansion are superimposed spectra
from samples heat treated as shown. The Ni diffuses up into the
Au, and a small Ni peak occurs where the Ni spreads out over the Au
surface. Similarly, the Au diffuses down into the Ni, and then
spreads out along the Ni/SiO_2 interface forming a peak in the backscattering spectrum.

We see that the two diffusions are quite different. The Ni
diffuses up into the Au during the lowest heat treatment. The
quantity of Ni inside the Au film continues to increase as the heat
treatment is done at higher temperatures. The Au appears to have
diffused into the Ni at a low level even in the no-anneal spectrum
of Figure 13. This may be due to momentary heating of the Ni
during the Au deposition, or it may be Au filling pinholes in the
Ni film. For the initial heat treatments there is no increase in
the Au concentration in the Ni film. At 350°C the Au suddenly
penetrates the Ni. One guess would be that the Au diffusion into
Ni has a high activation energy.

Figure 14 shows a more detailed example of thin film interdiffusion. The Au/Fe film is similar to the preceding Au/Ni film. As
the inset to Figure 14 shows, we tilt the target to the beam in
order to increase the apparent thickness of the films to the beam.
This enlarges the region of interest of the spectrum. The relative
concentrations of Fe in Au, and Au in Fe are clearly shown. The
vertical ordinates have been changed from "^{4}He Counts" to actual
atomic concentrations. Note that because the cross-section of Au
is much greater than that of Fe, the scale on the left is far larger
than the one on the right. Once the energy scale on the abscissa
is converted to depths one can directly measure concentration profiles.

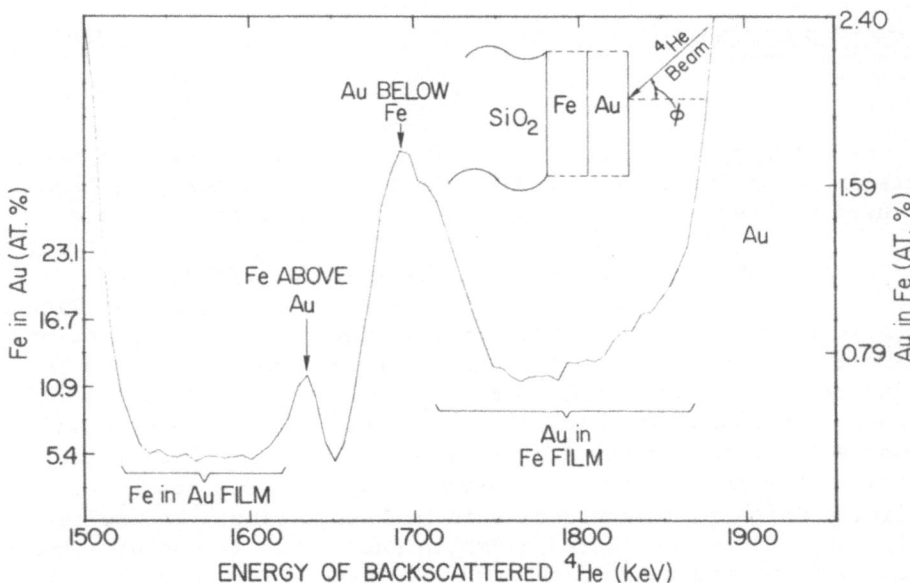

Figure 14) By tilting a target of Fe/Au the beam penetrates a
thicker target. This expands the region of interest between the two
films. The ordinate of the spectrum has been converted to atomic
percent of Fe in Au, and Au in Fe, so that concentration profiles
can be directly read.

        In conclusion, nuclear backscattering is simple and quantita-
tive. It is also non-destructive. The examples shown have been
ideally suited to backscattering analysis, and as will be shown in
the next chapter (Applications of Nuclear Backscattering) a few
problems can occur in nuclear backscattering. The primary problem
is spectrum overlap. If the peaks of two different elements overlap
one another, one must use detailed computer analysis to extract
meaningful profiles. And with more complex overlaps (say of three
elements) the accuracy of the technique can degenerate rapidly.

GENERAL COMMENTS ON NUCLEAR BACKSCATTERING

The example this chapter has used to illustrate nuclear back-scattering used ^{4}He ions at 2 MeV. Other ions and energies have been used with success, and we shall discuss a number of these.

He at an energy of 2 MeV is widely used for a number of reasons. One would like to use a high energy to separate as widely as possible backscattering from various elements. The separation is directly proportional to the ion's energy. One is limited by two factors: accelerator maximum energy and nuclear reaction resonances (non-Coulomb interactions). The reaction resonances introduce great complexity and require complex unfolding of the spectra to obtain meaningful material analysis. One loses the simplicity which is so attractive in backscattering. To nuclear physicists, ^{4}He is considered a "magic nucleus". It contains two protons and two neutrons, each with a spin up and spin down. The nuclear properties of ^{4}He are analogous to a closed electron shell: it is relatively inert. This means that it does not react with other nuclei very easily (react here means non-Coulomb inter-actions). With ^{4}He, we can use relatively higher energies and not be bothered by resonances. As shown in Figure 15, resonances occur in ^{4}He → Si at about 3.87 MeV. These results show the poor quality of semiclassical theory which indicates that an electrostatic barrier inhibits two-body nuclear-force reactions when the two bodys do not merge. That is, the interaction is merely Coulomb repulsion if:

(closest distance of approach) > (sum of radii)

or, more explicitly:

$$[(1+\csc \theta/2])(Z_1 Z_2)(\frac{M_1+M_2}{2M_2})(\frac{1.44}{E_o})] \quad > \quad (\frac{A_1^{1/3} + A_2^{1/3}}{0.24})$$

where the terms are defined as: Z_1, A_1, and M_1 refer to the projectile atomic number, mass number, and mass; Z_2, A_2, and M_2 refer to the target; E_o is the projectile's initial energy (MeV); and Θ refers to the center-of-mass scattering angle which may be described in terms of the laboratory angle θ as:

$$\Theta = \theta + \sin^{-1}[(M_1/M_2) \sin\theta].$$

For ^{4}He → Si at 4 MeV, the semiclassical distance of approach is greater than twice the sum of the radii, yet Figure 15 shows distinct resonances. The only safe approach is to study data published in the Nuclear Data Tables and find where the lowest energy resonances occur. This problem is discussed in detail in References 11 and 12 for ^{4}He projectiles.

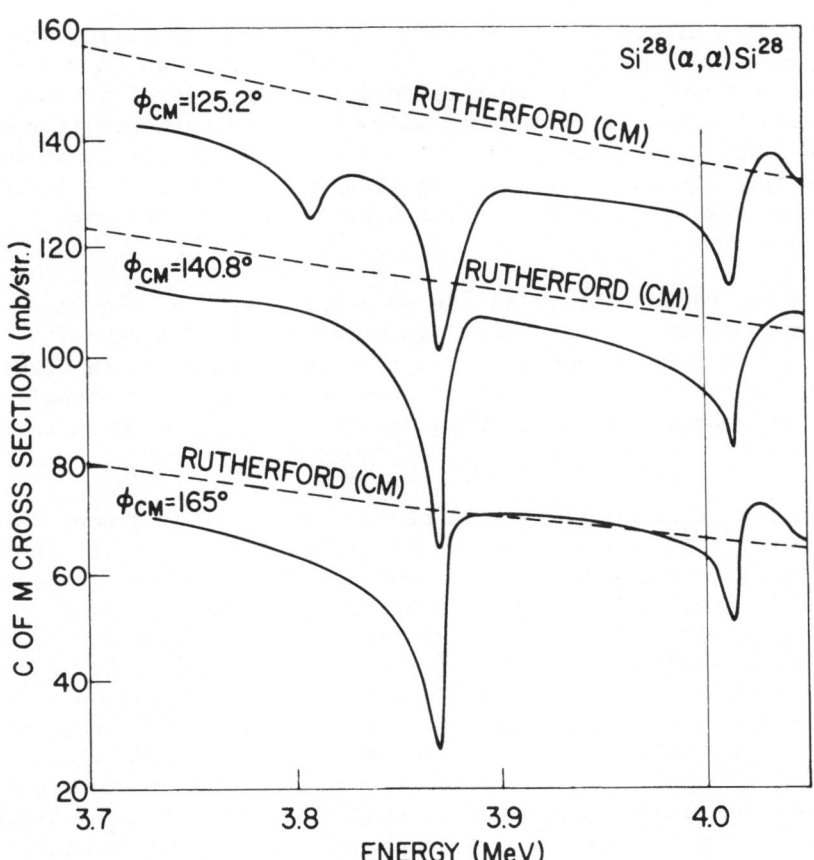

Figure 15) Comparison of Si^{28} (^{4}He, ^{4}He) Si^{28} experimental elastic cross-sections to Rutherford cross-sections (including recoil terms) for several angles. All angles and cross-sections are in center-of-mass coordinates. Actual data points were in steps of 2.5 keV and were replaced by smooth curves. This data was supplied by McEllistrem and Leung (Ref. 19) who are conducting a complete low energy analysis of this reaction.

Protons undergo nuclear reactions at energies of 100-300 keV for elements such as Al-Ge. Therefore, routine backscattering with protons is usually kept below this energy range. If the target contains only much heavier elements, the reaction threshold is higher, but the proton ability to differentiate different target masses is so poor that not much information can be gained from thick targets.

For protons with energies below 50 keV, there is significant interaction between the protons and the electrons of the target leading to beam neutralization. This creates serious problems of making analysis quantitative (see Ref. 13).

With both of these limitations, nuclear backscattering with protons appears limited. However, the limitations can be viewed as sources of powerful analytic capability. The use of protons to generate nuclear reactions can allow great sensitivity to some elements (see chapters on Nuclear Reactions and Lattice Location of Impurities). The interaction of protons with the target electrons allows some chemical binding information to be obtained (see Ref. 14).

A second reason for using ^{4}He as a probe is that the mass separation between elements goes approximately as $(2m)/(M+m)^2$ where M is the target atom mass and m is the ion mass (see Chapter on Formalism for rigorous details). Hence ^{4}He has considerably more recoil energy separation than protons for similar mass target elements.

Higher mass ion probes would be even more useful except for a single fact: there does not exist a simple detector which can analyze their energy with high resolution. Most laboratories have attempted to use ^{12}C beams for analysis, but except for very isolated analytic problems the poor detector resolution for ^{12}C nullifies the good recoil energy separation it achieves for similar mass target elements. This poor detector resolution can be eliminated by using a magnetic analyzer on the backscattered ^{12}C ions, but this technique is too complex to be described here. It should also be added that heavy ions such as ^{12}C can introduce significant damage and cause target sputtering considerably beyond that by H^+ or He^+ probes. For a comparison of He^+ and C^+ backscattering using solid state detectors see References 15 and 16.

To summarize the above choice of ions and energies: heavy probes and higher energies increase the mass analysis resolution of targets. Limitations are imposed by maximum accelerator energy, staying below nuclear reaction thresholds, and obtaining good detector resolution. These restrictions limit H^+ probes to below 200 keV and He^+ probes to below 4 MeV.

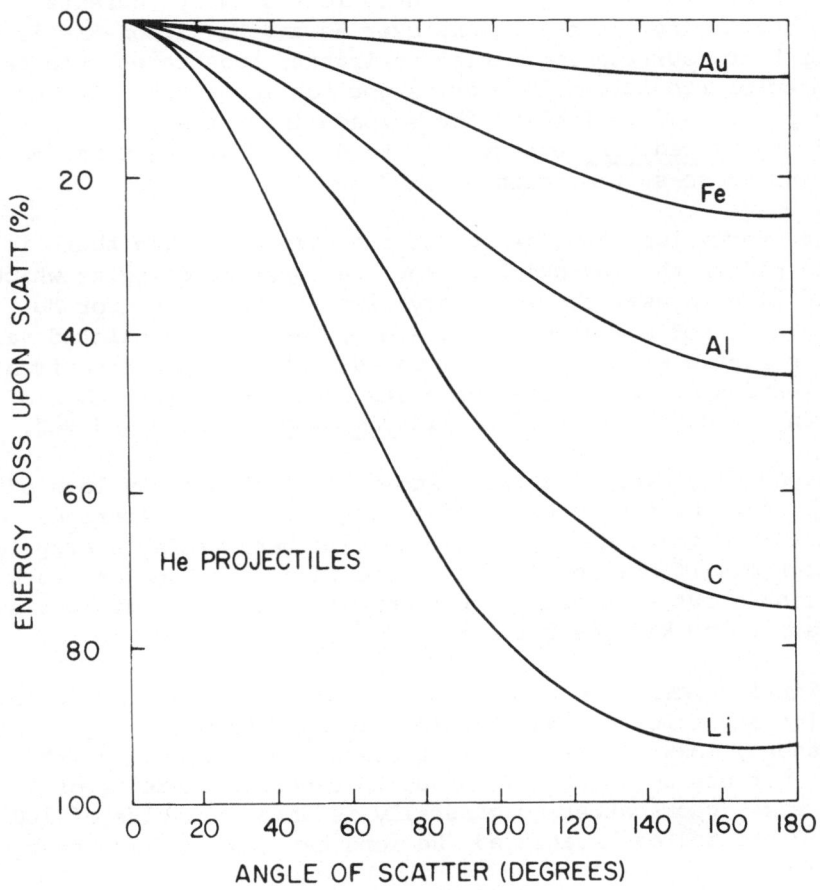

Figure 16) Comparison of the recoil energy of ^{4}He ions scattering from various targets at various laboratory angles. In order to separate elements of similar mass, the scattering should be at a back angle. This figure is reproduced from Ref. 20.

Two final considerations also enter into choice of ion and energy: cross-section for scattering, and energy loss in the target. The cross-section for scattering goes as $(Z_1/E)^2$ where Z_1 is the probe's nuclear charge, and E is its energy. For the special case of surface trace-element analysis it may be reasonable to use ^{12}C or ^{16}O as the probe. These ions greatly increase Coulomb scattering cross-sections over He and H. If one is looking only for heavy surface impurities on lighter substrates, the fact that detector resolution is poor is not too important. If one uses He as a probe, and is looking for surface impurities, one gains sensitivity by reducing energy. He at 0.5 MeV has 16 times more scattering cross-section than He at 2 MeV.

The energy loss considerations are simple because there is only one pig in the mud-puddle. For the ions and energies which we would like to use, there is very little data except for He^+. For this ion, for energies of 0.4-4 MeV, there are tabulated values for all target elements (Ref. 10) based mostly on the experimental work of Darden Powers. Also there are a considerable number of papers on the energy loss of He^+ in compounds (Ref. 9 and Ref. 17).

In contrast, very little is known about the energy loss of H^+, C^+, or D^+ for the energy range of interest to backscattering. One is left with a great deal of basic work in determining energy loss cross-sections if one would like to use these ions to get depth information about a target. For a compilation of recent references for H^+ at 50-150 keV see Ref. 18.

A final comment can be made on the choice of the laboratory scattering angle for nuclear backscattering. Shown in Figure 16 is the energy loss of ^4He ions scattering from various target atoms at various angles.[20] In order to separate elements of similar mass, backscattering generally is done at angles of 160-180 degrees where the backscattered ^4He ions have the maximum energy differences.

In the final chapter on nuclear backscattering we show briefly the formalism of nuclear backscattering analysis, and show where one can find the tables needed for detailed calculations.

Further discussion and many papers are included in Ref. 9.

APPENDIX

Numerical Examples

It is useful to have examples of numerical quantities used in the analysis of nuclear backscattering. The formulae to be used are discussed in detail in the chapter on Formalism of Nuclear Backscattering. We use that notation, and the energy loss values from Reference 10. For typical experimental conditions we assume:

> Projectile = ^{4}He
>
> Projectile Energy = E_o = 2 MeV
>
> Number of Projectiles = Q = 6.02 x 10^{12} (1μ C)
>
> Scattering Angle = θ = 170°
>
> Detector Solid Angle = Ω = 1 msr.
>
> Energy Channel Width = δE_1 = 5 keV/Ch.

For a target of Hf (see Fig. 4) on Si:

$$[\epsilon]_{Hf}^{Hf} = 210.6 \text{ eV}/(10^{15} \text{ atoms/cm}^2)$$

where $[\epsilon]$ is defined by Eq. (37). For a target of HfSi on Si we obtain:

$$[\epsilon]_{Hf}^{HfSi} = 153.8 \text{ eV}/(10^{15} \text{ atoms/cm}^2).$$

The reader is reminded that the epsilon in the square brackets is a backscattering stopping cross-section. That is, it includes both stopping on the inward and outward paths as defined in Eq. 12 in the chapter on Formalism of Nuclear Backscattering.

This epsilon, when it refers to stopping cross-sections in mixtures or compounds, has a superscript referring to the elements which make up the mixture, and a subscript which refers to the element from which the projectile scatters. For a mixture, we assume Bragg's Rule as defined in Eq. 38. Bragg's rule says that the stopping in a mixture is the same as the stopping in each individual element weighted by that element's abundance in the mixture.

The drop in Hf Counts/Channel (see Figure 17) from backscattering from pure Hf, to Hf in the compound HfSi, can be calculated

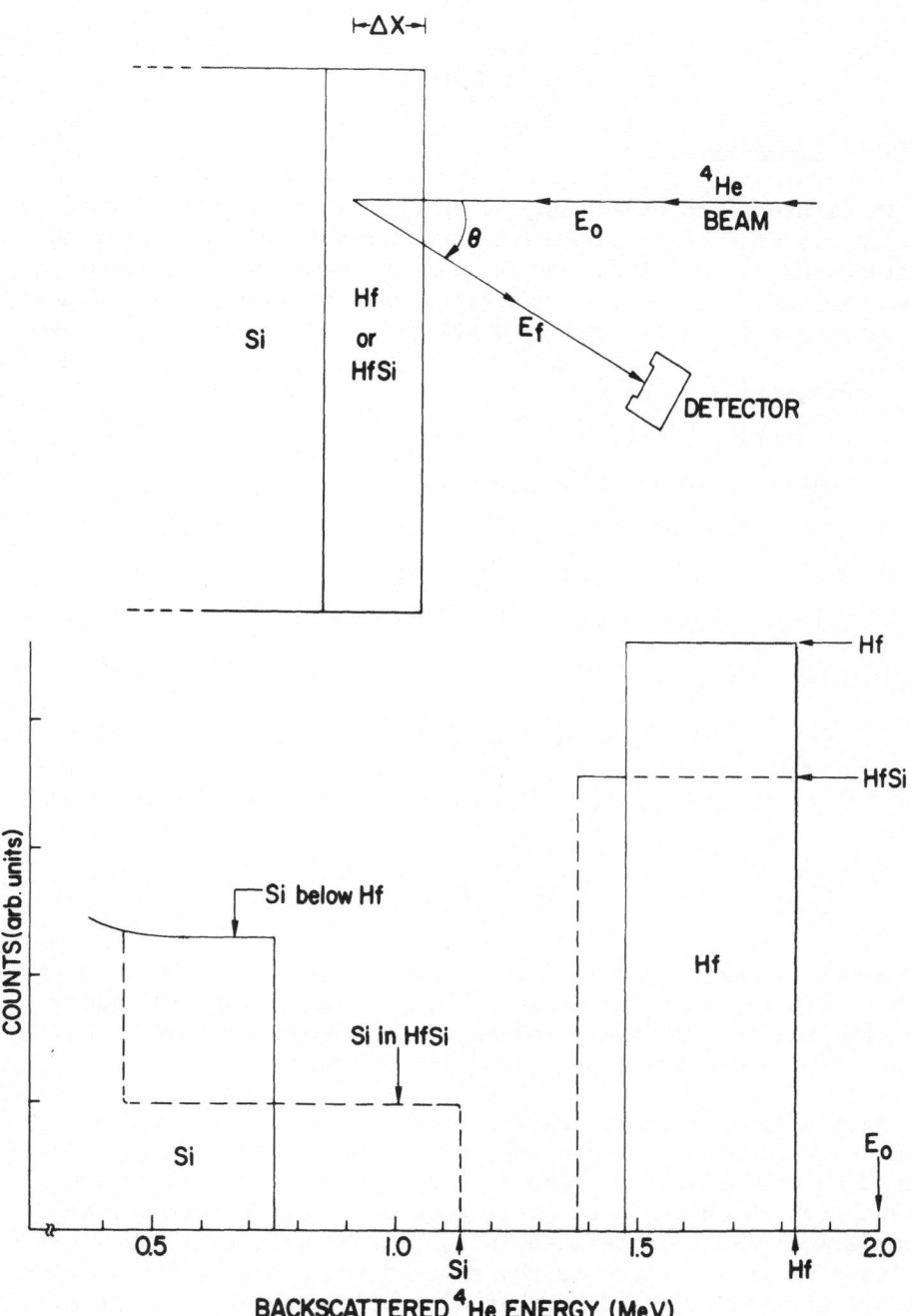

Figure 17) Simplified pictures of the formation of the compound
HfSi from a layer of Hf on a substrate of Si. To help identify
the compound it is necessary to calculate the Counts/Channel ratio
of Hf in pure form, to Hf in the compound. One may also calculate
the drop in the Si ledge. Also, one may calculate the Hf/Si Counts/
Channel ratio. Since more complex targets may not be as clear as
the one shown, one may need any of these calculations.

using Eq. 41:

$$\text{Ratio} = \left(\frac{153.8}{210.6}\right) \left(\frac{2}{1}\right) = 1.46 \ .$$

So we see that although the absolute concentration of Hf drops in half, the peak drops only to level of (1/1.46) = 68%. Similarly for Si:

$$[\varepsilon]_{Si}^{Si} = 92.6 \ eV/(10^{15} \ atoms/cm^2)$$

$$[\varepsilon]_{Si}^{HfSi} = 139.4 \ eV/(10^{15} \ atoms/cm^2)$$

and for $HfSi_2$:

$$[\varepsilon]_{Hf}^{HfSi_2} = 134.9 \ eV/(10^{15} \ atoms/cm^2)$$

$$[\varepsilon]_{Si}^{HfSi_2} = 123.8 \ eV/(10^{15} \ atoms/cm^2) \ .$$

The drop from pure Hf to Hf in $HfSi_2$ is from Eq. (41):

$$\frac{H_{Hf}^{HfSi_2}}{H_{Hf}^{Hf}} = .52 \ .$$

The reader may look back at Figure 6 to see actual spectra from layers of pure Hf, HfSi, and $HfSi_2$.

The ratio of the height of the Hf Counts/Channel peak to that of the Si peak can also be calculated using Eq. 41. Here we must use different values for the scattering cross sections in Eq. 41.

$$\frac{H_{Hf}^{HfSi}}{H_{Si}^{HfSi}} = 24.9; \qquad \frac{H_{Hf}^{HfSi_2}}{H_{Si}^{HfSi_2}} = 12.63.$$

Finally, one can calculate the ratio of areas of the Hf and Si in the areas corresponding to HfSi and $HfSi_2$. This is done by merely taking the ratio of scattering cross-sections and multiplying by the ratio of abundance. This technique is not as accurate because determining the area of the Si ledge is complicated by isotope effects.

$$\frac{A_{Hf}^{HfSi}}{A_{Si}^{HfSi}} = 27.5 \; ; \qquad\qquad \frac{A_{Hf}^{HfSi_2}}{A_{Si}^{HfSi_2}} = 13.76 \; .$$

For the final example of Fe/Au interdiffusion, Fig. 14, we can convert the raw data (counts/channel) to atomic concentrations by solving Eq. 41 for the ratio of abundance of the mixture's constituents, m/n. For every ratio of abundance we may calculate the Counts/Channel, and we may then directly change the spectrum ordinate to atomic concentration as shown in Fig. 14. For this example of Fe and Au mixture, we may construct a table such as:

Au in Fe		Fe in Au	
Counts	% Au Conc.	Counts	% Fe Conc.
3.62	0.2	.26	0.2
7.25	0.4	.52	0.4
10.9	0.6	.78	0.6
14.5	0.8	1.03	0.8
18.1	1.0	1.30	1.0
.	.	.	.
.	.	.	.
.	.	.	.
17 1	10.0	13.5	10.0

We assume the experimental parameters listed at the start of this Appendix. We see that we are far more sensitive to Au in Fe, mostly because of its higher scattering cross-section. Also as we increase to 10% impurities, the conversion of counts to concentration becomes more non-linear, due to changing values of $[\varepsilon]$ for the mixture.

REFERENCES

1. H. Geiger and E. Marsden, Proc. Roy. Soc., 82, 495 (1909).

2. E. Rutherford, Phil. Mag., 21, 669 (1911).

3. J. Stark and G. Wendt, Ann. d. Phys., 38, 921 (1912).

4. A. Turkevich, K. Knolle, R. A. Emmert, W. A. Anderson, J. H. Patterson, and E. Franzgrote, Rev. of Sci. Inst., 37, 1681 (1966).

5. A. Turkevich, E. Franzgrote, and J. Patterson, Science, 158, 635 (1967).

6. C. J. Kircher, J. W. Mayer, K. N. Tu, and J. F. Ziegler, Appl. Phys. Lett., 22, 81 (1973).

7. J. F. Ziegler, J. W. Mayer, C. J. Kircher, and K. N. Tu, J. Appl. Phys., 44, 3851 (1973).

8. J. F. Ziegler, J. E. E. Baglin, and A. Gangulee, Appl. Phys. Lett., 24, 36 (1974).

9. "Ion Beam Surface Layer Analysis", Ed. by J. W. Mayer and J. F. Ziegler, Elsevier Sequoia Co., Laussane, Switzerland (1974).

10. J. F. Ziegler and W. K. Chu, Atomic Data and Nuclear Data Tables, 13, 463 (1974).

11. J. F. Ziegler and J.E.E. Baglin, J. Appl. Phys., 42, 2031 (1971).

12. J. F. Ziegler and J.E.E. Baglin, J. Appl. Phys., 45, 1888 (1974).

13. A. van Wijngaarden, B. Miremadi, and W. E. Baylis, Can. Jour. of Phys., 49, 2440 (1971).

14. See references cited in the review article: R. E. Honig and W. L. Harrington, Thin Solid Films, 19, 43 (1973).

15. S. Peterson, P. A. Tore, O. Meyer, B. Sundquist, and A. Johansson, Thin Solid Films, 19, 157 (1973).

16. R. R. Hart, H. L. Dunlap, A. J. Mohr, and O. J. Marsh, Thin Solid Films, 19, 137 (1973).

17. J. F. Ziegler, W. K. Chu, J.S.-Y. Feng (to be published in 1975 in Appl. Phys. Lett.).

18. R. Behrisch and B.M.U. Scherzer, Thin Solid Films, 19, 247 (1973).

19. M. K. Leung, Ph.D. Thesis, University of Kentucky, 1972 (unpublished).

20. A. Turos and Z. Wilhelmi, Nukleonika, 13, 975 (1968).

MATERIAL ANALYSIS BY NUCLEAR BACKSCATTERING

Applications

J. W. Mayer and B. M. Ullrich

California Institute of Technology

Pasadena, California 91109

I.   INTRODUCTION

Applications of backscattering spectrometry as an analytical technique have primarily focused on depth microscopy - the ability to determine composition variation or impurity distribution as a function of depth below the surface of the sample. No other conventional microanalytical tool is capable of depth microscopy with depth resolution of a few hundred angstroms over depths of thousands of angstroms without recourse to sample erosion or layer removal. In addition, the technique provides quantitative analysis without recourse to standards of similar composition.

The majority of the applications of backscattering spectrometry have been directed toward layered structures containing but few elements or diffusion profiles. These would appear to be extremely narrow fields of investigation but backscattering spectrometry has had major impact on both technology and material science.[1-3] From a practical standpoint, the primary applications have been in two areas of materials technology: electronics and nuclear energy. In electronic materials, there have been studies of ion implantation, thermal and anodic oxidation, contact formation, thin-film deposition, dopant profiles, composition of thin film resistors, and metallization processes. In nuclear energy technology, studies have been made of cladding materials, first-wall erosion, helium and hydrogen incorporation, and corrosion. In the field of material science

there have been studies of thin film reactions, impurity solubility and diffusion, epitaxial layers, oxidation and corrosion,
and formation of superconducting or magnetic thin films. By
combining channeling with backscattering measurements, crystallographic information such as lattice site location of impurities or lattice disorder can be determined.[4] Applications
of channeling techniques have been primarily in the field of
ion implantation and growth of epitaxial layers.

        The applications listed above have been based on the
capability for microanalytical depth profiles without sample
erosion. In addition to these analyses of layered structures,
there have been studies of bulk composition of the outer few
microns of the sample surface. The most famous such analysis
was the Surveyor 5 "alpha-scattering experiment" which transmitted the first factual information on the composition of the
lunar soil.[5] These analyses agreed well with chemical analyses made later on moon soil returned to earth by the Appollo
missions. There have also been extensive studies of the composition of solids by the group at South Africa.[6] Such applications of mineral analysis of bulk composition are based on
the capability for quantitative microanalysis without the requirement for the use of the standards of a similar composition.
Backscattering techniques provide a very simple and direct
method for microanalysis of materials whose composition is uniform both laterally and in depth.

        There are areas where there has been little or no use
of backscattering techniques. Biological materials, except for
bone or shell structures, generally are not suitable due to
degradation caused by beam induced radiation damage. Organic
materials tend to decompose under vacuum or beam-irradiation
conditions. However, backscattering methods have been used to
measure the penetration of printing inks into paper[7] and the
composition of varnish on violins.[8] Chemical problems have
not received much attention because backscattering data do not
give information on chemical bonding. Particle energies and the
energy resolution of the detection system are much greater than
binding energies. Or course compositional information relating
to interfaces or cover layers could be obtained by backscattering methods. Mechanical aspects of materials have generally not
been considered as candidates for backscattering studies because
these properties are generally determined by bulk rather than
near surface composition. Even in this case, a study was made of
antimony segregation at fractures of steel.[9]

        One of the early applications of backscattering techni-
ques was the analysis of smog and microparticles collected on air

filters.[10]   There has been a revival of scattering techniques
in conjunction with ion-induced x-rays at the University of
California-Davis (Chapter by Cahill). In this case, detection
of forward scattered 4 MeV incident energy protons is employed
to determine the low mass constituents in thin filters (A<28)
while ion-induced x-rays identify the high mass-constituents
(A>20). The combination of the two techniques has been found to
be suitable for a wide range of materials ranging from air samples
to tree leaves.

In this chapter we would like to concentrate on applications in which depth analysis plays a key role. To put the
analytical capabilities in perspective we compare in Fig. 1,
techniques which utilize sputtering and those which rely on
energy loss to obtain depth profiles. Techniques (Auger electron
spectroscopy, secondary ion mass spectroscopy, and low energy
backscattering) which provide analysis of the composition at the
outermost layers of the sample must be combined with target
sputtering or some other means of layer removal to allow means
to obtain depth profiles. These surface sensitive analytical
tools have well-described in review articles[1] and commercial
instruments are available.

However, the sputtering process, itself, is not as well
characterized except in the case of homogeneous, monoelemental
targets. In polycrystalline targets there is the possibility
of accelerated erosion rates at the grain boundaries. With
alloys or compounds preferential sputtering effects of the constituent elements can lead to changes in the surface composition.
Methods which analyze the surface region such as Auger electron
spectroscopy (AES) will indicate deviations from the bulk composition that reflect the differences in sputtering yields.
However, a method such as secondary ion mass spectrometry (SIMS)
which collects the ejected particles can, after an initial
transient, reflect the original composition of the target. In
spite of these difficulties, layer removal by sputtering is
widely used in analytical problems which require information on
depth profiles.

Depth profiles generated by energy loss of charged
particles (Fig. 1b), is the subject of many sections of this
book. We comment that ion beam generated x-rays can give modest
depth information due to the strong energy dependence of the
excitation cross-section and that nuclear reactions, particularly those with sharp resonances, can be utilized to obtain
depth profiles. Both techniques have decided advantages in
certain applications which require high elemental specificity
(ion-induced x-rays) or sensitivity to low atomic number elements
(nuclear reactions).

Depth Concepts

A. Layer Removal by Sputtering

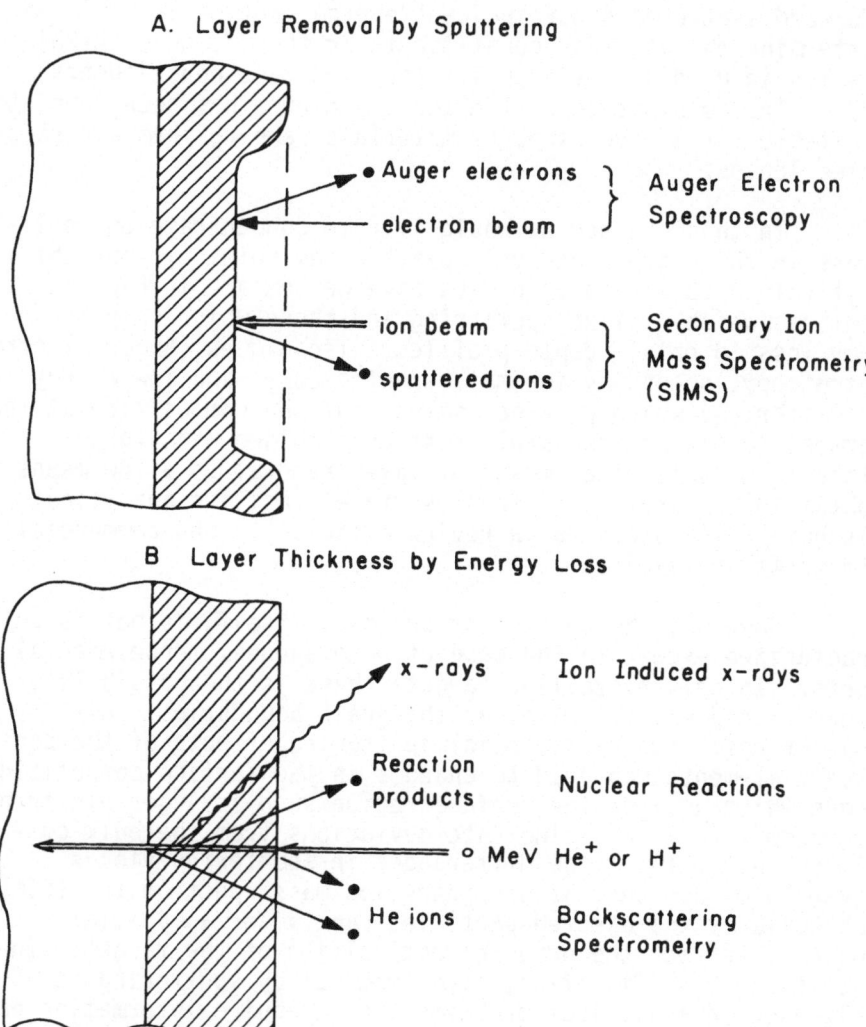

B Layer Thickness by Energy Loss

Figure 1.   Schematic representation of the two primary methods
            by which composition in depth has been obtained. In
            A, surface sensitive techniques, AES and SIMS, are
            used to analyze the outermost layers of the surface.
            Layer erosion by sputtering is used to expose deeper
            regions.   In B, energetic particles penetrate the
            layer of interest.   Depth profiles are determined from
            energy loss of particles on their inward and/or out-
            ward trajectories.

One of the major advantages of backscattering spectrometry is that it gives depth profiles in a fast, simple and direct fashion. To a first approximation, there is a linear conversion from the energy scale to the depth scale. The technique is quantitative and can be used to give absolute numbers for the number of impurities per cm^2 or layer thicknesses in terms of atoms per cm^2 (see Chapter by W.K. Chu). These features are important in technological or material science applications where the major emphasis is placed on the properties of the sample rather than the details of the analysis technique.

On the other side, there are limitations to backscattering spectrometry which must be born in mind when considering the application of the technique to specific problems. The sample depth that can be analyzed with MeV He ions is about one micron. Beyond this depth the total energy lost along the inward and outward paths becomes large and one must introduce corrections in the energy to depth conversion. There is a loss, then, in the simplicity of the technique. Greater analysis depths (up to about 10 microns) can be achieved with MeV protons with a concomitant loss in mass and depth resolution and hence loss in quantitative ability. Another solution is to utilize layer removal techniques. It is possible, for example, to mount a sputtering gun directly on the backscattering analysis chamber to permit analysis after incremental layers have been eroded away.[11]

Another limitation is that backscattering is not highly element specific and for heavy mass elements it is difficult to obtained good mass resolution. For example, it is difficult to distinguish between tungsten, tantalum, gold, or platinum with MeV He ions. Of course, in this case, ion induced x-rays can be used for element identification. The sensitivity to low atomic number elements is poor and nuclear reactions or resonance techniques must be introduced to obtain quantitative data on trace amounts of these elements. Of course if large concentrations (greater than a few atomic percent) of a low atomic number element are present, the concentration can be determined from its influence on the backscattering yield from higher mass elements. Such is the case for example, when analyzing the thickness of tantalum oxide layers on tantalum.[12]

To be ideally suited for backscattering analysis, samples should be laterally uniform across the dimensions of the beam (typically 1 to 2 mm). It is difficult to obtain meaningful depth distributions if the targets are nonuniform in lateral dimensions. Such nonuniformities can be determined directly by electron microprobes or scanning electron microscopy. Even optical microscopy can often identify potential problems due to nonuniform surface. In any event, it is desirable to have as much knowledge as possible of the lateral composition of the target.

To illustrate the problems of interpretation of laterally nonuniform targets, we show in Fig. 2 backscattering spectra
from Pb films deposited and heated on silicon substrates.[13]
The spectrum taken on the as-deposited sample (virgin) shows a
flat-topped peak and a sharp trailing edge for the Pb signal as
would be expected for a uniform film. After the sample was heat-
treated at 275°C for 10 min, the spectrum has a broad plateau
extending out to the energy position corresponding to Pb atoms
at the sample surface. The simplest interpretation of such a
spectrum is that the Pb and Si have intermixed and that Pb
penetrates greater than micron depths into the silicon. However,
the scanning-electron-microscope photograph in Fig. 3 shows that
the Pb film has formed islands and balls during the heat treatment and has not penetrated into the silicon. The backscattering spectrum in Fig. 2 is also consistent with this interpretation. Similar comparisons between backscattering spectra and
S.E.M. photographs of laterally nonuniform samples have been presented by Ottaviani, et al.[14]

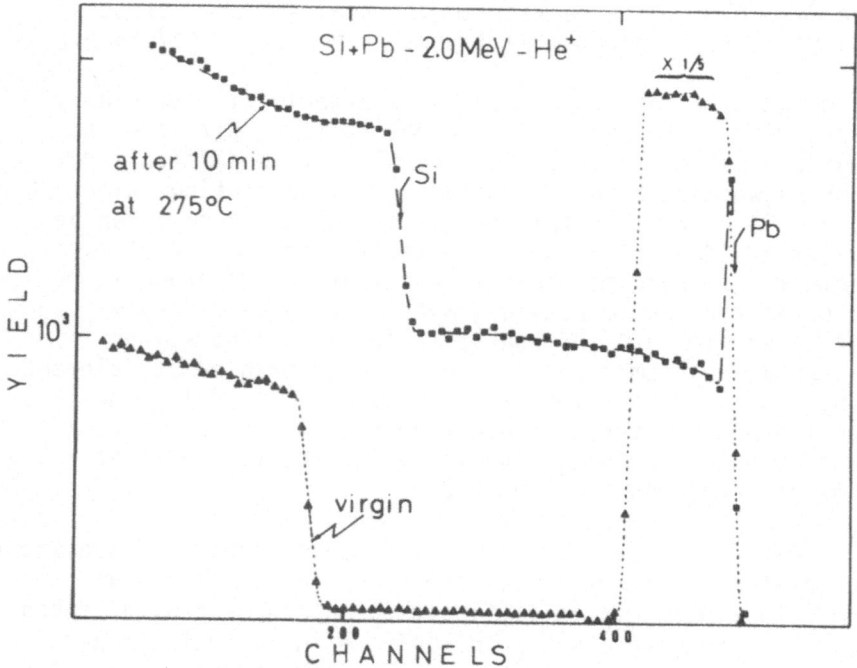

Figure 2.   Backscattering spectra from Pb layers deposited on Si
            targets before and after thermal anneal [Ref. 13].

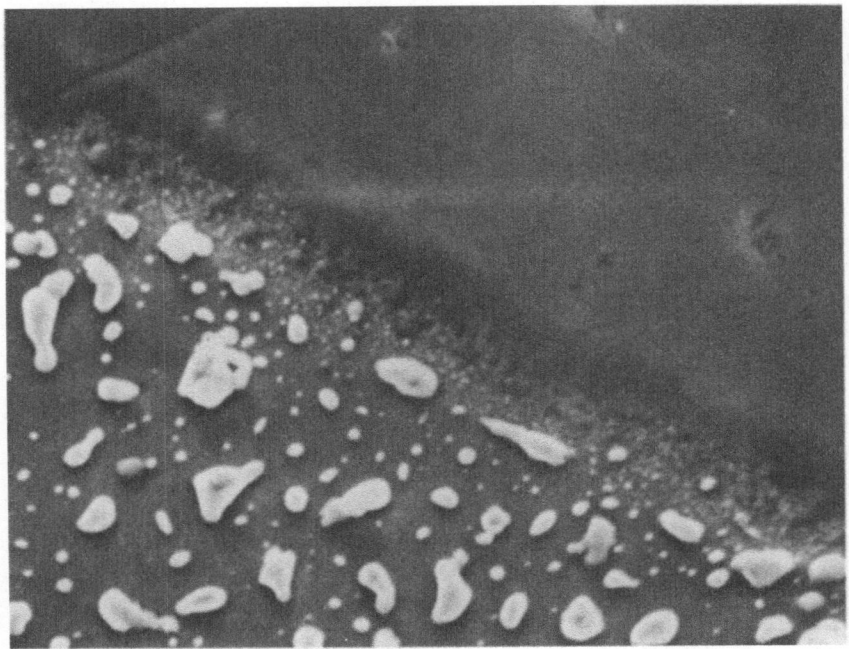

Figure 3.   S.E.M. micrograph of the annealed sample of Fig. 2.
The lower-left half of the figure shows the region
originally covered with Pb.[Ref. 13].

We have chosen to present examples of backscattering
based on the characteristics of the samples rather than on the
field of study. The sections are catagorized by: (1) ion
implantation, (2) deposited films, (3) thin film reactions, and
(4) bulk samples and diffusion profiles. The applications in any
section may deal with topics in electronic materials, CTR front-
wall cladding, basic materials science or applied studies of
material processing.

Most of the examples were drawn from studies using 1 to
2.5 MeV He ions. This is a convenient energy range for most low
energy accelerators and from an analytical standpoint it represents an ideal compromise between depth resolution, accessible
depth of analysis; and mass resolution. The relative advantages
and disadvantages of lower or higher beam energies and lighter or
heavier incident particles have been discussed in some detail by
Chu, et al.[15] We remark that changes in the beam parameters
only affect the quantitative aspects of the analysis not the
nature of the information that can be obtained. For example, 1
MeV protons rather than 1 MeV ^{4}He ions are required to monitor
thickness changes of one micron Au films. However, the fact that
Au film thicknesses can be monitored with ^{4}He ions by backscatter-

ing techniques can be demonstrated with 1000Å Au films. In the real world of practical applications, the sample configuration and layer thicknesses may be dictated by the problem rather than the capabilities of the accelerator.

In this chapter we will not discuss channeling as it is dealt with in detail by Picraux in a later chapter. It should be remembered, however, that channeling techniques are extremely powerful in obtaining quantitative data on lattice disorder and lattice location of impurity atoms.[4] In solid phase epitaxy, for example, channeling was one of the key measurements used to demonstrate epitaxial growth.[16]

II. ION IMPLANTATION

A natural application of backscattering spectrometry is the study of ion implanted layers. Implantation provides a method for changing the characteristics of the outermost layers of the sample and backscattering a method for measuring the change. In addition, the experimental techniques are virtually identical in some cases, the same accelerator is used for both backscattering analysis and implantation.

In the semiconductor field there is a symbiotic relationship between the two techniques. Backscattering and channeling measurements provided the necessary insight into implantation phenomena and the majority of groups currently applying backscattering spectrometry were initially concerned with the behavior of implanted layers.

The general features of the application of backscattering to implanted layers is illustrated (Fig. 4) in the backscattering spectrum of a silicon target implanted with 2×10^{15} As/cm^2 at 200 keV.[17] The integrated area of the As signal can directly give the number of As atoms/cm^2 if the experimental geometry, detector active area and integrated He ion current are known. Alternatively, if the stopping cross section, ε, in Si is known, comparison of the As signal area to the height of the yield from the silicon substrate gives the dose of implanted atoms. The energy shift ΔE of the peak in the As signal below the energy position corresponding to As on the sample surface gives the value for the projected range R_p; the FWHM of the As signal is directly related to the projected range straggling.[18]

There have been a number of such studies of range distributions in implanted layers from the standpoint of both basic atomic collision theory and applied dopant distribution evaluation in device structures.

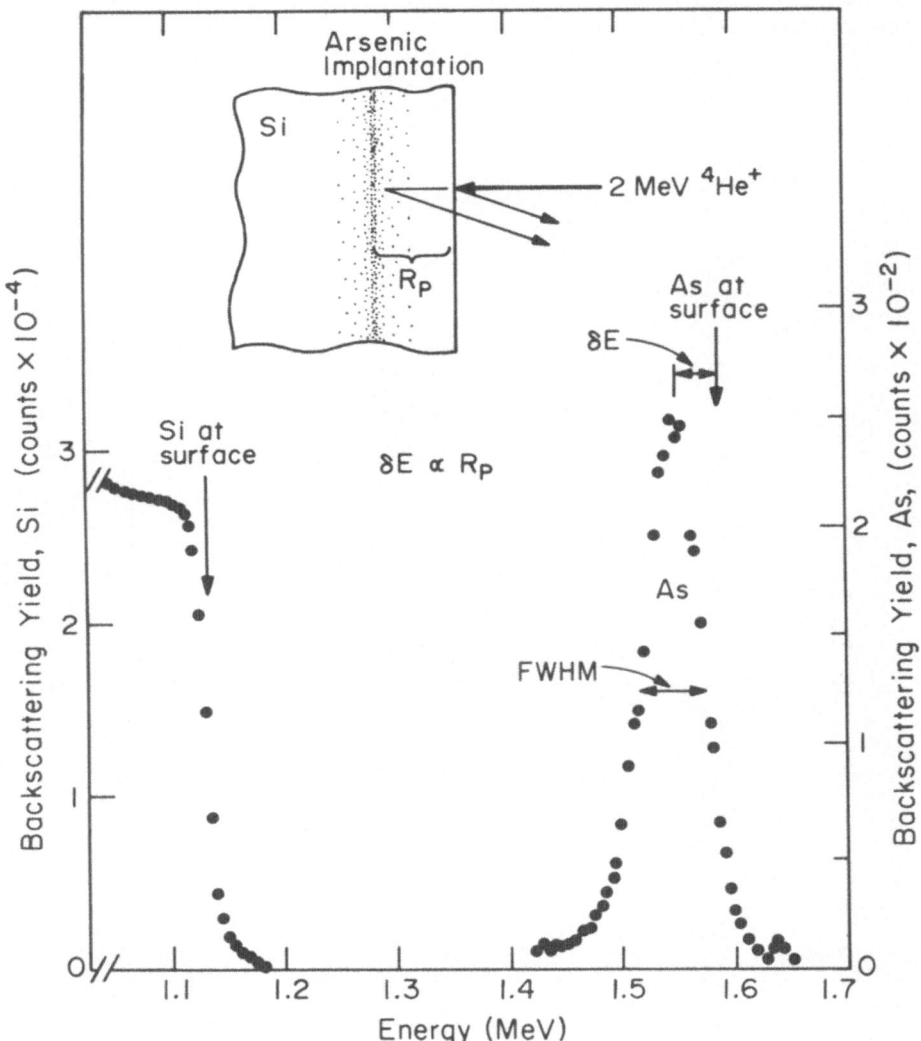

Figure 4.   Backscattering spectrum of 2 MeV He ions incident on
an As implanted Si sample.[Ref. 17]

The example shown in Fig. 4 represents one of the easiest
and most straightforward measurements: the mass of the implanted
dopant is nearly three times that of the substrate and the dopant
is located within a few thousand angstroms of the surface. A
counter example is shown in Fig. 5 for a light element (He) implanted in a heavy element (Ni). The distribution and total
number of the light element can be inferred from the decrease
"dips" in the heavy element signal at a position corresponding to

the depth of He in the film. The light element must be present
in concentrations exceeding a few atomic percent for the decrease
to be clearly identified.

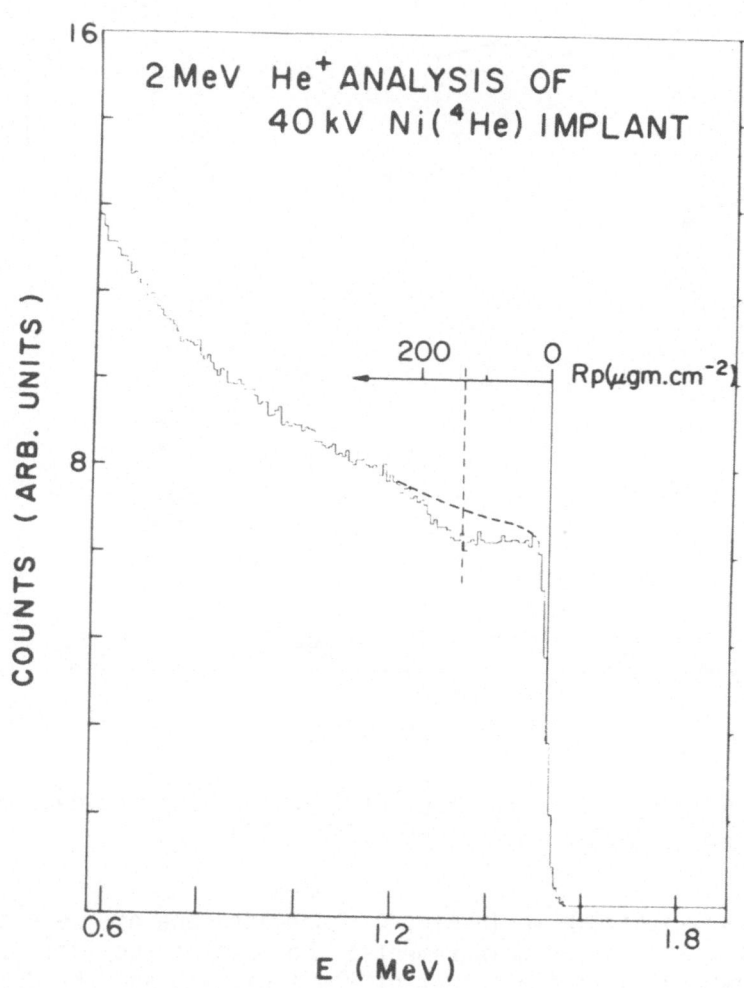

Figure 5. Backscattering spectrum from a Ni sample implanted with
           6 x 10^{17} ^{4}He ions/cm^2 at 40 keV. The "dip" in the Ni
           signal reflects presence of ^{4}He ions. The scale in the
           figure gives the depth scale in µg/cm^2. [From I.V. Mit-
           chell, Chalk River Nuclear Laboratories.]

The procedure of monitoring the valley in the signal from the heavy element rather than fighting the counting statistics and background substraction involved with extracting the signal of the light element has been used for many years. It was one of the techniques used by Powers and Whaling [19], for example, to obtain the range statistics for low mass ions in higher mass substrates. The disadvantage of the technique is that very high doses, often greater than a few hundred monolayers (i.e. greater than 2 to 4 x 10^{17}/cm^2) must be employed. At these high implantation doses, even the stopping power of the implanted species in the target can be influenced by the density changes introduced by the implanted species. Recently, Mitchell, et al. [20] have discussed a method whereby light elements can be detected by use of forward scattering and coincidence techniques.

For many light ions, resonance or reaction cross sections are sufficiently large so that implantation doses of the order of a monolayer can be easily detected. The application of nuclear reaction techniques is discussed in detail in the chapter by Wolicki, and several pertinent examples are cited. As an example of the use of resonance enhanced scattering cross section, we show in Fig. 6 a backscattering spectrum for 2 MeV protons on a micron

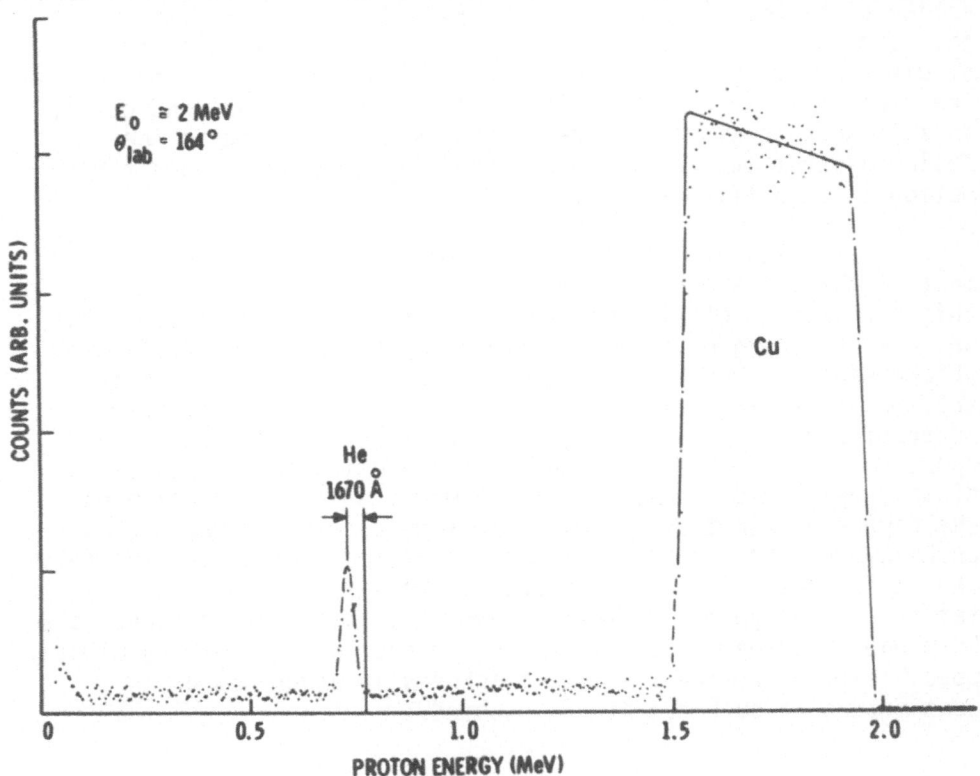

Figure 6.   Spectrum for 2 MeV protons incident on a Cu film implanted with He ions.[21]

thick Cu sample implanted with ^{4}He ions. [21] Although the implanted dose of He is nearly an order of magnitude less than that
in Fig. 5, the He signal stands out clearly above the background
level. For 2.5 MeV protons incident on ^{4}He ions, the scattering cross section ($\theta \approx 160°$) is a factor of 400 greater than that
predicted from the Rutherford formula. Since these cross section
enhancements are strongly dependent on energy and scattering
angle, some care is required when obtaining absolute values for
the dose of implanted ions.

The applications of ion implantation to metals, semiconductors and superconductors will be covered in detail in other
chapters. In summary we can reiterate that ion implantation and
backscattering analysis are natural partners. Implantation produces changes in the near surface region of the sample and backscattering provides analysis capability to determine these
compositional changes and impurity distributions.

III. THIN FILMS: GROWTH AND DEPOSITION

Thin film technology is another natural area of application of backscattering spectrometry. In fact in recent years
this area has been the subject of most of the backscattering
studies involving composition changes with depth. For convenience we will divide the subject into two broad areas and cover
in this section studies of deposition and growth and in the
following section, studies of interdiffusion and compound formation in thin film systems.

The most straightforward application is the measurement of the thickness in atoms/cm^2 of deposited metal films.
This measurement of film thickness is one of the basic elements
of backscattering depth microscopy. As shown in Fig. 7, film
thicknesses can be measured in multi-layer structures even if
the atomic masses of the different film elements are closely
adjacent.[22] In the lower portion of Fig. 7, the peak in the
spectrum near 1.3 MeV is caused by the overlap in signals (indicated by dashed lines) from the lower atomic mass Ni film on
the higher atomic mass Cu film. In some cases, for example Au
on W layers, the signals from the two elements overlap completely
and it is necessary either to change the sample to beam orientation (to change the effective layer thickness) or to change the
beam energy or particle. By these procedures it is nearly always
possible to determine the film thickness in atoms/cm^2 or in
centimeters if bulk density is assumed.

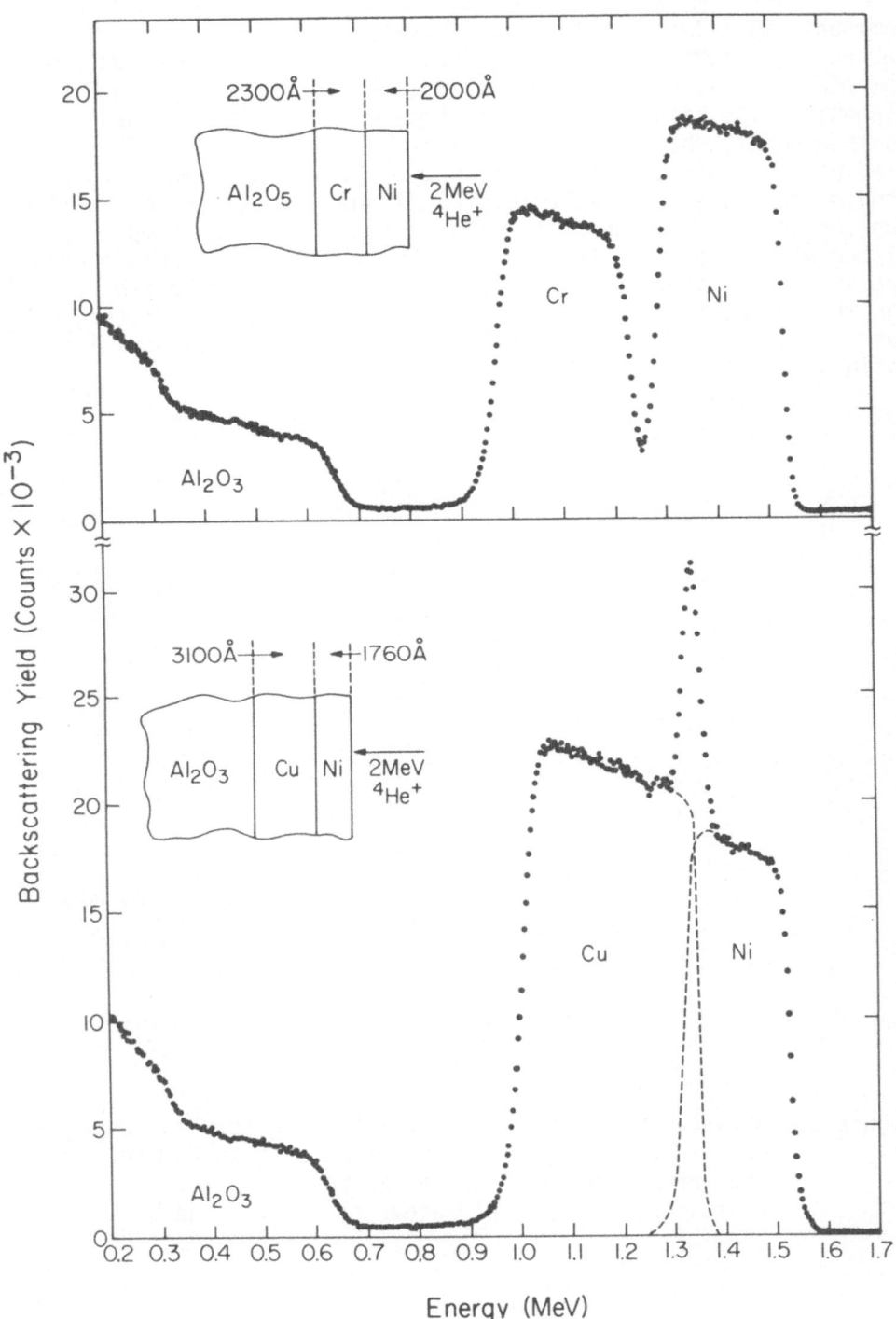

Figure 7. Spectra from metal layers on Al_2O_3 substrates.[Ref.22]

The same situation is not found when searching for contaminants in films. In favorable cases, when atomic mass differences are large and the background from the substrate does not
interfere, it is possible to make a clear separation of the
elements. The backscattering spectrum from a Ti-W layer deposited on Si, shown in Fig. 8, is such as example.[23] The
leading edges of both the Ti and W signals are at energies corresponding to those for scattering from Ti and W atoms at the surface
and the edges do not shift to lower energies when the sample is
tilted with respect to the beam. This indicates that both species
are at the surface. The signals from both elements are flat topped
and their widths are nearly equal indicating a uniform mixture of
both Ti and W in the deposited film, with an atomic composition
ratio of 38 at.%.

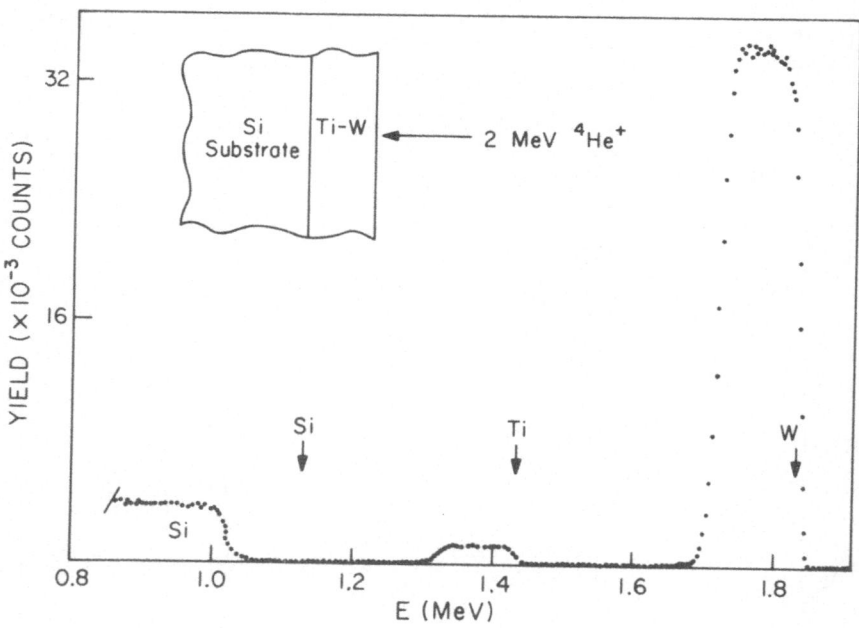

Figure 8.   Backscattering spectrum for 2 MeV ^{4}He ions incident
            on Ti-W film sputter deposited on a Si substrate.
            The arrows indicate the energy positions for
            scattering from surface atoms of Si, Ti, and W.
            [Ref. 23]

From backscattering kinetics alone it would not be possible to determine from the data in Fig. 8 whether the film contins W or Ta because the K factors differ by only 0.14%. To determine the identification of heavy mass elements such as Ta or W and Pt or Au other analysis techniques are required. A convenient method is ion-beam induced x-rays as discussed in a later chapter. This method has been used with great success in conjunction with backscattering to determine the composition of thin films and layered structures.

Identification of heavy elements is difficult because of the insensitivity of backscattering kinematics to differences in the atomic masses between heavy elements. For light elements such as carbon, oxygen or nitrogen the problem is often due to interference due to the signal from the substrate. Figure 9 gives the backscattering spectra for two different thicknesses of anodically grown tantalum oxide on tantalum.[12] The presence of a layer with reduced tantalum concentration is evidenced by the step in the tantalum spectrum. If it is assumed that the layer is tantalum oxide, then the height of step in the tantalum yield to that from the substrate signal would indicate a composition of Ta_2O_5. However, the oxygen signal is completely masked by the signal from the substrate. To identify the low mass signal in such a case either one must use nuclear reactions (discussed in the Chapter by Wolicki) or another technique such as secondary ion-mass spectroscopy (SIMS) or Auger electron spectroscopy (AES). In the present case, the analysis was performed with backscattering and SIMS. Backscattering was used to determine the thickness of the films while SIMS was used to determine the presence of oxygen and other low mass impurities.

As remarked before, SIMS and AES techniques when used with sputtering for layer removal are available in commercial units for analysis of depth profiles. The question that should be asked by users of low energy accelerators is what are the advantages of backscattering spectrometry over techniques employing sputtering. There are advantages: the depth scale can be determined directly, sample erosion during analysis is negligible, quantitative analysis is possible without recourse to standards, data acquisition is fast, and vacuum conditions are not as stringent as with surface analysis techniques. In spite of these advantages in the analysis of thin films and layered structures, nuclear analysis techniques are not used widely because of the ready availability of commercial AES or SIMS units. Also, the cost and operating expense of the commercial units is less than that of an accelerator. These considerations are not of paramount importance for groups using accelerators. For these groups, the problem is primarily to select a topic to investigate. In this respect, the proceedings from two conferences,

"Ion Beam Surface Layer Analysis"[1] and "Application of Ion Beams
to Metals"[2], provide an overview of the types of problems that
have been tackled by ion beam analysis techniques.

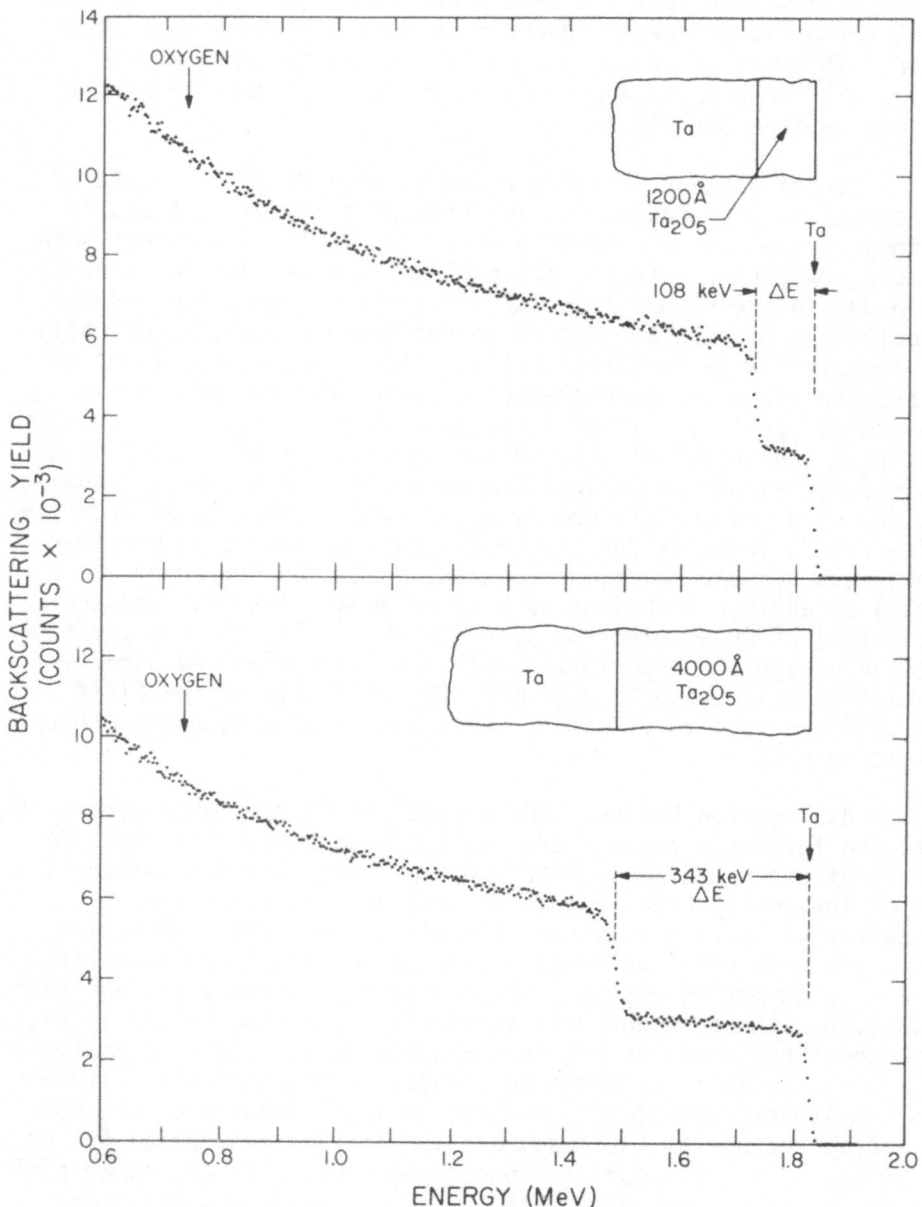

Figure 9.   Backscattering spectra for 2 MeV He ions incident
            on different thicknesses of Ta_2O_5 on Ta.[Ref. 12]

IV. THIN FILM REACTIONS: INTERDIFFUSION
AND COMPOUND FORMATION

When two metals or a metal and semiconductor are placed in contact with each other they will interdiffuse at temperatures well below the eutectic or melting point of any compounds. In fact with silicide forming systems, appreciable thicknesses of silicides are formed at temperatures less than one-half the melting point (in °K) of the silicide.[24] In all these thin film systems, the initial changes in composition occur at the interface without apparent changes at the sample surface. Backscattering spectrometry because of the short time required for data acquisition has been preferred for exploratory studies over the more lengthy and destructive analysis techniques employing sputtering for layer removal. In fact, backscattering analyses have been used in studies of thin film reactions to determine not only the temperature at which interdiffusion and compound formation occurs, but also to measure reaction kinetics and activation energies, to identify the diffusing species and the influence of impurities and to determine, in conjunction with channeling effect measurements, epitaxial growth conditions.[2]

The advantage of backscattering is that it gives information on thickness and atomic composition. However, knowledge of the atomic composition is often not sufficient to completely identify the phases that are present. In some cases as in the Au-Al system, two phases (Au_5Al_2 and Au_2Al) have compositions near one another; in others as with HfSi and $HfSi_2$, two phases are found intermixed in the same layer. It has been shown that glancing angle x-ray diffraction techniques have the sensitivity to determine the existence of compounds in layers a few hundred angstroms thick.[25] Many of the initial x-ray diffraction studies were made with the relatively sophisticated Seemann-Bohlin system; however, it has been shown that the simpler Read camera system is adequate in most cases.[26]

The combination of backscattering spectrometry and glancing angle x-ray diffraction techniques provides information on phase formation and reaction kinetics. It does not, however, give complete information on grain size of the polycrystalline layers that are nearly always found in thin film systems. For that information it is necessary to use electron microscopy or diffraction techniques. In this case, sample preparation is time consuming and it is often desirable to use backscattering analysis to first obtain a general idea of optimum film thicknesses and heat treatment conditions.

The general concepts involved in backscattering analysis of silicide layers has been presented in the chapter by Ziegler and in review articles.[15,24,27]   One aspect not covered is the use of implanted rare gas ions as diffusion markers.   The concept is shown schematically in the upper portion of Fig. 10 for Xe implanted in a Si sample which was subsequently covered with a

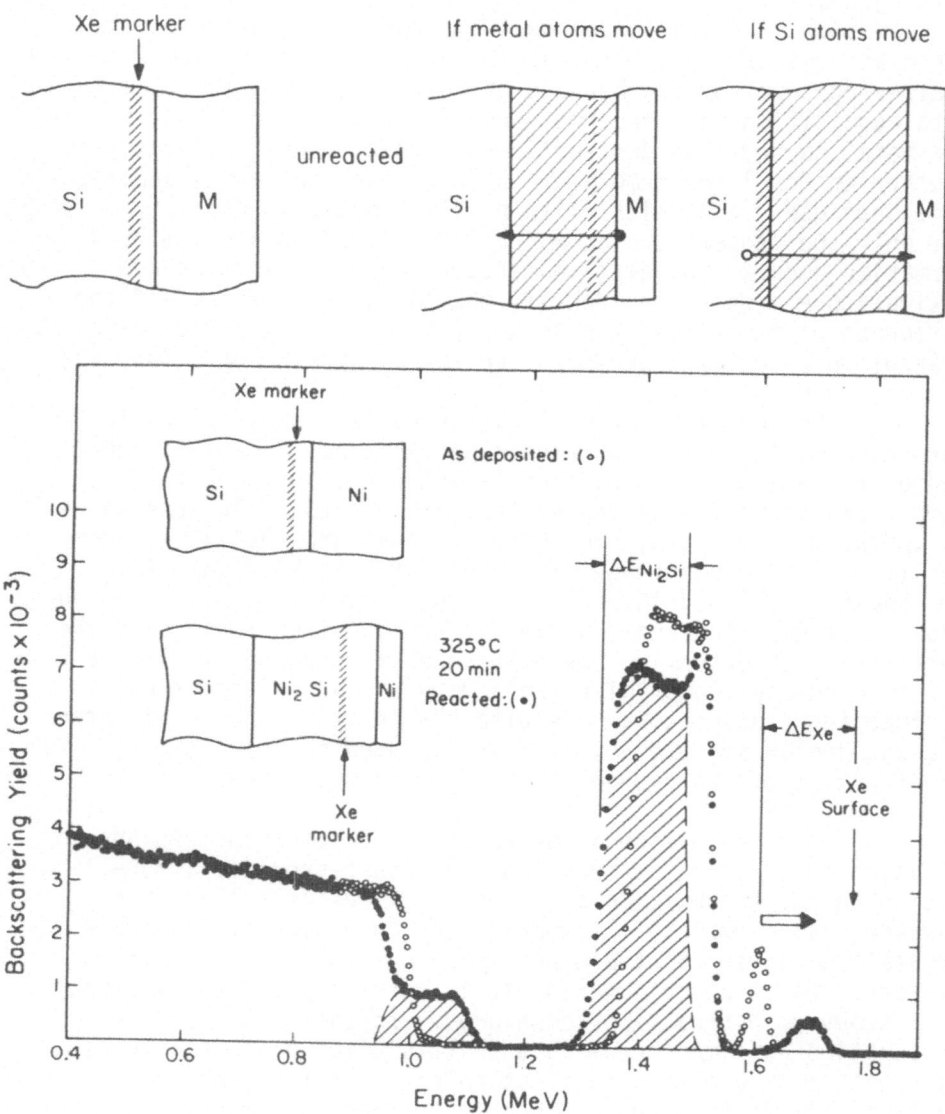

Figure 10.   The concept (above) and application of backscattering to diffusion marker studies.[Ref. 28]

metal layer.[28] Depending on whether the metal or silicon atoms are the diffusing species during silicide formation, the Xe marker atoms will be displaced toward or away from the surface. The example shown in the lower portion for the formation of Ni_2Si shows that Xe is displaced toward the surface. Measurements of the amount of displacement compared to the thickness of Ni_2Si show that Ni is the moving species. The idea of using implanted rare gas species as diffusion markers on a submicron scale has been used in studies of anodic oxidation as well as in silicide formation. It could also be applied to metal-metal thin film couples. Identification of the moving species in compound formation is one of the factors needed to establish a general model for compound formation in thin-film couples.

In systems such as Si-Al or Ge-Al exhibiting simple eutectics, diffusion occurs at temperatures well below the eutectic. It has been shown by electron microprobe measurements for Si-Al structures [29] and backscattering analysis for Ge-Al structures [30], that the solubility of the semiconductor in the Al film follows the same solidus solubility relationship found in bulk couples. As shown in Fig. 11 it is possible, then, to heat a Ge/Al couple to dissolve Ge in the Al and then slow cool to allow the Ge to regrow at the interface.[16] The regrown Ge layer is epitaxial on the underlying single crystal Ge substrate and is p-type due to incorporation of Al. The epitaxial layers grown under solid-phase conditions have been found to have good p-n junction characteristics and have been used to make nuclear particle detectors [31], transistors [32] and double injection diodes.[33] In this and in other work involving solid-phase epitaxy,[34] backscattering and channeling techniques were the key elements used in finding the optimum conditions for transport of the semiconductor through the metal film.

Metallization plays a large role in the success or failure of integrated circuit technology. In order to make electrical connection between the devices and the outside world, it is necessary to deposit metal layers for conducting paths. There are a host of problems associated with the metallization process, all of which stem from the face that thin film reactions occur at relatively low temperatures. For example, if Si is co-evaporated with Al to prevent dissolution of Si from the substrate, the excess Si in the Al can precipitate on the Si substrate. Also, gold does not adhere well to oxide layers and metals such as Cr or Ti are generally deposited as "glue" layers between the oxide and Au layer. However, Au and Ti react to form compounds. To prevent compound

SOLUBILITY OF GERMANIUM IN ALUMINUM

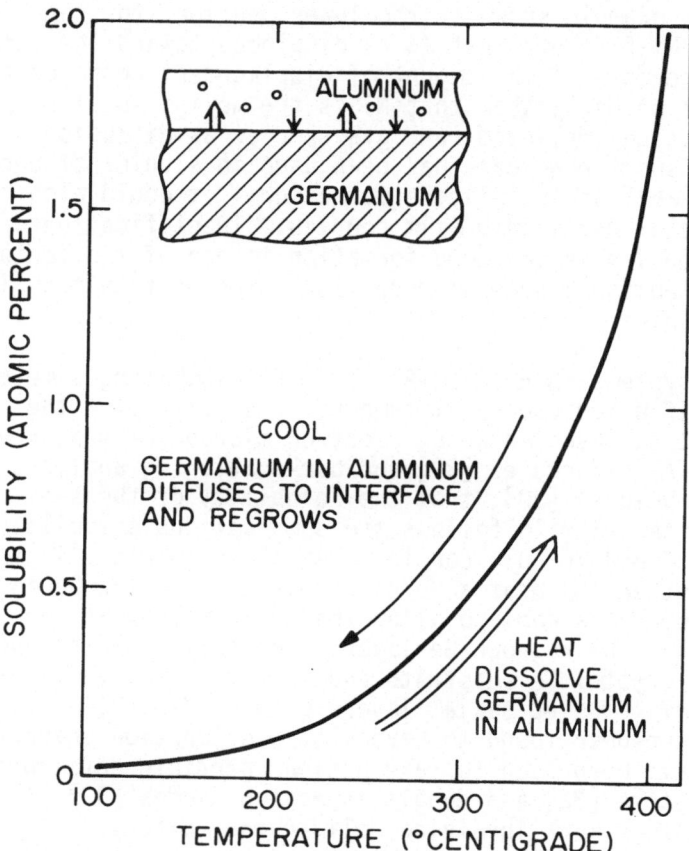

Figure 11.   Representation of the use of heating cycles to
produce Al-doped regions on Ge. [Ref. 16]

formation an intermediate layer is deposited as a barrier layer.
However, these barrier layers often only retard rather than
prevent interdiffusion. An example of the reactions observed
in the Ti/Au/Pd and Ti/Au/Rh systems are shown in Fig. 12.[35]
Here the positions of the Au and Pd as well as Au and Rh have
been inverted to show more clearly that interdiffusion occurred at temperatures as low as 312°C.

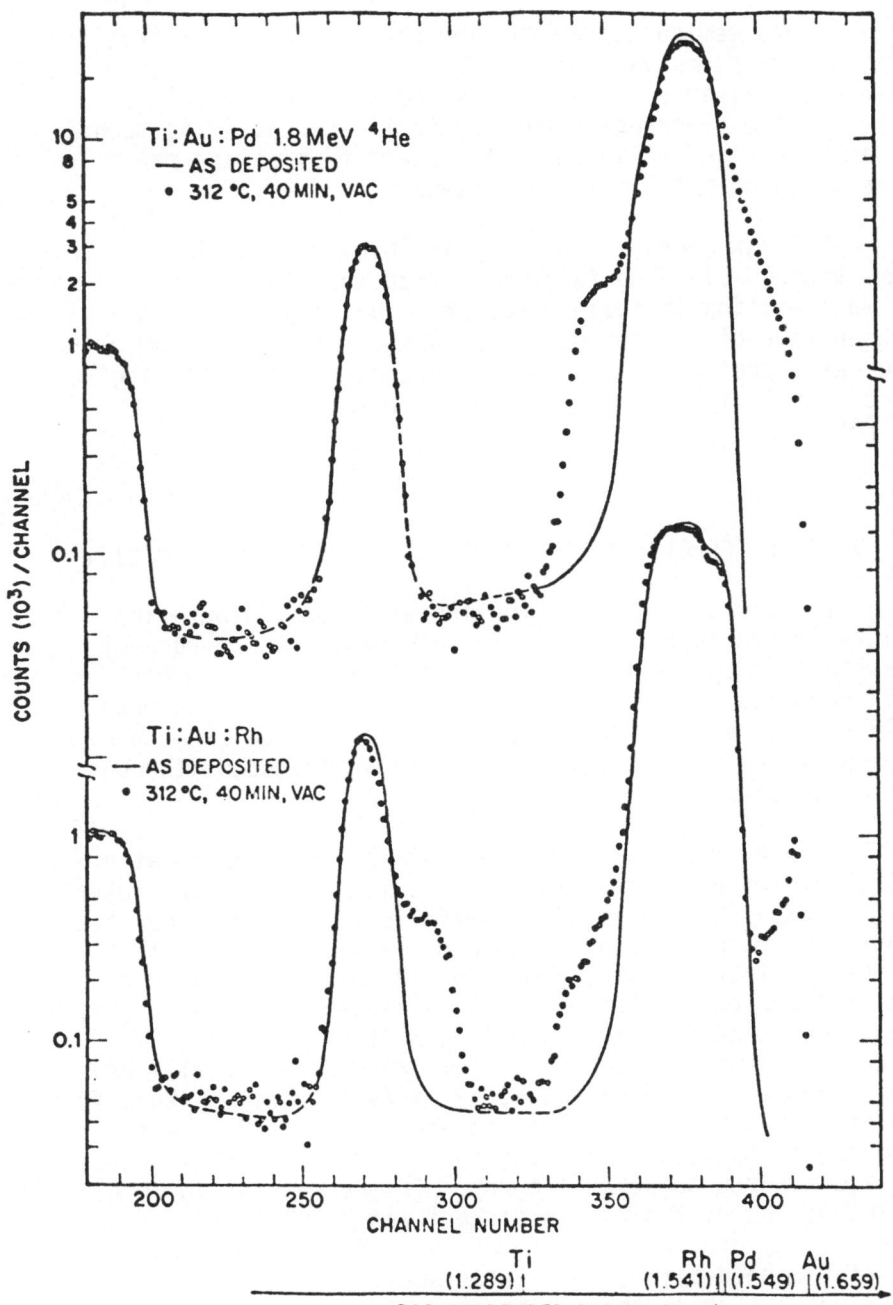

Figure 12. Backscattering spectra from thin films of Ti:Au:Pd (upper) and Ti:Au:Rh (lower) for as-deposited and heat treated conditions.[Ref. 35]

Every semiconductor company has its own metallization system. The systems vary widely from simple Al layers to multi-element layer systems. It is becoming more widely recognized in the electronic industry that backscattering techniques can be applied to investigate interdiffusion and reactions in these metallization systems.

Nuclear energy technology is involved with layered systems through cladding failures on reactor elements and front-wall erosion in fusion reactors. Backscattering studies have been made of sputtering rates during proton bombardment. In one case, protons of low energy were used for sputtering and protons of higher energy for measurement of erosion rates.[36]

V. BULK EFFECTS: COMPOSITION, DIFFUSION AND SOLUBILITY

In contrast to the situation with ion implantation or thin-film and layered structure technology, there has not been systematic applications of backscattering spectrometry to bulk samples. Oxidation and corrosion studies while made on bulk samples should more properly be considered as a thin film problems as the key parameters involve transport through a layered structure.

As a near surface, microanalytical tool, backscattering spectrometry has interesting possibilities. One can determine for example, the diffusion coefficient of heavy impurities in light substrates. This is shown for Zn diffusions into Al in the backscattering spectra of Fig. 13.[37] There is a clear change in shape of the Zn profile with increasing anneal time. It should be noted that the magnitude of the diffusion coefficient D that can be obtained from backscattering will be in the range between 10^{-11} and 10^{-15} cm^2/sec. This follows from the relation $D = x^2/t$ where the distance x varies from a few hundred angstroms to a micron and the heat treatment time, t, from a few minutes to several hours. In the case of Zn in Al the value of the diffusion coefficient was about 10^{-14} cm^2/sec and for a study of the diffusion of Cu in Be the measured values of D ranged between 4×10^{-12} and 10^{-15} cm^2/sec.[38] These values are below those that can be measured conveniently by sectioning techniques. In a practical sense, then, backscattering allows evaluation of diffusion processes at lower temperatures than commonly used.

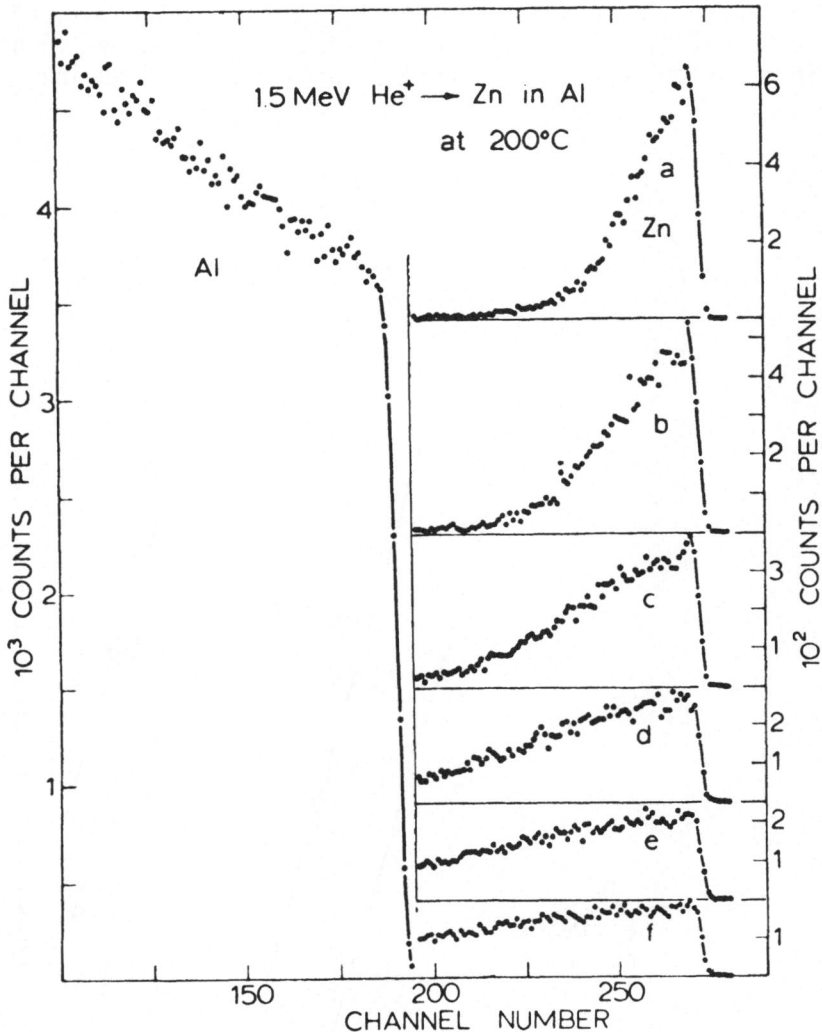

Figure 13.   Backscattering spectra illustrating the diffusion
of Zn in Al at 200°C. [Ref. 37]

          Another possibility is to study the solubility of
impurities in solids. There are several constraints imposed in
such studies: the solubility of the impurity should be above
$10^{18}/cm^3$ in order to be in the convenient range for backscattering analysis; the solubility should not be influenced
by proximity to the surface (although this limitation can be
overcome by etching or polishing the sample before analysis),
and precipitation, segregation or clustering effects should be
minimal. This last problem can be investigated in single crystal
substrates by use of channeling techniques. Such a procedure

was utilized, for example in the study of the solubility of Au in CdTe.[39] Figure 14 shows the backscattering yield of Au in atoms/cm^3 versus the partial pressure of Cd during

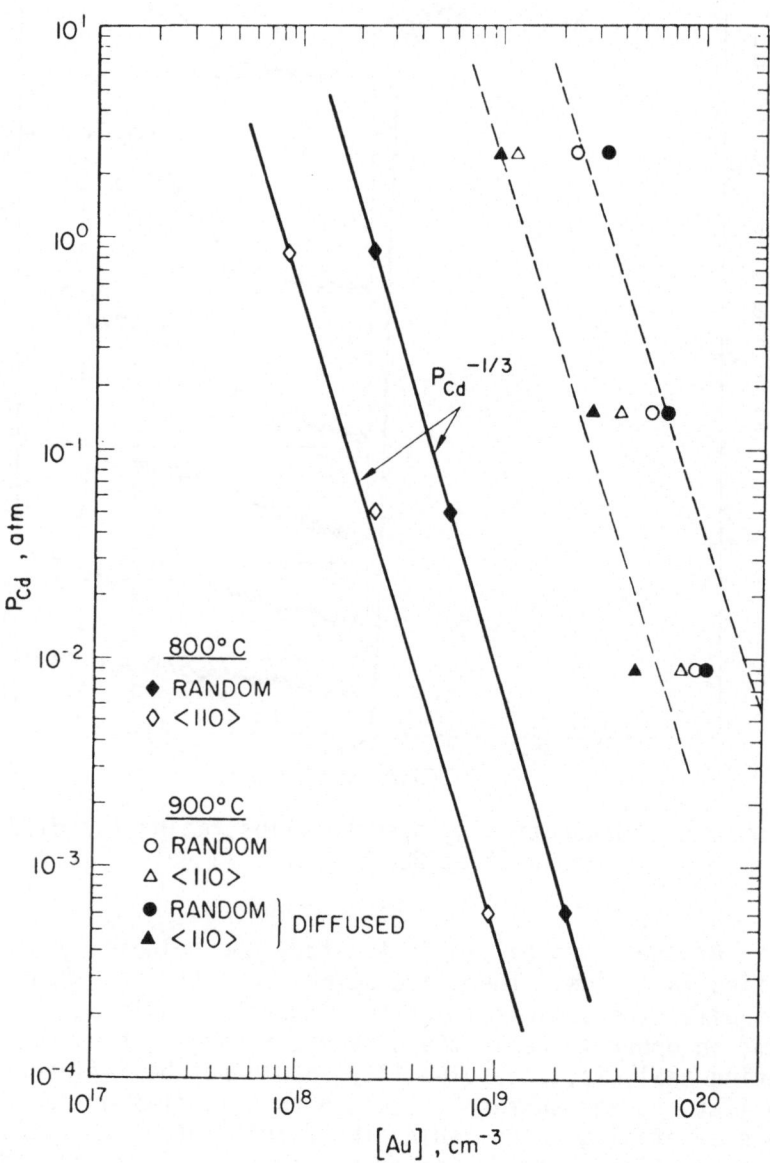

Figure 14.   Solubility of Au in CdTe as a function of the partial pressure of Cd during heat treatment. [Ref. 39]

anneal for measurements along "random" and <110> directions.
The difference between the random and <110> yields gives the
amount of Au on substitutional sites. The dependence of the
substitutional Au concentration upon the heat treatment
temperature and partial pressure of Cd agrees with theoretical
predictions. More importantly it shows that backscattering
and channeling effect measurements can be used to measure
impurity solubilities in compound semiconductors. Another
aspect of the same general problem is to use channeling effect
measurements to observe impurity motion off substitutional
lattice sites during different anneal cycles. This procedure was used, for example, to determine the substitutional
fraction of Te in GaAs.[40]

    To the extent that the first few thousand angstroms
are typical of the bulk of a target, backscattering can also
be used to analyze bulk materials. The spectrum is shown in
Fig. 15 for a magnetic bubble material grown on gadolinium

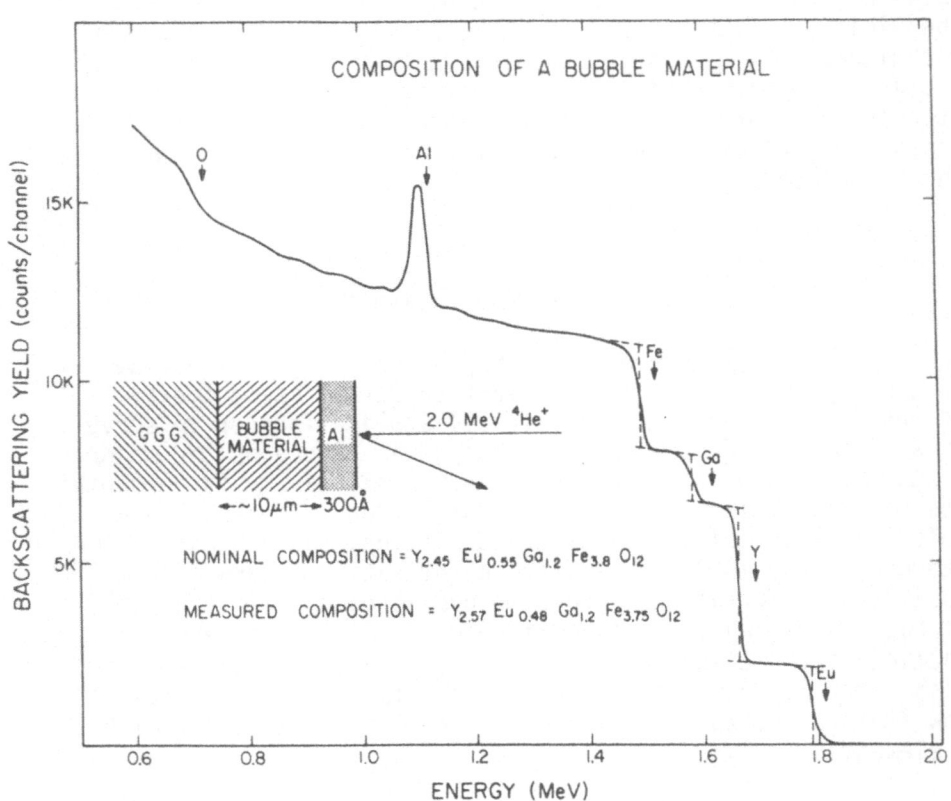

Figure 15.   Backscattering spectrum from "bubble" material.
              [Ref. 41]

gallium garnet.[41] Although the magnetic material is actually
a thick film, the layer is much thicker than the range of the
2.0 MeV ^{4}He beam, and thus effectively is a bulk material for
the purpose of analysis. A thin film of Al was deposited on
top to prevent charge buildup effects in the insulating material. The magnetic material was known to contain Fe, Gd, Y and
Eu in nominal amounts and to have an overall composition of
that of a garnet, X_8O_{12}. From the step height of the various
metal signals in the spectrum, the corresponding relative concentrations can be calculated by correcting for the cross
sections of the various elements and for differences in stopping power along the outgoing paths. The nominal and measured compositions are given in Fig. 15. The errors for such
determinations of composition are of the order of 5% in average cases and 1 to 2% in the best cases.

Note that without some previous knowledge of the elements contained in the sample, specific identification of the
elements would have been difficult in light of the mass resolution (about ± 2 amu in the vicinity of Fe and ± 9 amu in
the vicinity of Eu). In some cases spectrum differentiation
techniques [42] have been used to identify the elements from
the energy position of the resultant spikes (rather than
steps). However, for unambiguous identification of the elements, ion induced x-ray or electron microprobe analysis
should be used.

VI.   CONCLUDING REMARKS

The objective of this chapter was to give the flavor
of the type of applications of backscattering spectrometry.
It was not the intention to cite all applications or to serve
as a historical review. The emphasis was placed on the major
feature of backscattering spectrometry: microanalysis in depth;
that is depth microscopy. Of course, other features play a
role such as, rapidity of data acquisition, the nondestructive
feature of the analysis, and the quantitative nature of compositional information. Wherever possible, the limitations
of backscattering analysis were pointed out: the absence of
chemical information, the ambiguity of analysis when dealing
with laterally nonuniform targets, the lack of sensitivity to
low Z elements, and the difficulty in identification of high
Z elements.

When searching for new applications of backscattering,
the above advantages and limitations must be borne in mind. In
general, one looks for systems where near-surface analysis is

important. Examples are thin films, layered structures, diffusion profiles, ion implanted layers and dielectric layers. To take advantage of the capabilities of backscattering there should be a change in the composition during the process treatment such as interdiffusion between thin film elements or growth of oxide layers.

One should note that the majority of the applications of backscattering have been carried out with collaboration between the accelerator users and groups dealing with problems in materials. There are practical reasons for this in that sample preparation can be time consuming and supplementary measurements such as electron microscopy are often required. These supplementary measurements generally require equipment that is not immediately available to the accelerator users. From another standpoint, collaboration with groups dealing with materials provides ready access to the state-of-the art in a particular field.

Finally, it is desirable to design the scattering chamber so that measurements of ion induced x-rays or nuclear reactions can be made along with backscattering analysis. As will be obvious from the discussion in other chapters, these two techniques provide complementary data that may be crucial in the analysis of a given sample.

ACKNOWLEDGEMENTS

It is a pleasure to acknowledge the contributions to backscattering analysis of our Caltech colleagues: M-A. Nicolet, W-K. Chu, S.S. Lau and many others. To them we owe our thanks. The work was supported in part by ONR (L. Cooper).

REFERENCES

1. Proceedings of the conference on Ion Beam Surface Layer Analysis, Yorktown Heights, N.Y., June 1973; Thin Solid Films 19, 1-406 (1973).
2. Proceedings of the conference on Applications of Ion Beams to Metals, Albuquerque, N.M., October 1973;(Plenum Press, New York) 1974. Eds. S.T. Picraux, E.P. EerNisse and F.L. Vook.
3. Proceedings of the conference on Ion Implantation in Semiconductors and Other Materials, Yorktown Heights, N.Y., December 1972; Ed. B.L. Crowder,(Plenum Press, New York) 1973.

4.   "Channeling", Edited by D.V. Morgan (J. Wiley and Sons, New York) 1974.

5.   A.L. Turkevich, E.J. Franzgrote and J.H. Patterson, Science 165, 277 (1969).

6.   M. Peisach and D.O. Poole, in "Electron and Ion Beam Technology," Edited by R. Bakish, AIME (Gordon and Breach, New York) Vol. 2, 1195 (1969).

7.   L. Eriksson, G. Fladda and P.A. Johansson, International Meeting, "Chemical Analysis by Charged Particle Bombardment," Namur, Sept. (1971).

8.   P.A. Tove and W.K. Chu, private communication.

9.   M. Guttmann, P.R. Krahe, F. Abel, G. Amsel, M. Bruneaux and C. Cohen, Scripta Metallurgica 5, 479 (1971).

10.   S. Rubin, T.O. Passell and L.E. Bailey, Anal. Chem. 29, 736 (1957).

11.   A. Turos, W.F. van der Weg, D. Sigurd and J.W. Mayer, J. Appl. Phys. 45, 2777 (1974).

12.   W.K. Chu, M-A. Nicolet, J.W. Mayer and C. A. Evans, Jr., Anal. Chem. 46, 2136 (1974).

13.   S.U. Campisano, G. Foti, F. Grasso and E. Rimini, Proc. Conf. on Low Temperature Diffusion and Applications to Thin Films, Yorktown Heights, N.Y., August 1974 (to be published, Plenum Press).

14.   G. Ottaviani, D. Sigurd, V. Marrello and J.O. McCaldin, J. Appl. Phys. 45, 1730 (1974).

15.   W.K. Chu, J.W. Mayer, M-A. Nicolet, T.M. Buck, G. Amsel and F. Eisen, Thin Solid Films 17, 1 (1973).

16.   V. Marrello, J.M. Caywood, J.W. Mayer and M-A. Nicolet, Phys. Status Solidi a13, 531 (1972).

17.   T.W. Sigmon, W.K. Chu, H. Muller and J.W. Mayer, Appl. Phys. (in press).

18.   W.K. Chu, B.L. Crowder, J.W. Mayer and J.F. Ziegler, in Ref. 3, p. 225.

19.   D. Powers and W. Whaling, Phys. Rev. 126, 61 (1962).

20.   J.A. Moore, I.V. Mitchell, M.J. Hollis, J.A. Davies and L.M. Howe, J. Appl. Phys. 46, 52 (1975).

21.   R.S. Blewer, in Ref. 2, p. 557.

22.   W.K. Chu and B.M. Ullrich, private communication.

23.   J.M. Harris, private communication.

24.   J.W. Mayer and K.N. Tu, J. Vac. Sci. Technol. 11, 86 (1974).

25.   K.N. Tu and B.S. Berry, J. Appl. Phys. 43, 3283 (1972).

26.   S.S. Lau, W.K. Chu, J.W. Mayer and K.N. Tu, Thin Solid Films 23, 205 (1974).

27.   J.A. Borders and S.T. Picraux, Proc. IEEE 62, 1224 (1974).

28.   W.K. Chu, H. Krautle, J.W. Mayer, H. Muller, M-A. Nicolet and K.N. Tu, Appl. Phys. Lett. 25, 18 (1974).

29.   J.O. McCaldin and H. Sankur, Appl. Phys. Lett. 19, 524 (1971).

30. J.M. Caywood, Metall. Trans. $\underline{4}$, 735 (1973).
31. V. Marrello, T.A. McMath, J.W. Mayer and I.L. Fowler, Nuc. Inst. and Meth. $\underline{108}$, 93 (1973).
32. J.O. McCaldin and A. Fern, Proc. IEEE $\underline{60}$, 1018 (1972).
33. V. Marrello, T.F. Lee, R.N. Silver, T.C. McGill and J.W. Mayer, Phys. Rev. Lett. $\underline{31}$, 593 (1974).
34. C. Canali, J.W. Mayer, G. Ottaviani, D. Sigurd and W. van der Weg, Appl. Phys. Lett. $\underline{25}$, 3 (1974).
35. W.J. De Bonte, J.M. Poate, C.M. Melliar-Smith and R.A. Levesque, in Ref. 2, p. 147.
36. W. Eckstein, B.M.U. Scherzer and H. Verbeck, Rad. Effects $\underline{18}$, 135 (1973).
37. A. Fontell, E. Arminen and M. Turunen, Phys. Stat. Solidi a$\underline{15}$, 113 (1973).
38. S.M. Myers, S.T. Picraux and T.S. Prevender, Phys. Rev. B$\underline{9}$, 3953 (1974).
39. W. Akutagawa, D. Turnbull, W.K. Chu and J.W. Mayer, Solid State Comm. $\underline{15}$, 1919 (1974).
40. I.V. Mitchell, J.W. Mayer, J.K. Kung and W.G. Spitzer, J. Appl. Phys. $\underline{42}$, 3982 (1971).
41. M-A. Nicolet, private communication.
42. M. Peisach, Thin Solid Films $\underline{19}$, 297 (1973).

MATERIAL ANALYSIS BY NUCLEAR BACKSCATTERING

Formalism

Wei-Kan Chu*

California Institute of Technology

Pasadena, California 91109

From the previous two chapters and an earlier review[1] we have shown that backscattering is a very powerful technique for surface and thin film analysis. Many examples have been given to illustrate this method and its application. In this chapter we will give some essential concepts and basic formula about backscattering. Key formula and definitions are summarized and listed in the Appendix of this chapter.

There are only three basic concepts in backscattering and each gives an analytical capability:
1. Kinematic Factor → Mass Analysis
2. Differential Scattering Cross Section → Quantitative Analysis
3. Energy Loss → Depth Analysis
These three concepts serves as the building blocks for backscattering analysis. A series of formula can be derived from them to fit special problems, and some approximations and simplifications can be made for many special cases. We will discuss them in this chapter.

1. Three Basic Concepts in Backscattering

A. Backscattering Kinematic Factor → Mass Analysis

When a projectile of mass m with energy E_0 collides with a stationary target atom with mass M, there is momentum transfer from the projectile to the target atom. Assuming that there is no nuclear reaction introduced during the collision, the energies of the scattered projectile and recoil particle can be calculated from the conservation laws. The

*Permanent address: S.P. Division, IBM Corporation, Hopewell Junction, New York 12533.

kinematic factor, K, is defined as the ratio of the projectile energies after (E') and before collision, i.e., $K = (E'/E_0)$. This factor depends on the scattering angle θ in the lab system and the masses involved in the collision process, $K=K(M,m,\theta)$. Because in a given backscattering experiment, m and θ are predetermined, we use the notation K_M or K when there is no ambiguity. Throughout this chapter we will use the subscripts to denote the identity of the scattering center.

$$K = K_M = \frac{E'}{E_0} = \left(\frac{m\cos\theta + \sqrt{M^2 - m^2 \sin^2\theta}}{m + M}\right) \tag{1}$$

Equation 1 indicates the dependence of the backscattered energy on the target mass. When one measures E'/E_0, knowing m and θ, one can solve for M.

B. Differential Scattering Cross Section → Quantitative Analysis

The scattering process due to Coulomb interaction is a classical problem.[2] The differential scattering cross section given in laboratory coordinates can be written as:

$$\frac{d\sigma}{d\Omega} = \left(\frac{Z_1 Z_2 e^2}{2E\sin^2\theta}\right)^2 \frac{\{\cos\theta + [1 - (\frac{m}{M}\sin\theta)^2]^{\frac{1}{2}}\}^2}{[1 - (\frac{m}{M}\sin\theta)^2]^{\frac{1}{2}}} \tag{2}$$

In this equation, recoil correction is included, and Z_1 and Z_2 are the atomic number of the projectile and target atom. E is the energy of the projectile immediately before scattering and θ is the laboratory scattering angle.

The differential scattering cross section is directly related to the scattering probability and therefore it connects the backscattering yield to the quantitative analysis. The relation and formalism will be given in §3.

The average differential scattering cross section, σ, taken over a finite solid angle, Ω, spanned by the detector, is defined as

$$\sigma = \frac{1}{\Omega} \int \left(\frac{d\sigma}{d\Omega}\right) d\Omega \tag{3}$$

Deviations of the differential scattering cross section from the Rutherford formula do exist in some special cases. For extremely low energy projectiles scattered from heavy targets (e.g. 10 KeV ^{4}He$^+$ bombarding Au), a modification for electron screening needs to be made. For high energy ^{4}He ions (E > 2 MeV) or medium energy protons, elastic scattering from light mass elements can deviate significantly from the Rutherford formula because of nuclear reaction resonances. The enhancement of the yield due to nuclear resonance can be used to improve the sensitivity of the backscattering on light mass elements.

C. Energy Loss → Depth Analysis

When a projectile penetrates a target, it loses its energy throughout its trajectory to the electrons of the target atoms by ionization and by excitation, and it also loses its energy by nuclear collisions. When this projectile encounteres a hard collision, i.e. backscattering, it can change its trajectory into an outward direction. During its outward path it also loses energy to the target atoms until the particle emerges from the target. It is the existence of this energy loss phenomenon that enables one to determine the depth to which the projectile has penetrated, simply through measurements of how much energy loss it has suffered.

The energy loss of the projectile per unit path length depends on the energy of the particle and also on the projectile and the target. This information is generally obtained through experimental measurements. Compilations and reviews are available and sources of information will be given in the Appendix of this chapter.

Energy loss is normally expressed as dE/dx in the units of eV/Å, KeV/Å or MeV/μm, etc. The depth is often measured as mass per unit area, ρdx, or number of atoms per unit area, Ndx, where ρ and N are mass density and atomic density respectively. Therefore the energy loss is often expressed as $dE/\rho dx$ or dE/Ndx. The last term is called the stopping cross section ε. The stopping cross section ε carried the unit of eV-cm^2/atom sometimes shortened to eV-cm^2. The merit of using ε instead of dE/dx is that ε is on an atomic scale. When one deals with compound targets, the assumption of additivity of stopping cross section (Bragg rule) is made on the molecular and atomic scale rather than on the energy loss per physical thickness scale (dE/dx).

One other important fact is that ε for a given target is a well characterized fundamental parameter while dE/dx requires independent knowledge or assumption of the density value ρ or N. An implicit assumption that the density of the thin film sample is equal to that of the bulk sample of the same material has been made when dE/dx is used.

2.   Depth Scale in Backscattering Analysis [S]

Using two of the three concepts we learned from the pre-vious section, we will build up a simple formula for the depth analysis and we will also describe some approximations often used in such analysis.

A. Depth Scale in Backscattering Analysis

Let us consider the backscattering analysis of a well polished elemental target of atomic mass M, and atomic density N.   Figure 1 gives the notation and schematic backscattering energy spectrum for this backscattering example.

The incident energy of the projectiles is E_0 and the scattered particles have energy KE_0 when scattered from the surface.   When a particle is scattered at a depth t and emerges from the target into the detector, it will have lower energy because of the energy loss of the projectile in the target.   The incident and scattered angles are θ_1 and θ_2 with respect to the normal of the target, and the scattering angle is $\theta = 180° - \theta_1 - \theta_2$. The energy of the projectile at the depth t right before the scattering is E and it can be related to E_0 by

$$E = E_0 - \int_0^{\frac{t}{\cos\theta_1}} \frac{dE}{dx} \, dx \qquad (4)$$

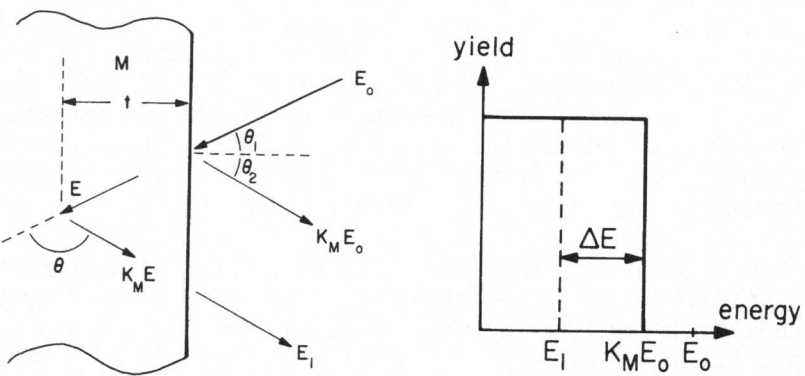

Figure 1.   Backscattering geometry, notation and schematic energy spectrum.

Equation 4 shows the energy loss dE/dx depends on the energy of the projectile, therefore it is also dependent on the depth x.

After scattering at depth t, the particle will lose its energy on its outward direction and it will have energy E_1 when emerging from the target.

$$E_1 = KE - \int_0^{\frac{t}{\cos\theta_2}} \frac{dE}{dx} \, dx \qquad (5)$$

The energy difference, ΔE, between the particles scattered from atoms on the surface and that scattered from atoms at a given depth t is defined from the spectrum given in Figure 1 as

$$\Delta E = KE_0 - E_1 \qquad (6)$$

and with Equation 4 substituted into Equation 5, Equation 6 becomes

$$\Delta E = K \int_0^{\frac{t}{\cos\theta_1}} \frac{dE}{dx} \, dx + \int_0^{\frac{t}{\cos\theta_2}} \frac{dE}{dx} \, dx \qquad (7)$$

Equation 7 has two terms, the first term comes from the energy loss in the incident path times the kinematic factor, and the second term comes from the energy loss in the outgoing path. The integration in Equations 4, 5 and 7 are line integrations and dE/dx is a function of energy at the given location inside the target.

From Equation 7 one can calculate ΔE vs t. Therefore, by measuring ΔE, one can get a direct measurement of the depth. Simple computer programs for this type of calculation can be written easily and are used in several laboratories. Since dE/dx is a very smooth function of E, Equation 7 can be simplified using many approximations. In general, for small energy loss where dE/dx does not change much there is a linear relation between energy loss and depth that can be expressed as

$$\Delta E = [S]t \qquad (8)$$

The symbol [S] is called the backscattering energy loss factor and it changes slowly as a function of E and t.

B. Surface Approxiation

When a film is very thin, one can assume that dE/dx of

the projectile in the thin film does not change and can be evaluated at the energy E_0 for its incoming path and KE_0 for its outgoing path. Equation 7 can be simplified into

$$\Delta E = (\frac{K}{\cos\theta_1} \frac{dE}{dx}\Big|_{E_0} + \frac{1}{\cos\theta_2} \frac{dE}{dx}\Big|_{KE_0})t \qquad (9)$$

From Equations 8 and 9 we have the backscattering energy
loss factor [S] at the surface defined as

$$[S] = \frac{K}{\cos\theta_1} \frac{dE}{dx}\Big|_{E_0} + \frac{1}{\cos\theta_2} \frac{dE}{dx}\Big|_{KE_0} \qquad (10)$$

Many of the backscattering experiments have normal incidence, i.e. $\theta_1 = 0$ and $\theta_2 = \pi - \theta$, so [S] becomes

$$[S] = K\frac{dE}{dx}\Big|_{E_0} + \frac{1}{|\cos\theta|} \frac{dE}{dx}\Big|_{KE_0} \qquad (11)$$

A parallel set of equations for ϵ instead of dE/dx can
be derived to give a backscattering stopping cross section factor
in the unit of $eV\text{-}cm^2$.

$$[\epsilon] = K\epsilon(E_0) + \frac{1}{|\cos\theta|} \epsilon(KE_0) \qquad (12)$$

The errors introduced by using the surface approximation
can be illustrated with the schematics shown in Figure 2.

Figure 2 gives the shape of a typical dE/dx or ϵ versus
energy curve. E_0 is the incident energy of the projectile. When
the projectile penetrates the target at a distance t, its energy
decreases to E. The values of energy loss dE/dx along the incident and outgoing paths as given in Equation 7, are accented
with two heavy bars on the dE/dx curve in Figure 2. The surface
approximation of Equations 9-12 assume that dE/dx can be evaluated at E_0 and KE_0. These values are given as two horizontal
lines on the dE/dx curve in Figure 2. One can see that when the
target is very thin, $E \approx E_0$ and the approximation becomes very
accurate.

To give a feeling about how accurate the surface approximation is, we calculated ΔE versus t for 2 MeV $^4He^+$ ions backscattering from Si along a random direction, the results are
given in Figure 3. The dashed line is the surface approximation
calculation of Equation 9 which has a constant slope given in
Equation 10, and the solid curve is a detailed calculation of

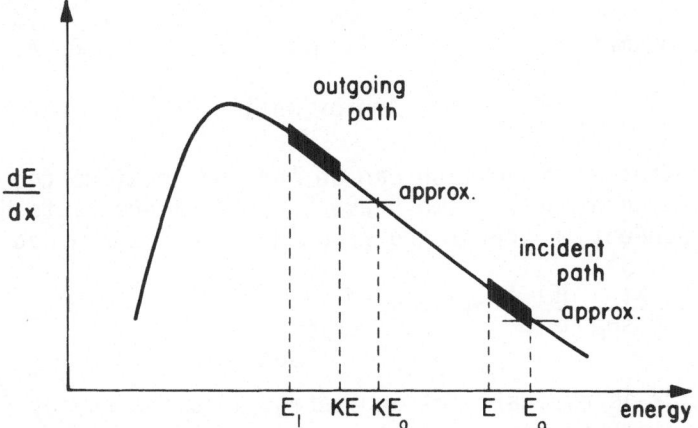

Figure 2. Schematic diagram of dE/dx versus E and regions used for backscattering analysis.

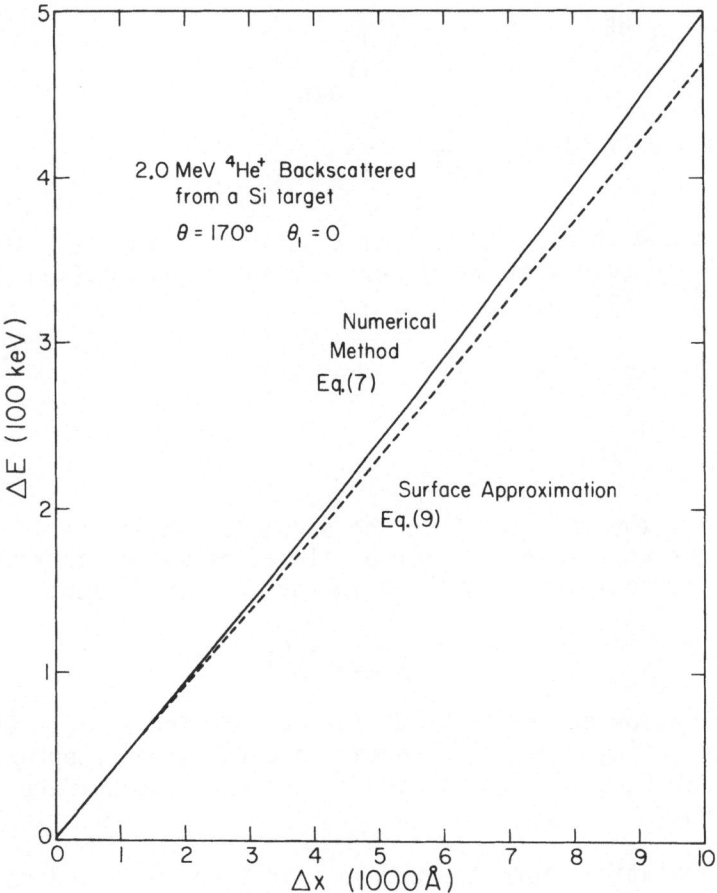

Figure 3. Energy to depth scale relation for 2 MeV ^{4}He$^+$ back-scattering from Si.

of Equation 7. One can see that there is a 5% error in using the surface approximation for the Si film at a depth of 8000Å.

C. Linear Approximation

A better approximation can be made by treating the incident and outgoing paths (two heavy bars in Figure 2) to be linearly dependent on energy and Equation 7 can be reduced to

$$\Delta E = \frac{Kt}{\cos\theta_0} \left.\frac{dE}{dx}\right|_{\overline{E}_{in}} + \frac{t}{\cos\theta_2} \left.\frac{dE}{dx}\right|_{\overline{E}_{out}} \tag{13}$$

Here the dE/dx is evaluated at an average incoming energy $\overline{E}_{in}$ and an average outgoing energy $\overline{E}_{out}$.

For normal incidence the linear approximation gives

$$[\overline{S}] = K \left.\frac{dE}{dx}\right|_{\overline{E}_{in}} + \frac{1}{|\cos\theta|} \left.\frac{dE}{dx}\right|_{\overline{E}_{out}} \tag{14}$$

$$[\overline{\varepsilon}] = K\varepsilon(\overline{E}_{in}) + \frac{1}{|\cos\theta|} \varepsilon(\overline{E}\ out) \tag{15}$$

Equations 14 and 15 are parallel to Equations 11 and 12. The bars above the E indicate that the linear average approximation has been applied. That is,

$$\overline{E}_{in} = \frac{1}{2} (E_0 + E)$$

$$\overline{E}_{out} = \frac{1}{2} (E_1 + KE) \tag{16}$$

Unfortunately, the value of E in the above two equations is unknown. It can be obtained either by iteration which starts from the surface approximation, or by a further approximation:

$$\overline{E}_{in} \simeq E_0 - \Delta E/4 \qquad \text{and} \qquad \overline{E}_{out} \simeq E_1 + \Delta E/4 \tag{16a}$$

This approximation generally holds for $\theta_1 \sim \theta_2$ and $K \sim 1$. Even for cases deviating from these conditions, the linear approximation used here with Equation 16a is much better than the surface approximation.

An analytical form to express E in terms of ε and $d\varepsilon/dE$ evaluated at E_0 and E_1 using Taylor series expansion has been given[3] and can be used for more precise analysis.

The linear approximation has been tested in 2 MeV ^{4}He ion backscattering and has been found to be very accurate even for thick films. For example, there is less than 0.2% error in depth calculation using linear approximation as compared to a detailed calculation given in Equation 7 for a gold film as thick as 8000Å.

3. Height of An Energy Spectrum

The shape and the height of an energy spectrum for backscattering contains quantitative information. This subject has been treated by Wenzel[4] and is well documented.[5,6] Several different versions of the formula exist and yet they all are based on the same concept. Since approximations have been applied in some cases to simplify the formulae. We will start with the most simple one.

A. Surface Approximation for Spectrum Height

Let us assume that we do backscattering from a thick elemental target. The schematic diagram for the scattering geometry, notations and energy spectrum are given in Figure 4. Let us focus our attention on the region near the surface of the target. All the notations have been defined earlier in Figure 1 except δE_1, δx and H, where δE_1 is the energy channel width of the detecting system and δx is a layer thickness which is related to δE_1 by Equations 9 and 11.

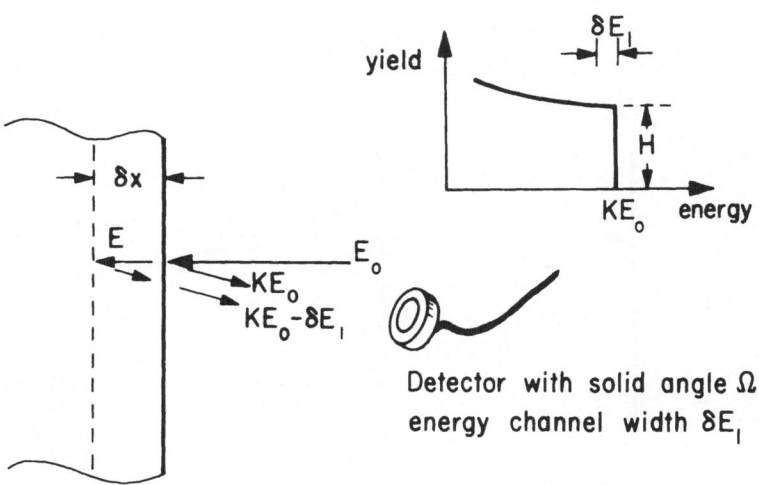

Figure 4.   Spectrum height at the surface.

$$\delta E_1 = KE_0 - E_1$$

$$\delta E_1 = [S]\delta x = [\epsilon]N\delta x \tag{17}$$

The height of the spectrum H that is the number of counts per channel, depends on the total number of projectiles, Q, incident on the target, the solid angle of the detecting system Ω, the average differential scattering cross section σ (Equation 3) evaluated at E_0 and the total number of target atoms per unit area $N\delta x$ which gives the energy difference corresponding to the energy channel width δE_1 detected, i.e.

$$H = Q\sigma\Omega N\delta x \tag{18}$$

By use of [S] or [ϵ] as defined in Equation 17, Equation 18 then becomes

$$H = Q\sigma\Omega N\delta E_1/[S] \tag{19}$$

or

$$H = Q\sigma\Omega\delta E_1/[\epsilon] \tag{20}$$

B. Thick Target Yield

The shape and height of a backscattering spectrum at a given depth is slightly more complicated than that at the surface. Figure 5 gives the notation and a schematic spectrum. From a derivation similar to the surface case we get

$$H(E_1) = Q\sigma(E)\Omega N \frac{\delta(KE)}{[S(E)]}$$

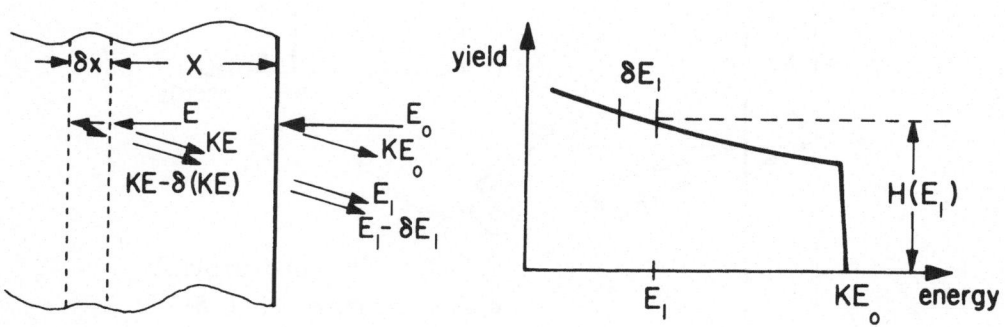

Figure 5.   Thick target yield.

or

$$H(E_1) = Q\sigma(E)\Omega \frac{\delta(KE)}{[\epsilon(E)]} \tag{21}$$

Equation 21, which gives the spectrum height at a given depth differs from Equation 20 which gives the spectrum height at the surface in several respects.

(i) δx which is inside the target instead of on the surface generates an energy width $\delta(KE)$ inside the target which in turn becomes δE_1 outside the target.

(ii) The differential cross section $\sigma(E)$ must be evaluated at an energy E at the depth x immediately before the scattering event took place.

(iii) The factors $[S(E)]$ and $[\epsilon(E)]$ are evaluated at a local energy E instead of E_0 in Equations 11 and 12. That is:

$$[S(E)] = K \frac{dE}{dx}\Big|_E + \frac{1}{|\cos\theta|} \frac{dE}{dx}\Big|_{KE} \tag{22}$$

$$[\epsilon(E)] = K\epsilon(E) + \frac{1}{|\cos\theta|} \epsilon(KE) \tag{23}$$

One has to be careful to notice the difference between $[S(E)]$ and $[\bar{S}]$. The factor $[\bar{S}]$ given in Equations 14 and 15 contains average energies $\bar{E}_{in}$ and $\bar{E}_{out}$ and is used to determine the depth at which scattering occurs.

The term $\delta(KE)$ in Equation 21 is not measurable but can be related to δE_1, since for uniform target,

$$\delta(KE) = \frac{\epsilon(KE)}{\epsilon(E_1)} \delta E_1$$

and Equation 21 becomes

$$H(E_1) = Q\sigma(E)\Omega \frac{\delta E_1}{[\epsilon(E)]} \frac{\epsilon(KE)}{\epsilon(E_1)} \tag{24}$$

Equation 24 or its equivalent has previously been derived by a different method by Power and Whaling,[6] by Siritonin, et al.,[7] by Brice[8] and by Feng, et al.[9] They have different notations, but all are basically the same.

One notices that in Equation 24, E needs to be solved from E_0 and E_1 in a separate calculation before Equation 24 can be applied. One can obtain E by two different methods. The first is the iteration method which starts from the surface approximation to establish the depth scale and from energy loss to obtain E. The second method is to express E in terms of ϵ and

$d\epsilon/dE$ evaluated at E_0 and E_1 using Taylor series expansion.[3]
Good agreement between the calculated thick target yield and a
backscattering spectrum has been verified on many occasions.[3,10]

C. Backscattering Yield of a Thin Film

Let us consider the energy spectrum of a thin film such
as presented in Figure 6.

A film is considered to be thin when $E_0 \gg \Delta E$ and the
energy loss does not change significantly when the energy changes
from E_0 to E and from KE to E_1. The surface approximation can
be used for a thin film giving

$$\Delta E = [S]t = [\epsilon]Nt$$

and

$$H = Q\sigma\Omega\delta E_1/[\epsilon]$$

One notices that the area A of the energy spectrum can be
calculated by taking the sum of counts per channel over the
channels in the spectrum. For a rectangular shape of spectrum,
we have

$$A = H \cdot \left(\frac{\Delta E}{\delta E_0}\right)$$

therefore

$$A = Q\sigma\Omega Nt \tag{25}$$

Equation 25 indicates that the area of the spectrum is
directly proportional to the thickness of the film. This relation is also true for extremely thin films such that the energy
spectrum becomes Gaussian rather than rectangular. For thicker
films, the fact that σ is a function of E and E changes over the
entire depth needs to be taken into consideration.

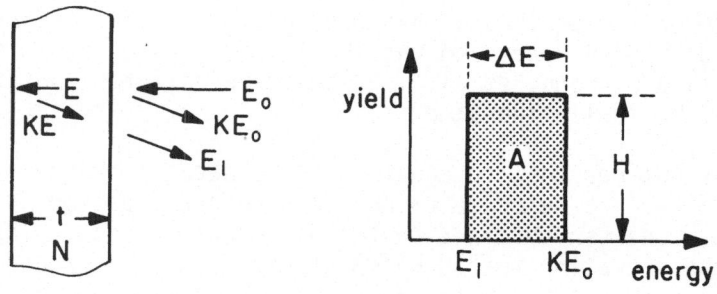

Figure 6.   Thin elemental film analysis.

4. Applications of Backscattering from Elemental Targets

So far, all the basic formulae and approximations have been given. We will illustrate some of the applications by combining or modifying some of the equations. Emphasis is on the method rather than on the example itself.

A. Surface Contamination and Ion Implantation

Let us analyze a heavy impurity on a light element substrate. The backscattering methods are not suitable for analyzing the case of trace contamination of light element on a heavy element substrate because the signals of the impurity appear at lower energies than that of the substrate. These particles usually cannot be detected because of the high background due to particles scattered from the thick substrate.

Figure 7 gives the notation and schematic spectrum of the surface impurity analysis. From the energy position of the impurity, $K_i E_o$, one can identify its atomic mass. From its area A_i, one can calculate the amount of contaminant Nt (in atoms per unit area) from Equation 25, if $Q\sigma$ and Ω are known. The total number of incident projectiles Q and the solid angle of the detecting system Ω require careful measurement. However, the product of Q and Ω can also be obtained from the surface spectrum height on the substrate portion of the spectrum:

$$Q\Omega = \frac{H_M [\varepsilon]_M}{\sigma_M \delta E_1} \tag{26}$$

Where subscript M means that the atomic mass M of the substrate has been used for the evaluation of stopping cross section, kinematics and scattering cross section; H_M is the surface height as indicated in Figure 7. From Equation 25, we can calculate the amount of impurity $(Nt)_i$ in number of atoms per unit area.

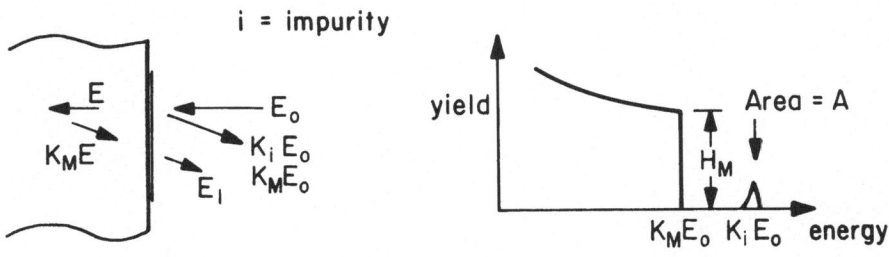

i = impurity

Figure 7.   Surface contaminant analysis.

$$(Nt)_i \quad A_i/Q\Omega\sigma_i \tag{27}$$

Substituting Equation 26 into Equation 27, to eliminate the experimental parameters, we have

$$(Nt)_i = \frac{A_i}{H_M} \frac{\sigma_M}{\sigma_i} \frac{\delta E_1}{[\epsilon]_M} \tag{28}$$

Equation 28 has been frequently used not only to estimate the amount of impurities on the surface but also to calculate the dose in a substrate in the ion implantation process. In the latter case, the implanted ions will appear at depth rather than on the surface. This introduces an ambiguity in distinguishing between light atoms at the surface and heavier atoms below the target surface. Identification is often made by tilting the target, since backscattering signals from species below the surface will shift to lower energies after tilting the sample. If the impurity atoms are located well below the surface, a correction to σ_i must be made to account for the energy loss along the inward path of the projectiles.

B. Doping Level of a Bulk Sample

The analysis of this case is similar to the analysis given above. A bulk sample containing heavy element impurities can be analyzed to give the identification and concentration of impurities. The mass identification is made from kinematics and the concentration is determined from the ratio of the spectrum height of the impurity to that of the bulk sample itself.

Figure 8 shows a schematic example of a bulk sample with atomic mass M doped with a heavy element impurity of atomic mass M_C.

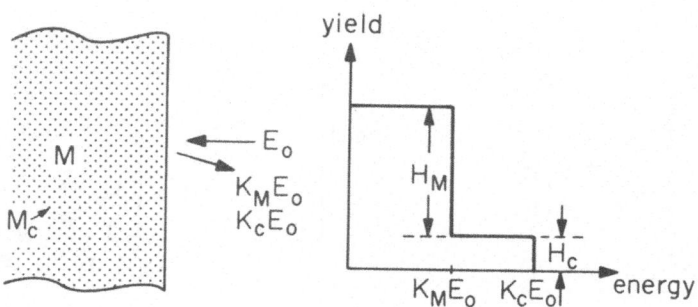

Figure 8. Impurity of heavy element in a bulk target.

The spectrum height at the surface can be written from Equation 19 as

$$H_M = Q\Omega\sigma_M N_M \frac{\delta E_1}{[S]_M^M} \tag{29}$$

$$H_c = Q\Omega\sigma_c N_c \frac{\delta E_1}{[S]_c^M} \tag{30}$$

Here the notation is slightly more complicated than that in Equation 19 where only one element is involved. Both subscripts and superscripts are used to denote the scattering atom and stopping medium respectively. H_M and H_c are the heights of the backscattering signals evaluated at the energies corresponding to scattering from the surface of the bulk element M and impurity element M_c, respectively. The σ's are the differential scattering cross sections and the N's are the atomic densities. The depth scale of $[S]_M^M$ is defined just as in Equation 10 or Equation 11 with $K = K_M$. The depth scale $[S]_c^M$ is similar to the above, except $K = K_c$, i.e.

$$[S]_c^M = K_c \left.\frac{dE}{dx}\right|_{E_0}^M + \frac{1}{\cos}\left.\frac{dE}{dx}\right|_{K_c E_0}^M \tag{31}$$

We assumed that the concentration of the impurity level is low enough so that the energy loss dE/dx of the projectile in the bulk material M is not changed by the presence of the impurity. If the concentration of the impurity gets higher, adjustment of the energy loss by the additivity rule becomes necessary. Use of the kinematic factor K_c in Equation 31 means that although the stopping material is the bulk material the scattering occurs between the projectile and the impurity atoms to give the spectrum height H_c.

By eliminating Q, Ω and δE_1 from Equations 29 and 30 we can obtain the impurity concentration, i.e.

$$N_c = \frac{H_c}{H_M} \frac{\sigma_M}{\sigma_c} \frac{[S]_c^M}{[S]_M^M} N_M \tag{32}$$

The impurity concentration can be measured and calculated from the spectrum height ratio H_c/H_M. The rest of the parameters are well defined and can be obtained. The values of $[S]_c^M/[S]_M^M$ is often close to unity for practical purposes. For example, the above ratio equals to 1.03 when a zinc impurity in silicon is analyzed using 2 MeV ^{4}He ion backscattering.

C. Film Thickness Measurement and dE/dx Measurements

It was obvious at the beginning of this chapter that energy loss is used to establish the depth scale, and therefore Equations 7, 10 or 13 can give a thickness measurement if dE/dx is known. On the other hand, if the thickness and density of the film are known by independent measurements, a backscattering measurement allows one to measure [S] or [ε]. By assuming the shape of the dE/dx curve, one can solve for dE/dx or ε from [S] or [ε]. Most of the available information on energy loss in thin films has been obtained either by backscattering or by trans- mission experiments. A few review articles and data compilation give a detailed listing of the sources of dE/dx meausrement; these will be listed in the appendix.

D. Yield Formula and dE/dx Measurements

In Section 3 we have related the backscattering yield to the depth scale as given for example by Equations 19, 20 and 21. Therefore, by measuring the spectrum height and other experimental factors carefully one can determine [S] and [ε] and therefore dE/dx or ε. Energy losses of protons and He ions have been obtained from the surface yield in several publications.[4,5,10-12]

All the above mentioned measurements require careful mea- surement of the total dose Q and solid angle of detection Ω. As demonstrated earlier, one can normalize Q and Ω by using a standard sample to measure the ratio of the two yields. Let us assume one measures backscattering yield from two different elemental targets, the ratio of the two heights at the surface from Equation 20 becomes

$$\frac{H_A}{H_B} = \frac{\sigma_A}{\sigma_B} \frac{[\varepsilon]_B}{[\varepsilon]_A} \tag{33}$$

Here the subscript A and B on H indicates two different measurements on elements A and B respectively. If the stopping cross section of element A, ε^A is known, from the measurement one can calculate $[\varepsilon]_B$. Leminen and Fontell[13,14] have measured the ε for proton and He ions in various metals using the spectrum height at the surface normalized to standards.

Feng et al.[9] have modified the above method of relat- ive measurement by depositing a thin film A on B (or B on A). By comparing the relative yield at the interface, they are able to measure the ration $[\varepsilon]_A/[\varepsilon]_B$ without requiring measurements of Q and Ω.

E. Differential Scattering Cross Section Measurement

In all the yield formula, the differential scattering cross section σ has been assumed to be Rutherford, i.e. as given by Equation 2. This is a very safe assumption when using 0.5 - 2 MeV ^{4}He$^+$ ions. At higher energies deviations from the Rutherford scattering cross section for ^{4}He$^+$ ions bombarding low Z elements due to nuclear reactions exist. A compilation at a few available results of alpha particle elastic cross section data on low Z elements has been reported by Semmler, et al.[15] A few examples of nuclear resonances for elastic scattering have been given elsewhere.[1]

For protons in the MeV energy region, deviations from Rutherford scattering cross section, very frequently occur, especially for low Z elements. A very complete compilation of the available measurements on cross sections for light nuclei reactions has been given by Ajzenberg-Selove and Lauritsen.[16] We have to emphasize however, that the information on scattering cross section is not complete; deviations from Rutherford formula should be expected when a low Z element is in the target. It is of interest to note that enhancement of the backscattering yield due to nuclear reaction resonances can be used in some cases to improve the sensitivity of the detection in backscattering.

If the stopping cross section ε is known, backscattering can be used to measure the scattering cross section σ, as is implied in Equation 20. If the amount of the target atoms is known one can measure the scattering cross section and without information about ε. This is shown in Equation 25.

5. Application of Backscattering to Compound Targets

A compound target can be studied by backscattering techniques in a manner similar to that used for elemental targets. The principles and concepts involved are identical. The only complication we have here is that we have to treat the energy loss and scattering kinematics differently. In an elemental target, the energy loss and scattering is in the same medium. In a compound target, the energy loss is calculated or measured from the stopping of the projectile in the compound while the scattering occurs between the projectile and a given atom of the compound molecule. We will illustrate this point by a few examples. The method given in this section is not limited to compound studies. The same method can be applied to alloy and mixture studies. We have to point out that backscattering analysis can only give the atomic composition distribution as a function of depth. It does not reveal the chemical information about the sample. Complimentary techniques such as X-ray dif-

fraction, are required for positive chemical identification. One
other restriction of the backscattering technique is that this
technique does not have lateral sensitivity. Lateral uniformative
is always assumes to hold; this can be verified by use of scanning electron microscopy.

A. Thin Film Analysis

Let us assume tnat we do backscattering from a thin
compound film on a light substrate or for simplicity a self-
supported compound film composed of element A and B with molecular composition A_mB_n (m and n are integers). This problem can
be generalized as the study of a mixture with composition A_mB_n
(where m and n are fractions with $m + n = 1$). The notations and
schematic backscattering spectrum are given in Figure 9.

In a manner similar to that for elemental film analysis,
a set of equations for compound (or mixture) film can be derived
for the depth scale and the spectrum height. The only difference
is that the notation is more cumbersome. We need the subscript
A or B to indicate that the scattering is from atom A or B and
we need the superscript A_mB_n to indicate that the stopping medium
is the compound (or mixture). We simply write down a set of
equations for compound (or mixture) thin film analysis which is
parallel to the elemental thin film analysis derived earlier.
First, let us assume that the film is thin enough so that a
surface approximation (Equations 8-12) can be applied. We have
in this case:

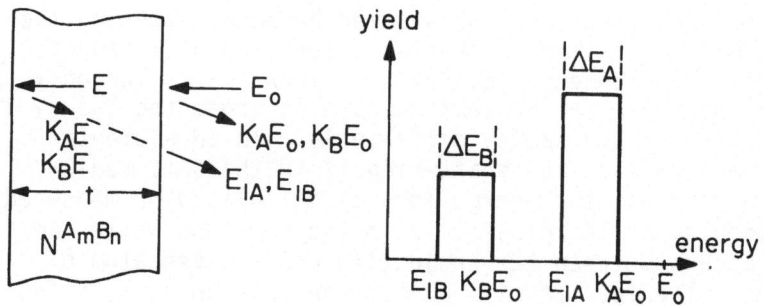

Figure 9.   Compound target analysis.

$$\Delta E_A = [S]_A^{A_m B_n} t = [\varepsilon]_A^{A_m B_n} N^{A_m B_n} t \tag{34}$$

$$\Delta E_B = [S]_B^{A_m B_n} t = [\varepsilon]_B^{A_m B_n} N^{A_m B_n} t \tag{35}$$

where $N^{A_m B_n}$ is the molecular density for the compound (number of molecules per unit volume) or the atomic density for the mixture and

$$[S]_A^{A_m B_n} = K_A \frac{dE}{dx}\Big|_{E_o} = \frac{1}{|\cos\theta|} \frac{dE}{dx}\Big|_{K_A E_o} \tag{36}$$

For $[S]_B^{A_m B_n}$ simply replace K_A by K_B in both terms of Equation 36; here dE/dx is the energy loss of the projectile in the compound or mixture. We can also write

$$[\varepsilon]_A^{A_m B_n} = K_A \varepsilon^{A_m B_n}(E_o) + \frac{1}{|\cos\theta|} \varepsilon^{A_m B_n}(K_A E_o) \tag{37}$$

where $\varepsilon^{A_m B_n}$ is the stopping cross section of the molecule in the compound or an weighted average stopping cross section of a atom in the mixture. From Bragg's rule, one can assume

$$\varepsilon^{A_m B_n} = m\varepsilon^A + n\varepsilon^B \tag{38}$$

where ε^A and ε^B are the stopping cross sections of the individual atoms A and B respectively.

Knowing the composition $A_m B_n$, one can always calculate $[\varepsilon]_A^{A_m B_n}$ (or $[\varepsilon]_B^{A_m B_n}$) and therefore by measuring the energy shift ΔE_A (or ΔE_B) one can calculate the film thickness from Equation 34 (or Equation 35).

The surface height of the spectrum H_A or H_B can be expressed similarly to Equations 19 and 20, except that

$$H_A = Q\sigma_A \Omega N_A \delta E_1 / [S]_A^{A_m B_n} \tag{39}$$

$$H_B = Q\sigma_B \Omega N_B \delta E_1 / [S]_B^{A_m B_n} \tag{40}$$

and the ratio of the above two equations gives

$$\frac{H_A}{H_B} = \frac{\sigma_A}{\sigma_B} \frac{m}{n} \frac{[S]_B^{A_m B_n}}{[S]_A^{A_m B_n}} \tag{41}$$

Here the ratio of the volume density N_A/N_B becomes the atomic ratio of the constituents, i.e. m/n. Equation 41 provides a way to determine the stoichiometry m/n from an unknown sample. The last term $[S]_B^{A_m B_n}/[S]_A^{A_m B_n}$ can be reduced to $[\varepsilon]_B^{A_m B_n}/[\varepsilon]_A^{A_m B_n}$. A simple exercise can demonstrate that this ratio is very insensitive to the composition and the systematic errors in the values of energy loss value used in the calculation. When a ratio is taken, the systematic changes in the numerator and denominator tend to cancel each other. This $[S]$ ratio or $[\varepsilon]$ ratio can easily be measured from $\Delta E_B/\Delta E_A$ since they are related by Equations 34 and 35. For element B lighter than element A, $\Delta E_B/\Delta E_A$ is less than unity, and it is typically between 0.9 and 1 depending on the atomic masses involved.

Parallel formalism for compound or mixture analysis using the linear approximation or numerical methods can be derived easily.

B. Thick Compound Targets

The composition of a thick compound or mixture target can be obtained from a height ratio as given in Equation 41. Because of the overlapping of the spectra due to scattering from different atoms, background subtraction can be a problem when extracting the height of the spectrum from each element is being extracted.

C. Analysis on Composition Varying Continuously with Depth

The analysis of the height ratio given in Equation 41 can be extended to thick targets; here the ratio m/n in Equation 41 becomes a variable and it is a function of depth. In this case one wants to obtain the composition ratio at a given depth by analysis of the spectrum height ratio of A and B from energy positions which correspond to a certain depth. The detailed formalism and a numerical method have been demonstrated by Brice.[8] Variation of the method by comparing the compound yield with a single element target has been given by Campisano, et al.[17]

Appendix 1. Notations

Target, Beam and Geometry:

m, Z_1 = mass and atomic number of the projectile (amu,-)
M, Z_2 = mass and atomic number of the target atom (amu,-)
    θ = scattering angle (°)
    Ω = solid angle of the detecting system (st. rad.)
    Q = total number of projectiles bombarding the sample (-)
dσ/dΩ = differential scattering cross section (cm^2)
    σ = averaged differential scattering cross section (cm^2)
    H = Height of a spectrum at a given energy E_1 (counts/channel)
    N = number of target atoms per unit volume (atoms/cm^3)
    ρ = density of the target (gm/cm^3)
  t,x = target thickness (cm)
    K = kinematics of the scattering between m and M at θ (-)

Energy of the particle and energy width

E_o = incident energy of the projectile before entering the
        sample (eV)
  E = particle energy at some depth in the target just before
      scattering (eV)
E_1 = emerging energy of a particle (eV)
$δE_1$ = channel width of the detecting system (eV)
δ(KE) = equivalent channel width at the scattering depth (eV)

Energy Loss

ε = stopping cross section = (1/N)dE/dx ($eV-cm^2$)
[ε] = stopping cross section factor = (1/N)[S] ($eV-cm^2$)
[S] = energy loss factor = ΔE/t (eV/cm)
$[S]_B^A$ = energy loss factor. Scattering from atom B in a matrix
      A (eV/cm)

Appendix 2. Formulae

Basic Formulae

$$K_M = \left(\frac{m\cos\theta + \sqrt{M^2 - m^2\sin^2\theta}}{m + M}\right)^2 \tag{1}$$

$$\frac{d\sigma}{d\Omega} = \left(\frac{Z_1 Z_2 e^2}{2E\sin^2\theta}\right)^2 \frac{\{\cos\theta + [1-(\frac{m}{M}\sin\theta)^2]^{\frac{1}{2}}\}^2}{[1-(\frac{m}{M}\sin\theta)^2]^{\frac{1}{2}}} \tag{2}$$

$$\varepsilon \equiv \frac{1}{N}\frac{dE}{dx}, \quad [\varepsilon] = \frac{1}{N}[S], \quad t = \Delta E/[S] \tag{8}$$

Elemental Films

$$[S] = K\left.\frac{dE}{dx}\right|_{E_0} + \frac{1}{|\cos\theta|}\left.\frac{dE}{dx}\right|_{KE_0} \quad \text{(surface approxi-} \atop \text{mation)} \tag{10-12}$$

$$[\overline{S}] = K\left.\frac{dE}{dx}\right|_{\overline{E}_{in}} + \frac{1}{|\cos\theta|}\left.\frac{dE}{dx}\right|_{\overline{E}_{out}} \quad \text{(linear approximation)} \tag{14-15}$$

$$\Delta E \text{ versus } t \qquad\qquad\qquad \text{(for thick target)} \tag{7}$$

$$[S(E)] = K\left.\frac{dE}{dx}\right|_{E} + \frac{1}{|\cos\theta|}\left.\frac{dE}{dx}\right|_{KE} \quad \text{(analysis at a given} \atop \text{depth)} \tag{21}$$

$$H(KE_0) = Q\sigma(E_0)\Omega\delta E_1/[\varepsilon] \qquad \text{(near surface)} \tag{18-20}$$

$$H(E_1) = Q\sigma(E)\Omega\delta E_1/[\varepsilon(E)]\times[\varepsilon(KE)/\varepsilon(E_1)] \quad \text{(in depth)} \tag{24}$$

$$\text{Area} = Q\sigma\Omega Nt \tag{25}$$

$$(Nt)_i = \frac{A_i}{H_M}\frac{\sigma_M}{\sigma_i}\frac{\delta E_1}{[\varepsilon]_M} \qquad \text{(surface impurity} \atop \text{Atom i in target M)} \tag{28}$$

Height Ratio

$$\frac{H_A}{H_B} = \frac{\sigma_A}{\sigma_B}\frac{[\varepsilon]_B}{[\varepsilon]_A} \qquad\qquad \text{(surface height)} \tag{33}$$

$$\frac{H_B}{H_A} = \frac{\sigma_B}{\sigma_A}\frac{[\varepsilon]_A}{[\varepsilon]_B}\frac{\delta(K_B E)}{\delta(K_A E)} \qquad \text{(interface height)} \qquad \text{(Ref. 9)}$$

Compound or Mixture Analysis - Bragg's Rule for compound and for
mixture

$$\varepsilon^{A_m B_n} = m\varepsilon^A + n\varepsilon^B \tag{38}$$

stopping cross section ϵ for:

(i) a compound molecule $A_m B_n$ (m and n are integers).
(ii) An average atom in a mixture $A_m B_n$ (m and n are fractions with m + n = 1).

Appendix 3. Sources for dE/dx Information

We list a few available compilation of energy loss information. References of individual measurements that the compilation is based on are not given here. The listing is in reversed chronological order.

1. Compilation by Chalk River Labs, D. Ward. H.R. Andrews, G.C. Ball, J.S. Forster, W.G. Davies, G.C. Costa and I.V. Mitchell (unpublished). They have compiled energy loss for helium ions and also for heavy ions in elements the latter being build around new experimental data. The Z_2-oscillatory structure of the stopping cross section is incorporated.
2. Compilation by J.F. Ziegler and W.K. Chu, Atomic Data and Nuclear Data Tables 13, 463 (1974). They have compiled stopping cross section of the ^{4}He ions in elements in the energy region 0.4 to 4 MeV. Most of the data is based on the measurements made at Baylor University. Interpolation of the unmeasured information are made including Z_2 oscillation. [S] factor and spectrum height for backscattering are also given.
3. Compilation by L.C. Northcliffe and R.F. Schilling, Nuclear Data Tables A7, 233 (1970). Tabulation of range and energy loss of proton, helium and all heavy ions in 12 solid elements, nine gaseous elements and three compounds are given.
4. Compilation by C.F. Williamson, J.-P. Boujot and J. Picard, Report CEA-R 3042 (1966).
5. Compilation by J.F. Janni, Air Force Weapons Lab. Report AFWL-TR-65-150 (1966).
6. Compilation by H. Bichsel, American Institute of Physics Handbook, McGraw-Hill, New York (1963), p. 8-22.
7. Review by W. Whaling in S. Flugge (Ed.), Hanbuck der Physik 34, Springer, Berlin (1958), p. 193.

A recent review with emphasis on the understanding of the energy loss of charged particles in solids is given by Peter Sigmund, Proceedings of the Advanced Study Institute on Radiation Damage Processes in Materials, Corsica, 1973.

References

1.   W.K. Chu, J.W. Mayer, M-A. Nicolet, T.M. Buck, G. Amsel and
     F. Eisen, Thin Solid Films 17, 1 (1973).
2.   Rutherford, Phil. Mag. 21, 669 (1911). This and other famous
     papers are published in R.T. Beyer, "Foundation of Nuclear
     Physics," Dover Publications, New York, 1949.
3.   W.K. Chu and J.F. Ziegler (to be published).
4.   W.A. Wenzel, Ph.D. Thesis, California Institute of Technology,
     1952.
5.   W.A. Wenzel and W. Whaling, Phys. Rev. 87, 449 (1952).
6.   D. Powers and W. Whaling, Phys. Rev. 126, 61 (1962).
7.   E.I. Siritonin, A.F. Tulinov, A. Fiderkevich and K.S. Shyskin,
     Vestnik MGU (Ser. Fiz. Astr.), 12 (1971) 541. Also see
     Rad. Effects 15, 149 (1972).
8.   D.K. Brice, Thin Solid Films 19, 121 (1973).
9.   J.S.-Y. Feng, W.K. Chu, M-A. Nicolet and J.W. Mayer, Thin
     Solid Films 19, 195 (1973).
10. W.K. Lin, S. Matteson and D. Powers (to be published).
11. W.K. Chu, J.F. Ziegler, I.V. Mitchell and W.K. Mackintosh,
     Appl. Phys. Lett. 22, 437 (1973).
12. R. Behrisch and B.M.U. Scherzer, Thin Solid Film 19, 247
     (1973).
13. E. Leminen, Ann. Acad. Sci. Fennicae A V1 386 (1972).
14. E. Leminen and A. Fontell, Rad. Eff. 22, 39 (1974).
15. R.A. Semmler, J.F. Tribby and J.E. Brugger, Report COO-712-
     89, US Atomic Energy Commission, Div. of Technical Infor-
     mation (Oct. 1964).
16. F. Ajzenberg-Selove and T. Lauritsen, Nuclear Physics.
17. S.J. Campisano, G. Foti, F. Grasso, J.W. Mayer and E. Rimini,
     Application of Ion Beam to Metal, by S.T. Picraux, ErNisse
     and F.L. Vook (Plenum Press, New York 1974) p. 159.

MATERIAL ANALYSIS BY MEANS OF NUCLEAR REACTIONS

Eligius A. Wolicki

Radiation Technology Division

Naval Research Laboratory, Washington, D.C. 20375

INTRODUCTION

The use of nuclear reactions for surface analyses is a field in which a wealth of interesting opportunities exists for small accelerator groups. The field is exciting and interesting both from the scientific standpoint of understanding the fundamental processes involved and from the specific applications which are still being developed. Important early contributions in this field may be found in references 1-9. Interest in the field, spurred perhaps in part by the advent of high resolution solid state detectors for charged particles and gamma rays and of improved data processing equipment, has increased to the point where entire conferences are now devoted to nuclear reaction analysis techniques which utilize charged particle bombardment (10-16).

The objectives of the discussion which follows will be to give the reader an overview not only of the present state of the art but also of anticipated future trends and potentials both for research and for applications. A further intent will be to present enough information so a reader can judge whether, for a given surface analysis problem, a nuclear technique may have significant advantage over other techniques. Toward this latter end, the text will include first a discussion of the general principles underlying the use of nuclear techniques and then a comprehensive list of examples selected so as to display the variety of ways in which these techniques can be applied. The examples will thus include not only those techniques which are in more or less routine use but also those which are still being investigated and developed in the laboratory.

If radioactivity is produced by the irradiation and detected afterward the method is called charged particle activation analysis; if radiations emitted instantaneously are detected, it is termed prompt radiation analysis. These two categories will be used for labelling major sections of the discussion which follows. (Since the letters NAA are commonly used to denote neutron activation analysis and are a great convenience, license will be taken in the remainder of this chapter to have the initials CPAA stand for charged particle activation analysis and PRA for prompt radiation analysis.)

The accelerated nuclear particles used to induce nuclear reactions will, with only one or two exceptions, be taken here to be protons, deuterons, tritons, helium-3, and alpha particles and the energies of interest will range from a fraction of an MeV up to about 5 MeV. This upper limit on the energy of the beam particles incident on a target is the factor which limits nuclear reaction techniques to the surface and near surface regions of the target; at 2.0 MeV, proton and alpha particle ranges in a metal are respectively about 50 and 6 μm.

Nuclear reaction techniques for analyzing surfaces consist of irradiating a sample with a beam of nuclear particles having sufficient energy so that nuclear reactions are produced and detecting the resulting characteristic radiations. For energies between about 0.5 and 5.0 MeV, the irradiating beam is most commonly produced by a Van de Graaff electrostatic accelerator; below 0.5 MeV Cockroft Walton accelerators are in common use. In these accelerators the ion beam travels in a vacuum and is generally analyzed magnetically so that but a single nuclear species with well defined energy is incident on the sample or as it is commonly called, the "target."

The irradiations are usually performed in vacuum. The emitted radiations may be detected with Geiger Mueller and proportional counters, scintillation counters, and solid state detectors, the pulses from the latter three types of detectors being analyzed typically with a multichannel pulse-height analyzer. The nuclear reactions which can be produced by the incident beam in the sample depend on the energy of the beam and the atomic species of both the beam and the sample. A schematic drawing of an ion beam incident on a sample and some of the emitted radiations produced is shown in Figure 1. Typically, nuclear radiations will be one or more gamma rays, a charged nuclear particle, a neutron or a beta particle.

In order for a nuclear reaction to occur, the incident beam particle has to have sufficient energy so that it can overcome the electrostatic potential energy which builds up as it approaches a target atom nucleus. This "Coulomb barrier" can be taken as

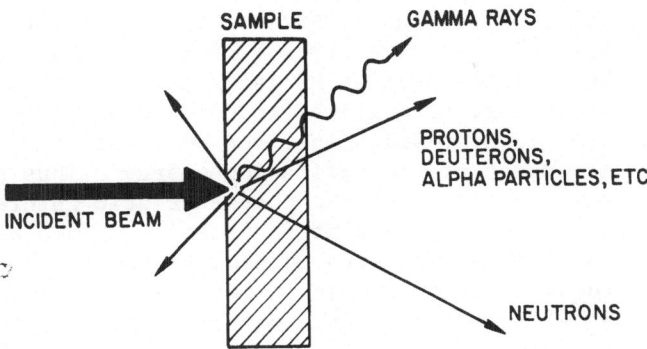

Fig. 1 A schematic diagram showing an ion beam incident on a target sample and the various nuclear radiations that can be produced.

proportional to Z_1Z_2/r where r is the radius of interaction and Z_1 and Z_2 are the atomic numbers of the incident and target nuclei respectively. Thus the barrier is higher for incident alpha particles than it is for protons and, for a given incident beam particle, it is higher, the higher the atomic number of the target nucleus. In general, for energies of 5 MeV and below, nuclear reaction analysis will be best suited for light elements; most of the examples which will be discussed in following sections are for $Z_2 \leq 15$.

Once a nuclear reaction has occurred the radiations which are emitted are characteristic of the excited nuclei produced in much the same way that optical radiation is characteristic of an excited emitting atom. It is the existence of a unique set of well defined energy levels in the atom or the nucleus that permits the identification of the emitted radiation with its source.

Symbolically a nuclear reaction may be written

$$M_1 + M_2 \longrightarrow M_3 + M_4 + Q \tag{1}$$

where M_1 is the incident nucleus, M_2 the target, M_3 is the emitted radiation, which may be either a nuclear particle or a gamma ray, M_4 is the residual nucleus and Q is the energy released (absorbed) in the reaction. Q is simply the difference between the total energy, at rest, of the interacting system before the reaction takes place minus that after the reaction has occurred. If the M's are taken to be masses,

$$Q = (M_1 + M_2)c^2 - (M_3 + M_4)c^2 \tag{2}$$

where c is the velocity of light. Nuclear reactions can be exoergic, in which case Q is positive, or endoergic, in which case Q is negative. If the residual nucleus M_4 is left in an excited energy state, the Q for the reaction will be reduced relative to the value which would obtain if the residual nucleus were left in its ground state by just the amount of the excitation energy. Thus the basis for the previous statement about the characteristic nature of the emitted radiation is seen to be that for a well defined beam energy and a fixed angle of detection with respect to the incident beam direction (the requirement on the angle being due to kinematic effects produced by the motion of the center of mass in the laboratory) the energy spectrum of M_3 will correspond to or be "characteristic" of the Q values possible to the reaction or, equivalently, to the excited states of the residual nucleus. In addition, even if the emitted particles M_3 are not observed the prompt gamma rays emitted (usually) in the decay of the excited M_4 nucleus may be observed and will also be characteristic of that nucleus.

In the case of charged particle activation analysis (CPAA) the identifying characteristics for the radioactive M_4 nucleus can be the half life for decay, the types of radiation emitted and the characteristic gamma rays emitted from the daughter nucleus of M_4. The half life is particularly useful when annihilation radiation following positron emission is being measured and when several different positron emitting species are produced in the target.

For the prompt radiation analysis (PRA) case, if M_3 is a gamma ray the nuclear reaction is called a direct capture reaction. The case $M_1 = M_3$ and $Q = 0$ is just the elastic scattering reaction which has been discussed in a previous chapter. (Although elastic scattering at low energies takes place usually via Coulomb forces without the involvement of nuclear forces, it can occur also via nuclear interactions; in such case the cross section for the scattering will usually be different from the value given by the Rutherford formula.) When $M_1 = M_3$ but $Q \neq 0$, the reaction is called inelastic scattering and, finally, when $M_1 \neq M_3$, it is commonly termed a rearrangement collision.

The characteristic spectrum of the nuclear radiation is measured with detectors having the property that the voltage pulse produced is proportional to the energy of the individual gamma ray or charged particle absorbed in the detector. The spectrum is produced by sorting the individual pulses according to pulse height (energy) with a pulse height analyzer. In such a spectrum, again in similarity to the atomic case, not only the energies but also the intensities of the various observed radiations are characteristic of the nuclear reactions which have been produced. Typically, gamma rays are detected with either a sodium iodide scintillation counter or a germanium lithium-drifted solid state detector, the trend being

toward use of the latter because of its greatly superior energy
resolution; the germanium lithium drifted detector however requires
continuous liquid nitrogen cooling. If detection efficiency is of
greater concern in an experiment than energy resolution then the
sodium iodide scintillation counter is usually used. Charged particles are detected most commonly with solid state detectors. For
neutrons with energies ranging from a fraction of an MeV up to a
few MeV time of flight methods are most often used for energy
measurement. Gamma ray and neutron detectors may be located outside the vacuum in which the target is irradiated but charged
particle detectors are usually located in the same chamber as the
target.

A typical target chamber arrangement for studying charged
particles emitted from nuclear reactions is shown in Figure 2. In
the chamber shown, the target is attached to a center post which
can be moved up and down externally so as to change from one sample
to another or from one position on the sample to another. Solid
state detectors are mounted on arms which can be moved, also externally, so as to position each detector at a desired angle. Measurements can be made with from one to four detectors, the larger number
of detectors being particularly useful for measuring how the intensities of the emitted charged particles vary as a function of angle.
The incident beam enters the target chamber through a set of collimating apertures so that its direction and the area being irradiated
on the target are well defined.

In contrast to the case of atoms, nuclear characteristics
usually differ markedly between two isotopes of the same chemical
element so the emitted radiations or reaction products are specific
not only to the chemical element but to a particular isotope of
that element; it is this property that provides the basis for the
many important applications of stable and radioisotope tracers.
Nuclear reactions have the disadvantage however that, with only a
few exceptions which do not pertain to the present discussion, they
are not affected by the state of atomic electrons and so do not
give information directly about the chemical bonds or the chemical
compound form of the elements in a sample.

The advantages of nuclear techniques are that they can have
excellent specificity, high sensitivity, and good accuracy. Good
accuracy, a characteristic which is especially important in trace
element work, is made possible because often little or no sample
processing is involved and the chances for introducing extraneous
elements are thereby reduced. For many materials the irradiations
are in addition nondestructive so that elemental analysis on a
particular sample does not preclude other subsequent tests or
measurements which may be desired. This latter statement may not
apply however for materials or devices whose performance is sensitive

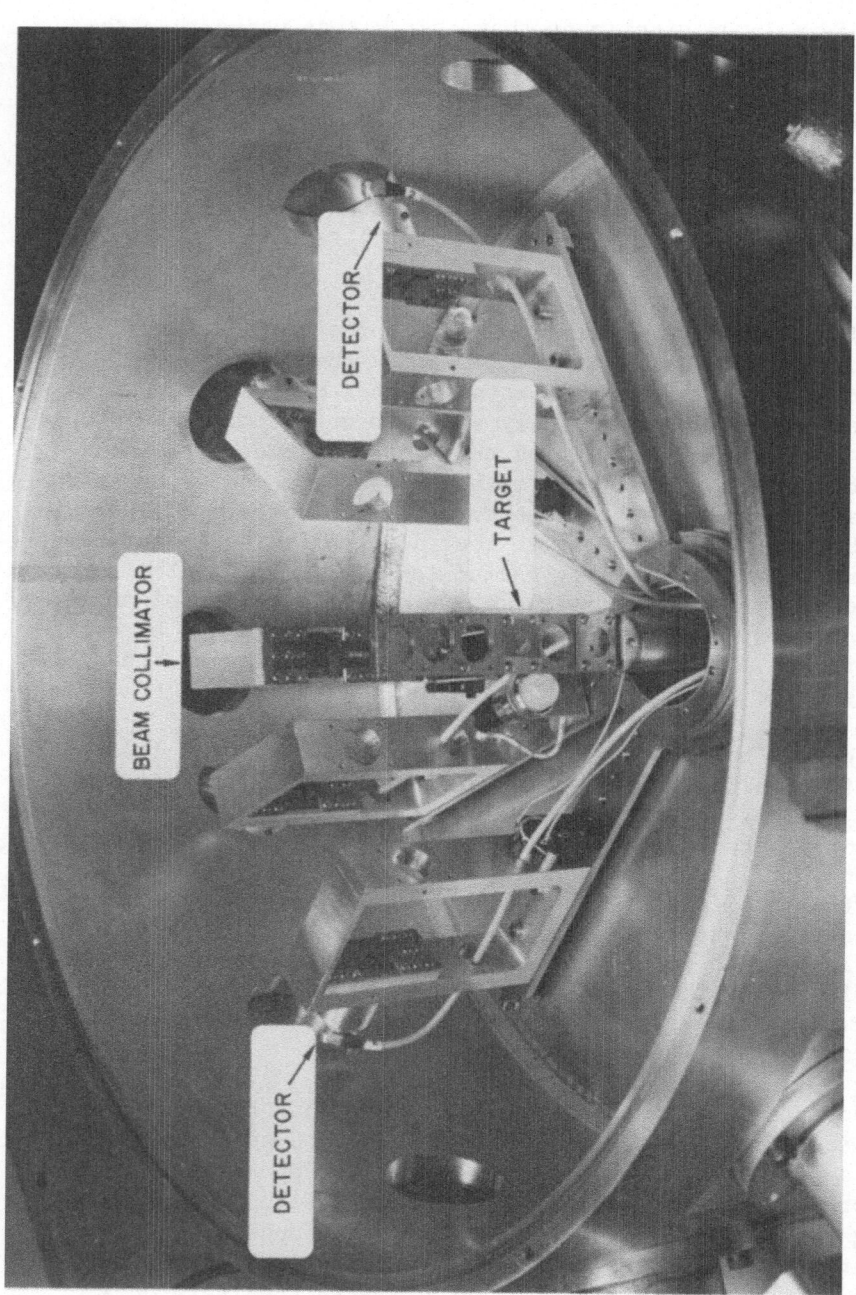

Fig. 2   A photograph showing the inside of a 45.7 cm diameter target chamber used at the NRL 5-MV Van de Graaff for backscattering and nuclear reaction analysis of solid materials.

to radiation damage. The stability of the target under bombardment
should not be taken for granted but should be carefully checked;
carbon buildup and target overheating are perhaps the two most
common problems which may be encountered.

The principal disadvantages of nuclear reaction analyses are
the high cost of the accelerator and the equipment required and a
low precision as compared to most standard analytical techniques.
The disadvantage of high cost does not always apply however because
nuclear techniques lend themselves to automation. Thus if a large
number of samples is being routinely analyzed with a given measurement technique, and particularly if multielement analysis in a
single irradiation is possible then the cost per element per sample
may turn out to be lower and on occasion even so much lower that the
nuclear technique may become the technique of choice because it is
the least costly.

The low precision is due to statistical errors which are
associated with counting random events and to calibration problems.
The difficulty in obtaining high precision caused by statistical
counting errors is easily illustrated. Because the standard devia-
tion for a Poisson distribution of $\underline{n}$ events is $\sqrt{\underline{n}}$, 10^4 counts must
be accumulated before a precision of 1% is reached and 10^6 events
must be detected for 0.1% precision. Conversely if only 10^3 or
10^2 counts are obtained the standard deviations are ±3.3% and ±10%
respectively. The yields from nuclear reactions and the detector
absolute efficiencies in practice are usually such that irradiations which would produce 10^6 counts would be inconveniently long.
The quantitative calibration of nuclear reaction techniques is most
commonly achieved through the use of comparison standards. One of
the principal difficulties in the use of such standards however is
the dependence of the measurement not only on the quantity of an
element present but also on its depth distribution. The depth
dependence arises because the incident ion loses energy as it penetrates into the solid and because the reaction cross section, particularly at low bombarding energies, changes with energy. Because
of both of these factors, precisions which are relatively easily
achieved generally range from 1% to approximately 10%. Precisions
of 0.1% have been reported (for neutron activation analysis) (17,
18) but are rare. A detailed discussion of accuracy, precision,
and comparison standards may be found in reference 14.

Even though this chapter is directed at surface analyses it
may be in order, since the intent of the book is to discuss the
opportunities which exist today for work with small accelerators,
to make brief mention of the use of Cockroft Walton accelerators
for fast neutron activation analysis. A more detailed discussion
of this type of measurement may be found, for example, in references
19 and 20. In fast neutron activation analysis the $^3H(d,n)^4He$

reaction is used for producing neutrons, with deuteron energies in the range from 100 to 200 keV being quite satisfactory. A flux of 10^9 neutrons/cm^2/sec suffices to measure oxygen concentrations in a 50 gram steel sample down to 10 ppm(19). Sealed tube, reliable, and long lived sources are commercially available which can produce even higher fluxes, namely 10^{10} neutrons/cm^2/sec (19). Fast neutron activation analysis is especially useful for bulk determination of light elements at trace levels and for many of these elements it is superior to the more widely used thermal neutron activation analysis. Because the neutron penetration range is long, neutron activation is not particularly well suited to surface analyses. It is interesting to note also that for minor (i.e., from approximately 0.1 atomic percent to a few percent) and major constituent analysis the reliability and the convenience of a fully automated small accelerator fast neutron activation analysis system are such that in at least one laboratory this technique is the method of choice for fully a third of all the elements in the periodic table (21).

PRA techniques are less well developed than CPAA, and are judged to present relatively greater potential for research in the future and to be of greater interest therefore to the reader. For this reason somewhat greater emphasis will be placed on prompt techniques than on CPAA in the sections which follow.

CHARGED PARTICLE ACTIVATION ANALYSIS

In this method a sample is irradiated with a charged particle beam and nuclear reactions with the element whose analysis is desired are used to produce a radioactive isotope whose activity can subsequently be detected. Most commonly the sample is removed from the target chamber and placed in a low background detecting system where the emitted radiations, usually gamma rays, positrons, or beta rays, are measured. While the sensitivity of the measurement is often so high that it can be used for trace element ($\leq$ 0.1 at. %) analysis, its application for minor constituent analysis should not be overlooked. For surface analyses particularly, it may be true that the detected element may be at the minor constituent analysis at the surface even though it is at the trace level in the bulk of the sample and charged particle activation analysis (CPAA) may be useful even if its bulk sensitivity for the element in question may be somewhat low.

An excellent discussion and a comprehensive list of early experiments in CPAA have been given by Tilbury (12). The field received a particularly strong impetus from the systematic investigation of the use of ^{3}He for activation analysis by Markowitz and Mahony (8) and has grown rapidly since that time. Interest has been spurred by the two advantages listed earlier, namely high sensitivity and good accuracy. As an example, for ^{3}He bombardments, detection limits for many light elements are better than 100 parts per billion (22). References 15 and 16, reporting on the two most recent international conferences on the use of nuclear reactions for materials analysis may be particularly useful in acquainting the reader with the present state of development and with the principal laboratories that are working in this area.

Because nuclear reaction cross sections or, equivalently, detection sensitivities generally increase as the bombarding particle energy exceeds the Coulomb barrier mentioned previously, the majority of CPAA experiments reported in the literature has been performed with cyclotron accelerators and at energies above 5 MeV. It remains true however that sensitivities adequate for trace element analysis can be obtained below 5 MeV for the light elements, at least through fluorine, and for minor constituent analysis as far as perhaps nickel. Clearly therefore, the number of problems that can be addressed by small accelerators and CPAA is not small. The charged particles of greatest interest for small accelerators are protons, deuterons, tritons (readers interested in developing a triton beam capability are cautioned that tritium, being radioactive, requires special handling and safety arrangements), and ^{3}He particles; for low energy alpha particles only the $^{10}B(\alpha,n)^{13}N$ reaction produces a useful radionuclide. With ^{3}He the interesting possibility exists also of using the doubly ionized species in order to obtain twice the singly ionized energy. The advantage of increasing the ^{3}He energy will in such case have to be weighed against the decrease in the beam intensity which can be produced. Still another potential use for ^{3}He^{++} is to eliminate the possibility of contamination in the ^{3}He beam from $(^1H^2H)^+$, H_3^+, and $^3H^+$ (this latter species may be present in systems which have been exposed to tritium). These ion species have the same mass to charge ratio as ^{3}He$^+$ and will not be separated out by the beam analyzing magnet.

An ever present problem which must be carefully considered for each experiment that is planned is that of interference from elements in the sample other than those whose determination is desired. This problem is of course not unique to nuclear reaction techniques. It may be worth noting however that surface analyses are especially susceptible to contamination from the ambient atmosphere and nuclear reaction measurements may in addition be affected by contaminants present in accelerator vacuum systems; carbon is the most common offender in this regard, fluorine also can be a problem. A cold trap arrangement which has been found to be very effective in preventing condensibles from depositing on the target being bombarded (23) is shown schematically in Figure 3. The important features of this arrangement are that the target be surrounded as completely as possible by a surface at liquid nitrogen

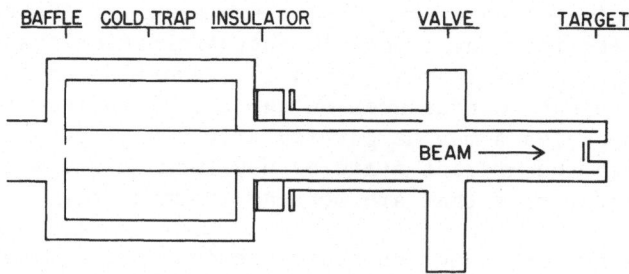

Fig. 3   A schematic diagram of a liquid nitrogen cold trap geometry
         for preventing carbon buildup during irradiations.

temperature (note that the cold tube extends past the target plane)
and that the solid angle subtended by the target into the vacuum
system be as small as possible. With this arrangement and a CPAA
detection efficiency which could have detected 10^{-2} atomic layers
of carbon, a five minute ^{3}He irradiation produced no observable
carbon. An atomic layer is taken here to be simply $N^{2/3}$, where N
is the number of host matrix atoms per cubic centimeter. It may
be worth noting at this point that in radioactivation analysis
extraneous impurities introduced after an irradiation do not affect
the measurement, provided of course that they are not radioactive
themselves.

     Under the assumptions that, for a specific analysis problem,
the literature has been consulted and that CPAA has been selected
as a candidate for the problem, the following general guidelines
are offered for reducing interferences with a desired measurement.
First of all if the radioactive decay lifetime associated with a
desired element is longer than those associated with interfering
elements it is possible to reduce the interference simply by delay-
ing the start of the measurement. In addition, the cross section
for a particular desired nuclear reaction will often be different
from that of possible interfering reactions and in general will
have a strongly different dependence on the energy of the incident
beam particle. It may be possible therefore to select a bombarding
energy so as to minimize the interference. Coulomb barrier effects,
for example, will, as the bombarding energy is decreased, tend to
reduce the cross section for higher atomic number (Z_2) elements
faster than for lower Z_2 elements, and will thus produce a relative
reduction of the higher Z_2 interference. An even more effective
reduction can be achieved if one of the interfering reactions has
a negative Q-value or energy release since such a reaction will not
proceed at all until the incident beam energy exceeds a threshold
value, E_{th} which is given by

$$E_{th} = -Q(M_2 + M_1)/M_2. \tag{3}$$

The reason that $E_{th} > |Q|$ is simply that, because of momentum con-
servation, the energy of the center of mass motion is not available
to the reaction. It is possible therefore to eliminate interference
from a negative Q reaction entirely by setting the bombarding energy
below its threshold value. Again, because nuclear reaction cross
sections can change markedly as the bombarding particle species is
changed it may be possible to reduce a particular interference by
choice of bombarding particle. Even when the interference cannot
be entirely removed, measurements at different energies and with
different bombarding particles may still be used to evaluate the
magnitude of the interference so that the desired measurement can
be obtained. Finally, if it proves necessary and provided the half
life for decay of the desired characteristic radiation is long
enough, a chemical separation can be performed to eliminate interferences. It is important to point out that, as was mentioned
previously, the final measurement in this case is not affected by
the introduction of extraneous impurities in the chemical separation
procedure. For trace element analysis, for example, at levels below
1 ppm this can be a very significant advantage.

The variation of reaction cross sections as a function of
bombarding energy, while useful for eliminating or evaluating interferences, at the same time complicates the problem of making quantitative measurements. The reason for this is that the incident beam
particle loses energy as it penetrates into the sample and the magnitude of the cross section varies therefore as a function of depth
into the sample. In this situation, if no assumptions are made about
the makeup of the target, a single measurement, in principle, cannot
yield the quantity of the measured element. In practice, reasonable
assumptions can be made about the depth distribution of the element
in question or, if they cannot, subsidiary experiments involving, for
example surface removal techniques or irradiations at different
energies are required. CPAA techniques in general are applied most
easily where the element being measured is present with uniform
density either throughout the bulk of the sample or in a layer of
known thickness. For such samples this technique usually has the
advantage of much higher sensitivity than prompt radiation measurements. Prompt radiation measurements however have the advantage
that, because they can yield depth distribution information, they
are not so dependent on assumptions about the sample.

There exist nuclear reactions whose reaction cross sections as
a function of incident energy exhibit a peaking or resonance behavior when the incident energy is just that needed for forming
the compound nucleus $(M_1 + M_2)$ in a particular narrow energy state.
In some cases the resonance is so sharp that the reaction cross
section changes by more than an order of magnitude when the incident
energy changes by only a few parts in a million. Such reactions
are important when prompt radiation measurements are made, particularly for depth distribution measurements, but are not commonly

used for CPAA. Resonant reactions will be discussed in detail
therefore under prompt radiation analysis (PRA).

In CPAA the amount, N, of an element present in a sample is
determined by the radioactivity produced, which in turn depends on
the number or yield of nuclear reactions produced by the irradiation.
For non resonant reactions the cross section changes relatively
slowly with energy. In this case, when the target thickness is
small compared to the penetration range of the incident beam particle, the reaction yield, Y, at incident energy E, may be written

$$Y = nN\Delta x\sigma(E) \tag{4}$$

where Y is the number of reactions produced, n is the number of
incident beam particles, N is the number of reactant target atoms
per cm^3, Δx is the target thickness in centimeters and $\sigma(E)$ is the
reaction cross section in cm^2 at energy E. If the target is thicker
than the range of the incident beam the formula becomes

$$Y = nN \int_0^R \sigma(x)dx \tag{5}$$

where R is the penetration range of the incident beam particles in
the target, $\sigma(x)$ is the reaction cross section as a function of x,
the penetration distance in the target, and N has been assumed to
be independent of x. The dependence of the beam particle energy
on x, required to convert $\sigma(E)$ to $\sigma(x)$, may be found from range
energy tables (24) for the beam particle and target material in
question. Once $\sigma(x)$ is known the integral can be evaluated numerically and the quantity N can be obtained.

Ricci and Hahn (22, 25) have greatly simplified this calcula-
tion by introducing a thick target cross section, $\bar{\sigma}$, that is approximately independent of target material and is constant for a given
nuclear reaction and a fixed energy interval, E to 0. This thick
target cross section is defined as:

$$\bar{\sigma} = \frac{1}{R} \int_0^R \sigma(x)dx \tag{6}$$

A substitution for the integral in equation (5) gives for the thick
target reaction yield then

$$Y = nN\bar{\sigma}R \tag{7}$$

Now if the integral is evaluated for a given nuclear reaction either
numerically or by measurement on a known composition thick target
then the thick target yield from a different target material but at
the same bombarding energy E, will be given by

$$Y' = n'N'\bar{\sigma}R' \tag{8}$$

where R' is the range of the beam in the second target. The primes
on Y', n', and N' simply denote that these quantities can all be
different in the second irradiation. If equation 7 pertains to
the comparison standard and equation 8 to the unknown sample we
can write for the desired quantity N'

$$N' = N_{std}(Y'/Y_{std})(n_{std}/n')(R_{std}/R') \qquad (9)$$

The reaction yield Y is of course not measured directly but requires
the usual corrections for detector efficiency, counting interval,
and effects due to the radioactive decay both during irradiation
and afterward. Detailed discussion of these corrections will not
be given here but may be found for example in (26) or in any of a
number of books on radiation detection.

Charged Particle Activation Analysis - Examples

The literature is sufficiently extensive so that it should
be possible for the reader interested in pursuing some specific
problem to find discussions of problems which will be related to
his own. A comprehensive bibliography on activation analysis with
extensive cross indexing by chemical element, sample material,
analysis technique, and author has been prepared by Lutz et al.
(27). It is an excellent starting point for any literature search
in this field. As an introduction to the literature and to display
the various ways in which CPAA may be used, a few selected articles
are abstracted briefly in the text which follows. Most, but not
all, of the measurements which will be discussed were performed at
energies of about 5 MeV or below and apply directly therefore to
small accelerators as the term is being used here. A few measurements which were performed at energies above 5 MeV have also been
included because many applications which require a given nuclear
reaction do not require the higher sensitivity provided by the
higher bombarding energies. A discussion of the measurement at
higher energies then is quite pertinent to the lower energy
measurement.

The application of ^{3}He-induced activation to analysis of oxygen
and other elements has been discussed in the paper of Markowitz and
Mahony (8) previously mentioned. For oxygen, the $^{16}O(^3He,p)^{18}F$
reaction was used and the radioactive ^{18}F isotope was detected;
^{18}F is a positron emitter with a 110 minute half-life. The radiations detected were the 511 keV gamma rays produced by the annihilation of the positrons. This reaction leads to extremely high
sensitivities for oxygen determination. For a 1 μA beam of
7.5 MeV ^{3}He ions incident on a 12.5 μm Al foil, an ultimate sensitivity of 1 part per billion (ppb) was estimated; for higher intensity

beams and water cooled target samples (to prevent melting) sensitivities of fractions of a ppb were indicated. Procedures for
determining the overall detection efficiency are described; precision and accuracy were estimated to be ±5 and ±10% respectively.
Analyses for oxygen were performed in beryllium, thorium, and mylar
samples. Suggestions were also made for evaluating possible interferences from sodium and fluorine as a result of the $^{23}Na(^{3}He,2\alpha)^{18}F$
and $^{19}F(^{3}He,\alpha)^{18}F$ reactions. This article also lists all of the
common nuclear reactions of ^{3}He ions with nuclides from ^{6}Li through
^{48}Ca along with product half-lives and reaction Q values.

An interesting application of CPAA is to combine it with the
use of stable isotope traces in order to study dynamic processes.
Thus Ritter et al. (28) have studied the room temperature oxidation
of silicon single crystals by exposing silicon samples for various
times to a mixture of 20% ^{18}O and 80% nitrogen at a pressure of one
atmosphere. The amount of oxidation was then measured through the
^{18}F radioactivity produced by the $^{18}O(p,n)^{18}F$ reaction; a proton
energy of 4.2 MeV was used since this reaction has a threshold
energy of 2.57 MeV. The activity was detected in a coincidence
arrangement which required that both of the 511 keV gamma rays
produced when a positron is annihilated be detected. This coincidence requirement greatly reduces background but, because the
two annihilation photons travel in opposite directions, does not
significantly reduce the detector acceptance solid angle; overall
sensitivity is therefore substantially increased. In the experiments cited, the sensitivity was such that .01 atomic layers
($\sim 3.6 \times 10^{-4} \mu g/cm^{2}$) were measurable. Carosella and Comas (29) used
this reaction and a somewhat different exposure technique to measure
oxygen sticking coefficients on silicon single crystals as a function of fractional surface coverage in the 0.03 to 0.97 atomic
layer range.

Benjamin et al. (30) have used 2.6 MeV deuterons for activation analysis of C, N, and O contamination on high purity Be, Mo,
and W surfaces. A particularly interesting technique was the
separation of the vacuum chamber for the irradiations from the Van
de Graaff accelerator vacuum system by a 2.5 μm thick molybdenum
foil to avoid carbon contamination. The reactions used and the
corresponding product half-lives were $^{12}C(d,n)^{13}N$, 9.96 min;
$^{14}N(d,n)^{15}O$, 2.07 min; and $^{16}O(d,n)^{17}F$, 1.1 min. The authors found
that the relatively long ^{13}N half-life makes the carbon analysis
insensitive to the presence of oxygen and nitrogen. The detection
limits for nitrogen and oxygen, however, depend on the concentrations of the other two impurities. Achievable detection limits
were estimated to be 1 ppm for carbon, 2 ppm for oxygen, and 5 to
10 ppm for nitrogen.

Still another way of using CPAA has been described by Holm et.
al. (31). In that work the microscopic distribution of oxygen and

carbon in metals was observed by bombardment of metallographically polished samples with 6.5 MeV ^{3}He ions and autoradiography of the activated areas. Although the technique was not optimized for ultimate resolution the authors were able to observe the outlines of metal grains with diameters of 12.5 μm in samples having bulk carbon concentrations of about 250 ppm.

Holt and Himmel (32) have combined stable isotope tracers and autoradiography following charged particle activation with microdensitometry measurements to produce a powerful way of obtaining the depth distribution of oxygen in thick oxide layers on metals. The measurements were performed in order to study the mechanisms of the scaling of metals at high temperatures. The authors exposed samples of iron and zirconium to an atmosphere enriched to 30% ^{18}O for roughly 1 minute at a temperature of 1050 °C and at a pressure of ~ 130 Torr. This exposure produced a 20 μm thick oxide layer which was used as a marker layer. The ^{18}O was then pumped out and the metal was oxidized in normal oxygen but at the same temperature and pressure until an oxide layer approximately 300 μm thick was formed. The sample was cooled slowly and beveled at a shallow angle to expose the entire cross section of the scale with a magnification of about 30. The beveled area was irradiated with 5 μA, 2.7 MeV proton beams for times of the order of 30-45 minutes. At this beam energy, which is only slightly above the 2.57 MeV threshold of the ^{18}O(p,n)^{18}F reaction, the ^{18}O is preferentially activated to ^{18}F and there are no interfering reactions. Contact autoradiographs were prepared within 30 minutes to one hour after irradiation. A microdensitometer measurement was then made on the autoradiograph to show the depth profile of the ^{18}O. Optimum depth resolution obtained was about 5 μm. Figure 4 shows the results obtained for an iron sample. It is immediately apparent that the ^{18}O marker layer, formed originally on the surface of the metal, has remained at the iron iron-oxide interface during the subsequent thickening of the scale in a normal ^{16}O atmosphere. The authors describe a number of other measurements. Since the technique and not the specific results are of prime interest here the reader is referred to the original publication for further details.

Tritons can be used to analyze for oxygen in the presence of fluorine because they do not induce any radioactivity which can interfere with the ^{18}F produced by the ^{16}O(t,n)^{18}F reaction. Discussions of the use of tritons for oxygen analysis in metal surfaces have been given by Wilkniss and Horn (33) and by Barrandon and Albert (34). In the latter investigation, excitations in the energy range from 0.5 to 3.0 MeV were used. The investigations included a study of the variation of surface oxygen on pure zirconium and aluminum as a function of chemical and mechanical polishing. The method can be used, with no chemical separations being necessary, for aluminum, iron, chromium, nickel, and zirconium. Sensitivities in the range from 100 to 10^{-3} μg/cm^2 were reported.

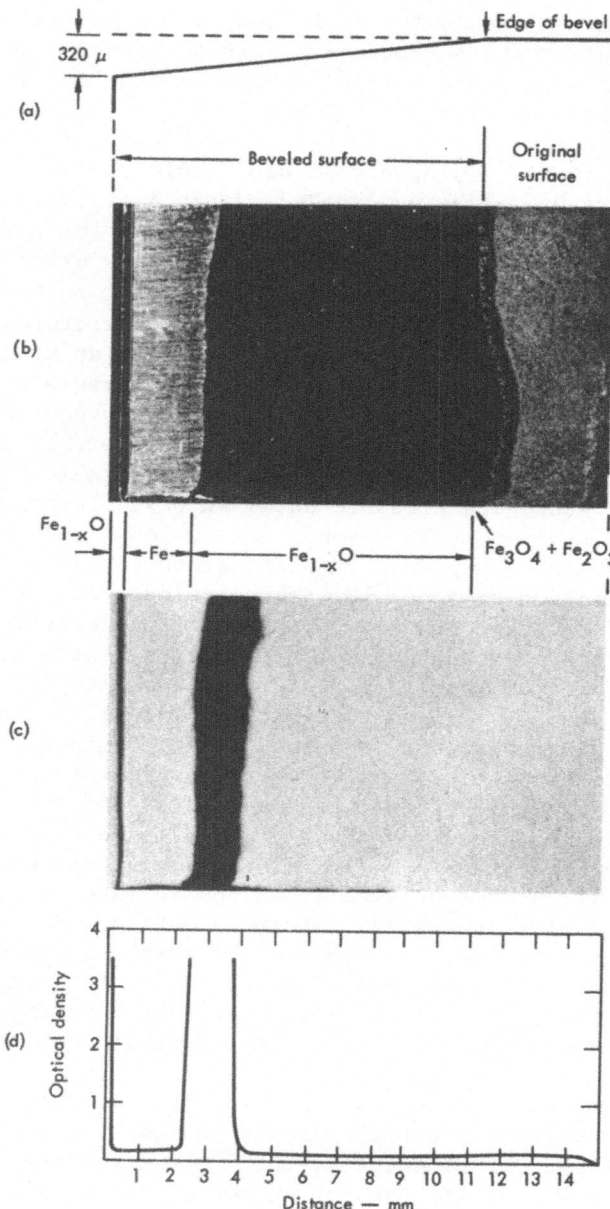

Fig. 4 The ^{18}O distribution in a marked zone-refined iron specimen
oxidized for 45 min. at 1050 °C in 1 atm O_2. a.) Plan view
of beveled surface; b.) low power photograph of beveled surface; c.) corresponding autoradiograph after proton activation; d.) microdensitometer scan of autoradiograph. (Reproduced by permission of the Journal of the Electrochemical
Society (32).)

Butler and Wolicki (23) performed activation analyses for oxygen and carbon on the surfaces of highly purified gold and platinum (99.999% pure). For oxygen the $^{16}O(He^3,p)^{18}F$ reaction was used and for carbon, $^{12}C(He^3,\alpha)^{11}C$; ^{11}C has a 20.5 minute half-life and is also a positron emitter as is ^{18}F. Irradiations were performed for approximately 5 minutes with one to two microampere beams of 4.5 MeV singly ionized ^3He particles. The 2 minute ^{15}O activity due to the $^{16}O(He^3,\alpha)^{15}O$ reaction was also observed. Sensitivities for oxygen and carbon were estimated to be 10^{-3} and 10^{-4} atomic layers ($\sim 4.1 \times 10^{-5} - 10^{-6} \mu g/cm^2$) respectively.

The $^{37}Cl(d,p)^{38}Cl$ reaction has been used by Knudson and Dunning (35) to detect trace amounts of chlorine on aluminum. The sample was bombarded with 5.0-MeV deuterons and the resulting gamma ray activity was measured with a 12.7 cm-diameter by 12.7 cm thick well-type NaI(Tl) scintillation detector. Carbon, nitrogen, and oxygen did not present any interference to the observation of ^{38}Cl. Measurements on 125-μm-thick commercial aluminum foil indicated considerable interference with the chlorine gamma-ray spectrum as observed with a NaI(Tl) detector. Later measurements with a high-resolution Ge(Li) solid-state detector showed that the interference was caused by the presence of manganese and gallium in the commercial foils. Radioactive ^{28}Al, produced by the $^{27}Al(d,p)^{28}Al$ reaction, could interfere with the ^{38}Cl. Fortunately, however, because the 2.27-min half-life of ^{28}Al is much shorter than the 37.2-min half-life of ^{38}Cl, counting can be delayed until the ^{28}Al activity has decayed to a tolerable level. A sensitivity of 0.2 $\mu g/cm^2$ was estimated. A substantially higher sensitivity should be achievable for chlorine on high-purity, high-Z metals.

If controlled surface removal by successive layers is performed after a charged particle activation and if the remaining activity is measured after each layer is removed, it is possible to measure the depth distribution of trace elements in the sample. This method, although more laborious than the prompt radiation techniques which are discussed in later sections, may still be useful in some applications. Krasnov et al. (36) have used this method for studying oxygen depth distributions in Ge and Nb samples. Samples were irradiated with 7.5 MeV ^3He ions. After irradiation the surface layers were removed, incrementally, by mechanical polishing with diamond paste on a cast-iron lap. The thickness of the removed layer was determined by weighing the samples before and after abrasion. Mechanical removal of the layers was chosen because in electrochemical etching the irradiated and non-irradiated parts of the surface may have different etching rates. On the average, each polishing removed a layer approximately 1.5 μm thick. The authors also considered the effect of the recoil of the ^{18}F atom, due to the nuclear reaction, on the depth distribution of the ^{18}F.

When surface removal techniques are used for measuring impurity depth distributions, care must be exercised to ensure that all components of the surface are being uniformly removed. Enough evidence of anomalous behavior has been obtained recently so that all surface removal procedures, whether mechanical, chemical, sputtering etc., should be suspect unless evidence exists or is obtained that the removal process is well understood. Dunning et al. (37) have found, for example, that ion beam sputtering mass spectrometry measurements can give grossly incorrect profiles for aluminum implanted into SiC. Similarly, Mackintosh and Brown (38) have found that during anodic oxidation of aluminum certain chemical species will move before the advancing oxide front and that anodic oxidation and chemical stripping, a commonly used technique, can then give erroneous impurity depth profiles. PRA depth profile measurements, to be described later, are particularly well suited for checking surface removal techniques because they are nondestructive. It is possible therefore to see if, in one and the same sample, a reproducible depth profile is being obtained. Reproducibility is then good evidence that the measurement does not disturb the impurity distribution.

Thus far the discussion has dealt exclusively with elements present only in trace amounts in the sample. Pretorius et al. (39) have reported on the use of CPAA for analyzing minor constituents in geological samples. Using 5.5 MeV deuteron irradiations they showed that for several compositional varieties of tourmaline the elements Al, Mg, B, Mn, Na, Fe, and Li could readily be determined as long as the concentrations were greater than approximately 0.1% by weight. A water cooled rotating sample holder which could accommodate up to 12 samples was used and samples and comparison standards were irradiated simultaneously to eliminate the need for total beam and beam intensity measurements. After irradiation the samples were placed in an automatic sample changer and analyzed by gamma-ray spectrometry with a Ge(Li) detector. The entire procedure was such that it could be made highly automatic and could be used to handle large numbers of samples on a routine basis.

PROMPT RADIATION ANALYSIS

In prompt radiation analysis (PRA) the presence of an element in a sample is detected through the nuclear radiations emitted instantaneously from nuclear reactions produced in the target by the irradiating beam. PRA techniques in practice, if not in objectives, resemble quite closely the way in which nuclear physics research is performed on the properties and characteristics of nuclear reactions, and the great variety of ways in which prompt techniques can be used to obtain information about a target sample is directly a reflection of the richness of nuclear phenomena and properties. Comprehensive summaries and discussions of nuclear structure and nuclear reaction properties are given for atomic mass numbers 5 through 44 in references 40 through 43; reference 44 gives a cumulative index for data on nuclei in the mass number range from 1 to 261. An excellent comprehensive summary and a bibliography on PRA have been prepared by Bird et al. (45). A recent set of tables, graphs, and formulas, compiled by Picraux et. al. (46) specifically to facilitate the use of low energy nuclear reactions for analysis, should also prove most useful.

The principal objectives of the text which follows will be to display the variety indicated above and, by so doing, to suggest that the "parameter space" in which these experiments are performed is large enough so that many opportunities for still further innovations and development exist. References 14, 15 and 16, containing more examples and greater detail than can be discussed here should prove helpful for planning surface analysis experiments. Particular attention is called to a summary article by Amsel (47) because it includes an interesting bibliography of applications that is listed according to such subject areas as oxidation phenomena, electrochemistry, metallurgy, biology, and others.

The kinds of beams, target chambers, detectors, and electronic pulse processing equipment are all the same as used in nuclear research; only the types of targets, the kinds of data, and the interpretation of the data are different. Thus a small accelerator laboratory equipped for nuclear research will be in an excellent position to undertake work with these techniques.

Detection limits for PRA can be quite good. Typically they are not as good, however, as can be achieved under ideal conditions with CPAA. Chemical separations are precluded although, of course, PRA can be performed on chemically separated sample constituents. Specificity and accuracy of PRA can be high. Precision suffers from the same statistical counting limitations and calibration difficulties that were mentioned earlier. Even though chemical separations are not possible, the problem of interferences is sometimes simpler to solve for the case of PRA than for CPAA because of the greater diversity of reactions and reaction types which can be considered for a specific problem.

One of the important advantages of PRA and the backscattering techniques discussed in the previous chapter is that they can frequently be used to measure the depth distribution of elements in the surface or near surface regions of the sample. The dependence of the characteristics of the emitted radiations on depth is due to the energy loss suffered by the incident ions as they penetrate into the sample and also to the energy losses suffered by charged particles emitted from the reaction as they emerge from within the sample. In some cases depth resolutions better than 200 Å can be achieved (48). PRA with narrow nuclear resonance reactions is

particularly powerful in this regard, the use of such resonances
for depth profiling having first been applied by Amsel and Samuel
(9). One recent resonance measurement, which will be discussed in
detail later, has reported a depth resolution of 20 Å (37).

Optimum depth resolution in the cases of both nonresonant and
resonant nuclear reactions depends on an accurate knowledge of
charged particle energy loss in solids and of the energy loss
straggling that is caused by the statistical nature of energy loss
processes. Discussions of energy loss and energy loss straggling
theories may be found in chapter 1 and in reference 48 and except
for a brief mention of energy loss straggling will not be repeated
here. It is worth noting however that, because a need exists for
improving depth resolutions to as good a value as can be achieved,
these questions are an important and common current interest in the
fields of both backscattering and PRA.

Another advantage of prompt radiation analysis is that spatial
information in the plane of the surface can be obtained about chemical composition, the spatial resolution of such measurements being
determined by the diameter of the incident ion beam. The excellent
work of Cookson and Pilling (49) on the use of small-diameter beams
for microprobe analysis deserves particular mention; specifically
these authors have developed a 3-MeV, 6-nA, proton beam less than
4 μm in diameter. Pierce (50) has discussed sample analysis which
utilizes a scanning of this beam across the sample either mechani-
cally or by electrostatic beam deflection.

Nuclear reactions may be categorized according to three basic
mechanisms, each of which may exhibit either resonant or nonresonant
dependence on the incident particle energy. The first of these is
the simple capture of the incident particle by the struck nucleus
to form a compound nucleus which then decays by the emission of one
or more gamma rays. The second is the inelastic scattering of the
incident particle, part of the incident energy being absorbed by
the struck nucleus which is then left in an excited state. The
third mechanism is that of rearrangement collisions. This term is
applied to all reactions in which the particles which separate after
a collision are different, i.e., $M_1 \neq M_3$, from those which entered
into the collision. For many reactions, in all three of the above
categories the emitted radiations have discrete and well-defined
energies characteristic of the spectrum of excited states in the
residual nucleus M_4 (and, strictly speaking, also of energy levels
in M_3 if it is itself a complex nucleus).

A clarification may be in order regarding the terms resonant
and nonresonant as used here to classify nuclear reactions. Strictly
speaking any nuclear reaction which exhibits a cross section maximum
that corresponds to a level in the compound nucleus, whether broad
or narrow, may be termed a resonance reaction. In this broad sense,

the term resonant would include the majority of low energy nuclear
reactions and would thereby lose some of its usefulness for distinguishing between types of applications. To avoid this situation
as well as questions about reaction mechanisms which, while most
fundamental, are not essential to the purposes of the present discussion, a working definition for the term resonant will simply be
taken here to include only those cases where the resonance property
of the reaction has purposely been used to provide some advantage
to the measurement.

Nonresonant Nuclear Reactions - Gamma Rays Observed

Nonresonant direct capture reactions are usually used for
identifying and measuring elements in a sample but not for depth
distribution measurements. The reason for this is that the majority
of gamma rays produced is due to crossover and cascade transitions
between well defined states of the compound nucleus and the gamma
rays do not depend therefore on the energy of the incident beam.
Energy conservation requires, of course, that the gamma ray transition from the initial state formed by the reaction to the first
well-defined level of the compound nucleus depend on the incident
energy, but this gamma ray is usually masked in the total spectrum
of gamma rays produced.

A very interesting and instructive exception to the above
situation is the experiment performed by Joy and Barnes (51) in
which use was made of a gamma ray that does depend on beam energy
specifically for depth distribution measurements. The reaction
used was $^{16}O(p,\gamma_2)^{17}F$ and the relevant part of the ^{17}F energy level
diagram is shown in Figure 5. For this reaction it is found that
the gamma ray labelled γ_2 has an energy which varies linearly with
proton energy with little change in intensity. If recoil effects
are neglected, the energy, $E(\gamma_2)$, of the gamma ray in MeV is given by

$$E(\gamma_2) = .600 + (16/17)E_p - .4955 \qquad (10)$$

where E_p is the incident proton energy in MeV, 0.600 MeV is the Q
of the reaction, and the energy of the first excited state in ^{17}F
is .4955 MeV. For reasonable depth resolution the method requires
a high resolution Ge(Li) detector. The detector used had a full
width at half maximum (FWHM) resolution of 2 keV at 1.33 MeV and a
volume of 35 cm^3. As an additional aid in calibrating the energy
scale the beam energy was chosen so that well established background
lines, a 1.461 MeV gamma ray from naturally occurring ^{40}K and a
1.632 MeV gamma ray from a contaminant $^{19}F(p,\gamma)^{20}Ne$ reaction, brack-
eted the γ_2 energies. The spectrum obtained from the bombardment
of an oxidized tungsten target with 1.514 MeV protons is shown in
Figure 6. The broad region of the spectrum just above the ^{40}K gamma
ray is γ_2. The width of γ_2 is due to the thickness of the oxide

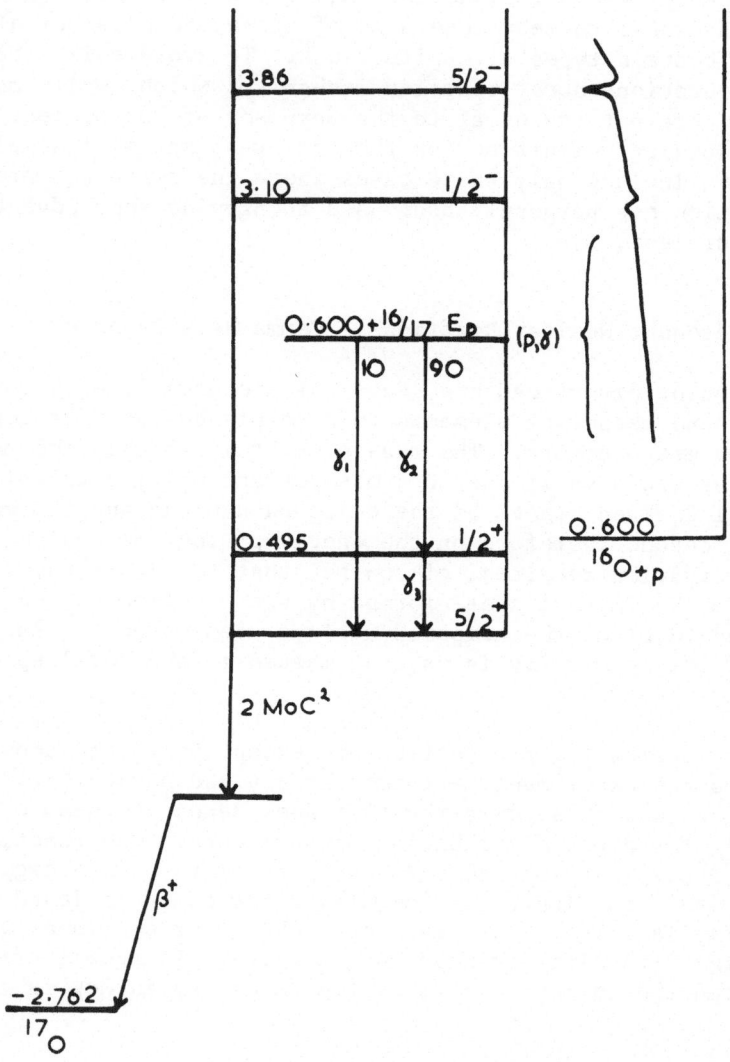

Fig. 5 Energy level diagram for the ^{17}F nucleus. (Reproduced by
permission of Nuclear Instruments and Methods (51).)

layer and the energy loss suffered by the protons in traversing it.
If the oxide layer were very thin the γ_2 peak would be approximately
as narrow as the two background peaks. Assuming that the tungsten
oxide had the composition WO_3 and using proton stopping powers from
reference (24) gave an oxide thickness of 433 ± 30 $\mu g/cm^2$. Although
the authors did not attempt to calculate the depth resolution of the
technique it would seem reasonable to expect that an unfolding of
the γ_2 shape should give a depth resolution at least as good and

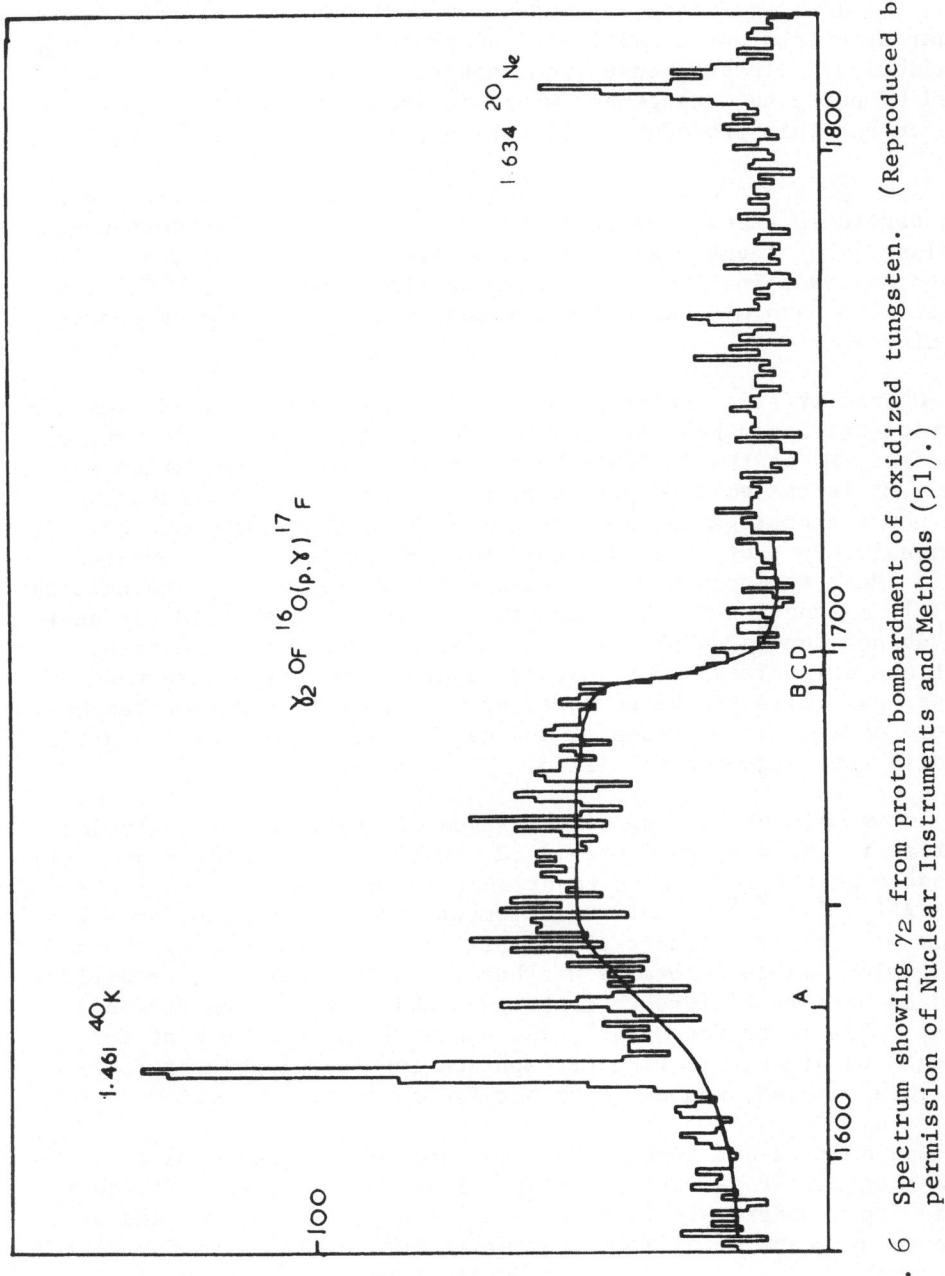

Fig. 6   Spectrum showing γ_2 from proton bombardment of oxidized tungsten.   (Reproduced by permission of Nuclear Instruments and Methods (51).)

perhaps even better than 43 $\mu g/cm^2$, this value corresponding to a
thickness of 600 Å. Unfortunately the cross section for the production of γ_2 is very small ($\sim 10^{-29} cm^2$) so the measurement cannot
be made unless the gamma ray background from competing reactions is
low. The low cross section means also that obtaining a sufficient
number of events for a meaningful shape unfolding will entail long
irradiations. Interference from background radiation can be reduced by using an arrangement where γ_2 and γ_3 are counted in coincidence but this procedure will not increase the yield for γ_2.

The $^{12}C(p,\gamma)^{13}N$ and $^{13}C(p,\gamma)^{14}N$ direct capture reactions have
been carefully studied by Close et al. (52). These authors have
concluded that these reactions can be used to perform isotopic
carbon analyses in times, including sample preparation, which are
competitive with mass spectrometry and nuclear magnetic resonance
techniques.

Characteristic prompt gamma rays can also be observed from the
reaction which may be written symbolically as $M_2(M_1, M_3\gamma)M_4$ where,
as before, M_1 is the incident beam particle, M_2 is the target nucleus, M_3 is the emitted particle, M_4 is the residual nucleus and
γ is now a prompt gamma ray. This reaction, mentioned only briefly
previously, in fact usually occurs when M_4 is left in an excited
state. Most commonly however, observations are made on the emitted
particle M_3 and not on the gamma rays. Because the gamma ray energies do not depend on whether the target is thick or thin these
reactions are suitable for thicker layers than is the case when
charged particles are being observed. Again, the reaction can be
either nonresonant or resonant and can be either inelastic scattering or a rearrangement collision.

An example of the use of this type of reaction for analyzing
material is the work of Coote et al. (53). These authors used the
inelastic proton scattering reactions, $^{19}F(p,p'\gamma)^{19}F$, $^{23}Na(p,p'\gamma)^{23}Na$
and $^{27}Al(p,p'\gamma)^{27}Al$ to make measurements on fluorine, sodium and
aluminum in black volcanic glass obsidian. The purpose of the investigation was to determine whether inelastic scattering reactions
could be used to differentiate between obsidians coming from different volcanic source sites. The nondestructive nature of the
nuclear technique, allowing its application to valuable or irre-
placeable samples, was an important factor in its selection.

One hundred and twenty obsidian samples from seven major
archaeological areas were analyzed. They were carefully flaked to
expose fresh moderately flat surfaces about 1 cm^2 in area and were
fastened by means of silicone rubber cement onto an aluminum plate.
To prevent the buildup of charge during irradiation, electrical
contact was made from the obsidian to the holder with silver conducting paint. The irradiations were performed with 0.6 μA of
2.2 MeV protons and the gamma rays were detected with a 5 cm^3 Ge(Li)

detector. The detection system had a resolution of 2.5 keV full width at half maximum (FWHM) at 350 keV. A counting time of 5 minutes for each sample was adequate for the measurement desired. The samples showed no deterioration due to the irradiation.

A typical spectrum obtained from an irradiation is shown in Figure 7. Gamma ray peaks from ^{19}F may be seen in the figure at 110 and 197 keV, from ^{23}Na at 439 keV, and from ^{27}Al at 842 and 1013 keV; attempts to measure still higher energy gamma rays were not made.

An interesting feature of these measurements is the simplification that occurs because only the ratios between elements in a sample are needed for its identification. Thus while the standard deviation for an individual element was about 18%, estimated to be perhaps due to inaccurate beam current integration, the ratio of fluorine to sodium had a standard deviation of only 2%, a value consistent with counting statistics. For the final analyses the area of each peak was compared to the sodium peak. Each gamma ray spectrum was processed automatically in about 20 seconds with a PDP-8 computer programmed to calculate significant peak areas.

The possibility of weathering effects on the obsidians, particularly the leaching out of sodium, was tested by making measurements at six energies ranging from 1.7 to 2.9 MeV, in this way sampling effectively from different depths, on both a weathered and a non-weathered surface of the same specimen. The gamma ray yield showed an approximate fourth power dependence on the bombarding energy and it was expected that the change in energies

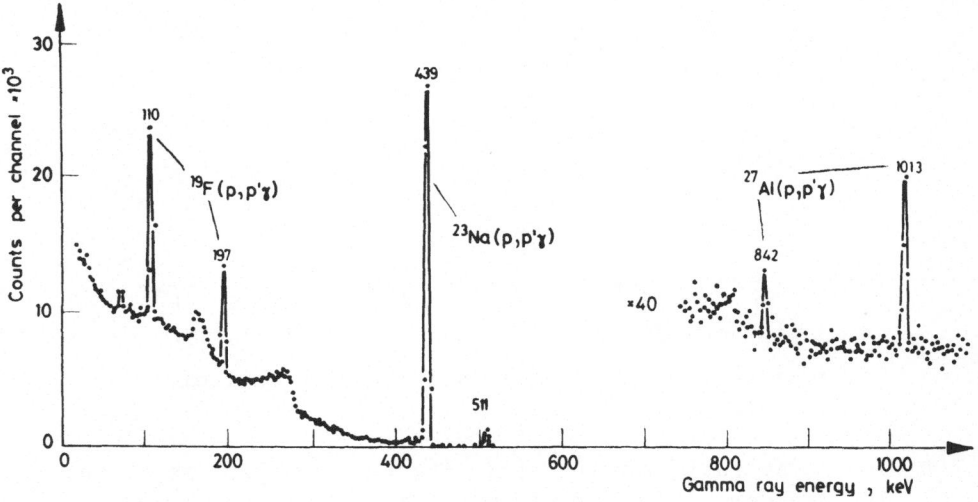

Fig. 7   Typical 512-channel gamma ray spectrum from an obsidian
         sample. (Reproduced by permission of the Journal of
         Radioanalytical Chemistry (53).)

might produce large differences in the ratio for the weathered
as compared to the unweathered surface. However, both sets of
ratios followed each other closely as a function of energy showing
that the weathering effect is not significant.

The ratios obtained for fluorine to sodium and aluminum to
sodium are shown in Figure 8. The Al/Na ratio was useful but in-
ferior to the F/Na ratio. The results show that many obsidian
source materials can be identified successfully on the basis of
their fluorine to sodium ratios. The principal advantages of the
technique were stated to be its nondestructive nature, its ability
to measure light elements, its speed, and the relative ease of
sample preparation. Even though measurements were made at the few
percent, level sensitivities were estimated to be 10-50 ppm. Inelastic scattering was judged to be a particularly good method for
determining fluorine.

Deconninck and co-workers (54, 55, 56) have undertaken a syste-
matic study of (p,γ), $(p,p'\gamma)$ and $(p,\alpha\gamma)$ reactions on a number of
nuclei in order to provide the data base which is needed if these
reactions are to be used for analyzing materials.

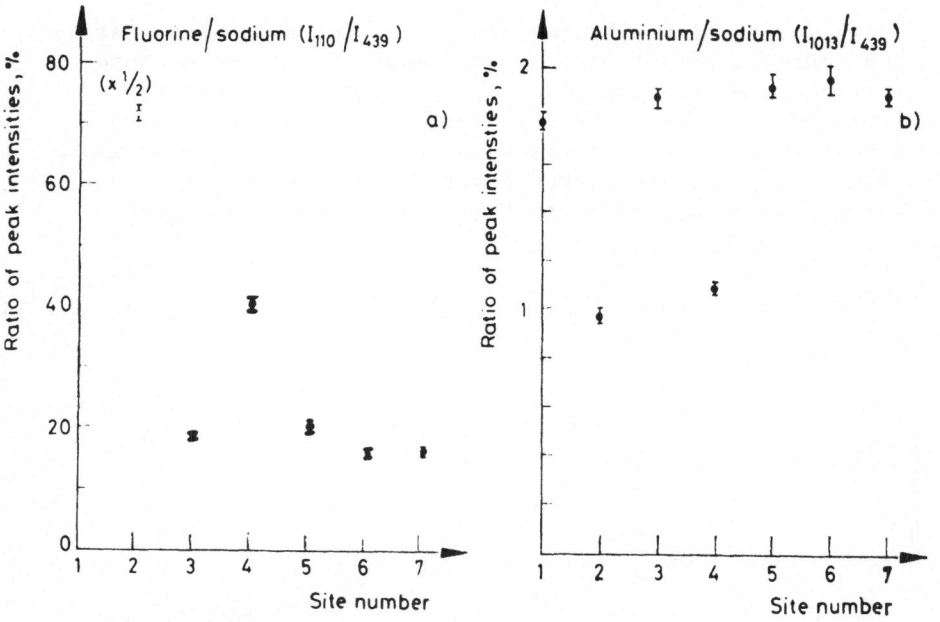

Fig. 8a   The F/Na ratio obtained for obsidian samples from
          various sites.

Fig. 8b   The Al/Na ratio for the same sites. Error bars represent
          the standard deviation of the mean. Reproduced by per-
          mission of Journal of Radioanalytical Chemistry (53).

Reference 54 presents an overview of the approach and a summary
of the principal gamma rays observed for the eleven elements, F, Na,
Al, P, Cr, Mn, Se, Rh, Ag, Pt, and Au. Since the work was performed
with protons in the energy range from 1.0 to 3.0 MeV it may be surprising to see the elements Se, Rh, Ag, Pt, and Au included in the
above list. Coulomb barrier effects, mentioned previously, would
be expected to make the probability of a nuclear interaction with
these nuclei very small. Indeed this is the case and the explanation of the gamma rays observed from these nuclei is that the nuclear
energy levels are excited not through nuclear reactions but by a
mechanism called Coulomb excitation which involves the interaction
of the Coulomb fields of the two approaching nuclei at separation
distances which are well outside the sum of their nuclear radii.
The use of Coulomb excitation for materials analysis may be of particular interest as an area for additional research in that very
few studies in it have been made. In this connection, some comparisons of cross sections for Coulomb excitation and compound nucleus
formation given by Stelson and McGowan (57) may be useful. For
3.0 MeV protons these authors have calculated, for example, that
even for Mo (Z = 42) the Coulomb excitation cross section is over
100 times as large as that expected for compound nucleus formation.

A principal objective in references (54, 55, 56) is to develop
a method which will be useful for quantitative analysis not only of
thin samples for which the yield is given simply by equation (4) but
also for thick targets. In order to avoid the integration over target thickness which is required, as shown for example in equation
(5), the authors have proposed an approximate formula which is similar but not identical to equation (9). The formula in the present
case, expressed in the notation of equation (9), is

$$N' = N_{std}(Y'/Y_{std})(n_{std}/n')(S'/S_{std}) \tag{11}$$

where N' is the number of reactant (to be measured) atoms per cm^3
in the sample, N_{std} is the number of atoms per cm^3 of the same
element in the comparison standard, Y' and Y_{std} are the gamma ray
yields from the sample and the standard respectively, and n' and
n_{std} are the corresponding incident numbers of beam particles. S'
and S_{std} are the stopping powers of the beam in the host material of
the sample and the standard, calculated at the energy where the
thick target excitation curve has half the cross section value it
has at the incident beam energy. The accuracy of the assumptions
underlying equation (11) have been tested with a nickel aluminum
standard containing 1.96% of aluminum and gave, for the aluminum
content, 1.96 ± 0.06%. Limits on the accuracy of the method were
judged to be due to uncertainties in the knowledge of the stopping
powers of elements and compounds. One of the assumptions made in
this method is that both the sample and the standard are homogeneous
in composition. For heavy elements, presumably those which are
excited by Coulomb excitation, sensitivity in general is estimated

to be about 1000 ppm; for the light elements F, Na, and Al sensitivity is estimated to be in the low ppm range.

The data base clearly required for the use of this method is to have available the thick target excitation curves for each reaction. From such curves, for a given bombarding energy, the energy at which the thick target yield is just one-half can be found and the stopping power required in equation (11) can be calculated. Reference 54 gives thick target excitation curves specifically for the reactions ^{23}Na(p,p'γ)^{23}Na (439 keV gamma ray), ^{107}Ag(p,p'γ)^{107}Ag (325 keV gamma ray), and ^{109}Ag(p,p'γ)^{109}Ag (311 keV gamma ray).

In this same program, Deconninck and Demortier (55) have reported an extensive series of measurements on the 1778 keV gamma ray from the ^{27}Al(p,γ)^{28}Si reaction, the 843 and 1013 keV gamma rays from the ^{27}Al(p,p'γ)^{27}Al reaction, and on the 1368 keV gamma rays from the ^{27}Al(p,$\alpha\gamma$)^{24}Mg reaction. Thick target excitation curves from pure aluminum targets were obtained in the incident energy range from 500 to 2500 keV. The gamma rays were measured with a Ge(Li), 26 cm^3, detector at 90° and at 0° to the incident beam direction.

The authors include detailed tables of gamma ray intensities as a function of incident energy with absolute errors estimated to be less than 15% and relative errors to be less than 5%. Also given are a discussion and a table listing the most likely interfering reactions and gamma rays from other chemical elements in the sample. The authors summarize their findings on how to minimize interference for a number of types of samples. Interferences from Na, Mg, Li and P are judged to be easily identified. For the rapid identification of aluminum in a sample the recommended procedure is to observe the 843 and 1013 keV gamma rays at 2 MeV and then to decrease the proton energy to 1.5 MeV to see if the yield of these two gamma rays becomes negligible, as it should if the origin is aluminum.

Quantitative analyses on samples in which aluminum is homogeneously distributed, at least in a 20 μm thick surface layer, are again performed by use of a comparison standard and equation (11). The accuracy of this formula for aluminum measurements is estimated to be 1%.

The applications of the technique are discussed for aluminum in stainless steel, in uranyl nitrate, in rock samples containing both Al and Na, and for a finely powdered mixture of W and Al$_2$O$_3$ in the proportion of 10^4 to 1. The sensitivity of the method, for measurement times not exceeding 60 minutes, is estimated to range from 10 to 1000 ppm, depending on the sample.

A third set of experiments in this program was reported by

Demortier and Bodart (56) and was a study of the 2237 keV gamma
ray from the $^{31}P(p,\gamma)^{32}S$ reaction, the 1266 keV gamma ray from
$^{31}P(p,p'\gamma)^{31}P$, and the 1778 keV gamma ray from the $^{31}P(p,\alpha\gamma)^{28}Si$
reaction. Again, detailed thick target excitation curves are ob-
tained and the observed results are presented in both graphic and
tabular form. Interferences are discussed and a table listing
interfering reactions and gamma rays is given. The target in these
studies presented special problems in that it was made of pure
compressed phosphorous powder. To keep it from melting it was
found necessary to keep the beam intensity no higher than 10 nA
and to rotate the target so as to distribute the heat delivered
by the beam. The front face of the target was also covered by a
thin evaporated gold layer to ensure good beam charge collection.
It was possible under these conditions to maintain beam on target
for several hours without causing observable destruction or evaporation.

It was concluded that when the sample consists mainly of
light elements it will be advantageous to irradiate at fairly high
energy and to detect phosphorous through the 1266 keV gamma ray.
When the sample consists mainly of heavy elements, a lower proton
energy should be used. Sensitivities for the determination of
phosphorous in different types of samples varied from 50 to about
500 ppm. Quantitative measurements were made, as before, on the
basis of equation (11).

Studies have been performed by Peisach (58) on the gamma emitting nuclear reactions produced by triton irradiation of oxygen.
The reactions studied were $^{16}O(t,n\gamma)^{18}F$, $^{16}O(t,p\gamma)^{18}O$, and
$^{16}O(t,\alpha\gamma)^{15}N$. Measurements were performed on gas targets whose
pressure or, equivalently, thickness could be varied. Typically
the thickness was such that 3.0 MeV tritons lost about 70 keV of
energy in the gas. Excitation curves were obtained for bombarding
energies between 1.1 and 2.8 MeV in steps of about 100 keV.

If the gamma rays are labelled according to the convention
s(a,b) where s is the prompt light particle emitted and the gamma
ray is emitted by a transition in the residual nucleus from level
a to level b (0 is the ground state, 1 the first excited state
etc.) then the gamma rays for which excitation curves were obtained
are the following: n(1,0), 940 keV; n(2,0), 1043 keV; n(3,0), 1095
keV; p(1,0), 1982 keV; α(1,0) α(2,0) (unresolved), 5300 keV. The
largest yields were obtained for the n(1,0) and n(2,0) transitions.
Although low compared to n(1,0), p(1,0) is considered useful for
analytical purposes because it falls in a region of the spectrum
where the continuum intensity is low.

A series of measurements was made on oxide layers on aluminum
foil and on copper and steel samples. For 1mC of 1.9 MeV tritons,
sensitivities were estimated to be 0.13 $\mu g/cm^2$ for iron and copper,
and 0.45 $\mu g/cm^2$ for aluminum.

Nonresonant Nuclear Reactions—Nuclear Particles Observed

Thus far the PRA examples discussed have involved the observation of gamma rays emitted from nonresonant nuclear reactions. With but the one exception, the measurements described were most useful for homogeneous samples or for homogeneous surface layers and did not contain information about depth distribution. Gamma ray observations can be used to obtain depth distribution information, but as will be seen in later sections, resonance reactions are most commonly used for this purpose and the depth distribution information is contained not in the energy of the gamma ray but in the dependence of the cross section on the incident particle energy.

When charged particles or neutrons emitted from a reaction are observed, the measurements can be used not only for quantitative measurements on homogeneous samples but, because the outgoing particle energy depends on the energy of the incoming particle, M_1, also for obtaining depth distribution information. Charged particles, in addition, lose energy as they emerge from the sample thus also giving information about the depth in the sample at which the reaction occurred.

General principles and examples will be discussed first for the simplest case, namely that in which charged particles from nonresonant nuclear reactions are used to analyze thin homogeneous layers. The discussion will then be extended, still for nonresonant reactions, to thicker layers and then to depth distribution measurements for nonhomogeneous layers. Resonant reactions will be treated last, the discussion in this last category including observations on both gamma rays and charged particles as well as some comments on the special energy loss straggling questions which arise for small energy losses. References 2, 45, 59 and 60 will also be found useful for additional information and excellent general discussions of analyses based on prompt nuclear particles.

For the nuclear reaction shown schematically in Figure 9, the conservation of total energy and momentum lead to the following nonrelativistic expression for the energy E_3 of the particle M_3 emitted from the reaction

$$
E_3 = \frac{E_1 M_1 M_3}{(M_3 + M_4)^2} \left\{ 2\cos^2\theta + \frac{M_4(M_3 + M_4)}{M_1 M_3} \left(\frac{Q}{E_1} + 1 - \frac{M_1}{M_4} \right) \right.
$$

$$
\left. + 2\cos\theta \left[\cos^2\theta + \frac{M_4(M_3 + M_4)}{M_1 M_3} \left(\frac{Q}{E_1} + 1 - \frac{M_1}{M_4} \right) \right]^{1/2} \right\} \quad (12)
$$

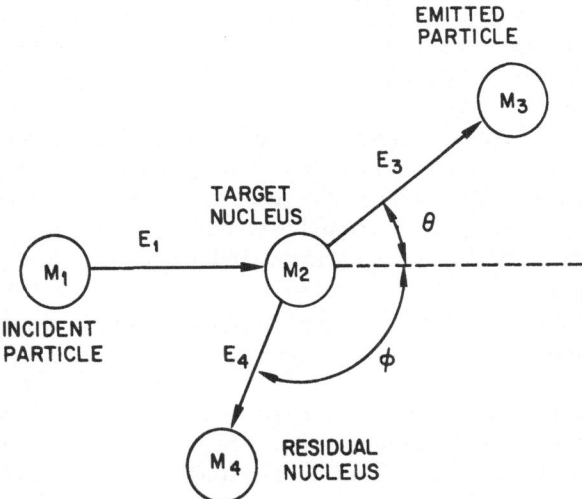

Fig. 9   A schematic diagram of a nuclear reaction.

where E_1 is the energy of the incident particle, M_1 and M_4 are the masses of the incident and residual nucleii respectively, Q is the energy released in the reaction, given by equation 2, and θ is the angle in the laboratory between the directions of the incident particle and the emitted particle.

Equation (12) shows that for a given reaction, incident energy E_1, and observation angle θ, the energy spectrum E_3 will depend on the Q values possible to the reaction. In a manner quite analogous to optical spectroscopy, if the residual nucleus can be left in any of several sharply defined energy states and if the target is thin so the incident energy has a single well-defined value, the energy spectrum of the emitted particles will exhibit a series of narrow peaks, each of which corresponds to one of the excited states of the residual nucleus. Such a spectrum will then be highly specific to the reaction which has occurred and can be used to identify the presence of a given elemental isotope. The positions of the peaks in the energy spectrum are easily calculable from Eq. (12). The intensity ratios of the various peaks in the spectrum are also characteristic of the reaction and provide additional useful information for identifying the reaction responsible.

The yield formula for thin targets is similar to Eq. (4),

$$Y_3(\theta) = nN\Delta x\sigma(\theta,E)\Omega \tag{13}$$

Here $Y_3(\theta)$ is the number of particles M_3 observed at angle θ by a detector subtending a solid angle Ω at the target, $\sigma(\theta,E)$ is the differential cross section in cm^2 per steradian at energy E and angle θ, and Δx, n, and N are, as before, the target thickness in centimeters, the number of beam particles, and the number of reactant atoms per cm^3 respectively.

A particularly simple example which illustrates how a nuclear reaction can be used for analyzing thin targets is an experiment performed by Wolicki and Knudson (61). The object of the measurement was to detect thin films of sulfur nondestructively on a copper nickel alloy in the presence of carbon and oxygen. The ^{32}S(d,p)^{33}S reaction was selected for the measurement because the Q value (6.419 MeV) for this reaction is considerably higher than those for the (d,p) reactions on carbon and oxygen.

Initially, a 4.0 MeV deuteron beam was used to bombard the samples. At that energy however, the intensity of protons from reactions with the copper and nickel isotopes was so high that the protons due to the thin sulfur film could not be observed. The deuteron energy was then reduced to 2.0 MeV and, because of the different Coulomb barriers involved, this reduction produced a very large preferential reduction of the (d,p) yields from nickel (Z = 28) and copper (Z = 29) relative to those from sulfur (Z = 16). The protons from the sulfur then became easily observable. Because the yield of beam particles scattered backward from a thick target is very large it is usually necessary to place a foil in front of the detector to stop these particles. In the present case a 12.5 μm foil of nickel was adequate for this purpose.

The spectrum of protons obtained with a test sulfide film less then 100 Å thick is shown in Figure 10. The target chamber shown earlier in Figure 2 was used in these experiments and the solid state detector was located at 135° with respect to the incident beam direction. The arrows labelled with the symbols p_0 through p_8 show the energies calculated from Eq. (12) for protons emitted from the ^{32}S(d,p)^{33}S reaction (corrected for energy loss in the 12.5 μm nickel foil). The symbol p_0 labels the proton group which corresponds to the ^{33}S nucleus being left in its ground state, p_1 corresponds to the first excited state of ^{33}S and so on. Note that the outgoing particle energies are narrowly defined because the beam energy has a well defined value and the target film is thin. As the target film becomes thicker the proton peaks in the spectrum get broader. For moderately thick targets the groups may be resolvable but for still thicker targets the broadened peaks will overlap and make identification difficult.

Table I shows the actual energies calculated and the corresponding energy levels in ^{33}S. To show how the energies change also as a function of angle, a column has been included which gives the energies calculated for 45°. In some cases interference from other reactions can be minimized by using this property to change the position a given particle group relative to a group from another reaction.

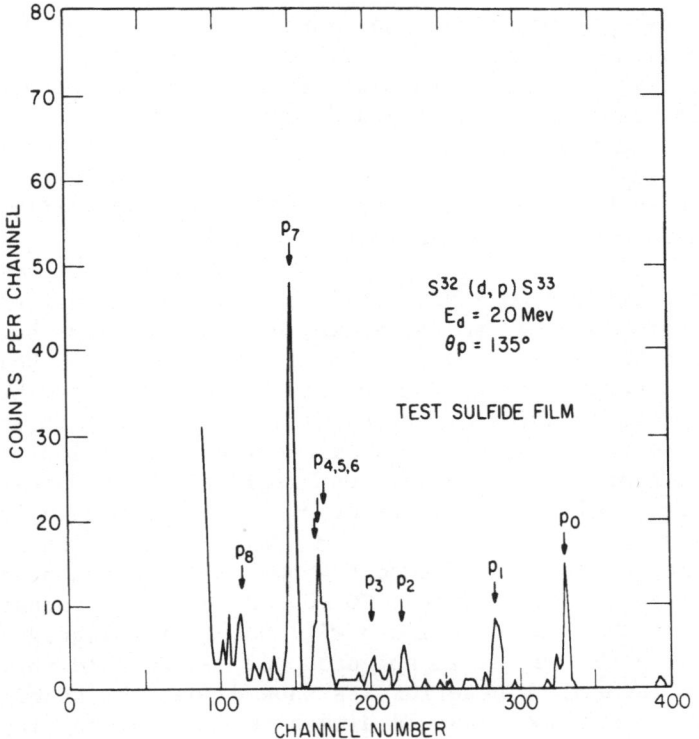

Fig. 10   Spectrum of protons obtained from an irradiation of a thin
          sulfide film with 2.0 MeV deuterons.   Reproduced by per-
          mission of the International Journal of Applied Radiation
          and Isotopes (61).

Table I

Values of E_3 calculated for the $^{32}S(d,p)^{33}S$ reaction from
Eq. (12) for E_1 = 2.0 MeV, θ = 45° and 135° and Q = 6.419 -
(^{33}S excitation energy in MeV)

Proton Group	E_3 (MeV) $\theta = 45°$	$\theta = 135°$	^{33}S Excitation Energy (MeV)
0	8.292	7.816	0
1	7.464	7.013	.84
2	6.350	5.934	1.97
3	6.015	5.610	2.31
4	5.461	5.076	2.87
5	5.402	5.019	2.93
6	5.363	4.981	2.97
7	5.116	4.743	3.22
8	4.512	4.163	3.83

The correspondence of all the peaks with the expected energies provides a positive identification of a thin sulfur film on the irradiated sample. Not only the energy positions of the peaks but also the relative heights (areas) and, for some reactions, the widths of the peaks are all characteristic of the reaction that is responsible. The sensitivity of the measurements for thin films was estimated to be easily 10^{-1} $\mu g/cm^2$ with the possibility existing for considerable improvement beyond this value.

Weber and Quaglia (62) have used (d,p) reactions not only to determine carbon, oxygen, and nitrogen in metal surfaces but also to study the behavior and stability of targets under ion bombardment. For the case of oxygen on copper the measurements showed, for example, that even for a current density on target as low as 25 nA/mm^2 the oxygen content showed a decay as a function of total integrated charge. The decay was most rapid at the beginning and, asymptotically, could exceed a factor of two. The oxygen layers studied were between 0.1 and 1.0 $\mu g/cm^2$ thick. Oxygen on silicon, for reasons not understood, showed a growth in oxygen content up to an integrated charge of about 600 μC. Results for carbon confirmed the fact that carbon buildup will occur if a liquid nitrogen trap surrounding the target is not used but can be held to very low levels if one is used. These results showed that the stability of a target under bombardment should always be checked and, in addition, that effects occurring at the start of an irradiation may not be the same as those occurring at a later time.

Deuteron induced reactions have been used on somewhat thicker targets by Olivier and Peisach (63). In these experiments boron was analyzed in ore and glass samples by (d,p) reactions produced by 2.7 MeV deuterons. Samples of the materials to be irradiated were ground to a fine powder with a particle size of about 0.5 μm and centrifuged onto tantalum discs to produce a layer of about 300 $\mu g/cm^2$. The proton peaks resulting from these targets, while broader than those observed from thin targets (as was the case for example in the $^{32}S(d,p)^{33}S$ measurements described previously) were still sufficiently well defined so that the boron p_O peak area could be determined. A point of interest in these experiments is that in order to take advantage of a maximum in the angular distribution of the ground state proton group, the protons were measured at 30°. At the 20° target angle required for this geometry the 3mm diameter beam then irradiated an elliptical area on the target having a major axis of about 8.8 mm. Again, the solid state proton detector had to be covered with a thin foil to stop the scattered deuterons from entering the detector. It was found also that beam currents higher than 1.0 μA resulted in a decrease in the boron count during irradiation so beam currents were kept below this value. The excitation curve for the boron p_O group at 30° showed that the yield was relatively flat between 2.0 and 3.0 MeV. This flat characteristic eliminates the need for integration over target thickness and simplifies the boron analysis considerably.

The spectra obtained from samples of tourmaline and borosilicate glass are shown in Figure 11. For the tourmaline sample, a separate irradiation of a test sample containing magnesium and boron in equal parts was used to place an upper limit of about 0.6% for the interference to the $^{10}B(p_o)$ from $^{25}Mg(p_o)$. The $^{14}N(p_o)$

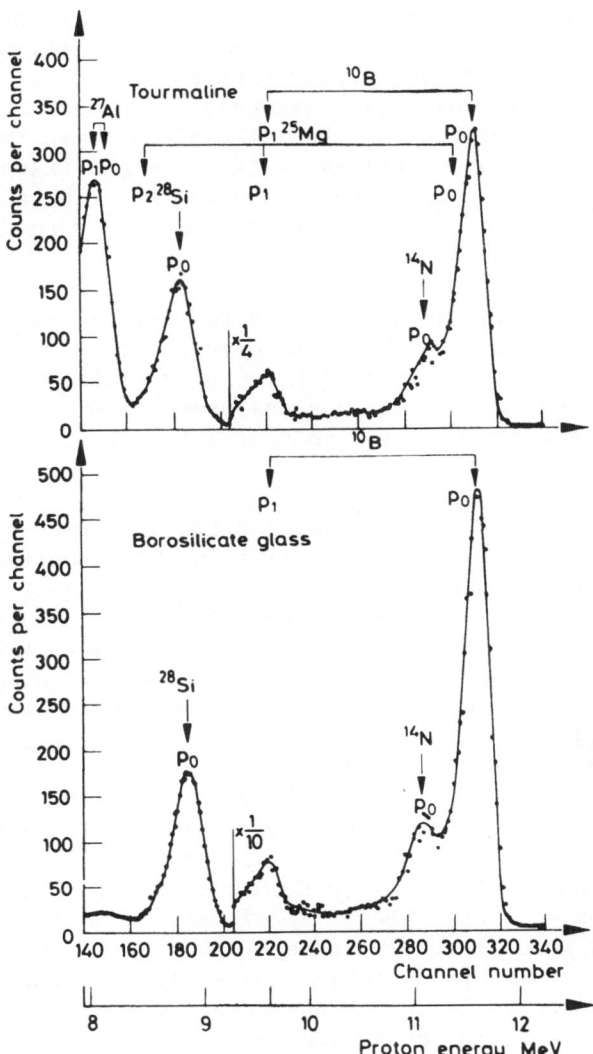

Fig. 11 Energy spectra of prompt protons obtained from the deuteron irradiation of thin deposits of tourmaline and borosilicate glass on tantalum. Reproduced by permission of the Journal of Radioanalytical Chemistry (63).

peak, while sizeable, was well enough separated from the $^{10}B(p_o)$
peak and its shape in addition sufficiently known so that its contribution to the latter could be reliably evaluated. Possible
interferences from ^{47}Ti, ^{33}S, ^{87}Sr, and ^{50}V, were checked against
but were not observed. For irradiation times of only 2 to 3 minutes,
the accuracy of the method in general was found to be about $\pm 5\%$
for boron contents ranging from 45% to about 3%.

In a related set of experiments, Peisach and Pretorius (64)
have analyzed various glasses by simultaneous spectrometry of backscattered alpha particles and prompt protons from (α,p) reactions.
The alpha particle backscattering was used to measure the heavy
elements and the prompt protons the light elements in the glass.
Samples were prepared as described in the previous example (63)
and, again, had a thickness of about 300 $\mu g/cm^2$. The alpha particle bombarding energy was 4.0 MeV and the protons were detected at
135° with a solid state detector.

The light elements which were measured were B, Na, and Al.
The (α,p) reactions on these elements have positive Q values and
produce protons with relatively high energies. The Coulomb barrier
for alpha particles kept reaction cross sections with higher atomic
number elements sufficiently low so that interferences from these
elements were not a problem.

Typical proton spectra obtained for two different types of
glass are shown in Figure 12. The peaks in this spectrum show
relatively little overlapping. There are also enough isolated
peaks so that peak shapes for peak stripping or subtraction procedures can be obtained if necessary. The measurements were calibrated by irradiating a comparison standard whose thickness was
comparable to that of the samples so that the variations in the
(α,p) cross section as a function of depth in the target affected
the standard and the sample equally.

Beam currents used were between 100 and 150 nA. With these
relatively small beams, irradiating times of about 25 minutes per
sample were adequate for analyzing B, Na, and Al present in amounts
which ranged from 0.1% up to about 15% in the various glasses. It
is worth noting, because doubly ionized alpha particle beams of
100 nA magnitude can be produced in many small accelerators, (provided that the ion source helium is sufficiently pure) that (α,p)
reactions can be investigated at 4.0 MeV, as was done in these experiments, with 2 MV accelerators.

An interesting way of eliminating an interfering group in a
reaction particle spectrum has been demonstrated by Pretorius et
al. (65, 66). These authors noted that for the nuclear reactions
$^6Li(p,\alpha)^3He$, $^6Li(d,\alpha)^4He$, and $^7Li(p,\alpha)^4He$, the residual nucleus M_4
could be observed as easily as the outgoing particle M_3 (in two of

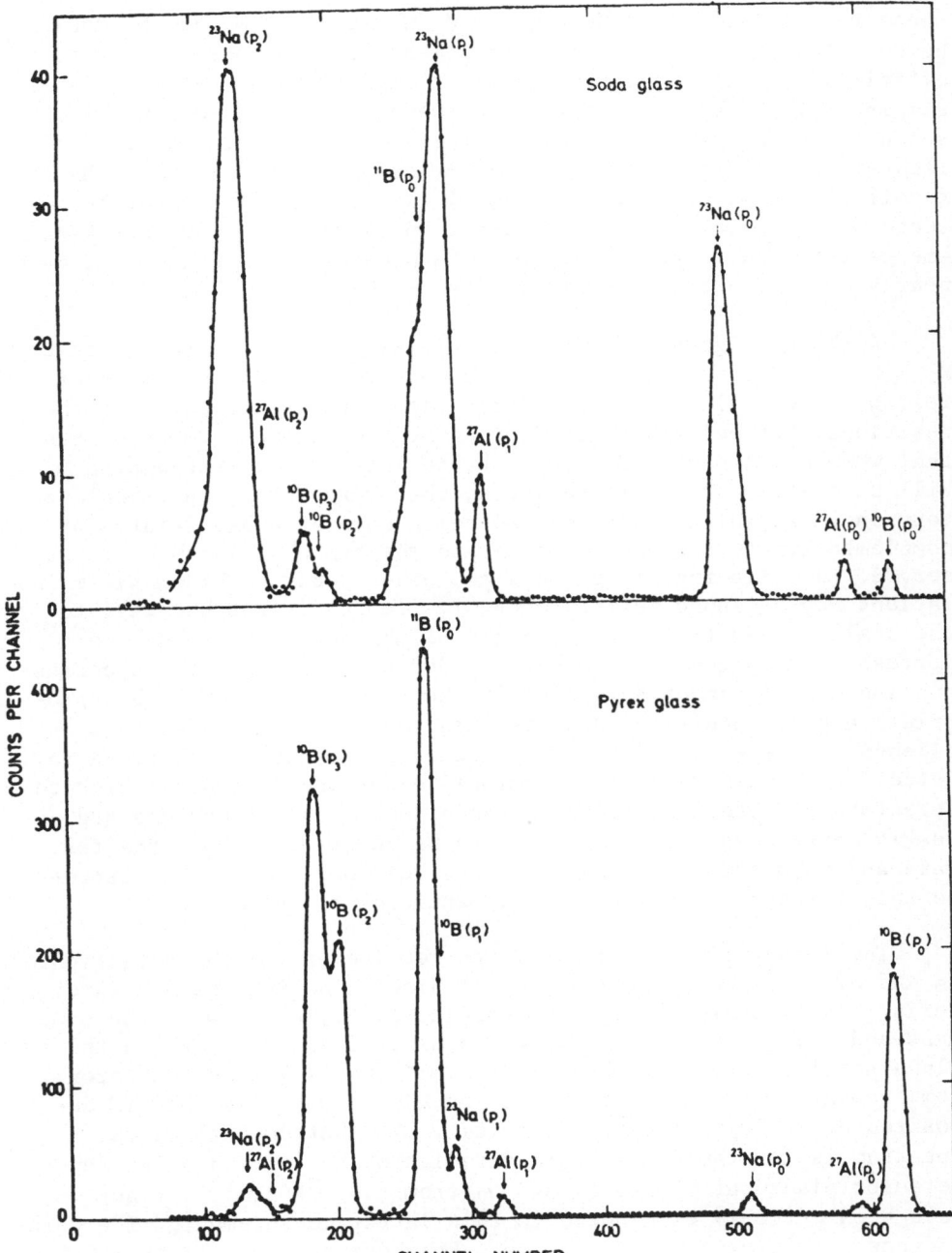

Fig. 12 Proton spectra obtained by alpha particle irradiation of soda-and pyrex-glass samples. Reproduced by permission of the Journal of Radioanalytical Chemistry (64).

these reactions in fact $M_3 = M_4$) and further that M_3 and M_4 could be detected in a coincidence arrangement with no loss of detection efficiency. The reason for the first part of this statement is simply that the energy partition between M_3 and M_4, because of conservation of momentum, depends on their difference in mass, the lighter particle receiving the larger energy. Thus if $M_4 \gg M_3$ the recoil energy of M_4 will be so small that it usually cannot be detected. Commonly, except for very thin targets, M_4 does not even emerge from the target. When $M_4 \approx M_3$ however the energies are more nearly equal and either particle may be observed.

Although Figure 9 does not accurately correspond to the case $M_3 \approx M_4$, it may be useful as a reference for the discussion that follows. For relatively high Q reactions, as is the case for the reactions here being discussed, conservation of momentum requires that when M_3 is observed at a backward angle the corresponding M_4 will be emitted at a forward angle, the energy of M_4 and the laboratory angle at which it is emitted being both uniquely related or complementary to the energy of M_3 and the angle at which it is observed. Details regarding these and other reaction-kinematic calculations may be found in reference 67. To put it another way, to the angle and subtended solid angle of the detector for M_3 there correspond a unique angle and subtended solid angle for the corresponding M_4. An important point is then that detectors with appropriate angular positions and solid angles can be operated in coincidence for detecting particles M_3 and M_4 without the decrease in detection efficiency that is commonly experienced when coincidence detectors are used on cascading radiations whose directions are only weakly correlated. For particles to be observed in both the forward and backward directions requires, of course, that the targets be thin compared to the ranges of particles M_3 and M_4.

The imposition of a coincidence requirement on the detection of M_3 and M_4 provides a powerful way of discriminating against interfering reactions which may be producing particles M_3 with the same mass and energy but not M_4. The top part of Figure 13 shows the alpha particle spectrum observed at 135° from a 4.0 MeV deuteron bombardment of a thin (between 10 and 200 $\mu g/cm^2$) film of LiF deposited on a 20-30 $\mu g/cm^2$ carbon foil. The bottom part of the spectrum is for the same length irradiation but with a coincidence detector placed at 50.4°; α_0 and α_1 from the $^{19}F(d,\alpha)^{17}O$ reaction have been entirely eliminated in this latter measurement. The coincidence resolving time was 0.2 μs. Beam currents varied from 0.2 to 1.0 μA and irradiation times were from 3 to 10 minutes per sample. Reference 66 reports an application of this technique and the (p,α) reaction to determine the quantities of isotopic ^6Li and ^7Li.

Still another way of using prompt radiation is that of using small diameter beams to scan a sample surface and to obtain thereby a measurement of the spatial distribution of an element or elements

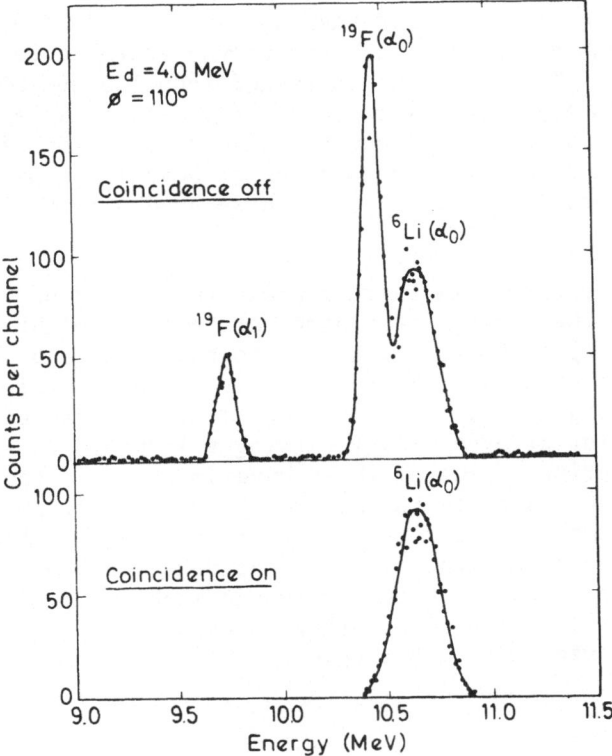

Fig. 13   Alpha particle spectra obtained by deuteron irradiation
of LiF showing the elimination of background and interferences by coincidence measurement of complementary
particles. Reproduced by permission of Radiochemical
Radioanalytical Letters (65).

in the plane of the target. The method is analogous to the electron
microprobe technique. High energy ion beams with diameters under
10 μm were first developed by Cookson and Pilling (49). These
authors also used a 5 μm diameter proton beam together with backscattering to demonstrate how scanning in two dimensions could be
applied. A discussion of the general principles of the technique
and the most recent experiments by Pierce et al. on the use of
this same proton beam together with prompt radiations from nuclear
reactions for analyzing samples may be found in reference 50.

The specific objective in these latter experiments was to
measure the depth distribution of carbon in steel samples by cutting
the sample in a plane perpendicular to the original surface and then
scanning the microbeam along the cut plane in a direction also perpendicular to the original sample surface.

The reaction used for the analysis was the $^{12}C(d,p)^{13}C$ reaction, the deuteron energy was 1.3 MeV, and ground state protons were detected at 135°. The scan lengths ranged from 100 to 400 μm. Simultaneous x-ray measurements were used for locating the edges of the specimen during a scan. The quantitative calibration of the measurement was obtained from the irradiation of several comparison standards under conditions identical to those used with the unknown speciments.

Because the target is thicker along the direction of the incident beam than the range of the beam particles, it might be expected that the proton peak $C^{12}(p_O)$ would be so broad as to be unresolvable. If the cross section for this reaction as a function of energy, or equivalently as a function of distance along the beam path in the target, were flat or even slowly varying this would in fact be true. Another interesting aspect of these experiments therefore was the use of the broad resonance in the $^{12}C(d,p)^{13}C$ reaction at about 1.3 MeV to reduce the effective thickness of the target. The results obtained are shown in Figure 14. At the deuteron bombarding energy of 1.3 MeV, the carbon p_O group, although much broader than the groups obtained from the thin carbon film on quartz, could be resolved and analyzed with the single channel analyzer set as shown.

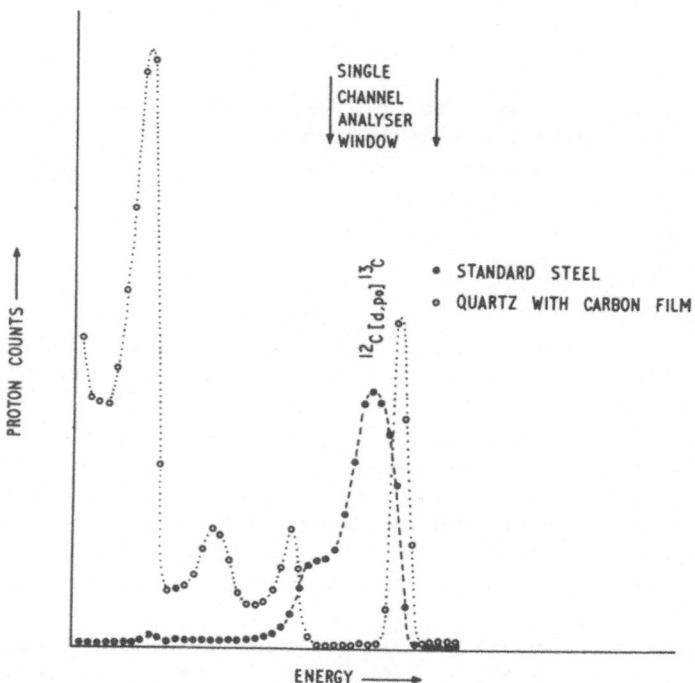

Fig. 14   Proton spectra obtained by deuteron irradiations of thin and thick carbon layers. Reproduced by permission of Nuclear Instruments and Methods (50).

A comparison between the microbeam scan and a mechanical profiling
result for carbon in steel are shown in Figure 15.

The production and use of beams with diameters as small as
4 μm requires special equipment and test instrumentation whose
development will be a major undertaking for any laboratory that
wants to acquire this capability. It should be noted however that
useful microbeam measurements can also be made with beam diameters
of about 50 μm and that such beams can be achieved with just a good
standard magnetic quadrupole focusing system and a limiting beam
aperture near the target. Price and Bird (68) have developed such
a proton beam and have used it together with the $^{18}O(p,\alpha)^{15}N$ reaction to study the distribution of oxygen in zirconium welds and in
titanium samples.

For the zirconium welds, which were made in an atmosphere consisting of 95% argon and 5% normal oxygen, irradiation scans were
performed in a direction perpendicular to the weld direction and
also along the weld on a 10° beveled and polished surface which was
cut through the weld. The spatial resolution of the method was
estimated to be better than 63 μm. The proton bombarding energy
was 870 keV which is just above a resonance at 845 keV in the
$^{18}O(p,\alpha)^{15}N$ reaction and which therefore reduced the effective
target thickness just as was discussed in the preceding example.

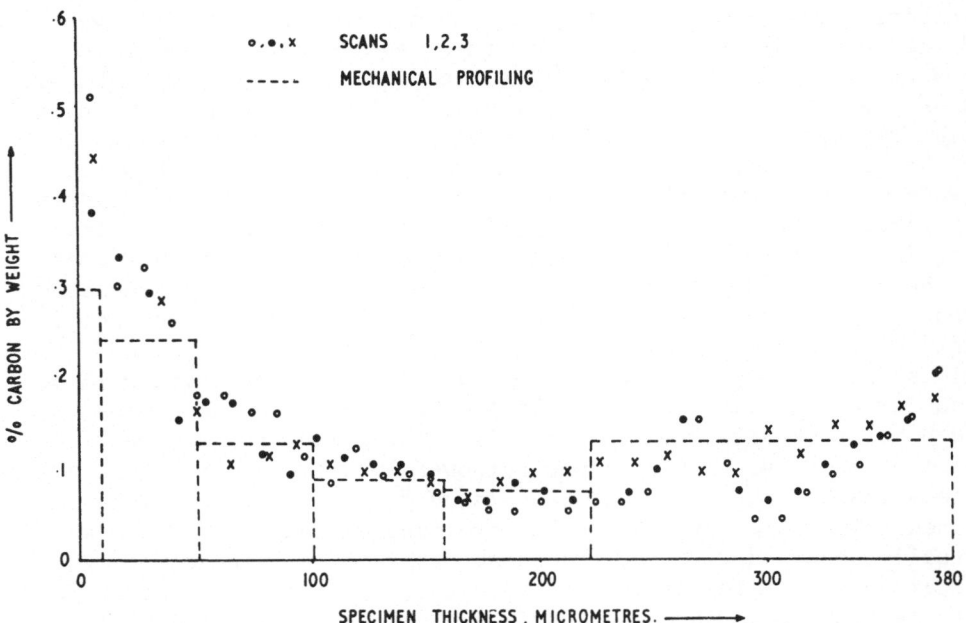

Fig. 15 Carbon profiles obtained for steel specimens by microbeam
scanning and by mechanical profiling. (Reproduced by permission of Nuclear Instruments and Methods (50).)

Beam currents of about 0.1 μA were used. The surface of the weld
was not treated in any way before the start of the scan measurements. The measurements showed that, although there was some interference from the $^{15}N(p,\alpha)\,^{12}C$ reaction, there was enough ^{18}O in the
weld from ordinary oxygen so measurements on the oxygen distribution
could be performed.

The $^{18}O(p,\alpha)\,^{15}N$ reaction has been used by Lightowlers and co-
workers (69) to measure the total bulk oxygen concentration in GaP
layers grown by liquid phase epitaxy from a melt which was doped
with ^{18}O enriched Ga_2O_3. The layers grown were from 40 to 50 μm
in thickness and the ^{18}O was assumed to be homogeneously distributed in the layer.

The principal objective of the study was to measure the solid
solubility of oxygen in zinc- and oxygen-doped GaP. The main purpose of the detection system for the ^{18}O-alpha particles therefore
was to have as large a solid angle as possible consistent with
keeping sufficient resolution so that interfering alphas from boron
and from the ^{18}O fraction of normal oxygen contained in the oxide
layer on the surface of the GaP could be resolved. Of particular
interest to the present discussion is the detection system that was
developed for these investigations. Equation (12) shows that the
energy of the emitted particle, for fixed Q and E_1, decreases as
θ gets larger. A detector which subtends a large solid angle will
therefore receive alpha particles with a spread in energies even
if the target is thin. This effect is usually called kinematic
broadening and is calculable. The technique developed by Lightowlers
et al. was to use an annular detector at 180°, with a 3 mm center
hole, through which the incident beam passed, and to place in front
of this detector a circularly symmetric, graded foil as shown in
Figure 16 which was thinnest at angles closest to 180° and thicker
for smaller angles. The grading of the foil thicknesses was then
set, in six steps, so that the energy lost by the alpha particles
in the foil would just undo the effect of the calculated kinematic
broadening. With this arrangement and a subtended detector solid
angle of 0.42 steradians, which is very large compared to the
values usually employed and for which kinematic broadening will be
also large, the authors were able to obtain an alpha peak whose
width was primarily due to energy loss straggling and not to kinematic broadening. The peak was narrower by a factor of 3 than the
kinematically broadened peak calculated for the solid angle mentioned. The calculated kinematic broadening along with the peak
actually observed with the graded absorber foils in place, the
latter peak labelled as zero solid angle, are shown in Figure 17.
Interferences from boron and surface oxygen are discussed along
with subtraction and evaluation procedures made possible by this
detector arrangement. Proton energies used were generally about
700 keV.

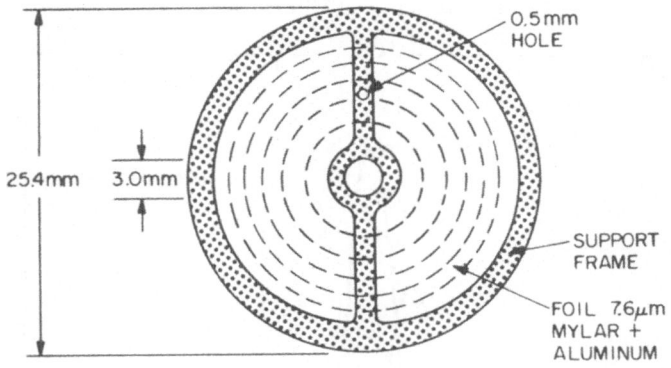

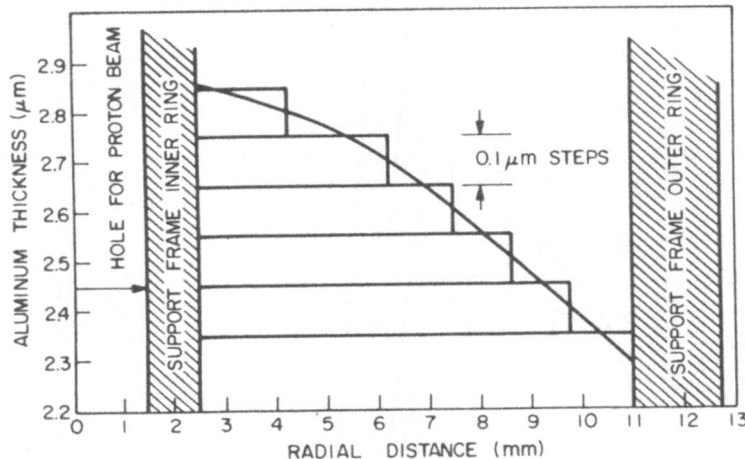

Fig. 16   Graded thickness foil covering an alpha particle detector.
The smooth curve shows the thickness of Al needed, in addition to 7.6 μm of mylar, as a function of radial distance
to make all alpha particles originating from reactions of
685 keV protons with ^{18}O at the surface of the target reach
the detector with a residual energy of 1.3 MeV.   Reproduced
by permission of the Journal of Applied Physics (69).

       Thus far the discussion for nuclear reactions which emit parti-
cles has dealt with examples in which the element to be analyzed had
a uniform distribution, as measured along the incident beam direction.   The next level of complexity for analysis with prompt particles therefore is to consider measurements of the depth distribution
of elements in a sample.

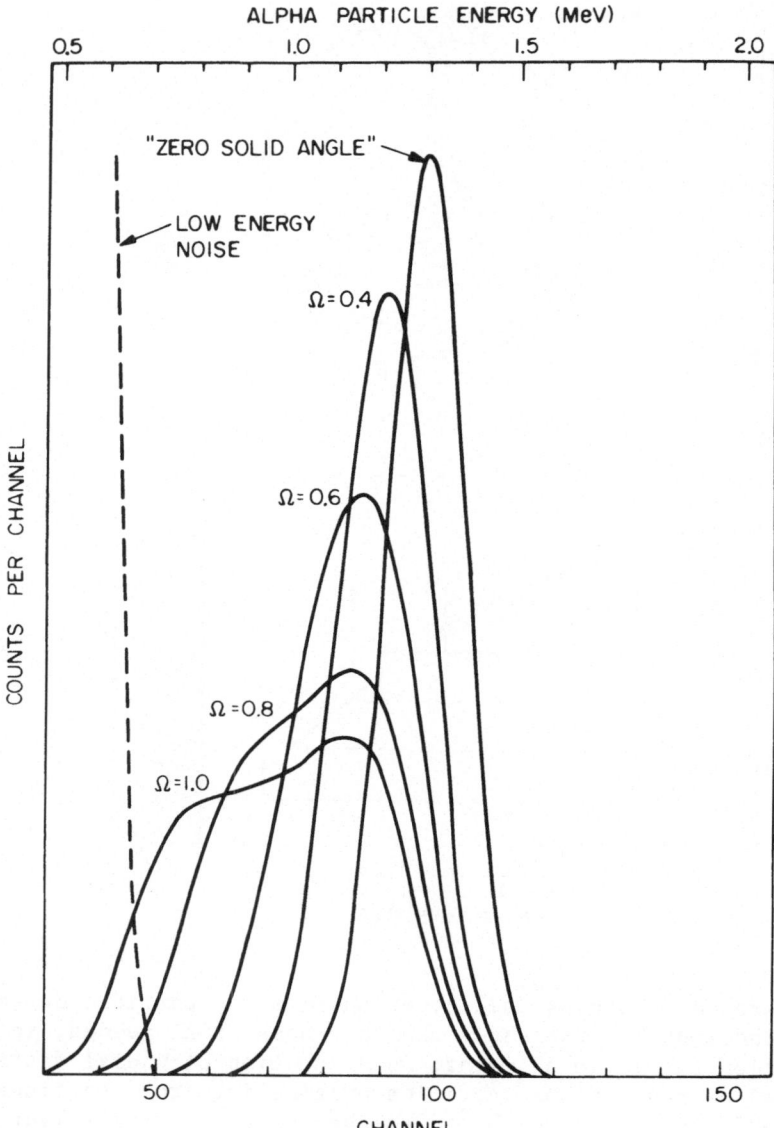

Fig. 17 Calculated kinematic broadening as a function of detector
         solid angle for alpha particles produced by the reactions
         of 685 keV protons with surface ^{18}O. The spectrum labelled
         "zero solid angle" is the experimental spectrum obtained
         with the graded foil and a solid angle of 0.42 sr. Repro-
         duced by permission of Journal of Applied Physics (69).

    As mentioned previously, the energy lost by the incident beam
particles as they penetrate into the target and the energy lost by
emitted charged particles as they emerge from the target both provide information about the depth in the sample at which the nuclear
reaction occurred. The depth resolution which can be achieved in

a given measurement depends on such factors as the stopping power
and energy loss straggling for both the incident beam and the outgoing particle, on kinematic broadening, on detector resolution,
on reaction cross section dependence on energy, and also on other
factors. The reader is referred to Chapter 1 for a detailed discussion of these factors and of the determination of depth resolution. The examples to be described should serve to show however
some of the ways in which nonresonant nuclear reactions can be used
for depth distribution measurements.

The recent experiments of Turos et al. (70) on oxygen depth
profiles provide a particularly good illustration of this method
in that the effects of energy loss straggling, detector resolution,
and experimental geometry are taken into account. In addition, these
authors made a systematic evaluation of the improvement in depth
resolution that could be obtained by tilting the target so the beam
entered it at an oblique angle; the target geometry for these measurements is shown in Figure 18. The importance of a thorough analysis of these effects may be judged from the excellent results obtained. Depth resolutions between 200 and 400 Å were obtained and
the method was found to be applicable to films ranging in thickness
from 2000 Å to 12000 Å.

The $^{16}O(d,\alpha)^{14}N$ reaction was selected for these oxygen measurements for two reasons. One was that this reaction is particularly
simple at low deuteron energy, emitting only a ground state alpha
group. For alpha particles corresponding to the first excited state
of the ^{14}N nucleus, the reaction has a negative Q value of .829 MeV
(and therefore a threshold energy) and will not occur for a deuteron
energy below 933 keV. The second reason was that for low deuteron
bombarding energies the α_0 energy at large angles is low and the
stopping power or energy loss per unit length therefore is relatively
high and provides improved depth resolution.

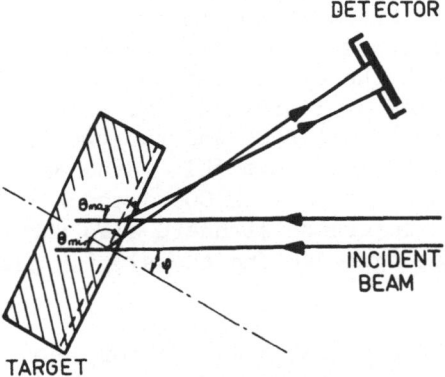

Fig. 18   Geometry of the incident beam and emitted particle tra-
          jectories. (Reproduced by permission of Nuclear Instru-
          ments and Methods (70).)

The studies were performed with 50 to 100 nA beams, 900 keV deuterons, and a solid state detector (25 keV resolution) at 145°; α_0 has an energy of 2.63 MeV under these conditions. In order to avoid interference from the $^{16}O(d,p_0)^{17}O$ reaction the detector depletion depth was not allowed to exceed 26 μm. While the alpha particles were stopped in this thickness and deposited their full energy in the detector, the protons deposited only a portion of their energy and were displaced thereby to lower energies in the particle energy spectrum. The samples used were single crystals of n-type silicon polished by a conventional mechanical and chemical procedure and oxidized at 1200 °C in a wet oxygen atmosphere.

Figure 19 shows the energy spectrum which was observed for a 6000 Å thick SiO$_2$ layer with the beam at normal incidence to the target ($\phi = 0°$) and the detector at 145°. This figure shows clearly the advantage of making measurements with particles having a high stopping power. Thus while the proton groups from the $^{16}O(d,p)^{17}O$ are quite narrow and cannot be used for depth profile measurements, the α_0 group is quite wide and can be so used.

Figure 20 shows the further widening of the α_0 peak that can be achieved if the target plane is tilted through $\phi = 30°$ as shown in Figure 18. For the 30° tilt angle it is interesting to note that a calculation of the alpha particle energy which assumes simply that the stopping powers for both the beam and the emitted alpha particles are constants, gives the same value as an exact calculation for depths in the target up to 0.8 μm. The target film in these irradiations was 4000 Å thick. The solid curves in this figure are theoretical calculations which have incorporated the reaction cross section variation with energy, the variation of the alpha particle's stopping power with energy, the detector resolution, and the energy spread in the alpha energies due to kinematic broadening, to different path lengths traversed in the target as a result of the geometry of the experiment, and to energy loss straggling. The energy spread function was assumed to be Gaussian and to be equal to the root of the sum of the squares of the energy spreads contributed from the sources listed above. Although the concentration profile in these calculations has to be obtained by iteration, the excellence of the fit obtained for a uniform 4000 Å SiO$_2$ and the two different tilt angles, shows that the method is very powerful. A comparison between alpha particle spectra obtained from a 12,000 Å thick SiO$_2$ layer and from a thick melted-quartz reference standard also showed excellent agreement along the high energy edge and the sloping plateau of the alpha peaks.

Somewhat surprisingly, calculations for depth resolution as a function of the tilt angle ϕ showed that, with the detector at 145°, the optimum tilt angle in this experiment was between 40 and 45°. The main factor which keeps the resolution from improving still further as the beam is brought into the target at an ever greater

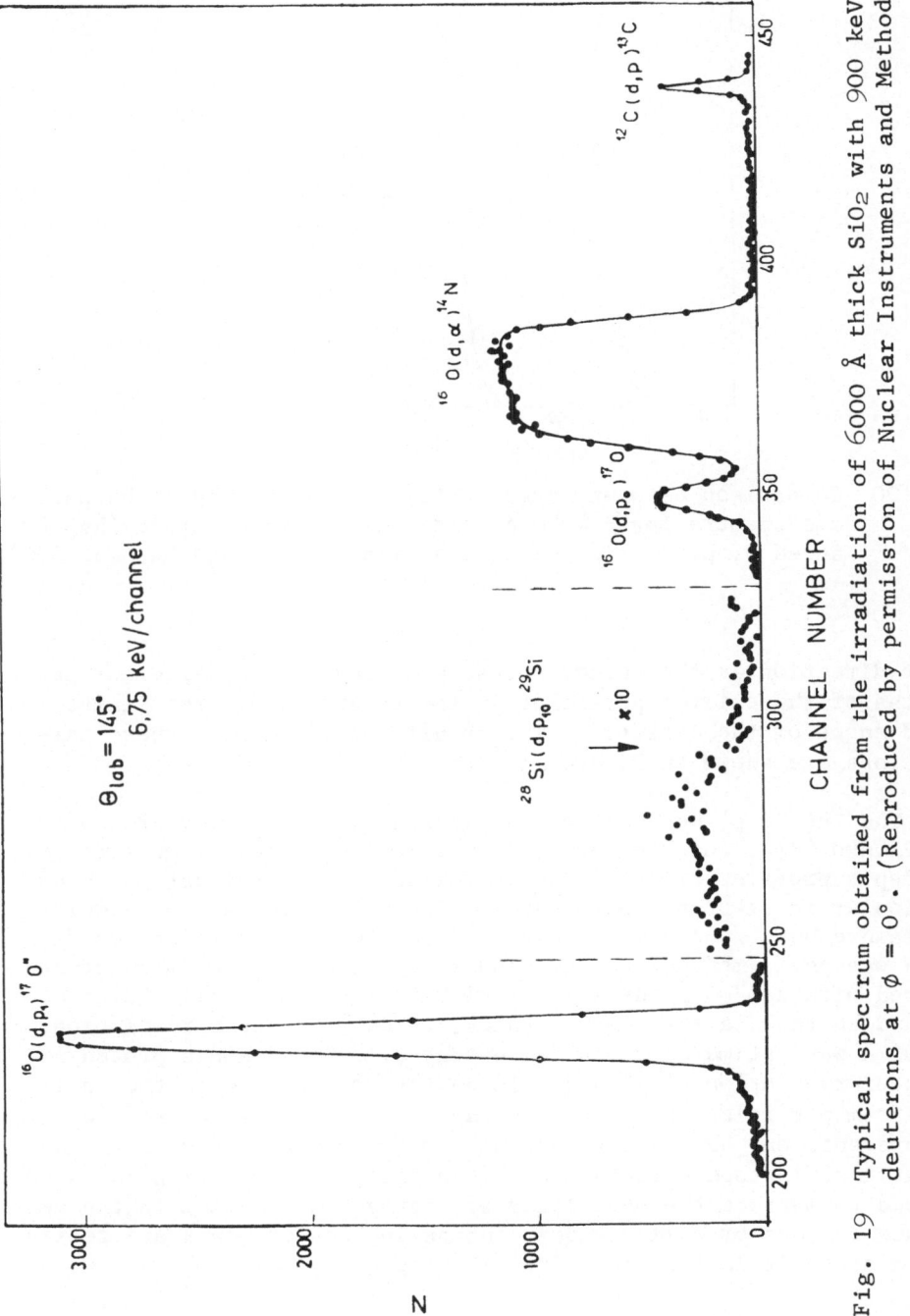

Fig. 19 Typical spectrum obtained from the irradiation of 6000 Å thick SiO₂ with 900 keV deuterons at $\phi = 0°$. (Reproduced by permission of Nuclear Instruments and Methods (70).)

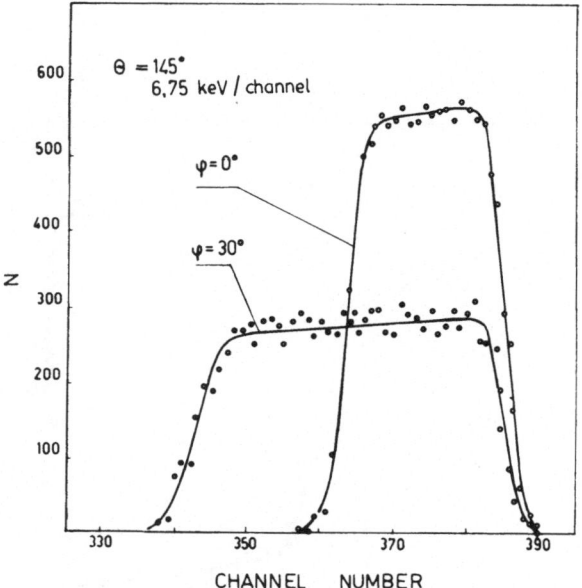

Fig. 20   Comparison between experimental and calculated alpha parti-
          cle spectra for a 4000 Å thick SiO₂ surface layer. (Repro-
          duced by permission of Nuclear Instruments and Methods (70).)

slant direction is the energy spread produced by the different path
lengths of the emitted particles in the target due to the finite
solid angle of the detector. The results obtained from these calculations are shown in Figure 21.

The $^2H(^3He,p)^4He$ reaction has been used recently by Pronko
and Pronko (71), and Langley, Picraux, and Vook (72) to measure
the depth profiles of deuterium in solids. This application is of
particular interest because hydrogen depth profiles are difficult
to measure by nonnuclear techniques. In the work of reference 72,
which was performed on samples consisting of a molybdenum substrate
covered with a 1400 Å thick layer of ErD₂, a 1600 Å thick layer of
Cr, and another layer, ~1000 Å thick, of ErD₂, a depth resolution
of 500 Å was estimated for a ^3He energy of 800 keV and a proton
detector resolution of 10 KeV. If deuterons are used as the incident beam particles this reaction can be used for depth profile
measurements on ^3He. A further point worth noting is that, to the
extent that isotope effects are unimportant, this reaction can also
be used to measure the mobilities of hydrogen and helium in the near
surface regions of solids and, if channeling measurements are feasible, also their lattice locations in single crystals.

An interesting suggestion for obtaining depth profiles has
been made by Palmer (73) in connection with use of the $^{18}O(p,\alpha)^{15}N$

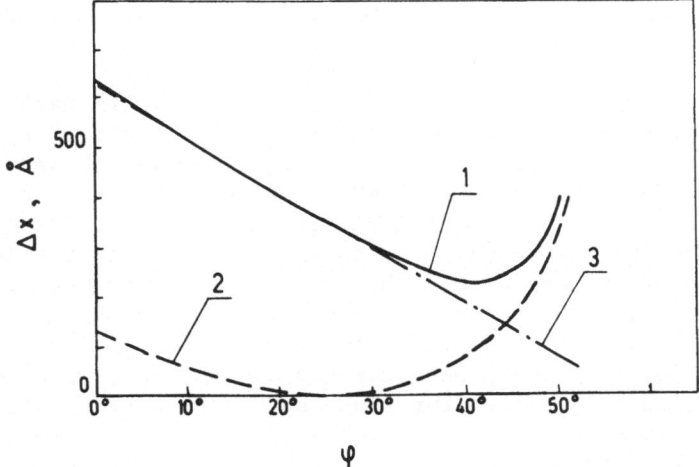

Fig. 21 Variation of depth resolution as a function of tilting angle
         φ calculated for a 2000 Å thick SiO₂ layer. Curve 1 is the
         full calculation; curve 2 is the contribution of geometri-
         cal factors and curve 3 is the contribution of detector
         resolution and straggling. (Reproduced by permission of
         Nuclear Instruments and Methods (70).)

reaction for studying oxygen diffusion in quartz. He stated that
if an alpha particle spectrum from an unknown sample is compared
channel by channel with an alpha particle spectrum obtained at the
same bombarding energy from a standard in which oxygen is uniformly
distributed and which has a stopping power the same as that of the
unknown sample, then the ratio (properly normalized and with background subtracted) will give the unknown profile directly.

    When neutrons are emitted from a nuclear reaction the energies
are given by equation 12, as for charged particles, and the neutron
energy spectrum may therefore also be used for analyzing materials.
Neutron emitting reactions for such application have received less
attention however than other reactions because neutrons are more
difficult to detect with reasonable energy resolution.

    Neutrons from (d,n) reactions were first used for analysis of
carbon, nitrogen, and oxygen, in steel by Möller and co-workers (74).
Prompt neutrons have also been used for analysis by Peisach (75).
In reference 74, the samples were irradiated by 4 ns long pulses of
3 MeV deuterons and the energy spectrum of the neutrons was measured
by their time of flight over the 382 cm distance to the detector.
The method was sensitive to about 0.1 $\mu g/cm^2$ and the depth resolution was estimated to be about 4500 Å at the surface of the sample.
Irradiations could be performed in as little as 5 minutes. The

method was recommended for depth profiling but not for the analysis
of bulk impurities.

Similar measurements with the $^{12}C(d,n)^{13}N$ reaction have been
reported more recently by Lorenzen et al. (76). Deuteron energies
used were in the range from 2.5 to 5.3 MeV, pulse length was 3 ns
and the repetition frequency was 1 MHz. To keep the heating in the
samples low, only currents up to 1 μA were used. The emitted neutrons were measured at 0° and 20° with the time of flight technique.
The depth resolution was found to be better for lower deuteron
energies. For 2.5 MeV deuterons it was estimated to be 1 μm at
the surface and .4 μm at a depth in the sample of 10 μm. The relatively large depth for which measurements are feasible with this
method is particularly noteworthy since such depths are usually not
practical when emitted charged particles are detected. Beyond 10 μm
the energy loss straggling of the deuterons was the limiting factor
on depth resolution.

An interesting additional feature of the measurements was the
simplified way in which the data were used to obtain depth profiles.
Recognizing that uncertainties in the reaction cross section $\sigma(E,\theta)$,
which was known only to about 20%, would produce unsatisfactory results if an equation similar to equation (13) were used to obtain
the depth profile $N(x)$, these authors used instead a ratio of the
neutron energy spectra, channel by channel, from the unknown sample
and from a standard in which carbon was homogeneously contained. As
was mentioned previously (73), for a standard in which the stopping
power of the deuteron beam varies with energy in the same way as in
the unknown sample, the ratio of the two spectra, after background
subtraction and normalization, will automatically produce the concentration profile. The results obtained in this way are shown in
Figure 22. The spectrum obtained from the unknown iron sample is
shown, after subtraction of a neutron background, as the solid line
in part a. The results similarly obtained for a standard sample of
iron containing 0.85% homogeneous carbon are shown in part b. The
ratio, showing a 4 μm thick carbon concentration gradient at the
surface of the unknown sample is shown in part c. For 2.5 MeV deuterons the detection limit for carbon, defined as the amount of carbon that gives a neutron yield 3 times greater than the statistical
error in the background at the same neutron energy, was estimated to
be 100 ppm at the surface and 600 ppm at a depth of 10 μm.

Fig. 22   Concentration profile of a 4 μm gradient of carbon in iron
          studied with the time-of-flight technique. (Reproduced
          by permission of Journal Radioanalytical Chemistry (76).)

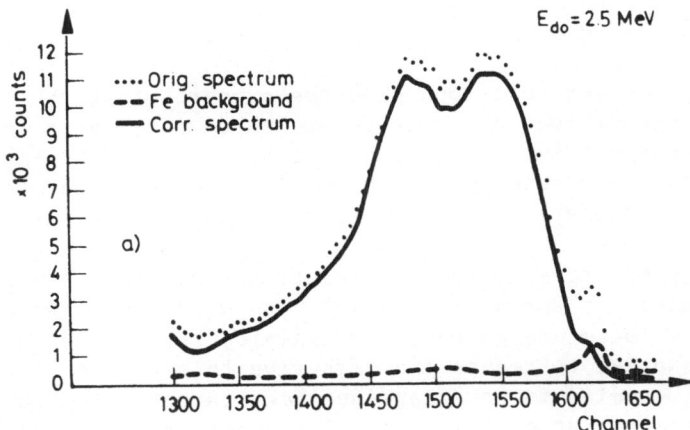

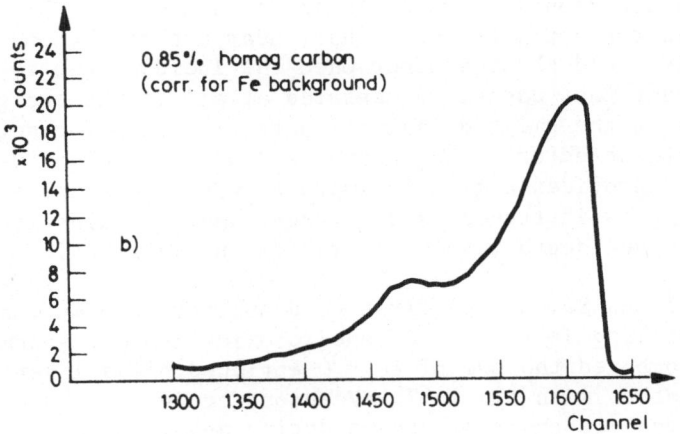

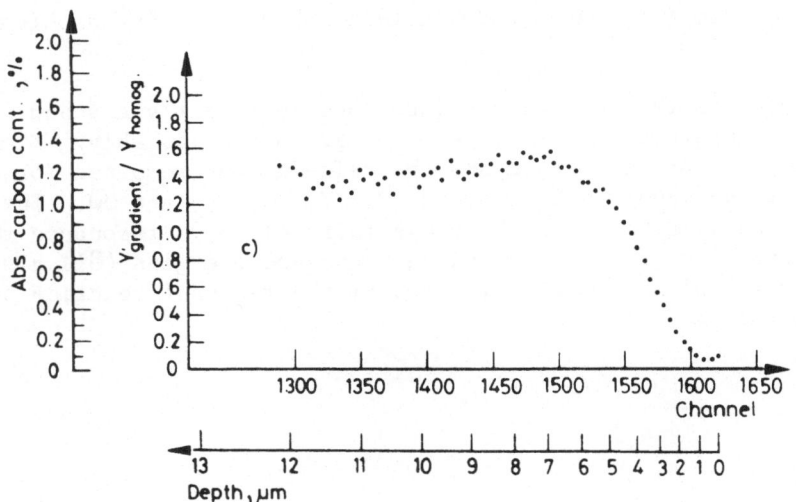

Resonant Nuclear Reactions

Many nuclear reactions have the property that the reaction yield exhibits one or more sharp peaks or "resonances" as a function of bombarding energy. Such a resonance is measured experimentally by varying the incident beam energy in small increments and measuring the quantity of radiation emitted per unit beam fluence at each energy.

The properties of nuclear reaction resonances are useful in several ways for surface analysis. In two experiments already discussed, the beam energy was, for example, set just at a resonance energy in order to reduce the effective thickness of surface that was being studied in an otherwise thick target. The resonance energy, being characteristic of a particular reaction, can also be used in some cases to identify an element in a sample when interferences exist off-resonance from other elements. The resonance technique which is of greatest interest in surface analysis however is that of measuring the depth distribution of an element in a sample. The important point in this latter application is that, near a sharp resonance, easily measurable yield changes occur when the incident energy is changed by an amount that cannot be measured at all if the energy of the beam particles or the emitted reaction particles is being detected with solid state detectors. The narrow width of a nuclear reaction resonance thus provides a greatly improved way of measuring the change in beam energy as it traverses the target and equivalently therefore provides improved depth resolution for depth distribution measurements.

Amsel and Samuel (9) first used nuclear resonance reactions for depth profiling in a study of anodic oxidation mechanisms. These authors combined the use of an ^{18}O enriched oxide layer as a marker layer together with the 1167 keV resonance in the $^{18}O(p,\alpha)^{15}N$ reaction to study the transport of oxygen during anodic oxidation and they used the $^{27}Al(p,\gamma)^{28}Si$ resonance at 992 keV to study the behavior of a thin film of aluminum evaporated onto tantalum which was then anodically oxidized.

Interest in this method has increased in recent years and the technical literature on the subject has grown considerably. Summaries and discussions of the method and the calculations required for data analysis may be found in references 60, 77, 78, 79, and 80. The excitation curves in reference 46 and a table of (p,γ) resonance reactions arranged in order of ascending resonance energies (81) also should prove useful. In simplified form the resonant reaction cross section can be written

$$\sigma(E) = \sigma_0 \frac{\Gamma^2/4}{(E-E_R)^2 + \Gamma^2/4} \tag{14}$$

where E_R is the resonance energy, Γ is the full width of the resonance at half maximum, and σ_0 is the peak cross section at resonance.
The width of the resonance as it is measured depends on the intrinsic
width of the resonance, on the energy homogeneity or energy distribution of the incident beam, on the thickness of the target, and on
the Doppler broadening which is produced by the thermal motion of
the reactant atoms in the target. The intrinsic width of the resonance corresponds to the width of the energy level in the compound
nucleus $(M_1 + M_2)$ which is responsible for the resonance; it needs
to be evaluated only once therefore for a particular resonance. The
Doppler broadening can be calculated to a useful accuracy from the
Debye temperature of a material. At the surface of the target the
beam energy distribution depends on the quality and characteristics
of the accelerator beam analyzing system. As the beam penetrates
into the target its energy distribution becomes broader as a result
of the energy loss straggling which occurs because of the statistical
nature of the energy loss process.

A formula which has been widely used for calculating energy
straggling in a single element material is that derived by Bohr (82)

$$\Delta E_s = 2Z_1 e^2 \left[\pi Z_2 N \Delta x\right]^{1/2} \tag{15}$$

where ΔE_s is the standard deviation of the energy loss straggling
distribution (ΔE_s = FWHM/2.355), e is the electronic charge, Z_1 and
Z_2 are the atomic number of the incident particle and target atoms
respectively, N is the number of target atoms per cm^3, and Δx is
the thickness of target traversed. Unfortunately, one of the assumptions made in the derivation of equation 15, namely that the velocity
of the projectile particle is faster than the velocities of the orbital electrons in the target, breaks down for many cases of practical
interest. The formula provides a simple guideline therefore but its
validity for specific cases of interest needs to be checked. An excellent discussion of the most recent data and calculations have been
given by Chu and Mayer (83). Included in this reference are the results of energy straggling calculations, also by Chu and Mayer (84),
which are based on the Hartree-Fock-Slater atomic-charge distribution.
Some selected results from these calculations are given in Table II
as ratios of ΔE, the Chu-Mayer value, to the Bohr value, ΔE_s, given
by equation 15. Because the energy straggling calculations show a
Z_2 oscillatory structure similar to that observed in the stopping
cross section (85), interpolation between the values given in Table
II is not advisable. The results are included here simply to show
the magnitude of the deviations from ΔE_s which can be expected.

For regions where the Bohr formula is valid and additivity may
be assumed, the formula for the energy spread in a compound (or alloy)
$A_p B_q$, due to straggling, becomes

$$\Delta E_s = 2Z_1 e^2 \left[\pi \Delta x N(AB) \left(p Z_A + q Z_B\right)\right]^{1/2} \tag{16}$$

Table II

Ratios of the energy straggling ΔE, calculated for incident
alpha particles on the basis of a Hartree-Fock-Slater atomic-
charge distribution (84), to the ΔE_s given by equation 15

Target Element	$\Delta E/\Delta E_s$						
	ALPHA PARTICLE ENERGIES IN MEV						
	4.0	3.0	2.0	1.6	1.2	0.8	0.4
Li	1.03	1.02	.994	.973	.942	.892	.807
C	.980	.966	.948	.939	.926	.908	.817
Al	.939	.902	.843	.810	.767	.713	.636
Cr	.881	.859	.808	.771	.721	.649	.534
Ag	.780	.752	.697	.660	.612	.542	.434
Au	.682	.647	.594	.562	.519	.458	.365

where $N(AB)$ is the number of "molecules" per cm^3 and Z_A and Z_B are
the atomic numbers of the elements A and B respectively. When the
Hartree-Fock-Slater atomic model is used, the additivity becomes
more complicated because of the involvement of a numerical integration over Z_A and Z_B (83). The validity of additivity for energy
straggling in compounds remains to be tested experimentally so until
such time as measurements are available, equation 16 should be used
only as a general guideline.

It is of interest to note that, typically, except for regions
very near the target surface, depth resolution is determined by
ΔE (or ΔE_s) and that when this condition obtains, the depth resolution is often simply taken to be $\Delta x = \Delta E_s (dE/dx)^{-1}$.

When all of the contributing factors are known, an experimentally observed resonant yield curve $Y(E)$ can be unfolded to give the
depth distribution $N(x)$ of the reactant atoms in the target, the
depth resolution being best near the surface, where energy loss
straggling effects are small and progressively worse with increasing
depth because of increased straggling. If no simplifying approximations are possible, this being the case when the best possible depth
resolution is required, then the calculations involve triple integrations and the use of a large digital computer program (77).

Figure 23 may be used to show in simplified form how a nuclear
resonance reaction, the $^{27}Al(p,\gamma)^{28}Si$ reaction as an example, can
provide depth distribution information about an element in a sample.
The top part of the figure shows a sample of SiO_2 in which there
are two sharply defined layers of aluminum, one at the very surface

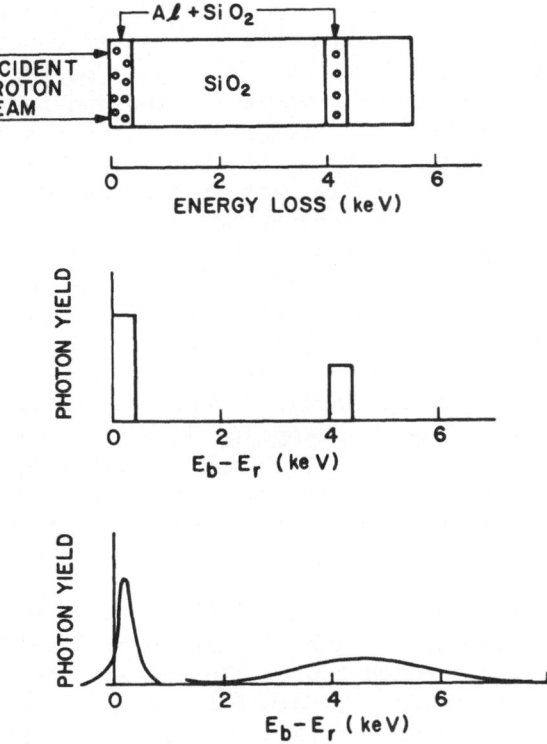

Fig. 23   Schematic diagrams showing a SiO₂ sample containing two
thin layers of Al and the corresponding idealized and
real curves for the $^{27}Al(p,\gamma)^{28}Si$ reaction.

and one below the surface with slightly fewer aluminum atoms in it.
For convenience, distance along the sample is expressed in keV; the
conversion is simply $\Delta x = \Delta E (dE/dx)^{-1}$ where the value for stopping
power or energy loss per unit length, dE/dx, is taken to be that
for protons of energy E_R in SiO₂ (the amount of aluminum is assumed
to be so small as not to affect the stopping power). The middle
part of the figure then shows the results which would be obtained
for the idealized case where the intrinsic width of the resonance,
Doppler broadening, beam energy distribution and energy loss straggling are all taken to be zero. The yield curve corresponding to
the surface layer has a flat top which shows that the aluminum
distribution is uniform throughout this layer. The width of the
yield curve corresponds to the width of the layer and the area under
the curve is proportional to the number of aluminum atoms per cm²
in the layer. For the deeper lying layer, the bombarding energy, at
which the resonant gamma ray yield is observed, has shifted to
higher energies by 4 keV. The flat top of the curve and the width

again correspond to a uniform aluminum distribution and the layer
thickness and the smaller area show that this layer contains fewer
aluminum atoms per cm^2 in it. In the bottom part of the figure
finally are the yield curves which would be produced in the real
case. The curves now reflect the real width of the resonance,
Doppler broadening, inhomogeneity in the beam energy, and beam
energy loss straggling. It is this last factor that has so markedly
broadened the yield curve corresponding to the deeper layer. The
areas under these two curves will still however equal the corresponding areas of the curves in the middle of the figure for equal
beam fluences. Even without a full data analysis which includes all
of the broadening factors mentioned, it is clear that these last
curves already provide at least some information about the depth
distribution of aluminum in the sample. Provided that energy loss
straggling is not unduly large, a complete data analysis, as will
be discussed later, can establish the depth distribution of the element being measured with a resolution which is in many cases superior
to that which can be achieved by either backscattering or nonresonant
prompt charged particle spectrometry.

The experiments of Bernett et al. (86) are a simple example of
the use of resonance reactions which did not involve full calculations
for the broadening effects which occur. In these experiments the object was to measure the amount of α-alumina and magnesia polishing
compounds, having mean grit diameters of 0.3 μm and about 1 μm respectively, which remained in metals of various hardnesses after a standardized polishing and rinsing procedure.

The results obtained for the alumina powder with the 992 keV
resonance in the ^{27}Al$(p,\gamma)^{28}$Si reaction are shown in Figure 24. For
the gamma ray measurements, a 7.62 cm thick by 7.62 cm diameter
NaI(Tl) scintillation counter was used; only gamma rays corresponding
to the transition which leaves ^{28}Si in its 1.78 MeV state were counted
($\sim$11 MeV). The beam energy spread after magnetic analysis was approximately 1 keV. The top part of the figure shows the yield curve obtained for Zr. The resonance was observed at the right energy for
the grit particles to be located right on the surface of the sample
and the width of the yield curve corresponds closely to the 0.3 μm
grit diameter. From the area under the curve, proportional to the
number of Al atoms/cm^2, an Al surface density of 0.15 μg/cm^2 was
calculated. The sensitivity of the technique was estimated to be
about 7 $\times$ 10^{-3} μg/cm^2.

The middle figure shows the results of a similar measurement
for gold. This curve has a greater area and is broader than the
first. These features show, respectively, that more polishing compound has been retained in the sample and that some of the particles
are embedded in the gold. Finally, the bottom curve shows that in
copper there is still more polishing compound retained in the sample
and that the grit particles are embedded still more deeply. In the

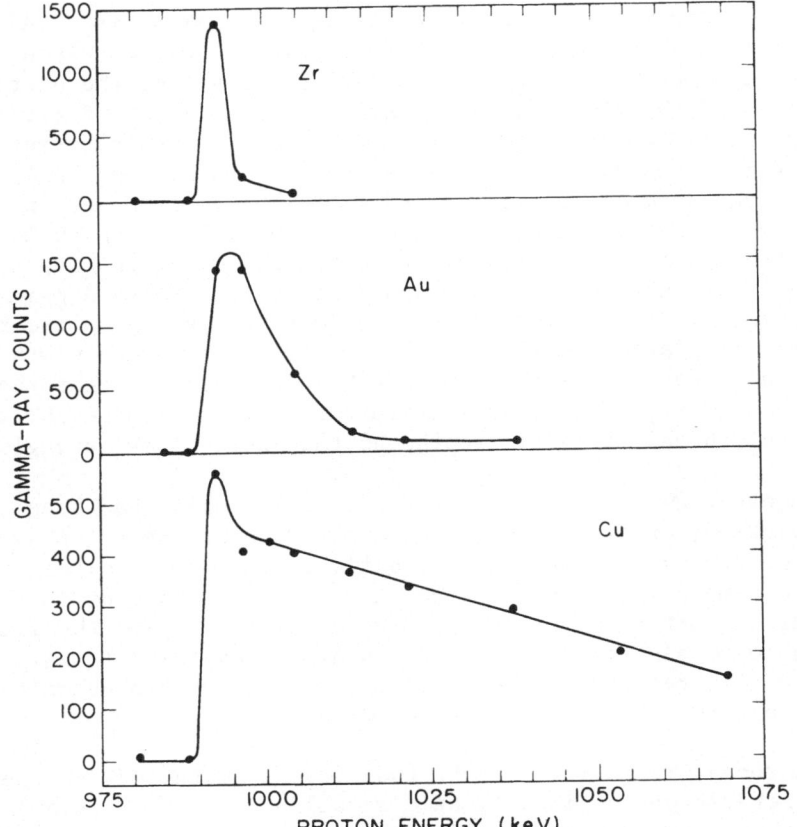

Fig. 24   Yield curves for the $^{27}Al(p,\gamma)^{28}Si$ reaction for Al_2O_3
polishing compound retained in Zr, Au, and Cu surfaces.
(Reproduced by permission of the Journal of Applied
Physics (86).)

latter case the interesting finding was that some of the particles
are buried at least 1 μm deep and are covered over by the copper,
even though the polishing was performed by hand on a relatively
slow-moving polishing usrface. The thickness of copper covering
the Al_2O_3 particles is evidenced by the fact that the reaction yield
is still appreciable at 1080 keV, an energy which is nearly 90 keV
above the resonance energy.

The measurements were performed for the magnesia powder with
the 1.548 MeV resonance in the $^{26}Mg(p,\gamma)^{27}Al$ reaction. Similar
results were obtained but the sensitivity was somewhat lower.

Measurements of fluorine contamination on and below the surface
of Zircaloy were performed by Möller and Starfelt (87) by means of a
resonance at 1375 keV in the $^{19}F(p,\alpha\gamma)^{16}O$ reaction. The fluorine

present in a sample can be observed through the gamma rays, and for
thin samples, also through the alpha particles. In the present case
the gamma rays were used and both the total amount and the depth
distribution of the fluorine were measured for a number of Zircaloy
samples which had undergone different treatments. Characteristic
gamma rays with energies in the range from 6.1 to 7.1 MeV were detected and the beam energy was varied from 1300 to 1500 keV in the
experiments. A set of linear equations was derived which could be
solved, with the aid of a digital computer, to obtain the depth
distribution of the fluorine corresponding to the observed gamma
ray yield curve. The effects of energy loss straggling on the results were calculated and found to be equal to the 11 keV width of the
resonance at a depth of about 500 $\mu g/cm^2$. At smaller depths therefore the depth resolution of the measurement was determined by the
resonance width and at larger depths by the energy loss straggling.

Altogether 29 samples were studied. The samples had been
ground, pickled, or electropolished, and had undergone autoclave
treatments varying in temperature and time. It was found that the
unoxidized samples contain fluorine only on the surface or in a surface layer thinner than 0.1 μm. In the oxidized sample, the fluorine
was found to be distributed down to depths of several microns. The
yield from this reaction is sufficiently high so that a quantity of
fluorine less than 0.01 $\mu g/cm^2$ can be detected.

More recently, Maurel et al. (88) have used the 872 keV narrow
(Γ = 4.2 keV) resonance in this same reaction, namely $^{19}F(p,\alpha\gamma)^{16}0$,
to study the fluorine contamination of tantalum by various polishing
procedures and its behavior during subsequent anodic oxidation. Samples were subjected to various treatments to reduce the amount of
fluorine present in the samples before anodic oxidation and the location and the effects of the fluorine on the growth and characteristics of the oxide layer were investigated. The measurements showed
that the ^{19}F is contained, after anodic oxidation producing an oxide
layer about 1400 Å thick, in a layer less than 50 Å thick near the
oxide-metal interface. The influence of ^{19}F on the oxide growth laws
was found to be only slight.

The $^{19}F(p,\alpha\gamma)^{16}0$ reaction has been used by Mandler et. al. (89)
and Porte et. al. (90) to study the depth distribution of fluorine in
tooth enamel after various treatments. Mandler and co-workers used
the 6 keV wide 672 keV resonance and varied the proton energy from
672 to just below 835 keV, the energy for the next resonance. This
range permitted the examination of tooth enamel down to a depth of
2.4 μm. The resolution of the method was estimated to be 0.07 μm
near the surface. At a depth of 2 μm, the depth resolution was estimated to be about 0.15 μm, this value being due to energy loss straggling.

For these measurements the proton beam was collimated to a diameter of 5 mm and kept to a value not exceeding 50 nA. Because teeth

are excellent insulators, charging was a serious problem. It was
alleviated by coating the teeth with a 200 Å layer of gold. A Ge(Li)
detector with 1.8 keV resolution at 1.33 MeV was used to observe
6.13 MeV gamma rays. Even with small beams and relatively low detector counting efficiency, times of 20 minutes provided a detection
sensitivity for fluorine of about 1000 ppm. A practical lower limit
for detection was estimated to be less than 100 ppm. Absolute fluorine concentrations were obtained by comparing measured gamma ray
intensities with calibration runs made on either teflon $(CF_2)_x$ or
CaF_2. The results obtained are shown in Figure 25.

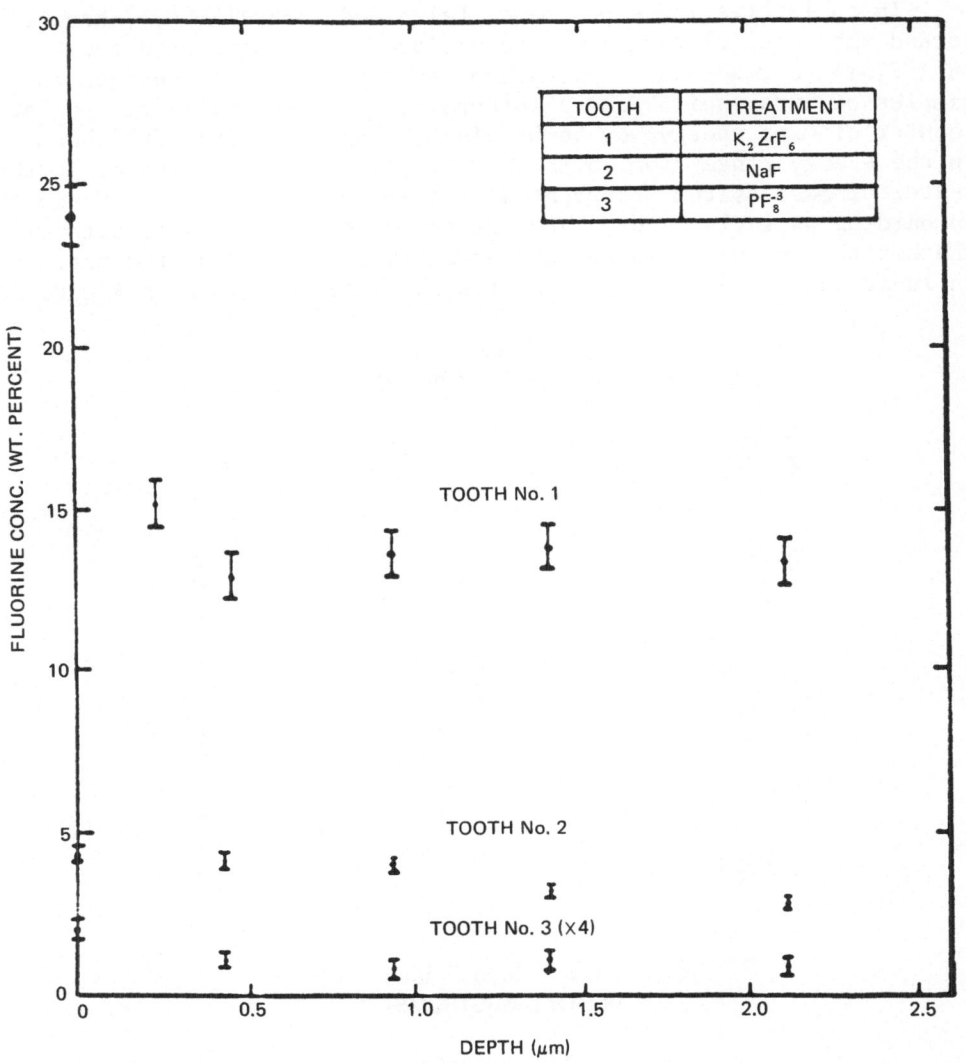

Fig. 25   Measured fluorine concentration profiles for three treated
          teeth. (Reproduced by permission of Thin Solid Films (89).)

Golicheff and Engelmann (91) have used the 4.5 keV wide 837 keV resonance in the $^{19}F(p,\alpha\gamma)^{16}O$ reaction to study fluorine content and depth profiles in metallic samples. Sensitivities better than 5×10^{-4} $\mu g/cm^2$ and depth resolutions of 1000 Å in aluminum were estimated; measurements were made to depths of about 0.6 μm. Data analysis methods for the depth distributions are presented which include energy loss straggling effects. A related paper (92) provides a useful tabulation of reaction resonance energies, peak cross sections, widths, and the energies of emitted gamma rays for proton induced reactions on Li, Be, B, C, N, O and F.

In an interesting experiment, Leich and Tombrello (93) have reversed the roles of target and beam elements and have used the $^{1}H(^{19}F,\alpha\gamma)^{16}O$ reaction to study the depth profiles of hydrogen in samples of lunar soil. These authors used the resonance at 830 keV (center of mass energy) by bombarding the samples with $^{19}F^{4+}$ beams in the energy range from 16 to 18 MeV. The 830 keV resonance in the center of mass system occurs at an energy of 16.6 MeV when ^{19}F is the bombarding particle. The object of the experiments was to determine whether the hydrogen gas concentrations were located at the surface in lunar soils. Some of the results obtained are shown in Figure 26.

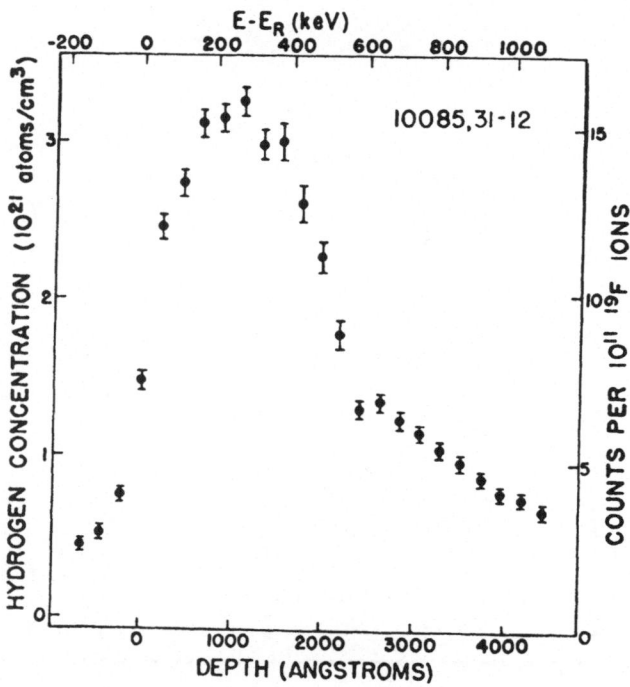

Fig. 26   Hydrogen concentration profile in lunar soil sample 10085, 31-12 as measured with the $^{1}H(^{19}F,\alpha\gamma)^{16}O$ reaction. (Reproduced by permission of Nuclear Instruments and Methods (93).)

The depth resolution of the measurement was estimated to be 50 Å and the depth which could be studied 3 μm. Because of the increased stopping power of the beam and the changed reaction kinematics, these values are a considerable improvement over those that can be achieved with this reaction when fluorine is bombarded with protons.

Although this particular reaction is not within the range of small accelerators as the term is being used here, it is worth mentioning because the same idea can be applied to other reactions which can be reached with small accelerators. The resonance energy E_R* with the heavy particle M used as the bombarding particle, in terms of the laboratory resonance energy E_R observed when the lighter particle m is the bombarding particle, is simply

$$E_R* = E_R M/m \qquad\qquad (17)$$

Thus there are resonances for (p,γ) reactions with ^7Li, ^{11}B, and ^{14}N that lie well within the range of small accelerators and also have the narrow widths required for depth profiling. Boron (from BF_3 gas) and nitrogen beams can in addition be produced easily by a radio frequency discharge ion source; lithium is more troublesome but can be made.

The $^{18}O(p,\alpha)^{15}N$ resonance reaction has been used by a number of investigators for studying ^{18}O profiles and transport mechanisms. The studies of Amsel and co-workers (9,60) have already been mentioned. Other references of interest include Whitton et al. (94), Calvert et. al. (95), and Neild et. al. (78). Whitton and co-workers used the 633 keV resonance to study the depth distribution of ^{18}O implanted into GaP at energies of 20, 40 and 60 keV to local concentrations as high as 3×10^{21} atoms/cm^3. Of particular interest in this reference is the detailed description of the mathematical convolution methods that are used to obtain the depth distributions of ^{18}O from the measured yield curves. From the depth distribution histograms which are presented, depth resolution may be estimated to be about 200 Å; the profiles calculated extended out to about 2500 Å.

Calvert and co-workers (95) used the 1.766 MeV resonance to study the self diffusion of oxygen in oxides. Samples were prepared by annealing single crystals of rutile (TiO_2) in ^{18}O at 1152°C, for maximum times extending up to 178 hours. Measurements were made to depths of up to 4 μm and the depth resolution was estimated to be 0.1 μm. The depth profiles obtained agreed with the diffusion law.

Neild et al. (78) developed a convolution procedure for determining depth profiles which uses the experimentally measured yield curve in tabular form, and takes the initial beam energy spread and energy loss straggling into account. Straggling is assumed to be given by the Bohr formula (equation 15).

The method of calculation was tested against samples prepared
in three different ways. In the first sample a buried step function
distribution of ^{18}O was obtained by oxidizing a chromium sample first
in $^{18}O_2$ at 850° for 3 hours and then in $^{16}O_2$ at the same temperature
for a further period. In the second an error function distribution
was obtained by diffusing ^{18}O from the gas phase into a small single
crystal of TiO_2 for 5 days at 1000 °C. In the third sample a thin-
film-diffusion function was obtained by depositing a thin layer of
$Cr_2{}^{18}O_3$ onto a small single crystal of Cr_2O_3 and diffusing this layer
for 20 hours at 1247 °C.

The experimental measurements were made with the 1.766 MeV resonance of the $^{18}O(p,\alpha)^{15}N$ reaction. Figure 27 shows a comparison between the experimental measurements obtained for the step function
sample and a yield curve calculated on the basis of a step function
distribution. For the edge of the ^{18}O distribution closest to the
surface (about 0.21 μm below the surface) a depth resolution of about
.01 - .02 μm was obtained. For the deeper edge (about 1.43 μm below
the surface) the depth resolution was about .02 - .04 μm. It is important to point out that the excellence of the fit in Figure 27 was
made possible because the calculations took into account all of the
broadening effects which have been mentioned.

Dunning et al. (37) have shown that near the surface of a sample, where broadening due to straggling is still small, the depth

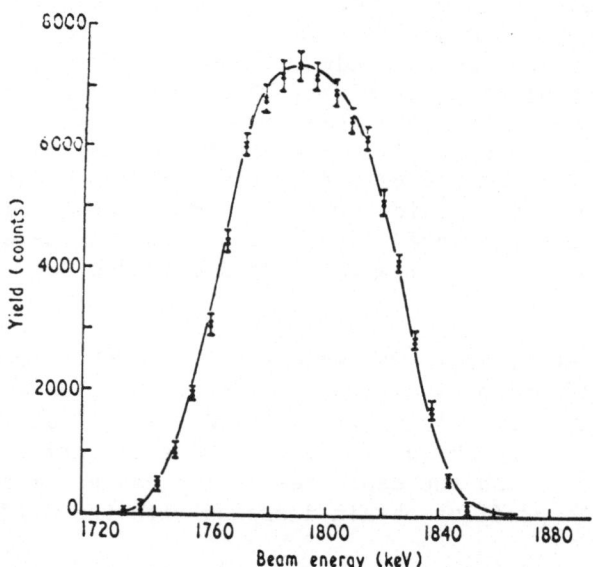

Fig. 27   Best fit calculated for a step function oxygen depth pro-
          file. (Reproduced by permission of the Journal of Physics
          D: Applied Physics (78).)

resolutions which can be obtained with a narrow resonance, a small
beam energy spread, and with accurate calculations which take into
account all broadening effects, can be as small as 20 Å (37). In
that investigation the narrow 992 keV resonance in the $^{27}Al(p,\gamma)^{28}Si$
reaction (the intrinsic width of this resonance is about 100 eV
(96)) was used to study the depth profile, after annealing, of Al
implanted into SiC.

The formula used to calculate the observed gamma ray yield
$Y(E_p,t)$ was

$$Y(E_p,t) = n\varepsilon \int_0^t \int_{E_i} \int_E N(x)g(E_p,E_i)f(E_i,E,x) \; \sigma(E,E_R)dEdE_idx \qquad (18)$$

where $Y(E_p,t)$ is the observed gamma ray yield at an incident energy
E_p from a target of thickness t, n is the number of incident beam
particles as before, ε is a factor which takes into account the
absolute detector efficiency, $N(x)$ is the concentration of the reactant atoms at depth x below the surface, $g(E_p,E_i)dE_i$ is the probability for an incident proton to have its energy in an interval E_i to
$E_i + dE_i$, $f(E_i, E,x)dE$ is the probability for an incident proton with
energy E_i to have its energy in the interval E to E + dE at depth x
in the target, and $\sigma(E,E_R)$ is the cross section at energy E. The
integration of this formula requires a large digital computer. A
description of the fortran programs required to perform the integration may be found in reference 77.

For proton energy losses greater than about 12 keV the beam
energy distribution due to straggling, contained in $f(E_i,E,x)$, is
Gaussian, with a variance equal to ΔE_s^2, where ΔE_s is given by equation 15. For energy losses smaller than this amount the beam energy
distribution due to straggling is asymmetric and is given by the
theory of Vavilov for the transport and energy loss of heavy charged
particles in thin targets (97). A tabulation of the Vavilov distribution may be found in reference 98. A graphic illustration of
how the beam energy distribution due to straggling changes from the
Vavilov asymmetric distribution for small energy losses to a symmetric distribution for larger energy losses is shown in Figure 28.
The parameter $\bar{\Delta}$ in the figure is proportional to the thickness of the
target. For depth profile calculations near the surface it is essential that the Vavilov energy loss straggling distributions be used.

The results obtained by Dunning and co-workers (37) for the profile of Al implanted at 60 keV energy into SiC and then annealed for
15 minutes at 1400 °C are shown in Figure 29. The open circles are
the experimentally measured yield curve, the open triangles are the
yield curve calculated on the basis of equation (18) and the solid
line is the Al depth profile that was required to produce the

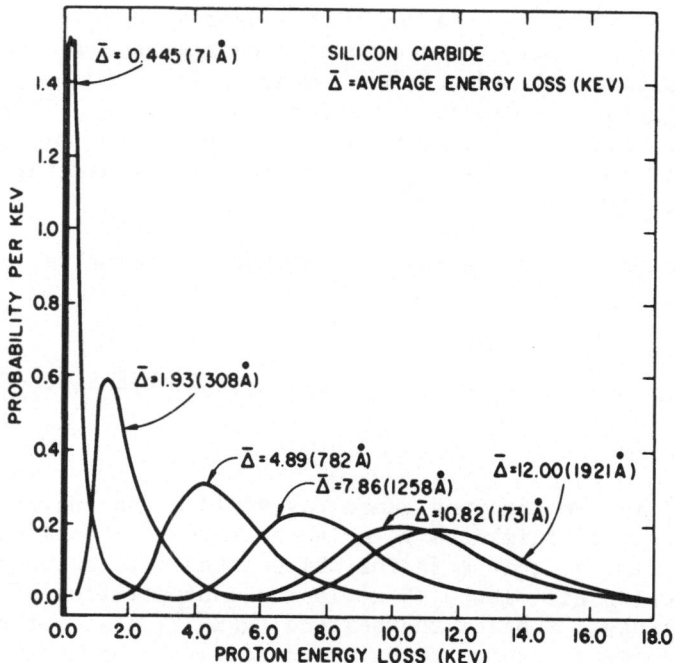

Fig. 28   Energy loss distributions for protons in SiC. The distribution corresponding to the 71 Å layer has been truncated.
(Reproduced by permission of Thin Solid Films (37).)

calculated, open triangles, curve. It is clear that the annealing
has caused the implanted Al to migrate to the SiC surface; the width
of this Al layer is less than 20 Å. The excellence of the fit between the experimental and calculated curves was made possible because all broadening factors were taken correctly into account and
because, for small energy losses the Vavilov distribution for energy
loss straggling was used. It appears likely that once certain integrals are calculated on a large computer and tabulated, equation 18
can be calculated on intermediate size computers such as might be
used on-line with small accelerators. Work in this direction is in
progress at the time of this writing (99).

                                    SUMMARY

     The use of nuclear reactions for surface analysis, clearly a
fertile area for new developments during approximately the last 10
years, continues to be a promising area for significant scientific
development and innovation. The physical quantities which may be
expected to drive future development are detection sensitivity,
specificity, accuracy, and, for prompt radiation analysis, spatial

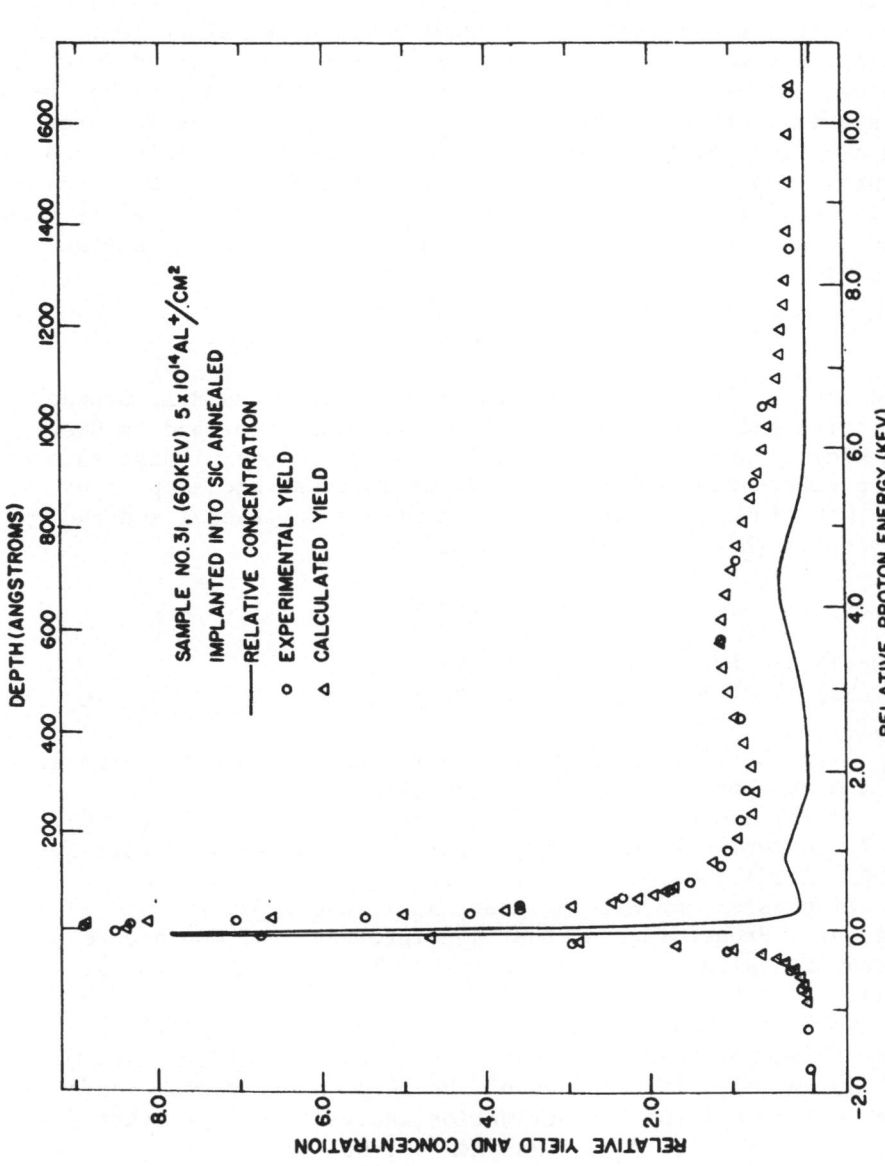

Fig. 29 Profile, and experimental and calculated yields for 60 keV Al implanted into SiC to a fluence of 5×10^{14} atoms/cm^2 and then annealed for 15 min. at 1400 °C. (Reproduced by permission of Thin Solid Films (37).)

and depth resolution. The following few examples may serve to illustrate some of the specific areas that are judged to be interesting
for future research. Thus in charged particle activation analysis,
emphasis on improving quantitative accuracy and multielement analysis
capability should prove useful. In prompt radiation analysis, energy
loss and energy straggling should receive increased attention. In
particular, energy straggling corrections should be applied to both
the incident and emitted charged particles in order to provide improvements in depth resolution. The use of magnetic spectrometers
to obtain better energy resolution for the emitted particles also
will improve depth resolution. Finally, the use of nuclear techniques for studying techniques which require surface removal is judged
to be an area that can impact significantly on the entire surface
analysis field.

ACKNOWLEDGEMENT

The author wishes to thank Drs. S.T. Picraux and B.L. Crowder
for editorial and technical reviews of the manuscript and to Drs.
K.L. Dunning, J.K. Hirvonen, and J.W. Butler for useful discussions.
It is a pleasure also to acknowledge the interest and support of
Dr. J.M. McElhinney for ion beam applications in general and the
preparation of this chapter in particular.

REFERENCES

1. E. Odelblad, Acta. Radiol. 45, 396 (1956).
2. S. Rubin, T.O. Passel, and L.E. Bailey, Anal. Chem. 29, 736
    (1957).
3. R.A. Gill, U.K. Atomic Energy Research Establishment, Harwell,
    England, Report No. AERE C/R 2758 (1958).
4. R.F. Lippel and E.D. Glover, Nuc. Inst. & Meth. 9, 37 (1960).
5. J.A.B. Heslop, U.K. Admiralty Materials Laboratory, Report No.
    A/58(M) April 1961.
6. J.W. Winchester and M.L. Bottons, Anal. Chem. 33, 472 (1961).
7. P. Albert, Proceedings of the 1961 International Conference on
    Modern Trends in Activation Analysis, Texas A&M University,
    College Station, Texas, Dec. 15 and 16, 1961.
8. S.S. Markowitz and J.O. Mahony, Anal. Chem. 34, 329 (1962).
9. G. Amsel and D. Samuel, J. Phys. Chem. Solids, 23, 1707 (1962).
10. J.P. Guinn, ed., Proceedings of the 1965 International Confer-
    ence on Modern Trends in Activation Analysis College Station,
    Texas, Apr. 19-22, 1965, Texas A&M U., College Station, Texas,
    1965.
11. Proceedings of the First Conference on Practical Aspects of
    Activation Analysis with Charged Particles, Grenoble, France,
    June 23, 1965, Euratom, EUR 2957 d-f-e.
12. R.S. Tilbury, "Activation Analysis with Charged Particles," Nat.
    Acad. Sci., Nat. Res. Council Nuclear Science Series, NAS-NS 3110,

Clearinghouse for Federal Scientific and Technical Information, Springfield, Va., 1966.

13. H.G. Ebert, ed., Proceedings of the 2nd Conference on Practical Aspects of Activation Analysis with Charged Particles, Liege, Belgium, Sept. 21-22, 1967, Euratom, EUR 3896 d-f-e.

14. J.R. DeVoe, ed., Modern Trends in Activation Analysis, NBS Special Publication 312, Vols. 1 and 2, Government Printing Office, Wash., D.C. 1969.

15. G. Deconninck, G. Demortier, and F. Bodart, eds., Chemical Analysis by Charged Particle Bombardment, Proceedings of the International Meeting, Namur, Belgium, 6-8 Sept. 1971, Elsevier Sequoia S.A., Lausanne (1972).

16. Proceedings of the International Conference on Modern Trends in Activation Analysis, C.E.N. Saclay, France 2-6 Oct. 1972, J. Radioanal. Chem. 16, #2 (1973).

17. F.F. Dyer, L.C. Bate, and J.E. Strain, Anal. Chem. 39, 1907 (1967).

18. R.H. Marsh and W. Allie, Jr., Ref. 14, pg. 1285.

19. E.E. Wicker, "Activation Analysis," Chapter 3 in Determination of Gaseous Elements in Metals, L.M. Melnick, L.C. Lewis, and B.D. Holt, eds., Wiley Interscience, New York 1974.

20. P. Kruger, Principles of Activation Analysis, John Wiley & Sons, Inc., 1971, pg. 308.

21. J.F. Cosgrove, Ref. 14, pg. 457.

22. E. Ricci and R.L. Hahn, Anal. Chem., 37, 742 (1965).

23. J.W. Butler and E.A. Wolicki, "Surface Analysis of Gold and Platinum Disks by Activation Methods and by Prompt Radiation from Nuclear Reactions," in Ref. 14, Vol. II, pg. 791.

24. C. Williamson, J. Borojot, and J. Picard, Tables of Charged Particle Energy Losses, CEA-R3042, 1966.

25. E. Ricci and R.L. Hahn, Anal. Chem., 39, 794 (1967).

26. C.E. Crouthamel, F. Adams, and R. Dams, Applied Gamma-Ray Spectrometry, Pergamon Press, New York (1970).

27. G.J. Lutz, R.J. Boreni, R.S. Maddock, and W.W. Meinke, eds., Activation Analysis, A Bibliography, Tech. Note 467, Pts. 1 and 2 National Bureau of Standards, Wash., D.C., 1971.

28. J.C. Ritter, M.N. Robinson, B.J. Faraday and J.I. Hoover, J. Phys. Chem. Solids, 26, 721 (1965).

29. C.A. Carosella and J. Comas, Surface Sci., 15, 303 (1969).

30. R.W. Benjamin, L.D. England, K.R. Blake, I.L. Morgan, and C.D. Houston, Trans. Amer. Nucl. Soc., 9, 104 (1966).

31. D.M. Holm, J.A. Basmajian, and W.M. Sanders, "Observations of the Microscopic Distribution of Oxygen and Carbon in Metals by He^3 Activation," Rept. LA-3515, Los Alamos Scientific Laboratory, 1966.

32. J.B. Holt and L. Himmel, J. Electrochem. Soc. 116, 1569 (1969).

33. P.E. Wilkniss and H.J. Born, Int. J. Appl. Rad. Isot. 18, 57 (1967).

34. J.-N. Barrandon and Ph. Albert, "Determination of Oxygen Present

at the Surface of Metals by Irradiation with 2 MeV Tritons,"
in Ref. 14, Vol. II, pg. 794.

35. A.R. Knudson and K.L. Dunning, Anal. Chem., 44, 1053(1972).

36. N.N. Krasnov, I.O. Konstantinov, and V.V. Malukhin, J. Radioanal. Chem. 16, 439 (1973).

37. K.L. Dunning, G.K. Hubler, J. Comas, W.H. Lucke, and H.L.
Hughes, Thin Solid Films 19, 145 (1973); see also Ref. 48,
pg. 145.

38. W.D. Mackintosh and F. Brown, "Movement of Ions During Anodic
Oxidation of Aluminum," in Applications of Ion Beams to Metals,
S.T. Picraux, E.P. Eernisse, and F.L. Vook, eds., Plenum Press,
New York (1974), pg. 111.

39. R. Pretorius, F. Odendaal, and M. Peisach, J. Radioanal. Chem.
12, 139 (1972).

40. T. Lauritsen and F. Ajzenberg-Selove, Nuc. Phys. 78, 1 (1966).

41. F. Ajzenberg-Selove and T. Lauritsen, Nuc. Phys. 114, 1 (1968).

42. F. Ajzenberg-Selove, Nuc. Phys. 152, 1 (1970); 166, 1 (1971);
A190, 1 (1972).

43. P.M. Endt and C. Van der Leun, Nuc. Phys. A214, 1 (1973).

44. D.J. Horen et al., Nuclear Data Sheets 11, ii (1974).

45. J.R. Bird, B.L. Campbell, and P.B. Price, Atomic Energy Review
12, 275 (1974).

46. S.T. Picraux, G. Amsel, and L. Feldman, "Low Energy Nuclear
Reaction Tables," in Catania Working Data: A Compilation of
Tables, Graphs and Formulas for Ion Beam Analysis, to be
published.

47. G. Amsel, J. Radioanal. Chem. 17, 15 (1973).

48. J.W. Mayer and J.F. Ziegler eds., Ion Beam Surface Layer Analysis, Elsevier Sequoia S.A. Lausanne, Switzerland (1974).

49. J.A. Cookson and F.D. Pilling, "A 3 MeV Proton Beam of Less Than
Four Microns Diameter," AERE-R-6300, Harwell, 1970.

50. T.B. Pierce, J.W. McMillan, P.F. Peck and G. Jones, Nuc. Inst.
& Meth. 118, 115 (1974).

51. T. Joy and D.G. Barnes, Nucl. Inst. & Meth. 95, 199 (1971).

52. D.A. Close, J.J. Malanify and C.J. Umbarger, Nuc. Inst. &
Meth. 113, 561 (1973).

53. G.E. Coote, N.E. Whitehead, and G.J. McCallum, J. Radioanal.
Chem. 12, 491 (1972).

54. G. Deconninck, J. Radioanal. Chem. 12, 157 (1972); also Ref. 15,
pg. 157.

55. G. Deconninck and G. Demortier, J. Radioanal. Chem. 12, 189 (1972);
also Ref. 15, pg. 189.

56. G. Demortier and F. Bodart, J. Radioanal. Chem. 12, 209 (1972);
also Ref. 15, pg. 209.

57. P.H. Stelson and F.K. McGowan, Phys. Rev. 110, 489 (1958).

58. M. Peisach, J. Radioanal. Chem. 12, 251 (1972); also Ref. 15,
pg. 251.

59. O.V. Anders, Anal. Chem. 38, 1442 (1966).

60. G. Amsel, J.P. Nadai, E. D'Artemare, D. David, E. Girard, and
J. Moulin, Nuc. Inst. & Meth. 92, 481 (1971).

61. E.A. Wolicki and A.R. Knudson, Int. J. Appl. Rad. Isot. 18, 429 (1967).
62. G. Weber and L. Quaglia, J. Radioanal. Chem. 12, 323 (1972); also Ref. 15, pg. 323.
63. C. Olivier and M. Peisach, J. Radioanal. Chem. 12, 313 (1972); also Ref. 15, pg. 313.
64. M. Peisach and R. Pretorius, J. Radioanal. Chem. 16, 559 (1973).
65. R. Pretorius, Radiochem. Radioanal. Lett. 10, 303 (1972).
66. R. Pretorius, P.P. Coetzee, and M. Peisach, J. Radioanal. Chem. 16, 551 (1973).
67. J.B. Marion and F.C. Young, Nuclear Reaction Analysis, North-Holland Publishing Company, Amsterdam, 1968.
68. P.B. Price and J.R. Bird, Nuc. Inst. & Meth. 69, 277 (1969).
69. E.C. Lightowlers, J.C. North, A.S. Jordan, L. Derick and J.L. Merz, J. Appl. Phys. 44, 4758 (1973).
70. A. Turos, L. Wieluński, and A. Barcz, Nuc. Inst. & Meth. 111, 605 (1973).
71. P.P. Pronko and J.G. Pronko, Phys. Rev. B9, 2870 (1974).
72. R.A. Langley, S.T. Picraux, and F.L. Vook, J. Nuc. Mat. 53, 257 (1974).
73. D.W. Palmer, Nuc. Inst. & Meth. 38, 187 (1965).
74. E. Möller, L. Nilsson, and N. Starfelt, Nuc. Inst. & Meth. 50, 270 (1967).
75. M. Peisach, Ref. 13, pg. 650.
76. J. Lorenzen, D. Brune, and S. Malmskog, J. Radioanal. Chem. 16, 483 (1973).
77. K.L. Dunning, "Fortran Programs for (p,γ) Yield Calculations Based on Vavilov's Theory of Energy Loss Distributions," NRL Report 7230, Mar. 1971.
78. D.J. Neild, P.J. Wise, and D.G. Barnes, J. Phys. D: Appl. Phys. 5, 2292 (1972).
79. K.L. Dunning and H.L. Hughes, IEEE Trans. on Nuc. Sci. NS-19, 243 (1972).
80. J.F. Chemin, J. Roturier, B. Saboya, and G.Y. Petit, Nuc. Inst. & Meth. 97, 211 (1971).
81. J.W. Butler, "Table of (p,γ) Resonances," NRL Report 5282, Apr. 1959.
82. N. Bohr, Mat. Fys. Medd. Dan. Vid. Selsk 18, No. 8 (1948).
83. W.K. Chu and J.W. Mayer, "Energy Loss and Energy Straggling," in Catania Working Data, to be published.
84. W.K. Chu and J.W. Mayer (unpublished)
85. C.C. Rousseau, W.K. Chu, and D. Powers, Phys. Rev. A4, 1066 (1970).
86. M.K. Bernett, J.W. Butler, E.A. Wolicki, and W.A. Zisman, J. Appl. Phys. 42, 5826 (1971).
87. E. Möller and N. Starfelt, Nuc. Inst. & Meth. 50, 225 (1967).
88. B. Maurel, D. Dieumegard, and G. Amsel, J. Electrochem. Soc. 119 (1972).
89. J.W. Mandler, R.B. Moler, E. Raisen, and K.S. Rajan, Thin Solid Films 19, 165 (1973).

90. L. Porte, J.-P. Sandino, M. Talvat, J.P. Thomas, and J. Tousset,
     J. Radioanal. Chem. 16, 493 (1973).
91. I. Golicheff and Ch. Engelmann, J. Radioanal. Chem. 16, 503
     (1973).
92. I. Golicheff, M. Loeuillet and Ch. Engelmann, J. Radioanal.
     Chem. 12, 233 (1972); see also Ref. 15, pg. 233.
93. D.A. Leich and T.A. Tombrello, Nuc. Inst. & Meth. 108, 67
     (1973).
94. J.L. Whitton, I.V. Mitchell and K.B. Winterbon, Can. J. Phys.
     49, 1225 (1971).
95. J.M. Calvert, D.G. Lees, D.J. Derry, and D. Barnes, J. Radioanal.
     Chem. 12, 271 (1972); see also Ref. 15, pg. 271.
96. R.O. Bondelid and C.A. Kennedy, Phys. Rev. 115, 1601 (1959).
97. D.V. Vavilov, Zh. Exper. Teor. Fiz. 32, 320 (1957), transl.
     JETP. 5, 749 (1957).
98. Studies in Penetration of Charged Particles in Matter, NAS-NRC
     publication 1133, 1964.
99. P.R. Malmberg, private communication.

LATTICE LOCATION OF IMPURITIES IN METALS AND SEMICONDUCTORS*

S. T. Picraux

Sandia Laboratories

Albuquerque, New Mexico 87115

I. INTRODUCTION

Many important physical properties of materials are controlled by the crystallographic location of impurities in solids. In addition, the interpretation of solid state experiments or the study of solids by theoretical calculations often requires a knowledge of the lattice location of impurities. Examples of experiments include spin resonance studies such as NMR and EPR, hyperfine methods such as Mossbauer effect and perturbed angular correlation, and internal friction measurements. Theoretical examples would include band theory calculations of electronic, optical and other properties due to the presence of impurities. While the interest in impurity location has been strong for many different types of systems, there have been few techniques available for direct location determination.

Ion channeling is a technique which can be used to directly determine the impurity location within a crystal lattice. It requires the generation of a monoenergetic collimated beam of particles to energies easily available on typical low-energy accelerators ($\sim$ 1 MeV). Under favorable conditions it allows determination of the impurity position to an accuracy $\sim$ 0.1 Å. Even though the channeling method for lattice location is far from an ideal technique, it has received a great deal of attention due to the importance of impurities in solids and the need for location data.

Other techniques for crystallographic determination of impurity location include neutron and x-ray scattering techniques. Usually

*This work was supported by the United States Energy Research and Development Administration, ERDA.

these techniques give only the periodicity of the impurity location
and high impurity concentrations ($\geq$ 1 atomic %) are required. In
contrast, ion channeling gives the location relative to the lattice
atoms and, typically, concentrations as low as 10^{-2} atomic % can be
studied. Another important difference is that ion channeling probes
only the near-surface region rather than throughout the bulk of a
solid. Thus the channeling technique is quite amenable to surface-
related methods of impurity introduction such as ion implantation.

The channeling effect has recently been comprehensively reviewed
by D. S. Gemmell.[1] In addition, a recent book edited by D. V. Morgan
on channeling[2] gives a detailed treatment both of the theory and
application of ion channeling; a chapter on lattice location by
J. A. Davies[3] is included. Another chapter in this book is on
nuclear lifetime measurements by ion channeling[4] and the analysis
for this application is closely related to that for lattice location.

Lattice location studies can be carried out either by directing
the incident beam of particles along axial or planar directions in a
crystal (channeling), or by observing a scattered or emitted beam of
particles along axial or planar directions (blocking). When both the
channeling and blocking effects are combined in a single measurement
the technique is referred to as double alignment. Almost all recent
lattice location studies have been carried out for the channeling
configuration. Some of the early measurements were carried out by
implanting radioactive emitters for the impurity atoms and detecting
the blocking patterns.[5-7] A disadvantage of this latter approach is
that a signal is obtained only from the impurity atoms, whereas when
an external beam of particles is used signals can be obtained simultaneously from the host and impurity atoms, a requirement for quantitative location measurements. Since according to the rule of rever-
sibility the channeling and blocking effects are the same phenomena,[8]
the blocking geometry will not be discussed further.

Although both positive and negative particles exhibit channeling
effects,[2] our discussions will be limited to positive ions. Blocking
measurements of β^- emitting impurities have exhibited location
effects.[6] However, electrons have not yet been demonstrated to be
as useful as positive ions for location studies.

This chapter will briefly discuss impurity detection methods
and the ion channeling technique. Lattice location analysis from
the channeling data will then be described and examples of applications to metals and semiconductors for various classes of impurity
sites will be given. Finally, the existing location data available
from channeling studies will be summarized, and limitations of the
technique will be discussed.

II. IMPURITY DETECTION

For impurity lattice location by ion channeling it is necessary to monitor close impact parameter collision events of the incident ions with both the lattice atoms and the impurity atoms. The means of detecting close impact parameter collisions are ion scattering, ion-induced x rays and ion-induced nuclear reactions. Each of these types of detection are subjects of earlier chapters and only the relative advantages of these techniques for dilute impurity detection will be summarized here.

Ion backscattering is the most versatile of the three techniques and is used primarily for higher-Z impurities in a lower-Z target. Incident He ions are usually the best compromise for good depth and mass resolution by backscattering, although heavier incident ions such as C and N can provide better sensitivity for impurities which are only slightly heavier than the host lattice atoms. Ion backscattering also has the advantage that the energy spectrum can be easily analyzed to give the atomic concentration as a function of depth. Thus the signal from the impurity and the host can be selected from the same depth within the crystal and this simplifies location interpretation of channeling data. Also, regions of the spectrum may be rejected, for example the surface region if some fraction of the impurities are tied up in a surface oxide layer. Figure 1 shows a schematic of the backscattering energy spectrum for He incident on Si which contains implanted Bi. The scattering signals from Si and Bi are well separated and portions of the scattering spectrum corresponding to Bi and Si at the same depth can be easily selected using single channel analyzers.

Ion-induced x-ray analysis provides a means to study an even wider range of impurity-host combinations. Although the absolute cross sections are large, the relative sensitivity to impurities over the host background currently tends to be lower by a factor $\sim$ 10. Also, since the x rays give a weighted signal over the near-surface region of a solid, it is sometimes difficult to directly compare impurity and host channeling signals from the same depth, or to reduce dechanneling effects which increase with depth. In the case of ion-induced x-ray studies care must be taken if outer shell x rays (e.g., L or M) are detected since x-ray production cross sections may still be appreciable[9] for certain channeled particle trajectories and this can complicate interpretation of channeling angular distributions.[10]

Ion-induced nuclear reaction analysis where either emitted particles or gamma rays are detected can also provide a convenient alternative to ion backscattering, particularly for low-Z impurities. The absolute cross sections for nuclear reactions are lower than for ion backscattering and in practice impurity concentrations need to

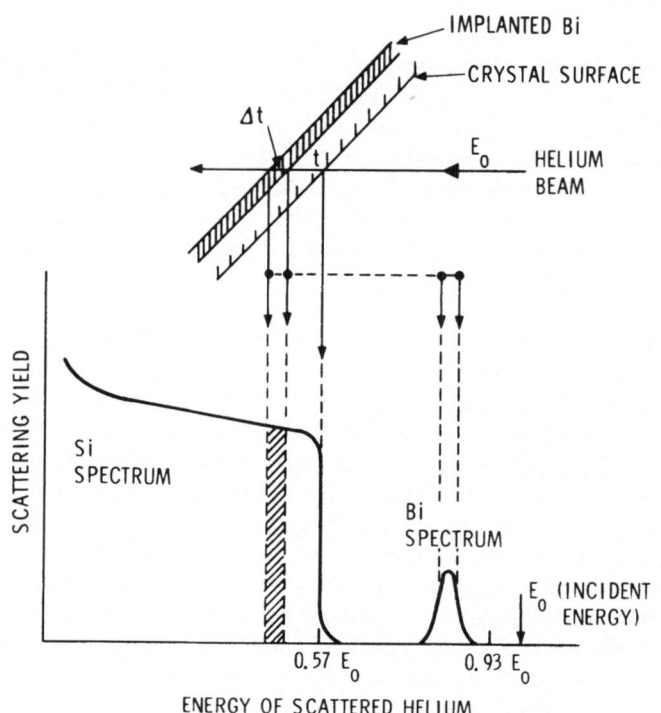

Fig. 1. Schematic of energy spectrum for He backscattering from
Si containing a buried Bi profile. The scattering regions corresponding to Si and Bi at the same depth as well as to Si at the
surface are indicated by arrows.

be greater by a factor ~ 1-10 than in backscattering studies. This
results not so much from detection limits, since background levels
can be quite low for high energy reaction products, but rather from
radiation damage effects by the analysis beam.

In practice the ion channeling technique requires relatively
high impurity concentrations for accurate lattice location measurements due to limitations of detection sensitivity. Most studies
have used impurity concentrations ~ 10^{19} to $10^{20}/cm^3$. The lower
concentration limit is set primarily by the need to minimize radiation damage by the analysis beam prior to completion of the measurement. Otherwise defect association or radiation damage to the
lattice may change the impurity location or the channeling behavior
during the measurement. In the extreme case this can prevent
measurement of the normal location of a dilute impurity.

Since ion channeling is a near-surface analysis technique, it
is frequently convenient to combine it with ion implantation for
the controlled introduction of impurities near the surface. Typical

implantation fluences required are $\geqslant 10^{14}/cm^2$, although studies
have been performed for implant fluences as low as $10^{13}/cm^2$.[11]
Also, ion implantation offers the possibility of introducing impurities above solubility limits without precipitation, and these metastable phases may show location properties characteristic of more
dilute systems.

III.   THE CHANNELING TECHNIQUE

III.1.   Channeling Concept

Axial or planar channeling occurs when a collimated energetic
beam of ions is incident on a single crystal along an axial or
planar direction. The rows or planes of atoms act as potential
barriers such that the ions undergo a series of correlated screened
coulombic collisions. These small angle collisions result in the
channeled ions being steered by the rows and planes towards the
central channel region between the rows or planes. As discussed by
Lindhard[12] in an early and definitive analytic treatment of the
channeling effect, the potential due to the atomic rows or planes
can be averaged in a continuum model to explain many of the basic
features of channeling. Using this continuum model the critical
angle of the incident ion beam relative to a row within which axial
channeling will occur is given by

$$\psi_{\frac{1}{2}} = \alpha \left(\frac{2Z_1 Z_2 e^2}{Ed} \right)^{\frac{1}{2}} , \tag{1}$$

provided the projectile is of sufficient energy such that

$$E \geqslant \frac{2Z_1 Z_2 e^2 d}{a^2} , \tag{1a}$$

where Z_1 and E are the atomic charge and the energy of the projectile, Z_2 and d the average atomic charge and atomic spacing along
the row, a is the Thomas-Fermi screening distance and α is a slowing varying function of vibrational amplitude of order 1. Typical
critical angles for ion channeling for which formula 1a is satisfied
are $\leqslant 2°$. The fraction of incident particles channeled and the
critical angles are greatest for low index (closely packed) directions since the "strength" of channeling is determined primarily by
the average charge density of the rows or planes.

The most important aspect of the channeling effect for lattice
location studies is the spatial distribution of the incident ion
beam which results during channeling. This can best be understood
with respect to Fig. 2. For a beam uniformly incident along a nonchanneling ("random") direction there is a uniform spatial

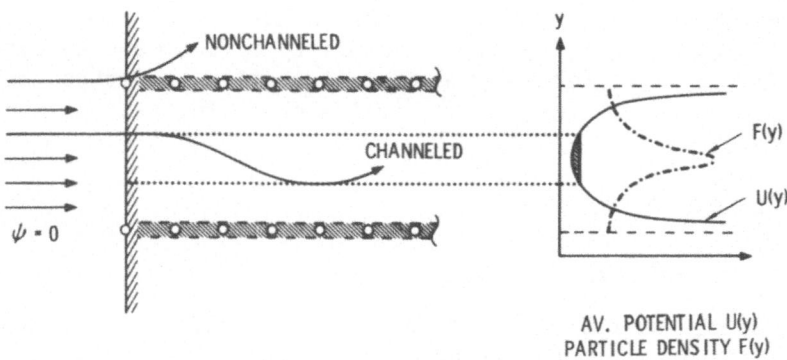

Fig. 2.   Schematic of channeling in a crystal with plot of representative average potential and particle density across the channel.

distribution inside the crystal.   However, when the beam is aligned
along a low index crystal direction as shown in Fig. 2 then the
majority of the ions ($\approx$ 95-99% for axes) are channeled and undergo
oscillatory motion between crystal rows or planes.   The channeled
particles have insufficient transverse momentum to penetrate the
potential barriers formed by the atom rows or planes and thus are
prevented from entering closer to the rows or planes than $\approx$ 0.1 Å.
This minimum impact parameter with the rows or planes is determined
by the Thomas-Fermi screening distance of the interaction potential
and by the vibrational amplitude of the atoms.   Close impact parameter processes, such as ion backscattering, are effectively forbidden for the channeled part of the beam so that when the incident
beam is aligned with a crystal axis there is a strong reduction
($\approx$ 95-99% for axes) in the yield of backscattered particles near
the surface.   The small remaining fraction of the incident beam
which enters close to the crystal rows or planes is scattered into
nonchanneling directions, giving rise to a uniform background level.

     The spatial distribution of particles during channeling results
in an appreciable enhancement of the particle density near the center
of the channels relative to that for beam incidence along a random
direction.   This can be understood from energy balance arguments
since to first order the total transverse energy, $E_\perp$, of a channeled
particle is a conserved quantity:

$$E_\perp = E\Psi^2 + U(x,y) \quad , \tag{2}$$

where the first term on the right-hand side of the equation is the
kinetic energy due to particle momentum transverse to the channel
and the second term is the potential energy resulting from the
screened coulombic interaction with the lattice atoms.   Here E is
the particle energy, Ψ is the angle of the particle with respect to

the crystal rows or planes, and U is the average potential along the
z direction at a position (x,y) in the channel and is normalized to
zero at the minimum in U. An example of an average potential contour plot for the $\langle 001 \rangle$ direction in Cu is shown in Fig. 3. The
circles at the four corners represent the Cu rows and a diagonal
slice through Fig. 3 would give the sort of plot shown in Fig. 2 by
$U(y)$ vs. y. To determine the spatial distribution of channeled
particles it is necessary to know the channel area available to a
given particle. The area available to a channeled particle which
enters the crystal at (Ψ_0,x_0,y_0) lies within a contour line of the
potential given by $U(y) = E_\perp = E\Psi_0^2 + U(x_0,y_0)$. From this simple
estimate the spatial distribution of particle density for channeling
under the assumption of a statistical equilibrium is given by

$$F(x,y,\Psi_0) = \int \frac{dA(x_0,y_0)}{A(x_0,y_0,\Psi_0)} , \qquad (3)$$

$$E_\perp(x_0,y_0,\Psi_0) \geq U(x,y)$$

where $F(x,y)$ is the particle probability density as a function of
position (x,y) in the channel and A is area of the channel within a
contour $U(x,y)$ such that $E_\perp(\Psi_0,x_0,y_0) \geq U(x,y)$.[12-14]

A cross section of a typical spatial distribution and average
potential for an axial direction is indicated schematically in
Fig. 2. As the channeled particles penetrate the crystal, multiple
scattering by electrons and the interaction with the vibrating lattice atoms increase the root mean square transverse energy, and in
a more detailed treatment this could be approximately accounted for
in Eqs. 2-3. This increase in transverse energy reduces the particle
density in the center of the channel (flux peak), smooths out the
spatial distribution with increasing depths and gives rise to dechanneling of the incident beam by increasing $E_\perp$ above the critical
value for channeling ($\approx E\Psi_1^2$). Also there are oscillations with
depth in the particle density at a given position (x,y) in the
channel, since statistical equilibrium is not reached until the
particles penetrate appreciable depths ($0.1 - 1.0~\mu m$) into the
crystal. These oscillations are greatest near the surface and calculations of such effects require more detailed treatments than
given above (such as Monte Carlo or equations of motion methods[15-18]).

By comparing the channeling effects resulting from the crystal
lattice atoms to those of impurity atoms as a function of incident
angle (channeling angular distributions), one can determine information about the impurity position. The impurity location information
is obtained primarily from the forbidden region around the rows
(planes) and the increased particle density (flux peak) at the center
of the channel. Since the details of the particle spatial distribution are depth dependent, the interpretation is simplified by select-
ing the host and impurity atom signals from approximately the same
depth.

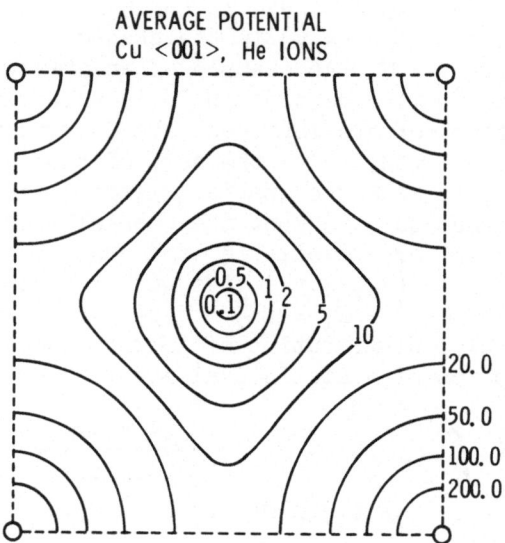

AVERAGE POTENTIAL
Cu <001>, He IONS

Fig. 3. Contour plot of average potential normalized to zero at the center of the channel for He particles for the ⟨001⟩ direction in Cu. The ⟨001⟩ rows are perpendicular to the plane of the paper and the numbers are in units of eV. From Van Vliet.[14]

III.2. Experimental Technique

The basic experimental technique for lattice location measurements involves the same requirements as those given in the previous chapters for either ion backscattering, ion-induced x rays or nuclear reaction analysis of solids. In addition, ion channeling studies require that the incident beam be collimated to a low angular divergence, typically ≤ 0.05°.

The primary additional piece of experimental equipment required is a goniometer which allows orientation of the crystal with respect to the beam direction. Usually one uses either an x-ray goniometer which has been modified for operation in a vacuum or a goniometer which has been specifically designed for channeling measurements such as one shown in Fig. 4.[19] Either a three- or two-axis goniometer is satisfactory provided the orientational accuracy is ≤ 0.05°. The 2-axis goniometer shown in Fig. 4 allows tilt motion around the vertical axis of the entire lower frame and 360° rotation of the crystal about an axis perpendicular to the plane of the paper. For a 2-axis goniometer of the type shown in Fig. 4 the tilt axis must be perpendicular to the beam direction, since a misalignment from this by $\Delta\theta$ will result in a solid angle $\pi\Delta\theta^2$ of orientation which is inaccessible for all settings.

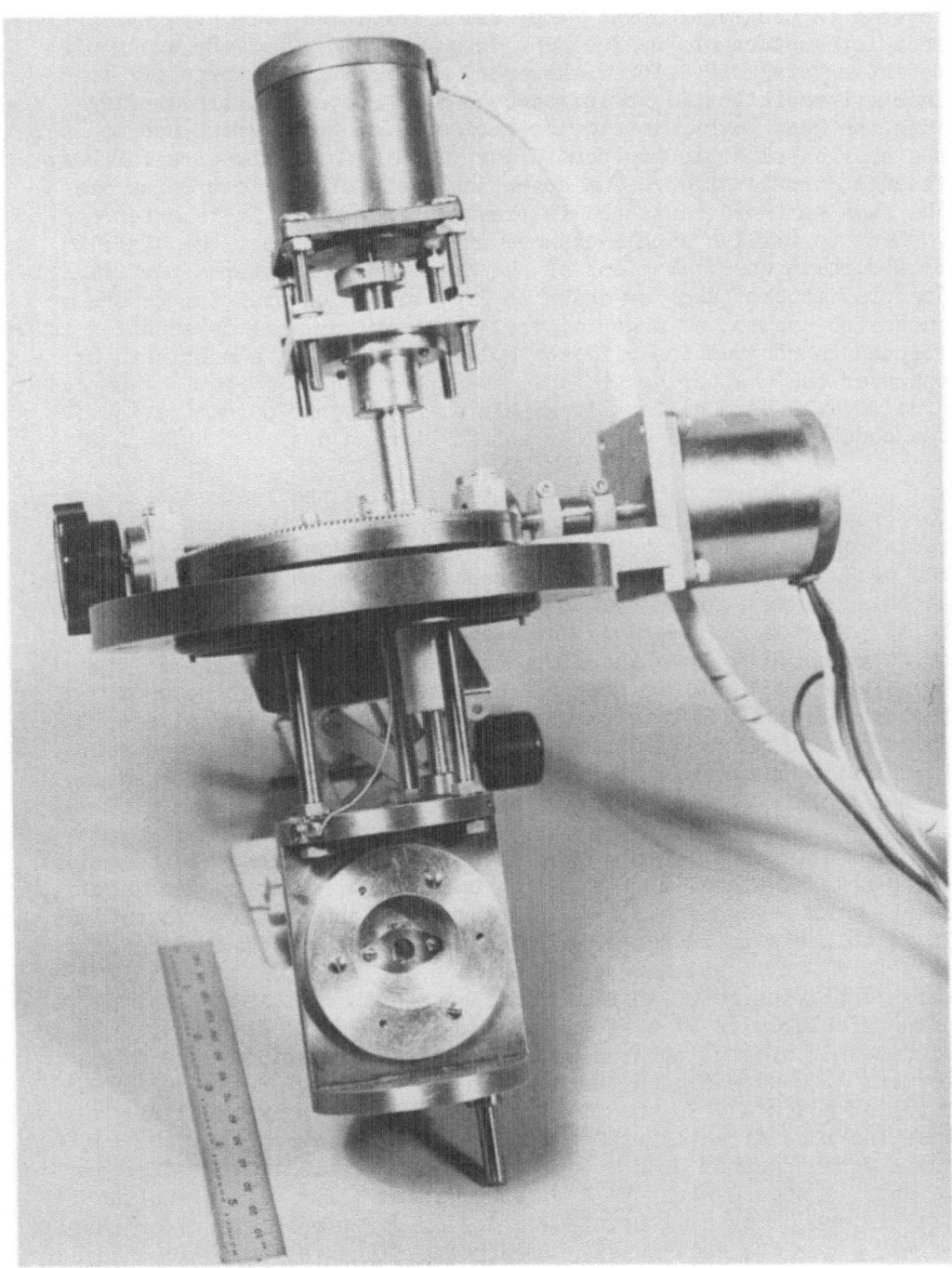

Fig. 4. Photograph of two-axis goniometer used for channeling measurements.

The goniometer is usually located within a Faraday cup or it faces a secondary electron suppression shield so that accurate current integration of the incident beam charge can be made along different crystal directions. However, an effective alternative to charge integration is to introduce a small rotating wire or flag into the beam path prior to its incidence on the crystal and monitor the charge collection or scattering yield off of this wire for beam fluence normalization. The detection systems and electronics are the same as those mentioned in previous chapters. It is often convenient to utilize single-channel analyzers to select the signals from certain energy regions of the spectrum of scattered particles. Stepping motors, such as shown in Fig. 4, can be used for remote or automated control of a goniometer. When the crystal orientation is stepped in conjunction with the multiscale mode of a multichannel analyzer the scattering yield versus angle can be directly recorded provided the beam current is stable or electronic normalization of the count rate by the beam current is carried out.

It is also convenient to be able to move the beam spot to check for lateral variations in the sample or to check for beam-induced radiation damage effects on the lattice location measurement. This can be done either by moving the beam position (by moving the beam-defining slits), or by having lateral sample movement capability built into the goniometer. When moving the beam collimating slits it is valuable to move the slit closest to the sample and to use a "dog-leg" electrostatic steering arrangement to bring the beam through this slit. This requires the use of two identical steering units between the collimating slits with the polarities reversed. In this way the crystal orientation relative to the beam direction is maintained while allowing lateral movements of the beam.

The initial step in a lattice location measurement is to orient the single crystal with respect to the beam direction. An example is shown in Fig. 5 for a 2-MeV He beam incident on a bcc Mo crystal. It is valuable to know the approximate orientation of a major crystal axis ($\pm 5°$) prior to alignment, although this is not essential. The alignment illustrated in Fig. 5 was done with a goniometer with two perpendicular axes of tilt (φ and ω). The crystal face was cut to within $\pm 3°$ of the $\langle 100 \rangle$ axis, so a series of $6°$ tilts were made forming the box shown on the left side of Fig. 5. A typical scan labelled A for $\omega = 30$ to $36°$ at $\varphi = 181°$ is shown in the upper inset as read directly from a multichannel analyzer using the multiscale mode together with stepping motors. Several planar dips can be seen in scan A with the two larger being $\{110\}$ planes and the smaller one being a $\{100\}$ plane. The larger planes are wider than normal since the scan passes through the planes at an oblique angle. By plotting the intersection of the planes on the stereogram shown on the left of Fig. 5 for a series of scans the location of the $\langle 100 \rangle$ axis is determined. This is confirmed by scan B shown in the lower right-hand inset of Fig. 3. The scans of Fig. 5 were carried

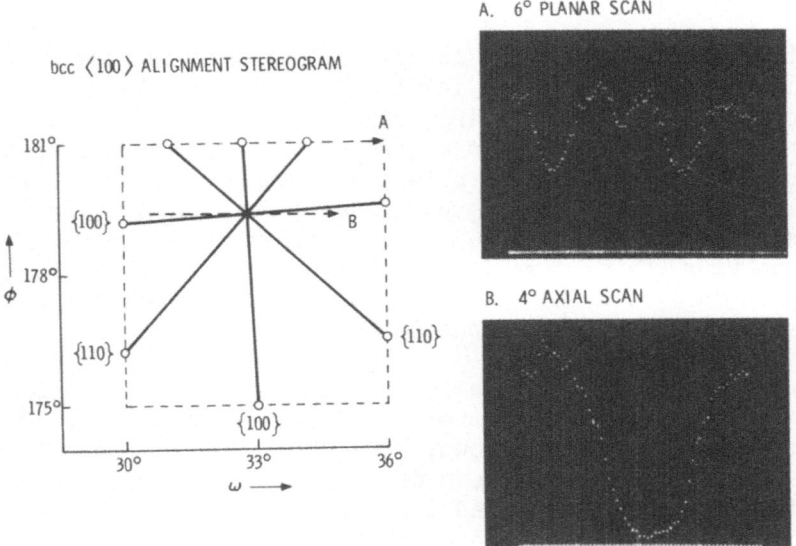

Fig. 5. Alignment stereogram together with some of the planar and axial angular scans used to align a (100) Mo crystal using 2-MeV He.

out with a 10 nA He[+] beam current at a scan rate of 4° per minute and accepting a scattering yield from an energy just below the surface scattering peak to an energy corresponding to the upper 2/3 of the Mo backscattering energy spectrum.

The minimum yield χ_{min} along axial directions should be of the order of the theoretical estimate for channeling prior to impurity introduction For low index axes this generally implies a near-surface $\chi_{min} \lesssim 5\%$. A good empirical estimate of the minimum yield just below the surface has been given by Barrett:[15]

$$\chi_{min} \simeq Nd\pi(3.0\rho^2 + 0.2a^2) \quad , \tag{4}$$

where N is the atom density of the crystal, d is average spacing along the crystal row, ρ is the mean vibrational amplitude and a is the Thomas-Fermi screening distance. When the lattice minimum yields are quite high it is difficult to interpret the results of a location measurement with confidence, except when the impurity and host channeling angular distributions coincide due to complete impurity substitutionality. High minimum yields may be indicative of appreciable disorder in the near-surface region of the crystal due to inadequate surface preparation or may indicate poor crystal quality due to imperfections or mosaic spread.

IV. LATTICE LOCATION ANALYSIS

Lattice location analysis is carried out by comparing the signals from the impurity and the lattice atoms for angular scans along
axial and planar directions. The location obtained along any given
channeling direction is the projection of the impurity position(s)
along that direction. Thus measurements along several directions
are usually required to unambiguously assign the lattice location.
When measurements are restricted to the minimum yield rather than
complete angular scans, a reliable interpretation of the impurity
location can only be made when large and equal attenuations are
obtained along the various channeling directions.

A two-dimensional illustration of the site differentiation process is given in Fig. 6 where the (110) plane from the cubic diamond
lattice structure for Si is shown. The $\langle 110 \rangle$ and $\langle 111 \rangle$ axial directions lie in this plane and the 3 different impurity positions shown
by the filled circle, the filled triangle and the x can be easily
differentiated by comparing the channeling along these two directions.
In the case of the substitutional impurity (filled circle) this would
give channeling behavior similar to the lattice atoms for both the
$\langle 110 \rangle$ and $\langle 111 \rangle$ directions. The tetrahedral interstitial site (x)
would exhibit the same behavior as the substitutional impurity for
the $\langle 111 \rangle$ direction but would give an enhanced yield (flux peak) for
the $\langle 110 \rangle$ direction. Finally for the other off-site position (filled
triangle) the channeling distribution along the $\langle 110 \rangle$ direction would
not be too different from that for the tetrahedral site since both
impurities are displaced similar amounts from the $\langle 110 \rangle$ row, whereas
along the $\langle 111 \rangle$ direction the on and off row positions would give
quite different channeling behavior. This process of looking along
different directions is sometimes referred to as the "triangulation
process" and simply reflects the fact that, in general, to locate a
position in a 3-dimensional lattice projections along more than one
direction are required.

A second important consideration in determining the location of
an impurity in the lattice is the lattice symmetry. For a given nonsubstitutional site the symmetry of the lattice may imply that there
are several equivalent positions within a unit cell corresponding to
the same site. Thus a projection along an axial or planar direction
will result in multiple impurity positions within a given channel.
In order to consider the consistency of an angular distribution
along a given channeling direction with the proposed site, one must
determine the projected positions within a channel and the relative
fractional occupation of each of these equivalent sites.

An example is shown in Fig. 7 for the tetrahedral interstitial
site in the bcc lattice for projections along the $\langle 100 \rangle$ axis and the
(100) plane. Circles represent the projected lattice rows for the
fraction of the unit cell shown and the lines indicate the projection

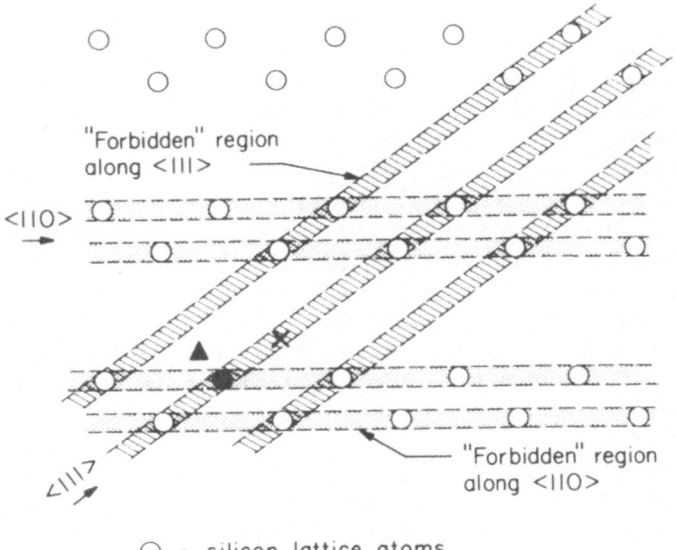

O - silicon lattice atoms

Fig. 6. The {110} plane of Si showing ⟨110⟩ and ⟨111⟩ channeling directions. The filled circle corresponds to a substitutional site, the x to the tetrahedral interstitial site and the filled triangle to another nonsubstitutional position. After Mayer, Eriksson and Davies.[20]

TETRAHEDRAL SITE

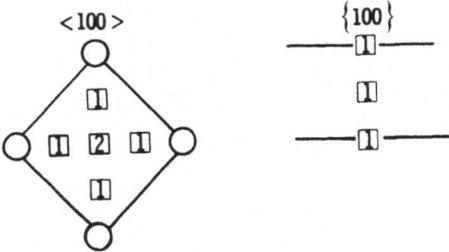

Fig. 7. Schematic showing the projected positions (squares) of the tetrahedral interstitial site in the bcc lattice for the ⟨100⟩ axial and the {100} planar directions. The circles represent the ⟨100⟩ rows for part of a unit cell, the lines represent planes and the numbers refer to the relative number of sites for that projected position.

of the planes. The square boxes represent the projected position of
the tetrahedral interstitial site and the numbers within the boxes
indicate the relative number of sites along that projection. Thus
for the ⟨100⟩ axis 2/3 of the sites are halfway between the center
and the atom rows whereas the other 1/3 of the sites are located in
the channel center. Correspondingly for the case of the planes,
2/3 of the sites lie in the (100) planes and 1/3 of the sites are
in the center of the planar channel. There are two methods to determine the number of equivalent sites for a given projection: 1. all
the equivalent sites within a unit cell are filled and the fractional
occupational number for each projected position along a given direction is determined; 2. a single site is projected along all equivalent directions, i.e., for the ⟨100⟩ direction this would be [100],
[010] and [001].

Unambiguous interpretation of channeling angular distributions
is difficult when there are more than 1 or 2 inequivalent projected
impurity positions (there are 2 for both the ⟨100⟩ and the (100)
examples of Fig. 7). Thus lattice location assignment becomes increas-
ingly difficult for low-symmetry interstitials or for impurities with
multiple-site occupation. In such cases planar channeling in conjunction with axial channeling can be particularly useful since frequently
planar directions can be selected such that a given interstitial site
will be completely in planes or completely in the channel center
between the planes. In this way a clear-cut elimination or confirmation of a given site can be obtained.

Double alignment can supply additional information in lattice
location studies. The technique involves having the beam incident
along a channeling direction and also detection of the emitted or
backscattered particles along a channeling direction. In general the
resultant yield involves a product of the channeling probabilities
along both the incoming and outgoing paths. For example, the axial
minimum yields are appreciably smaller, i.e., X_{min} < 1%, compared to
X_{min} ≈ 3% for single alignment channeling. Although double alignment has received little attention, it can be useful in the interpretation of site location for the case of multiple site occupation.
A simple example is the case of a large fraction of the impurities
being exactly on substitutional sites with the remaining fraction
distributed randomly throughout the lattice vs. the case of all the
impurity atoms being in substitutional or near-substitutional sites.
This will be illustrated in Section V.2 for the case of Bi in Si.
A second possible use of double alignment would be to determine with
greater accuracy the total fraction of impurities which are substitutional in the case of nearly 100% substitutionality. The substitutional fraction is determined by comparing the minimum yields of the
impurity and lattice atoms but is limited by counting statistics.
In a typical single alignment case one might determine that all the
atoms were on substitutional sites to within ± 3%; whereas in the

double alignment case the decreased minimum yields could allow one to determine if $\lesssim 1\%$ of the atoms were off substitutional sites.

Frequently a qualitative determination of the predominant lattice site occupied by an impurity can be obtained from channeling measurements along various directions without the aid of any calculations. In fact it is important to obtain a good qualitative indication of the approximate location from the data prior to detailed calculations, since certain features such as the magnitude of the yield for nonsubstitutional sites cannot yet be calculated with certainty. Simple calculational estimates can be applied to certain features of the angular distributions such as the minimum yield and angular width of dips, and the flux peak position for useful estimates of site location and fractional occupation. Detailed calculations can be valuable in determining more precisely a given location or fractional occupation of sites, particularly in the case of low symmetry or multiple site occupation.[13]

A complete calculational analysis requires that the particle probability density versus position across the channel be calculated as a function of incident angle of the beam with respect to the channeling direction. These spatial distributions are then used together with trial impurity sites to calculate the angular distributions for comparison to the measured angular distributions. This is illustrated in Fig. 8 by analytic calculations for the case of 3.5 MeV ^{14}N ions incident along the ⟨100⟩ direction of the bcc Fe lattice for 4 different sites.[13] The left side of the figure shows the projected positions, the center panel shows the spatial distributions relative to the channel center ($\rho=0$) for 3 different incident angles and the right panel shows the angular distributions predicted for the 4 different sites considered. In the case of the substitutional (S) site the angular distribution would be expected to be the same as that for the lattice atom. Site B would be a near-substitutional site where the angular distribution is narrowed but still shows a dip as for the lattice atoms. Sites nearer the center of the channel show flux peaking effects with a large central peak of typical width ~ 0.2° for position 0 at the channel center and weaker side-band peaks for site A displaced from the channel center.

There are two currently-used methods for calculating the channeled particle spatial distribution in order to obtain the angular distributions.[13,21] One is Monte Carlo calculations which are at this point the most reliable and also the most time-consuming to perform.[13-15,22] Monte Carlo calculations basically involve trying to simulate the true solid state situation as accurately as possible and performing the collision experiments directly on the computer. The second method is the analytical calculation based on the continuum model.[12-14,22,23] The analytic approach can be much simpler to perform and in general gives the qualitative features although in some cases it may be less accurate, particularly in

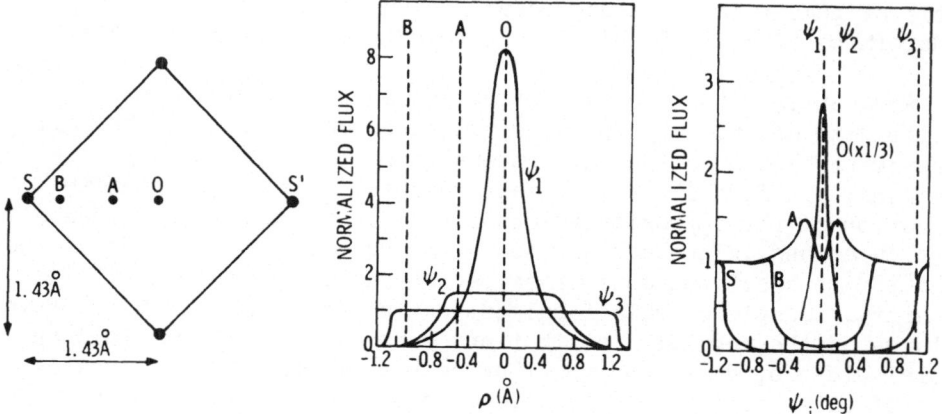

Fig. 8. Calculated particle spatial and angular distributions
by the multi-row analytic method for 3.5 MeV^{14}N incident along
the $\langle 100 \rangle$ axial direction in Fe. Left panel shows channel with
Fe $\langle 100 \rangle$ rows at the 4 corners and impurity positions 0, A, B and
S. Center panel shows spatial distributions across section SOS′of
the channel for incident beam at angle $\psi_1 = 0$, $\psi_3 = \psi_1$, and $\psi_1 < \psi_2 < \psi_3$
with respect to the $\langle 100 \rangle$ direction. Right panel shows angular
distributions for the four positions shown. From Alexander et al.[13]

terms of magnitudes of dips and flux peaks. In both methods it is
difficult to account for second order effects such as scattering by
the electrons, lattice atoms vibrating out from static axial or
planar positions, crystal defects or surface contamination. The
net result is that it is difficult to calculate the exact quantitative levels for the angular distributions observed in a given
experiment. The above complications are often included to some
approximate degree in Monte Carlo calculations but usually neglected
in analytic calculations.

In the analytical method there have been two basic techniques
used for location analysis, both based on the continuum model. One
approach has been a single-string or single-plane calculation, which
is a blocking calculation and can be done with either the continuum
model or a refinement of the continuum model referred to as the half-
way plane model.[12,23] This approach is useful for substitutional
and near substitutional impurities. A second approach for analytic
calculations has been a multiple-string calculation based on the
available channel area, which for a given channeled particle is
dependent on its transverse energy (see Eq. 3).[13,24] This approach
is particularly useful for interstitial impurities.

The analytic calculations generally assume statistical equilibrium.[12] It is known that statistical equilibrium is not reached near the surface of the crystal, i.e., for depths < 1000 Å, but this is where most lattice location measurements are made in order to reduce degradation of the channeling dips by dechanneling of the beam. However, since lattice location measurements tend to accept a signal from the impurity over a range of depths, i.e., typically over $\Delta x \sim$ 500-1000 Å, the oscillatory effects of statistical nonequilibrium are not too significant.

One additional and potentially useful analytic approach which has not yet been applied to location studies is the differential equation method.[16-18] Here the equations of motion under the continuum model are written down and solved directly for a representative number of initial conditions (x_0, y_0, ψ_0). At present this method appears to have the advantage of not assuming statistical equilibrium and the disadvantage of requiring appreciable computer time.

V. EXAMPLES

For all the channeling location examples to be discussed the impurity and host atom signals are obtained from events in which the impact parameter is much less than the minimum impact parameter of a channeled particle with the atom rows or planes (≈ 0.1 Å). Usually ion backscattering has been used. Also in all cases it is assumed that the signals from the impurity and host lattice atoms have been selected to correspond to the same depth within the crystal.

V.1. Substitutional Impurities

The substitutional impurity case is the easiest to interpret in channeling measurements of the lattice location. For a dilute substitutional impurity the channeling angular distributions for the impurity and the host lattice atoms should coincide, provided the thermal vibrational amplitudes for the impurity and host atoms are not too different. Excellent examples of this have been given for Au introduced into copper single crystals from the melt and by ion implantation.[25,26] The ⟨110⟩ channeling angular distributions for Au and Cu from a melt grown Cu-2 at.% Au single crystal are shown in Fig. 9.[25] The Au and Cu angular scans are seen to coincide, indicating that within the accuracy of the statistics 100% of the gold atoms are located on fcc copper lattice sites. Recently it has been shown that 100% substitutionality is also found for Au implanted into single-crystal Cu.[26] Figure 10 shows angular scans along the ⟨110⟩ axis for an implant of 10^{16} Au/cm^2; the scans labeled "surface" were taken just behind the surface scattering peaks and the scans labeled "gold distribution peak" were taken at the region of maximum Au concentration corresponding to a depth

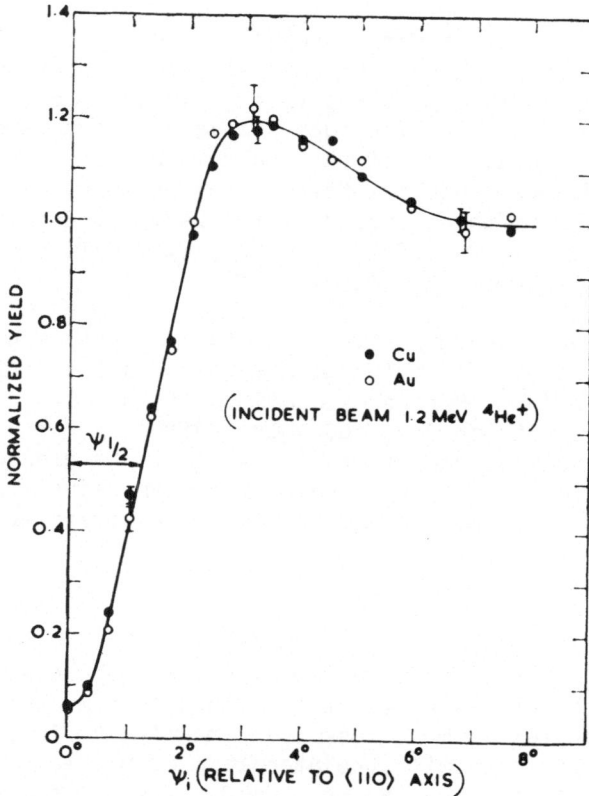

Fig. 9. Channeling angular distributions observed by backscattering from Cu-2 at.% Au for 1.2-MeV He incident along the ⟨110⟩ axis. From Alexander and Poate.[25]

≈ 800 Å below the surface.[27]   The agreement between the Au and Cu scans at each depth indicates 100% substitutionality for the implanted Au.   In addition, Fig. 10 demonstrates that even though the channeled particle spatial distribution and therefore the angular distribution undergo changes with depth, the impurity and lattice angular distributions will still coincide for 100% substitutional impurities.

In this particular example for the fcc crystal lattice structure, exact agreement between impurity and host along the ⟨110⟩ axis is sufficient to confirm that the impurity is entirely located on substitutional sites.   In general it may be necessary to make measurements along more than one direction.   An example would be the ⟨111⟩ direction in the diamond lattice where the tetrahedral interstitial site is always contained within the ⟨111⟩ rows so that 100% occupation of either this interstitial site or substitutional sites would give the same angular distribution.

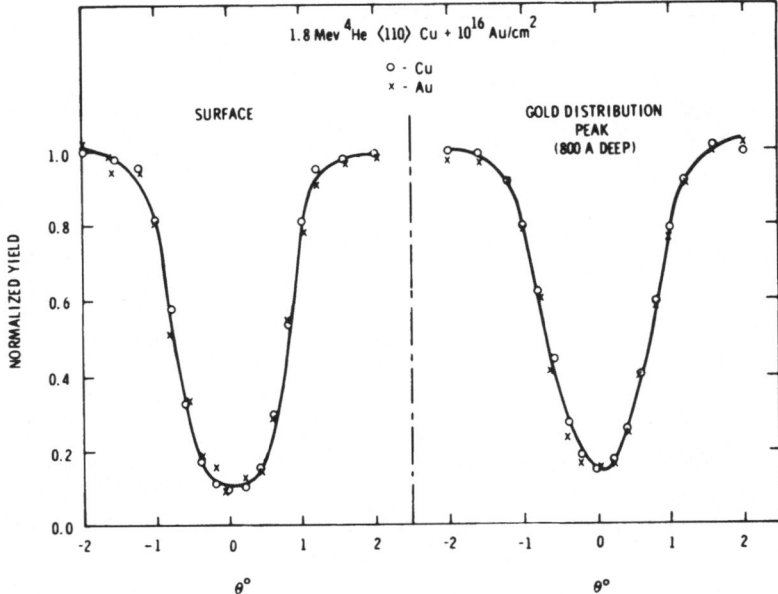

Fig. 10.   Channeling angular distributions observed by backscatter-ing from 10^{16} Au/cm^2 implanted at 200 keV into Cu for 1.8 MeV He incident along the $\langle 110 \rangle$ axis.   Scans were made from regions of the backscattering spectrum corresponding to just below the surface peak and to the maximum in the Au distribution at ~ 800 Å.   From Borders et al.[27]

An example of a highly-substitutional impurity for a semi-conductor is given in Fig. 11 for 1 x 10^{16} Te/cm^2 implanted into GaAs at 200°C.[28]   Angular scans for the impurity and lattice atoms nearly coincide for both the $\langle 111 \rangle$ and $\langle 110 \rangle$ directions indicating the Te is essentially all substitutional.   There is some lack of agreement in shoulder regions (tilt angle > ± 1° in Fig. 11); how-ever this region generally seems to be of less importance than the central region of the dip in determining the impurity location.

A third example demonstrates how in favorable cases for a com-pound material, ion channeling can be used to determine on which sublattice a substitutional impurity is located.[29]   Advantage is taken of the fact that the angular width over which channeling can occur is determined by the average nuclear charge along the atom row and is proportional to $(Z_2/d)^{\frac{1}{2}}$ as given in Eq. (1). Experimentally the critical angle ($\psi_{\frac{1}{2}}$) is determined from a channel-ing angular distribution by the half-width at a level halfway between the minimum yield and the random level.   For the diamond lattice structure of GaP the $\langle 111 \rangle$ rows contain equal numbers of Ga and P atoms; thus for a substitutional impurity the critical angle should coincide with that for the Ga and the P host atoms.   This is shown

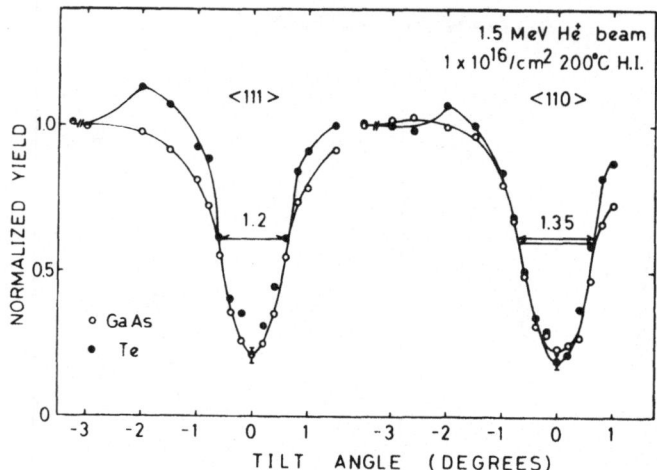

Fig. 11.   Channeling angular distributions observed by backscatter-
ing from 10^{16} Te/cm^2 implanted at 70 keV, 200°C into GaAs for 1.5
MeV He incident along the ⟨111⟩ and ⟨110⟩ axial directions.   From
Takai et al.[28]

to be approximately the case for Bi implanted in GaP in Fig. 12.[29]
The ⟨110⟩ direction contains pure rows of Ga and of P, each of
which can steer a channeled beam to a different critical angle
with the $\psi_{\frac{1}{2}}$ for P being smaller by a factor ≈ 0.8.   If the Bi is
located on the Ga sublattice the Bi critical angle will be similar
to that for Ga whereas if on the P sublattice it will be similar
to that for P.   As seen in Fig. 12 the Bi angular distribution for
the ⟨110⟩ direction is significantly narrower than for Ga and its
magnitude indicates that the Bi is located in nearly substitutional
sites on the P sublattice.[29]   As will be discussed in the following
section, the slight increase in yield and narrowing of the impurity
dip in the ⟨111⟩ direction is indicative of distortion of the Bi
from an exactly substitutional position.

Implanted impurity atoms are often less than 100% substitutional
and thus the impurity and host angular scans do not coincide.   Under
such conditions it is necessary to perform complete angular scans
along more than one direction for unambiguous location assignment.
One simplifying case is when a certain fraction of the impurities
are distributed randomly and thus give no angular structure, such
as can occur after precipitation of the impurity.   The randomly distributed component will give no orientation dependence, whereas the
substitutional component would contribute angular dips which are
proportional in magnitude to the substitutional fraction and which
would have the same angular width as that for the lattice.   The
fractional impurity concentration, f_s, lying along a given lattice
row or plane provided the impurity and host critical angles are the
same is

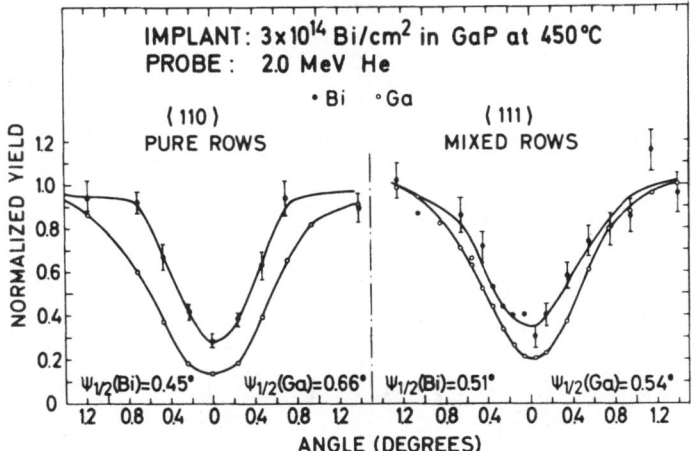

Fig. 12. Channeling angular distributions of Bi and the Ga sub-lattice observed by backscattering from 3 x 10^14 Bi/cm² implanted at 450°C into GaP for 2.0 MeV He incident along the ⟨110⟩ and ⟨111⟩ axial directions. From Merz et al.[29]

$$f_s = \frac{1 - \chi_i}{1 - \chi_h} \; ,$$

(5)

where χ_i and χ_h are the impurity and the host lattice minimum yields respectively. This determination of the fraction shielded by a row or plane is very useful for both substitutional and simple interstitial lattice site determinations. However, it must be cautioned that it cannot be applied if there is appreciable narrowing of the impurity critical angle or if the presence of flux peaks preclude accurate determination of the minimum yield.

An example of partial substitutionality is given in Fig. 13 for ⟨111⟩ channeling scans for Sb, Te, I, and Xe implanted in Fe.[30] A progressive increase in the impurity minimum yield can be seen and use of Eq. (5) gives a substitutional fraction which varies from $f_s \approx 0.9$ for Sb to $f_s \approx 0.4$ for Xe. As discussed, this assumes the impurities are either located on substitutional sites or randomly distributed. In this case there is also some narrowing of the angular width of the dip (≤ 15%) indicating that somewhat larger fractions of the impurities are substitutional and that they are displaced slightly from exactly substitutional sites. Such effects will be discussed in the following section. In this particular case deWaard and Feldman have analyzed these implanted impurities in Fe by both channeling and hyperfine measurements and concluded that impurity-defect association may be playing an important role.[30] Also in the case of I in Fe the nonsubstitutional fraction indicated

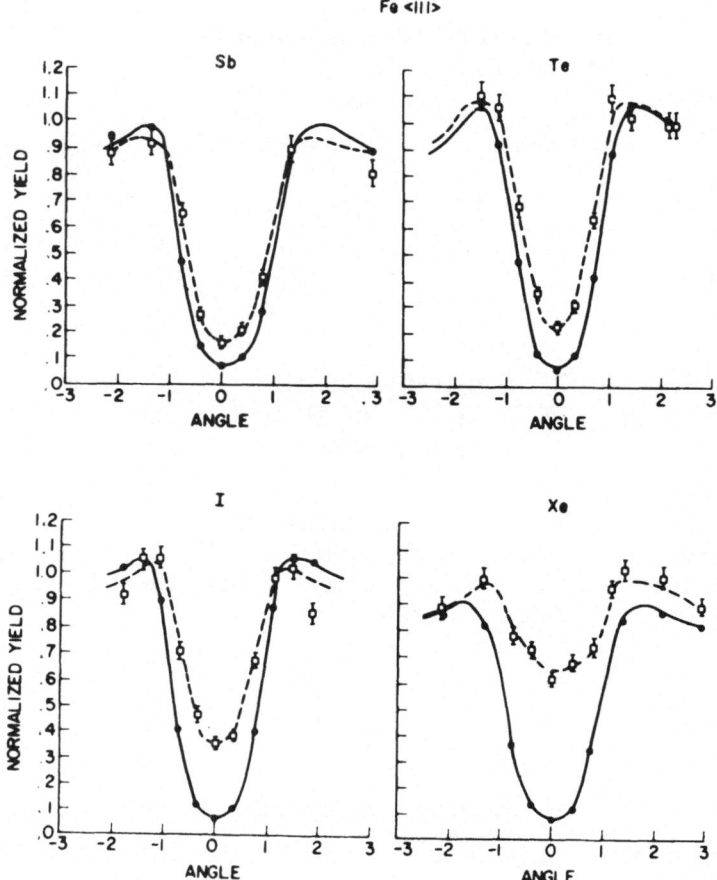

Fig. 13.   Channeling angular distributions for Sb, Te, I and Xe implants at 100 keV to fluences $5.5-8 \times 10^{14}/cm^2$ in Fe observed by backscattering for 1.8 or 2.0 MeV He incident along the ⟨111⟩ axis. From deWaard and Feldman.[30]

by the ⟨111⟩ scan is not simply in randomly-distributed sites since a different orientation dependence is observed along the {100} plane.

V.2.   Nearly Substitutional Impurities

A nearly substitutional impurity will be defined to have an appreciable time average probability of being located ~ 0.1-0.5 Å from a substitutional site. The channeling angular distribution for such impurities will be narrowed and the minimum yield will be increased over that for the host. This change can be observed along all channeling directions, depending upon the projected impurity position distribution along that direction. Nearly substitutional impurities include those with equilibrium displacements due to such effects as defect association and those with vibrational amplitudes appreciably larger than the host lattice. For most cases of practical interest the impurity atoms do not have a vibrational amplitude appreciably different from that for the host lattice atoms. For impurities of light atom mass relative to the host, large vibrational amplitudes may be observed, although in such cases the impurity is frequently located on interstitial rather than substitutional sites.

Reasonably accurate analytic estimates can be made of the effect of equilibrium displacement or large vibrational amplitude on the impurity channeling angular distribution by the analytic methods discussed in Section IV.[31-33] An approximate way to think of the narrowed width and increased yield for the angular distribution is to consider that when impurity atoms are located outside the minimum impact parameter for channeling (r_{min} is typically ~ 0.1 Å) they may be detected by channeled ions at angles less than the critical angle for channeling ($\psi_{\frac{1}{2}}$). A simple blocking estimate for the narrowing in axial channeling using the continuum model[12] with Lindhard's standard potential for a single row of atoms is given by

$$\psi_{\frac{1}{2}} = [U(r_{min})/E]^{\frac{1}{2}} \approx \left[\frac{1}{2} \ln \left(\frac{C^2 a^2}{r_{min}^2} + 1 \right) \right]^{\frac{1}{2}} \left(\frac{2Z_1 Z_2 e^2}{Ed} \right)^{\frac{1}{2}} , \qquad (6)$$

where $r_{min}^2 \approx \rho^2 \ln 2 + r_0^2$, r_0 is the equilibrium displacement, ρ the projected rms vibration amplitude, $C^2 \approx 3$ and the other terms as defined in Eqs. (1) and (2). A slightly more detailed treatment for the narrowing can be obtained using the half-way plane model.[12,23,32-34] An example of results of such a calculation of the angular distribution as a function of equilibrium displacement from a substitutional site is shown in Fig. 14 for the $\langle 110 \rangle$ direction in Si.[32] As can be seen the angular distribution narrows rapidly with increased displacement and in the limit the evolution of a flux peak can be seen from the calculations. The case of increasing vibrational amplitude without equilibrium displacement gives results similar to that for small displacements.[23,32] The half-way plane model does not accurately predict the minimum yield although it does show the correct trend. Modifications to minimum yield estimates like Eq. (4) can be used,[34,35] or for a better combined estimate of the minimum yield and angular narrowing, a multiple row calculation based on Eq. (3) can be used.[24,25]

Examples of near-substitutional impurities have been given in studies of Bi and Sb in Si.[32,33] In these studies it was concluded that some fraction or all of the impurity atoms were displaced by several tenths of an Å from substitutional sites. An example of

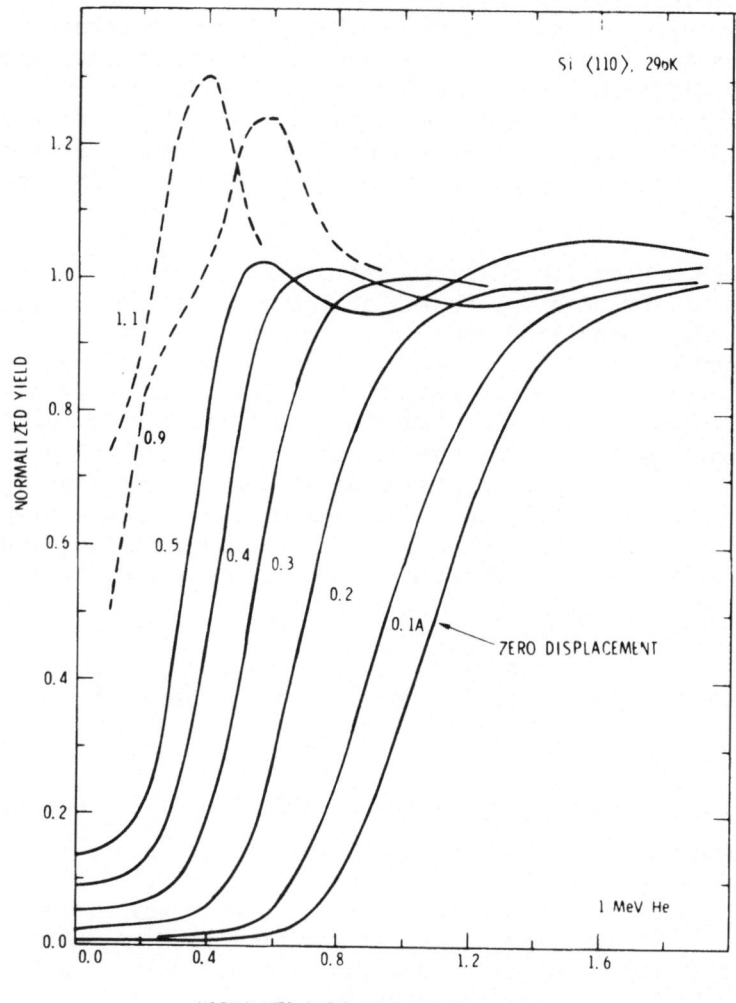

Fig. 14. Calculated angular distributions by half-way plane model as a function of displacement distance from the ⟨110⟩ row for 1 MeV He incident on Si. Angles normalized by characteristic angle $\psi_1 = (2Z_1 Z_2\ e^2/dE)^{\frac{1}{2}}$. From Picraux et al.[32]

the ⟨111⟩ angular distribution for Bi implanted in Si is shown in Fig. 15.[32] The Bi distribution is seen to be appreciably narrower than that for Si and to have a larger minimum yield, in qualitative agreement with calculations for nearly substitutional impurities. This effect was also observed along other channeling directions. Half-way plane calculations of the Bi angular distribution are shown by the dashed line and with corrections to the minimum yield by the dotted line. The curves for Fig. 15 were calculated for the Bi location which gave the best overall fit to all the angular distri-

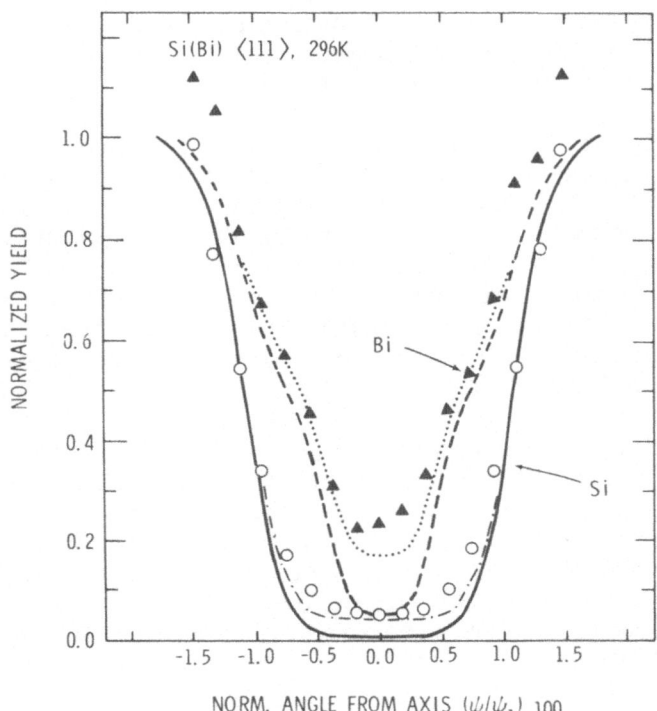

Fig. 15. Measured angular distributions for 2×10^{14} Bi/cm^2
implanted at 150 keV into Si and annealed to 650°C for 1 MeV He
incident along the ⟨111⟩ direction. The lines show the calculated
angular distributions for 50% of the Bi displaced 0.45 Å and 50%
substitutional without (solid and dashed) and with (dot-dashed and
dotted) minimum yield correction. From Picraux et al.[32]

bution data; that location was ≈ 50% of the Bi displaced ≈ 0.45 Å
from the Si lattice sites and ≈ 50% located substitutionally on Si
lattice sites. Another study of implanted Bi and Sb in Si has given
more uniform narrowing of the angular distributions consistent with
all the impurity atoms being displaced a small amount from substitutional sites.[33] In both cases the narrowing and increase in
minimum yield give a strong indication of significant fractions of
impurity atoms being displaced in their equilibrium position from
exact substitutional sites.

While there have been relatively few applications of double-
alignment channeling to lattice location studies, one potentially
valuable area of application is multiple-site occupation by impurity
atoms. Detailed analytic calculations of double-alignment angular
distributions have not been carried out to this author's knowledge
except for minimum yield estimates.[36] The double-alignment yield is
essentially of the order of the square of the single-alignment yield.

Thus if the impurity single-alignment yield is a sum of contributions from say two different sites with appreciably different angular distributions, then the double-alignment yield might help to further distinguish this from the single site case.

An example of double-alignment angular distributions is shown for Bi implanted in Si in Fig. 16.[32] The Si minimum yield decreases from 0.039 to 0.0059 from single to double alignment while the Bi minimum yield decreases from 0.164 to 0.050. Consistent with the discussion given above, this further indicates that the Bi $X_{min} \approx 0.16$ for single alignment cannot simply be due to this fraction of the Si atoms being displaced into random sites since from double-alignment measurements at least 95% of the Bi atoms are near (< 0.5 Å) Si lattice sites.

An additional use of double alignment would be to determine more accurately the total fraction of impurities which are in substitutional sites for cases of $> 95\%$ substitutionality as determined by single-alignment measurements. Counting statistics will limit the application of Eq. (5) in this range since typical host minimum yields are $X_h \approx 0.01-0.05$. However for double alignment the host minimum yield is further reduced to $\approx 0.001-0.01$, allowing an accuracy $\sim 1.$ to 0.1% for determining the substitutional fraction for the exactly substitutional case.

V.3. Interstitial Impurities

As has been discussed in Section IV, impurities near the center of a channel can lead to enhanced yields (flux peaks) near the center of the channeling angular distribution. This peaking effect along with the substitutional-like dips due to shielding of well-defined interstitial sites along certain axial or planar directions provides the primary means of locating interstitial impurities.

One of the earliest examples of a detailed lattice location study making use of the flux peaking effect has been given by Andersen et al. for Yb implanted in Si.[37] Channeling angular scans for this system along the ⟨100⟩, ⟨111⟩ and ⟨110⟩ axial directions are shown in Fig. 17. The flux peak along the ⟨110⟩ direction reaches a level ≈ 1.6 times the random yield and indicates that a significant fraction of the Yb is located near the center of the ⟨110⟩ channel. For the ⟨100⟩ direction, a fractional dip of the order of 30% is observed with the angular width similar to that for the Si lattice, suggesting that $\approx 30\%$ of the Yb is located along the ⟨100⟩ rows with the remaining fraction in some intermediate site between the lattice row and the center of the channel. For the ⟨111⟩ direction there is appreciable narrowing of the Yb channeling dip indicating a significant fraction of Yb atoms are located near ⟨111⟩ rows with a displacement of several tenths of an angstrom. From these axial scans as well as scans along planar

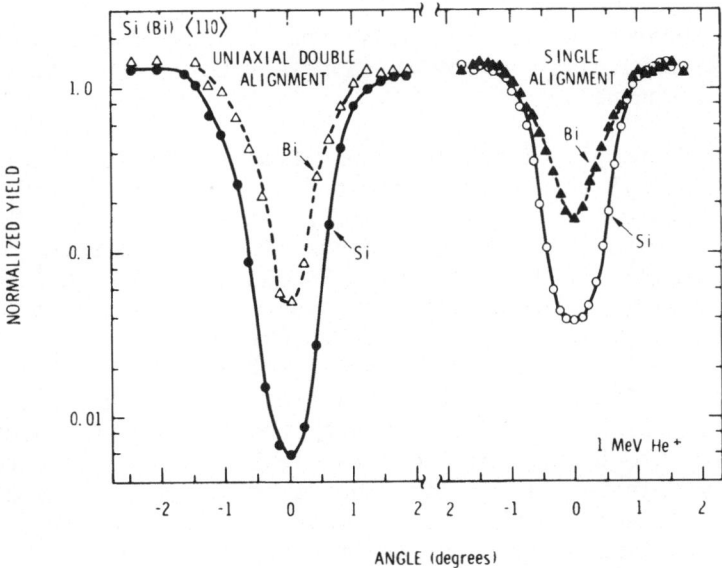

Fig. 16. Observed single- and uniaxial double-alignment angular distributions along the ⟨110⟩ axis for the same experimental conditions as given in Fig. 15. From Picraux et al.[32]

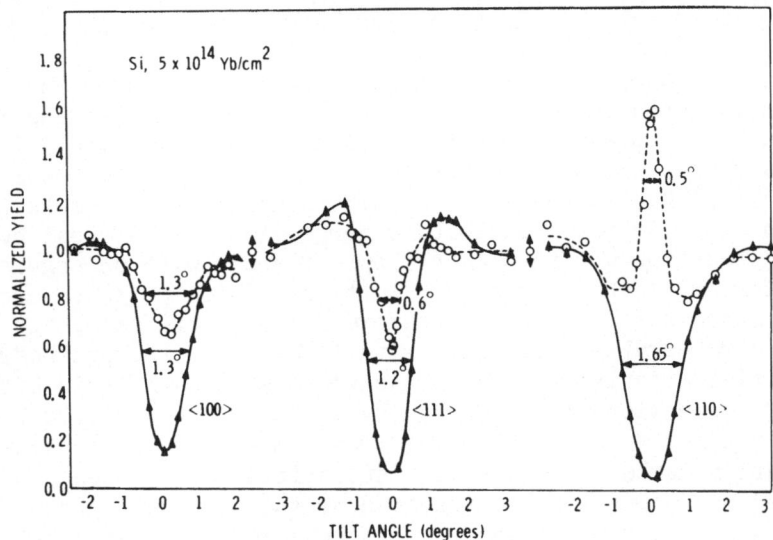

Fig. 17. Channeling angular distributions observed by backscattering from 5 x 10^{14} Yb/cm^2 implanted at 60 keV, 450°C into Si for 1 MeV He incident along the ⟨100⟩, ⟨111⟩ and ⟨110⟩ axial directions. From Andersen et al.[37]

directions Andersen et al. concluded that Yb was located in an
interstitial site given by moving along a ⟨100⟩ row to the center
of the ⟨110⟩ channel.[37]

A second example of the location of an interstitial impurity
by channeling is B in W. The implanted B was found to occupy the
octahedral interstitial position in the bcc W lattice.[38] This position gives rise to flux peaking effects along all three major axial
directions, ⟨111⟩, ⟨100⟩ and ⟨110⟩. In addition the ⟨100⟩ and ⟨110⟩
flux peaks reside in the envelope of a dip which is 1/3 of that for
the host lattice since 1/3 of the octahedral sites are contained
within the rows for a given ⟨100⟩ and ⟨110⟩ direction. However,
the most conclusive evidence that the B is located in the octahedral
interstitial position in W was given by the angular scan along the
{100} planar direction. For the octahedral position the sites
always lie within the {100} planes and thus the channeling dip for
the B should coincide with that for W as was observed. This study
pointed out that planar channeling directions can be very valuable
in distinguishing between possible lattice sites and also that
channeling dips along directions where the interstitial sites are
contained within rows or planes are the most definitive way to
obtain quantitative estimates of the fraction of impurity atoms
occupying a particular interstitial site. At present flux peaks
give valuable information for determining the location but have
not been useful for quantitatively determining the fraction of the
impurity in a particular interstitial site. This is due primarily
to the sensitivity of the shape and magnitude of the flux peak to
many experimental parameters as discussed in Section IV.

A third example of interstitial location by channeling is for
hydrogen isotopes in metals.[39] Figure 18 shows ⟨100⟩ scans for
deuterium (D) implanted into bcc crystals W and Cr. The D is
detected by the emitted protons from the ion-induced nuclear reaction $D(^3He,p)^4He$ and the W and Cr are detected by ion backscattering.
For both W and Cr the D exhibits a large central flux peak but for
Cr this peak lies within the envelope of a dip with an angular
width similar to that for the Cr lattice. Based on these and other
angular scans the D has been interpreted to be predominantly located
in the tetrahedral interstitial site in W and in the octahedral
interstitial site in Cr. The projected positions of this site along
the ⟨100⟩ direction are given by the squares in the insets where the
circles represent the ⟨100⟩ rows. The numbers in the squares give
the relative number of the given interstitial site in each of the
projected positions for the fraction of the unit cell shown. As
discussed in the B and W case, the octahedral site has 4 out of the
12 possible sites, corresponding to 1/3 of the D atoms, located
along ⟨100⟩ rows. Thus the observed narrow central flux peak due
to the D atoms in sites located in the center of the channel resides
within a broader channeling dip which is of similar width and ≈ 1/3
of the magnitude of the dip for the Cr lattice.

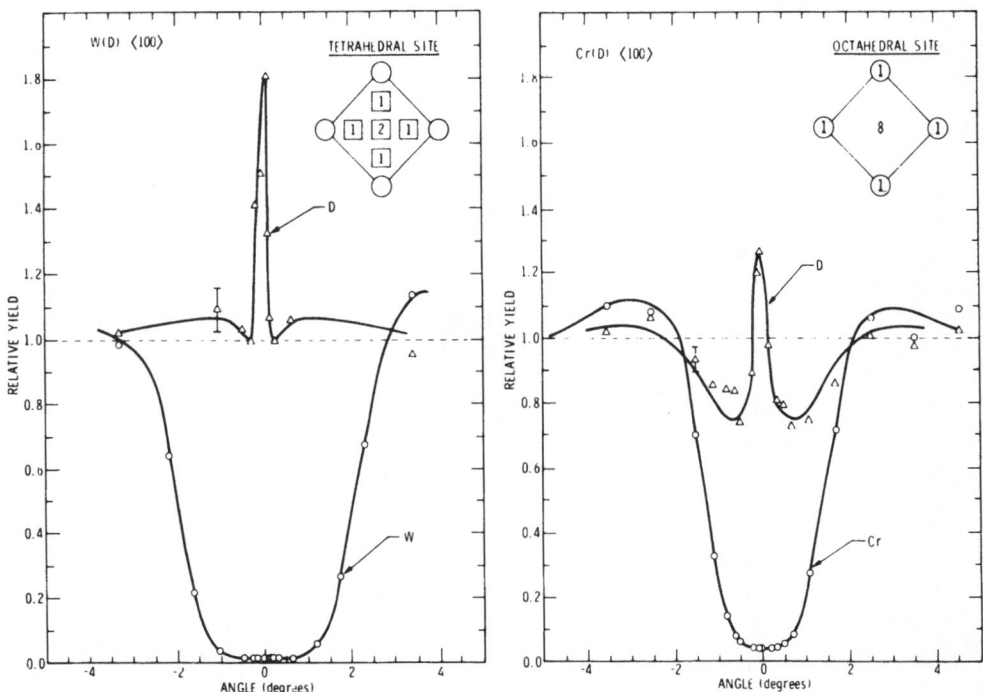

Fig. 18. Channeling angular distributions along the ⟨100⟩ axis observed by backscattering for the host lattice and by nuclear reaction for the impurity from $3 \times 10^{15} D/cm^2$ implanted at 30 keV into W and Cr using a 750 keV ^{3}He incident beam. Insets show octahedral and tetrahedral interstitial sites along the ⟨100⟩ direction as defined in Fig. 7. From Picraux and Vook.[39]

Impurity sites which have a projected position displaced from the center of the channel can result in multiple flux peaks as indicated by results of analytic calculations for position A in Fig. 8. The angular position ψ of these side-band flux peaks can be used to estimate the impurity displacement from the center of the channel by conservation of transverse energy arguments. Assuming statistical equilibrium, the impurity position will be on the contour line U(x,y) given by $E\psi^2 = U(x,y)$. Since the potential is much more slowly varying near the center of the channel than near the rows, the increase in ψ with distance r from the center of the channel is initially small. Thus relatively sharp central flux peaks (0.2 to 0.3° full width) would be expected even for displacements ~ 0.2-0.3 Å from the channel center.

The previous example of D in W shown in Fig. 18 has intermediate sites along the ⟨100⟩ directions sufficiently displaced to expect side-band flux peaks. While more detailed scans as shown by the curve labeled "initial" in Fig. 19 have given indications of such side-band peaks,[40] in general they are low in magnitude and

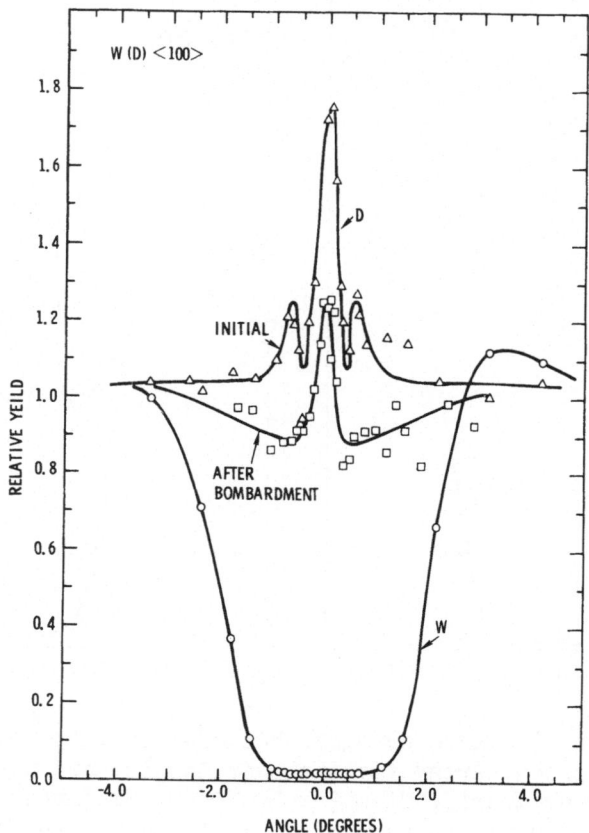

Fig. 19.   Channeling angular distributions for $1 \times 10^{15} D/cm^2$
implanted at 15 keV in W measured as in Fig. 18. The D scans are
shown as initially measured and after 750 keV ^{3}He irradiation to a
fluence of $4 \times 10^{16}/cm^2$. From Picraux and Vook.[40]

are difficult to observe experimentally, especially in the presence
of central flux peaks. Also the magnitude of side-band peaks may
depend fairly strongly on the azimuthal angle of the direction in
which an axial scan is made due to deviations of the channeled
particles from statistical equilibrium. Another example where
side-band peaks have been observed is Br implanted in Fe. Here
detailed measurements as well as analytic and Monte Carlo calculations indicated the Br was located 60% in a particular interstitial site and 40% substitutional.[13,41] Indications of multiple
peaking structure in channeling angular distributions also have
been observed for ⟨100⟩ scans of He implanted in W.[42] These results
have been interpreted in terms of multiple He atoms trapped at
vacancies which gives rise to the He located in well-defined sites
intermediate between the center of the ⟨100⟩ channel and the ⟨100⟩
rows.

V.4. High Impurity Concentrations

For sufficiently high concentrations of impurities located on discrete lattice sites the channeling behavior for the host lattice will be modified. The most important effects occur for interstitial impurities. As their concentration is increased interstitial impurities can give rise to strong multiple scattering of the channeled beam. This can broaden rapidly varying regions of the angular distribution such as flux peaks at quite shallow depths. At still higher concentrations impurities can directly modify the average atomic potential of the rows and planes, as well as form new rows.

The simplest high concentration case is that of a substitutional solid solution. While the critical angle may change due to the change in the atomic composition of the rows, this can be easily calculated from Eq. (1) and the angular scans will still coincide for the host and the impurity atoms. For example, for the $\langle 110 \rangle$ scans in Fig. 9 the concentration of Au in Cu was ≈ 2 at.% which is sufficient to just begin to broaden the dips.

For high concentrations of interstitial impurities the location interpretation of the data must be made with care since the angular distributions may be concentration dependent. An example of this is given by the work of Della Mea et al. in a study of the system TiO_x, where x was varied from 0.11 to 0.39.[43] Figure 20 shows the $\langle 0001 \rangle$ angular scans made within the $\{1\bar{2}10\}$ plane for the Ti and O signals as a function of O concentration from x = 0.11 to 0.28. For the $\langle 0001 \rangle$ direction the O atoms are known to lie in the center of the channel formed by Ti rows; the $\{1\bar{2}10\}$ plane is a mixed plane containing both O and Ti and thus the reduced yields with this planar dip should be approximately the same for both the Ti and O as seen in Fig. 20. For the $\langle 0001 \rangle$ axial direction there is no central flux peak even at the lowest O concentration studied of 0.11 atomic fraction since the observed yield rises only slightly above that for the random direction (100%). At the higher O concentrations of 0.20 and 0.28 one can see the formation of a dip in the O signal indicating that rows are being formed by the interstitial O of sufficiently high average potential to steer the incident beam and give rise to channeling effects. The large scattering yield to either side of the O dip is due to the strong scattering by the weaker O rows in the vicinity of the axial-to-planar transition.[43]

Another case where high impurity concentrations could give rise to a modification of the channeling behavior is if the impurity formed coherent compound precipitates in the single-crystal lattice. These regions, sometimes referred to as Gunier-Preston zones, would have to be sufficiently large to establish channeling behavior within the zones and have any orientation variation ("mosaic spread") small relative to the critical angle for channeling. Under such conditions the impurity channeling dip would have a critical angle

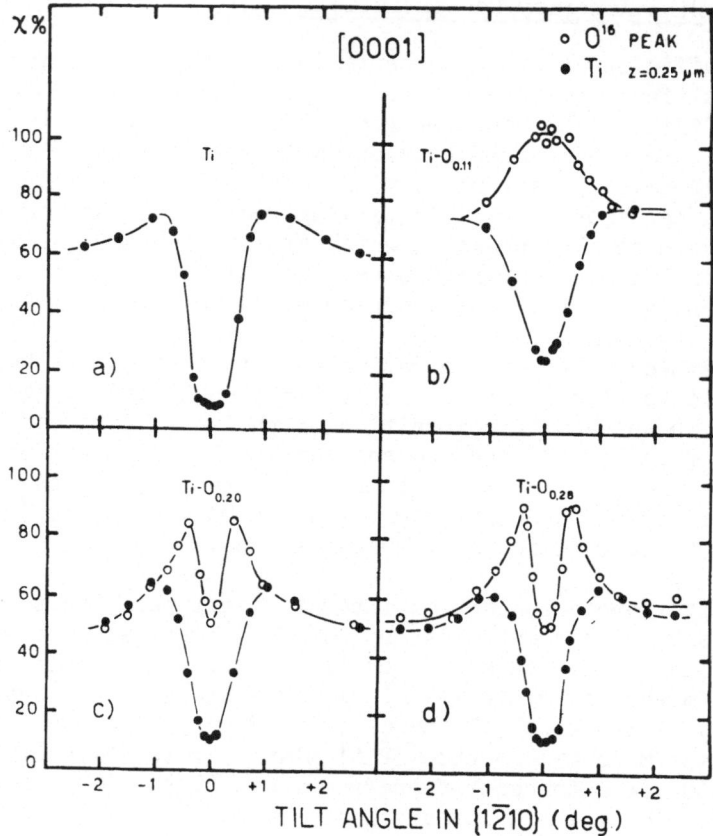

Fig. 20.   Channeling angular distributions along the $\langle 0001 \rangle$ axis for scans within the $\{1\overline{2}10\}$ plane in Ti and TiO$_x$ by backscattering from Ti and nuclear reaction product from O using a 1 MeV D incident beam.   From Della Mea et al.[43]

characteristic of the average potential of the atom rows in the precipitate rather than that of the crystal lattice.   Thus from Eq. (1) the impurity critical angle would be a measure of $(Z_2/d)^{\frac{1}{2}}$ thus providing crystal structure information about the precipitates. No effects of this nature have been reported to the author's know-ledge, although minimum yield measurements have been carried out as a function of Gunier-Preston zone formation in the AlCu system[44] and angular scans have been carried out on AlSb coherent precipi-tates formed by Sb implantation in Al.[45]

V.5.   Radiation-Induced Change in Impurity Sites

In the presence of radiation damage impurities may become asso-ciated with defects.   While the local environment of an impurity cannot be directly sampled by channeling, the influence of a defect can be observed in cases where it modifies the lattice location of the impurity. An example of this is given by recent studies of Mn in Al at 35°K where channeling measurements indicate that prior to bombardment > 96% of the Mn atoms are on substitutional sites.[46] However, irradiations with fluences of > 10^{15} He/cm^2 at temperatures 35-65°K give rise to the new channeling angular distribution shown for the $\langle 110 \rangle$ direction in Fig. 21. The indication of a flux peak on the envelope of a dip for the Mn signal suggests some appreciable fraction of the Mn atoms are located in discrete sites appreciably displaced from substitutional positions. Channeling angular scans along these and other directions have been interpreted as being con- sistent with the trapping of radiation-induced interstitial Al atoms giving rise to the formation of Al-Mn atom pairs in the $\langle 100 \rangle$ dumbell configuration.

A second example of radiation-induced change in lattice site is given by studies of As-doped Si.[47] Prior to bombardment, the As was nearly completely substitutional as determined by channeling angular scans. This is indicated by the backscattering energy spectra for measurements along $\langle 100 \rangle$ and random directions in Fig. 22. After bombardment by 1.8 MeV He ions to a fluence of ~ 1 x 10^{16}/cm^2 the aligned As signal increased to a normalized yield ~ 35%, indicating that as many as 30% of the As atoms have been displaced from substitutional lattice sites. Channeling angu- lar distributions along the $\langle 110 \rangle$ and $\langle 111 \rangle$ directions showed no change in the angular width from that for the host, as indicated in Fig. 23. This suggests that the impurity atoms displaced from substitutional sites are either not of sufficiently high symmetry or else not in sufficiently well-defined sites to give channeling structure to the scans.

A third example is shown in Fig. 19 for $\langle 100 \rangle$ angular scans of D implanted in W before and after He bombardment.[40] The data strongly suggest that He irradiations to fluences ~ 5 x 10^{16}/cm^2 result in the movement of the D from simple tetrahedral interstitial sites to fairly well-defined new interstitial sites which are dis- placed appreciably from the tetrahedral position. While the angular scans indicate that an appreciable fraction of the D has moved to this new lattice site after He irradiation the exact position of the site has not been unambiguously determined.

These examples demonstrate how impurity-defect interactions can be studied directly by channeling lattice location measurements. However, they also serve to caution one to adopt certain procedures in channeling location studies. To guard against modification of the impurity site prior to the measurement it is useful to move the analysis beam or the sample so that a new spot is sampled once the crystal has been aligned. Also, it is valuable to take the first data point directly aligned with the lattice row or plane (i.e., $\psi = 0°$)

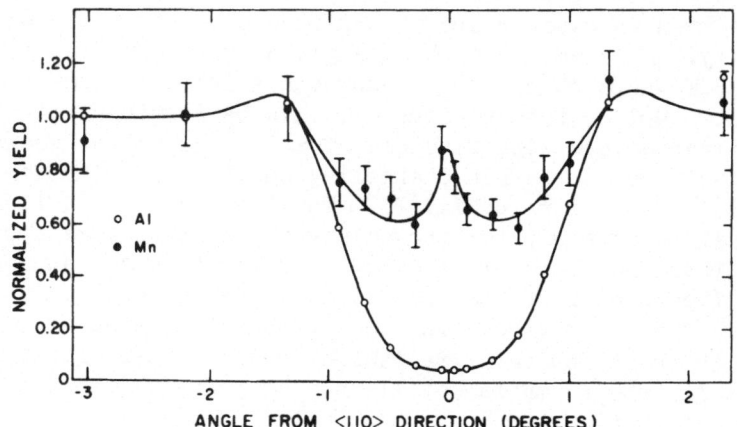

Fig. 21.   Channeling angular distributions along the ⟨110⟩ axis observed by backscattering of incident 1 MeV He from Al with 0.09 at.% Mn after irradiation at 65°K with ∼ 10^{16} He/cm^2 at 0.3 MeV.   From Swanson et al.[46]

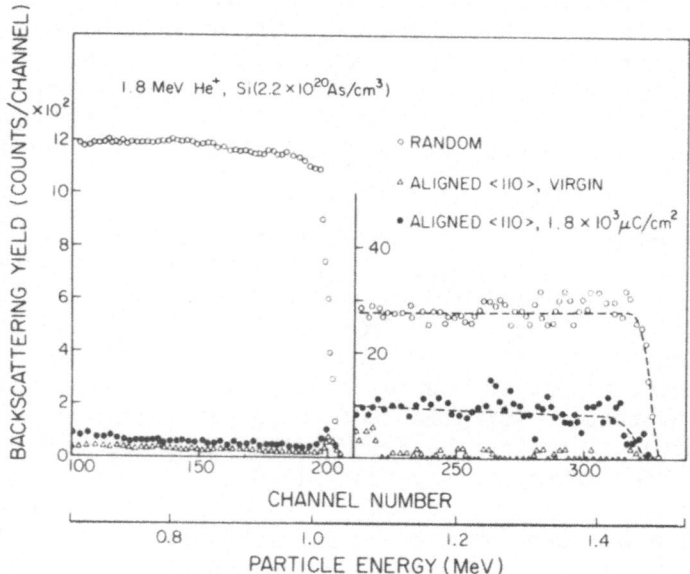

Fig. 22.   Backscattered energy spectra for 1.8 MeV He incident on As-diffused (2.2 x 10^{20} As/cm^3) Si along a ⟨110⟩ axial direction before (Δ) and after (●) irradiation to a fluence of ≈ 1 x 10^{16} He/cm^2 at 1.8 MeV, and along a random direction (o) From Haskell et al.[47]

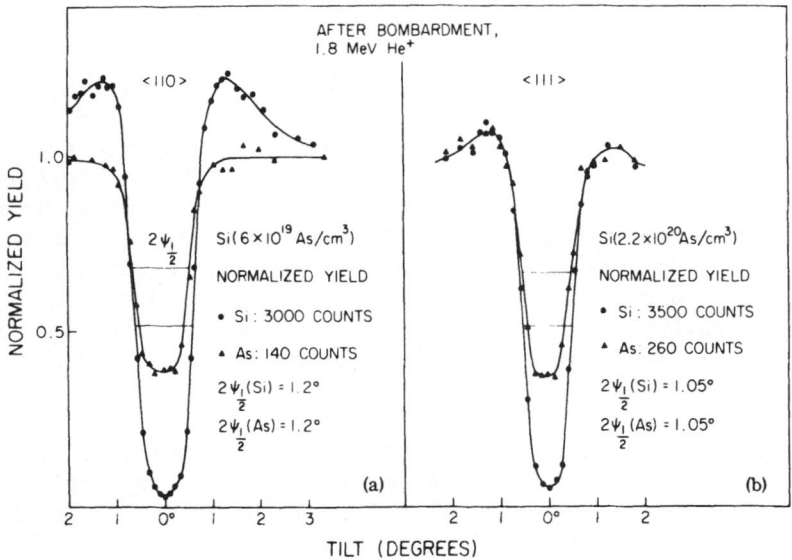

Fig. 23. Channeled angular distributions for 1.8 MeV He incident on as-grown (6 x 10^{19} As/cm^3) Si along the $\langle 110 \rangle$ axis and As-diffused (2.2 x 10^{20} As/cm^3) Si along the $\langle 111 \rangle$ axis, after 1.8 MeV He irradiation to a fluence $\approx$ 6 x 10^{15}/cm^2. From Haskell et al.[47]

and to repeat this point at the end of the angular scan to check for any changes during the measurement. The magnitude of a flux peak is very sensitive to small levels of radiation damage and may decrease somewhat during the course of an angular scan; however, any additional changes in the shape of the angular scan should be viewed as a possible indication of radiation-induced impurity movement.

VI. SUMMARY OF THE LITERATURE ON CHANNELING LATTICE LOCATION DATA

A summary of all the published ion channeling lattice location studies known to the author for dilute impurities in metals and semiconductors is given in Tables I and II, respectively. The method of introduction and detection of the impurity is listed and the observed location is indicated qualitatively. For substitutional and near-substitutional impurities S has been used for aligned fractions (see Eq. (5)) f_S > 50% and S-R for 25% < f_S < 50%. In some cases the assigned location appears to this author to be tentative. Also defect association, which can alter impurity location, can only be inferred in channeling measurements and so detailed experimental conditions may affect impurity location in such cases. For implanted impurities, particularly in semiconductors, the impurities will often only occupy well-defined sites after appropriate annealing either during or after implantation.

Thus it is important to consult the original references when considering reasons for the location of a particular impurity.

Early studies in semiconductors[20] and studies in metals[30] have been summarized previously. Unfortunately, in the case of semiconductors much work was done prior to a detailed understanding of flux peaking effects for nonsubstitutional impurities. As a result some of these measurements are only determinations of the minimum yields for the impurity and the lattice atoms along various axial directions. An example of this has been given in studies of Tl in Si where, although simple axial attenuation data indicate a mixture of Tl on substitutional and interstitial sites, detailed angular scans result in modification of the relative number of Tl assigned to these sites.[105]

A majority of the location studies have been done for the impurity introduced into the metal or semiconductor by implantation. In cases where impurities have been introduced by diffusion or growth from the melt, as well as by implantation, the agreement has been good. The most extensive investigations have beem made for impurities in Fe for the metals, and in Si for the semiconductors. While Si has received the most attention there has recently been an increase in interest in the metals.

VII. LIMITATIONS

One area of limitations in the application of ion channeling to lattice location studies is experimental in nature. The relatively low bulk sensitivity for impurity detection by ion beam analysis requires high concentrations of the impurity to be present. In many cases these concentrations exceed equilibrium solid solubilities. While such systems can frequently be formed by ion implantation at temperatures sufficiently low that the impurity cannot migrate away from the near-surface region or form precipitates, one cannot always be certain that measurements at these high concentrations are completely characteristic of the dilute limit.

A second experimental limitation is that of radiation-induced damage by the analysis beam. As demonstrated in Section V.5, association of the impurity with irradiation-induced defects may modify the impurity location. Even in the absence of these effects sufficiently high radiation fluences will degrade the channeling in the host crystal due to direct displacement of lattice atoms. In many cases damage effects may effectively set lower limits on the impurity concentrations which can be studied in lattice location determinations.

A third experimental limitation is the perfection of single crystals. Moderately large areas of single crystals are required

with a relatively low mosaic spread. Preferably the mosaic spread
should be a factor ~ 5 smaller than the characteristic angular
widths of features which are being studied. Host lattice minimum
yields will be increased for mosaic spreads of the order of a few
tenths of a degree and the slope of the sides of the dip will be
decreased. In the case of 100% substitutional impurities such
effects may not be drastic or in some cases even noticed. But in
the case of nonsubstitutional impurities where flux peaks with
characteristic widths of the order of 0.2° would be present, such
mosaic spreads may preclude precise location determination.

Another area of limitations relates to difficulties in making
detailed analytic estimates of the channeling effect. These difficulties relate primarily to accurate analytic determinations of the
particle flux density across the channel. While Monte Carlo calculations give the best available results, it is difficult to
accurately simulate the true experimental situation.

Numerous experimental conditions as discussed in Section IV
strongly influence the magnitude of central flux peaks in channeling
measurements and additional complications arise from such effects
as oscillations in the particle density as a function of depth at
a given position in the channel. Flux peaks may be broadened due
to high impurity concentrations as discussed in Section V.4 and
this is difficult to account for accurately in simple analytic
descriptions of ion channeling. In addition, some corrections to
the impurity yield may be necessary due to the difference in the
stopping power for random to aligned directions. This difference
can give rise to enhancement of the impurity signal over an angular
width of the order of that for the channeling dips (i.e., much
greater impurity flux peak widths than normally observed). Such
wide impurity flux peaks could occur, for example, for a uniformly
distributed impurity detected by an ion-induced nuclear reaction
with an energy dependent cross section. In this case even with the
impurity randomly distributed throughout the lattice there would
still be an enhanced yield along the channeling direction, since
the decreased dE/dx of the channeled beam would permit interaction
with a greater number of impurity atoms for a given energy loss ΔE
in a channeled direction. These effects on the flux peak imply
that careful consideration must be given to lattice location data
when observed flux peaks are appreciably broader than 0.2 to 0.3°.

A third area of limitations is more basic to the channeling
effect itself. Unambiguous lattice site determination may not be
possible in cases where the impurity occupies multiple sites or a
single site of low symmetry (i.e., there are not good channeling
directions for which large fractions of the impurities can be found
either near the rows or near the center of the channel). For such
cases it may only be possible to show consistency with proposed
locations based on independent information. An example of a low

symmetry site would be an impurity displaced along the $\langle 110 \rangle$ direction a sufficient distance that while it would appear substitutional along one of the $\langle 110 \rangle$ rows its equivalent projection along directions corresponding to the other five equivalent $\langle 110 \rangle$ directions would not give rise to any strong flux peaks or channeling dips (a position intermediate between A and B in Fig. 8). In such a case only 1/6 of the impurities would be contained in any given $\langle 110 \rangle$ row and one would observe an $\approx 17\%$ dip with no other strong angular features. In many cases when there is only a single site occupied by an impurity this site can be determined for a sufficient number of angular distribution measurements. However, if all the angular distribution measurements result in relatively poor dips ($< 50\%$) with no other strong features such as narrowing of the dips or flux peaks, then in practice it is quite difficult to unambiguously determine the location. Also it is possible that different combinations of multiple sites could give rise to similar angular distributions, so that while it is true that a given distribution of impurity sites should give rise to a unique angular distribution, the angular scans will not necessarily uniquely specify the distribution of impurity sites. An additional intrinsic limitation in the channeling technique is that usually small fractions of an impurity in one site (i.e., $\approx 5\%$) cannot be detected in the presence of the much stronger signal due to the majority of the impurity atoms being located on another site.

VIII. CONCLUSIONS

Current capabilities of the ion channeling technique for the study of impurity location in single crystals allow the determination of substitutional, nearly substitutional, and well-defined interstitial positions to accuracies ~ 0.1 Å. In addition semiquantitative information can usually be given about the relative fraction of impurities in a particular site. In the case of multiple site occupation unambiguous interpretation of the data is much more difficult and usually only semiquantitative or qualitative information can be given about the lattice location. There are numerous possibilities for new and important work in this field of research. One aspect is to improve and further develop the channeling technique for lattice location studies. Here there is a need for careful and systematic work. Improvements are needed in the analytic estimates of the spatial flux distribution of particles across a channel. This should allow more quantitative determinations of lattice locations for nonsubstitutional impurities. In addition, further Monte Carlo computer simulation studies will be needed to establish the analytic models on firm ground. Experimental studies of the flux peaking effect as a function of depth for uniformly-doped samples would give better confidence in calculations. Additional studies are needed on the influence of defects, of high impurity concentrations and of the azimuthal orientation of angular scans on flux peaking effects. There has been little evaluation of the usefulness of double alignment measurements. A careful evaluation of electron channeling for location studies would be of interest. The low radiation damage by electrons, for example, might give electrons some unique advantages although quantum mechanical effects and impurity detection capability may limit the general applicability.

Another aspect to be pursued would be to extend the lattice location measurements to new systems. Often there is great value to studying classes of systems. Examples of such studies which have been discussed here would include the impurities used for the electrical doping of Si, impurities for which there exists hyperfine measurements in Fe, hydrogen in metals, and certain classes of defect-associated impurities in Al. Examination of Tables I and II show that there has been relatively little work on non-implanted impurities. Also, while only location studies in metals and semiconductors have been included in this review, one should indicate that there have been only a very few studies in other materials. Finally, while this chapter has concentrated on the technique of ion channeling for lattice location studies in metals and semiconductors, we believe that the most important contributions to solid state science are given by correlations of channeling location measurements with other studies of the behavior of impurities in solids.

TABLE I - Summary of Location Studies in Metals

Host	Impurity	Method Intro.[a]	Method Detect.[b]	Location[c]	References
Be	Ag	I	BS	S	48
	In	I	BS	T	48
	Hf	I	BS	T	49
Al	B	I	(p,α)	R	38
	Mn	M,I	BS	S,D	46,50
	Cu	M	BS	S	44
	Zn	M	BS	S,D	50
	Ga	I	BS	S	51
	Mo	I	BS	S	51
	Ag	M,I	BS	S,D	46,50
	Cd	I	BS	R	51
	Sn	M	BS	S,D	50
	Sb	I,M	BS	R	45
	Cs	I	BS	R	51
	Au	I	BS	R	51
	Bi	I	BS	R	51
Ti	O	I,D,M	$(p,\alpha)(d,p)$	O	43,52,53
	In	I	BS	S	54
V	Ga	I	BS	S	55
	Se	I	BS	S	55
	Kr	I	BS	R	55
	In	I	BS	S	55
	Cs	I	BS	R	55
	Bi	I	BS	S	55
Cr	D	I	$(^{3}He,p)$	O	39
Fe	B	I	(p,α)	I	38
	C	I	(d,p)	O	56
	F	I	$(p,\alpha\gamma)$	R	57
	Ar	I	x ray	R	58
	K	I	x ray	R	58
	Ca	I	x ray	S	58
	Cu	I	BS	S	59
	Se	I	BS	S	60
	Br	I	BS	I-S	13,41
	Sn	I	BS	S	61
	Sb	I	BS	S	30,62
	Te	I	BS	S,D	30
	I	I	BS	S,D	30
	Xe	I	BS	R-S	30,63
	Ce	I	BS	I	62
	Yb	I	BS	S,D	64,65
	Au	I	BS	S	62

TABLE I (continued)

Host	Impurity	Method Intro.[a]	Method Detect.[b]	Location[c]	References
Fe	Tl	I	BS	S	66
	Pb	I	BS	S	66
	Bi	I	BS	S	66
Co	Se	I	BS	S	60
Ni	B	I	(p,α)	R	38
	F	I	$(p,\alpha\gamma)$	R	57
	Sn	I	BS	S	51
	Dy	I	BS	S	67
	Hf	I	BS	S	68
	Au	I	BS	S	26
Cu	D	I	(d,n)	I	69
	B	I	(p,α)	R	38
	Rb	I	BS	R	51
	Mo	I	BS	S	70
	Ru	I	BS	S	51
	Ag	I,M	BS	S,D	71,72
	Sb	I,M	BS	S,D	71,72
	I	I	BS	S	71
	Xe	I	BS	R	71
	Cs	I	BS	R	51
	Ce	I	BS	R	51
	Eu	I	BS	R	51
	Ta	I	BS	S	51
	W	I	BS	S	26
	Pt	I	BS	S	71
	Au	I,M	BS	S,D	25,26,70,73
	Hg	I	BS	S	71
	Tl	I	BS	S	71
	Pb	I	BS	S	71
	Bi	I	BS	S	51,71
Zn	In	I	BS	R	54
Zr	Au	I+D	BS	S,D	74
Nb	D	D	(d,p)	T	75,76
	Li	I	(n,α)	I	77
	O	D	(p,α)	O	78
Pd	Au	I	BS	S	26
Ag	Au	I	BS	S	26
Sn	Au	D	BS	S	79

TABLE I (continued)

Host	Impurity	Method Intro.(a)	Method Detect.(b)	Location(c)	References
Ta	Xe	I	β^- emit.	-	5
W	D	I	$(^3He,p)$	T	39,42
	^3He	I	(d,p)	D	42
	B	I	(p,α)	O	38
	Rn	I	α emit.	S	5
Pb	Au	M	β^- emit.	I	80

(a) Method of impurity introduction: I = ion implantation,
     D = diffusion, M = grown from the melt.

(b) Method of impurity detection: BS = ion backscattering,
     (a,b) = ion-induced nuclear reaction with a the incident
     particle and b the emitted particle, x ray = ion-induced
     x rays, α emit. and $\beta^\pm$ emit. correspond to detecting radio-
     active α, β^+ or β^- decay.

(c) Categories of lattice site location: S = $\geq$ 50% of the
     impurity on or near substitutional sites, I = interstitial
     with T = tetrahedral and O = octahedral interstitial sites,
     R = random sites, i.e., no strong orientation dependence,
     D = defect association.

TABLE II. Summary of Location Studies in Semiconductors[a]

Host	Impurity	Method Intro.	Method Detect.	Location	References
Si	Li	I	(p,α)	I,R	81
	B	I,D	(p,α)	I,S	33, 81-88
	N	I	(p,α)	R	89
	Na	I	(p,α)	I	90
	P	I	BS,β^-emit	S	81, 91, 92
	Zn	I	BS	I	93, 94
	Ga	I	BS,x ray	S-I	95-101
	As	I,D	BS	S,D	47,82,92,95,96,98,102,103
	Se	I	BS	S	93,94
	Rb	I	BS	R	104
	Zr	I	BS	I	105
	Cd	I	BS	I	93,94,98,106
	In	I	BS	S-I	82,95,96,98,101,107
	Sn	I	BS	S	98,108,109
	Sb	I,D	BS,x ray	S	33,47,82,95,96,98,102,110,111
	Te	I	BS	S	93,94,98,106
	Xe	I	BS	R-S	82,98
	Cs	I	BS	R	98,104
	Tm	I	BS	I	112
	Yb	I	BS,β^-emit	I	37, 113-115
	Hf	I	BS	I	105
	Au	D	BS	$S^{[b]}$,I	98,102,116
	Hg	I	BS	S-I	93,94,105,117
	Tl	I	BS	S-I	95,98,99,105,107,118,119
	Pb	I	BS	S	33
	Bi	I	BS	S	32,33,95,98,119
Ge	Li	D	(n,α)	-	120
	N	I	$(d,\alpha)(p,\alpha)$	R	120a
	P	I	x ray	S	121
	S	I	x ray	S-I	121
	In	I	BS	S	95,96,99,122-124
	Sb	I	BS	S	95,122-124
	Hg	I	BS	S-I	123
	Tl	I	BS	S-I	99,122-124
	Pb	I	BS	S	122,123
	Bi	I	BS	S	122-124
SiC	N	I	(p,α)	S	125
	In	I	BS	S	126
GaP	Bi	I	BS	S	29,127
GaAs	Cd	I	BS	S	128,129
	Sn	I	BS	S	130
	Sb	I	BS	S	131
	Te	M,I	BS	S	28,131-133

TABLE II (continued)

Host	Impurity	Method Intro.	Method Detect.	Location	References
InSb	Tl	I	BS	R	134
	Bi	I	BS	R	134
CdTe	Au	M,D	BS	S	135
	Tl	I	BS	R	134
	Bi	I	BS	S	134

(a) Symbols same as given by footnotes of Table I.

(b) Measurements in heavily P-doped Si where Au-P pairing is believed to occur.

REFERENCES

1. D. S. Gemmell, Rev. Mod. Phys. 46, 129 (1974).

2. Channeling: Theory, Observation and Applications, ed. by D. V. Morgan (John Wiley, 1973).

3. J. A. Davies, ibid., p. 391.

4. W. M. Gibson and M. Maruyama, ibid., p. 349.

5. B. Domeij, Nucl. Instrum. Methods 38, 207 (1965).

6. E. Uggerhoj and J. U. Andersen, Can. J. Phys. 46, 543 (1968).

7. H. Matzke and J. A. Davies, J. Appl. Phys. 38, 805 (1967).

8. E. Bogh and J. L. Whitton, Phys. Rev. Lett. 19, 553 (1967).

9. E. Laegsgaard, J. U. Andersen, and L. C. Feldman, Phys. Rev. Lett. 29, 1206 (1972).

10. J. A. Davies, L. Eriksson, N.G.E. Johansson, and I. V. Mitchell, Phys. Rev. 181, 548 (1969).

11. L. Eriksson, G. R. Bellavance, and J. A. Davies, Radiation Effects 1, 71 (1969).

12. J. Lindhard, K. Dan. Vidensk. Selsk. Mat.-Fys. Medd. 34, No. 14 (1965).

13. R. B. Alexander, P. T. Callaghan, and J. M. Poate, Phys. Rev. B9, 3022 (1974).

14. D. Van Vliet, Radiation Effects 10, 137 (1971).

15. J. H. Barrett, Phys. Rev. B3, 1527 (1971).

16. F. Abel, G. Amsel, M. Bruneaux, and C. Cohen (private communication).

17. J. Ellison, University of New Mexico, Dept. of Mathematics and Statistics Technical Reports #300 and #305, Sept. 1974.

18. Y. Hashimoto, J. H. Barrett, and W. M. Gibson, Phys. Rev. Lett. 30, 995 (1973) and Atomic Collisions in Solids, ed. by S. Datz, B. R. Appleton, and C. D. Moak (Plenum Press, N. Y., 1974) in press.

19. This goniometer was built from design of W. Augustyniak with modifications by J. Smalley.

20. J. W. Mayer, L. Eriksson, and J. A. Davies, <u>Ion Implantation in Semiconductors</u> (Academic Press, N. Y., 1970).

21. J. U. Andersen and L. C. Feldman, Phys. Rev. <u>B1</u>, 2063 (1970).

22. D. V. Morgan and D. Van Vliet, Radiation Effects <u>12</u>, 203 (1972).

23. J. U. Andersen, K. Dan. Vidensk. Selsk. Mat.-Fys. Medd. <u>36</u>, No. 7 (1967).

24. Y. Yokoyama, Masters Thesis, Nagoya University (1974).

25. R. B. Alexander and J. M. Poate, Radiation Effects <u>12</u>, 211 (1972).

26. J. M. Poate, J. A. Borders, W. J. DeBonte and W. M. Augustyniak, Appl. Phys. Lett. <u>25</u>, 698 (1974).

27. J. A. Borders, J. M. Poate, and W. J. DeBonte, Bull. Am. Phys. Soc. <u>19</u>, 257 (1974) and private communication.

28. M. Takai, K. Gamo, K. Masuda, and S. Namba, Jap. J. Appl. Phys. <u>12</u>, 1926 (1973).

29. J. L. Merz, L. C. Feldman, D. W. Mingay, and W. M. Augustyniak, <u>Ion Implantation in Semiconductors</u>, ed. by I. Ruge and J. Graul (Springer-Verlag, Berlin, 1971), p. 182.

30. H. deWaard and L. C. Feldman, <u>Applications of Ion Beams to Metals</u>, ed. by S. T. Picraux, E. P. EerNisse, and F. L. Vook (Plenum Press, N. Y., 1974), p. 317.

31. J. A. Davies, <u>European Conference on Ion Implantation</u> (Peregrinus, Stevenage, England, 1970), p. 172.

32. S. T. Picraux, W. L. Brown, and W. M. Gibson, Phys. Rev. <u>B6</u>, 1382 (1972).

33. D. Sigurd and K. Bjorkqvist, Radiation Effects <u>17</u>, 209 (1973).

34. K. Tachibana, Radiation Effects <u>19</u>, 135 (1973).

35. H. E. Roosendaal, W. H. Kool, W. F. Van Der Weg, and J. B. Sanders, Radiation Effects <u>22</u>, 89 (1974).

36. L. C. Feldman and B. R. Appleton, Appl. Phys. Lett. 15, 305 (1969).

37. J. U. Andersen, O. Andreasen, J. A. Davies, and E. Uggerhoj, Radiation Effects 7, 25 (1971).

38. J. U. Andersen, E. Laegsgaard, and L. C. Feldman, Radiation Effects 12, 219 (1972).

39. S. T. Picraux and F. L. Vook, Phys. Rev. Lett. 33, 1216 (1974).

40. S. T. Picraux and F. L. Vook, Ion Implantation: Science and Technology, ed. by S. Namba (Plenum Press, N. Y., 1975) in press.

41. R. B. Alexander and P. T. Callaghan, Phys. Lett. 45A, 379 (1973).

42. S. T. Picraux and F. L. Vook, Applications of Ion Beams to Metals, ed. by S. T. Picraux, E. P. EerNisse, and F. L. Vook (Plenum Press, N. Y., 1974), p. 407.

43. G. Della Mea, A. V. Drigo, S. Lo. Russo, P. Mazzoldi, S. Yamaguchi, G. G. Bentini, A. Desalvo, and R. Rosa, Phys. Rev. B10, 1836 (1974).

44. R. Hellborg, Physica Scripta 5, 219 (1972).

45. G. J. Thomas and S. T. Picraux, Applications of Ion Beams to Metals, ed. by S. T. Picraux, E. P. EerNisse, and F. L. Vook (Plenum Press, N. Y., 1974), p. 257.

46. M. L. Swanson, F. Maury, and A. F. Quenneville, ibid., p. 393 and Phys. Rev. Lett. 31, 1057 (1973).

47. J. Haskell, E. Rimini, and J. W. Mayer, J. Appl. Phys. 43, 3425 (1972).

48. E. N. Kaufmann, R. S. Raghaven, P. Raghaven, E. J. Ansaldo, and R. A. Naumann, Phys. Rev. B (to be published).

49. E. N. Kaufmann, K. Krien, J. C. Soares, and K. Freitag (to be published).

50. M. L. Swanson and F. Maury, Can. J. Phys. (to be published).

51. D. K. Sood and G. Dearnaley, AERE Report, Harwell (to be published).

52. R. B. Alexander and R. J. Petty, <u>Atomic Collisions in Solids</u>,
     ed. by S. Datz, B. R. Appleton, and C. D. Moak (Plenum Press,
     N. Y., 1974) in press.

53. G. Della Mea, A. V. Drigo, S. Lo. Russo, P. Mazzoldi,
     S. Yamaguchi, G. G. Bentini, A. Desalvo, and R. Rosa, <u>Atomic</u>
     <u>Collisions in Solids</u>, ed. by S. Datz, B. R. Appleton, and
     C. D. Moak (Plenum Press, N. Y., 1974), p. 791.

54. E. N. Kaufmann, P. Raghaven, R. S. Raghaven, K. Krien,
     E. J. Ansaldo, and R. A. Naumann, <u>Applications of Ion Beams to</u>
     <u>Metals</u>, ed. by S. T. Picraux, E. P. EerNisse, and F. L. Vook
     (Plenum Press, N. Y., 1974), p. 379.

55. G. Linker, M. Gettings, and O. Meyer, <u>Ion Implantation in</u>
     <u>Semiconductors and Other Materials</u>, ed. by B. L. Crowder
     (Plenum Press, N. Y., 1973), p. 465.

56. L. C. Feldman, E. N. Kaufmann, J. M. Poate, and
     W. M. Augustyniak, <u>Ion Implantation in Semiconductors and</u>
     <u>Other Materials</u>, ed. by B. L. Crowder (Plenum Press, N. Y.,
     1973), p. 491.

57. J. R. MacDonald, E. N. Kaufmann, W. Darcey, and R. Hensler,
     Radiation Effects (to be published).

58. J. R. MacDonald, R. A. Boie, W. Darcey, and R. Hensler, (to
     be published).

59. R. A. Boie, J. R. MacDonald, and J. M. Poate (to be published).

60. P. T. Callaghan, N. J. Stone, and B. G. Turrell, Phys. Rev.
     <u>B10</u>, 1075 (1974).

61. E. Bogh, Proc. R. Soc. A <u>311</u>, 35 (1969).

62. R. B. Alexander, N. J. Stone, D. V. Morgan, and J. M. Poate,
     <u>Hyperfine Interactions in Excited Nuclei</u>, ed. by G. Goldring
     and R. Kalish (Gordon and Breach, N. Y., 1971), p. 229.

63. L. C. Feldman and D. Murnick, Phys. Rev. <u>B5</u>, 1 (1972).

64. F. Abel, M. Bruneaux, C. Cohen, H. Bernas, J. Chaumont, and
     L. Thomé, Solid State Commun. <u>13</u>, 113 (1973).

65. R. B. Alexander, E. J. Ansaldo, B. I. Deutch, J. Gellert, and
     L. C. Feldman, <u>Applications of Ion Beams to Metals</u>, ed. by
     S. T. Picraux, E. P. EerNisse, and F. L. Vook (Plenum Press,
     N. Y., 1974), p. 365.

66. L. C. Feldman, E. N. Kaufmann, D. W. Mingay, and
     W. M. Augustyniak, Phys. Rev. Lett. $\underline{27}$, 1145 (1971).

67. G. A. Stephens, E. Robinson, and J. S. Williams, <u>Ion Implanta-
     tion: Science and Technology</u>, ed. by S. Namba (Plenum Press,
     N. Y., 1975) in press.

68. E. N. Kaufmann, J. M. Poate, and W. M. Augustyniak, Phys. Rev.
     $\underline{B7}$, 951 (1973).

69. H. Fischer, R. Sizmann, and F. Bell, Z. Phys. $\underline{224}$, 135 (1969).

70. D. K. Sood and G. Dearnaley, J. Vac. Sci. Technol. $\underline{12}$, 445
     (1975).

71. J. A. Borders and J. M. Poate (to be published).

72. M. L. Swanson, A. F. Quenneville, and F. Maury (to be
     published).

73. M. L. Swanson, A. F. Quenneville, and F. Maury, Phys. Status
     Solidi (to be published).

74. M. L. Swanson and L. M. Howe, J. Nucl. Mater. $\underline{54}$, 155 (1974).

75. H. D. Carstanjen and R. Sizmann, Ber. Bunsenges. Phys. Chem.
     $\underline{72}$, 1223 (1972) and Phys. Lett. $\underline{40A}$, 93 (1972).

76. G. A. Iferov, G. P. Pokhil, and A. F. Tulinov, JETP Lett. $\underline{5}$,
     250 (1967).

77. J. P. Biersack and D. Fink, <u>Applications of Ion Beams to
     Metals</u>, ed. by S. T. Picraux, E. P. EerNisse, and F. L. Vook
     (Plenum Press, N. Y., 1974), p. 307.

78. P. P. Matyash, N. A. Skakun, and N. P. Dikii, JETP Lett. $\underline{19}$,
     18 (1974).

79. J. W. Miller, D. S. Gemmell, R. E. Holland, J. C. Poizat,
     J. N. Worthington, and R. E. Loess, Phys. Rev. $\underline{B11}$, 990
     (1975).

80. P. N. Tomlinson and A. Howie, Phys. Lett. $\underline{27A}$, 491 (1968).

81. W. M. Gibson, F. W. Martin, R. Stensgaard, F. P. Jensen,
     N. I. Meyer, G. Golster, A. Johansen, and J. S. Olsen, Can. J.
     Phys. $\underline{46}$, 675 (1968).

82. J. A. Davies, J. Denhartog, L. Eriksson, and J. W. Mayer,
     Can. J. Phys. 45, 4053 (1967).

83. J. C. North and W. M. Gibson, Appl. Phys. Lett. 16, 126 (1970).

84. J. C. North and W. M. Gibson, Radiation Effects 6, 199 (1970).

85. G. Fladda, K. Bjorkqvist, L. Eriksson, and D. Sigurd,
     Appl. Phys. Lett. 16, 313 (1970).

86. J. A. Cairns, R. S. Nelson, and J. S. Briggs, Ion Implantation
     in Semiconductors, ed. by I. Ruge and J. Graul (Springer-Verlag,
     Berlin, 1971), p. 299.

87. Y. Akasaka and K. Horie, Ion Implantation in Semiconductors
     and Other Materials, ed. by B. L. Crowder (Plenum Press, N. Y.,
     1973), p. 147.

88. Y. Akasaka and K. Horie, J. Appl. Phys. 44, 3372 (1973).

89. J. B. Mitchell, P. P. Pronko, J. Shewchun, D. A. Thompson
     and J. A. Davies, J. Appl. Phys. 46, 332 (1975).

90. N. A. Skakun, N. P. Dikii, P. P. Matyash, and V. M. Korol,
     Sov. Phys.-Solid State 15, 123 (1973).

91. C. R. Allen and C. R. Thomas, Electron. Lett. 9, 475 (1973).

92. F. Fujimoto, K. Komaki, M. Watanabe and T. Yonezawa, Appl.
     Phys. Lett. 20, 248 (1972).

93. J. Gyulai, O. Meyer, R. D. Pashley, and J. W. Mayer,
     Radiation Effects 7, 17 (1971).

94. O. Meyer, N. G. E. Johansson, S. T. Picraux and J. W. Mayer,
     Solid State Commun. 8, 529 (1970).

95. J. W. Mayer, J. A. Davies, and L. Eriksson, Appl. Phys. Lett.
     11, 365 (1967).

96. J. W. Mayer, L. Eriksson, S. T. Picraux, and J. A. Davies,
     Can. J. Phys. 46, 663 (1968).

97. L. Eriksson, J. A. Davies, and J. W. Mayer, Radiation Effects
     in Semiconductors, ed. by F. L. Vook (Plenum Press, N. Y.,
     1968), p. 398.

98. L. Eriksson, J. A. Davies, N. G. E. Johansson and J. W. Mayer,
     J. Appl. Phys. 40, 842 (1969).

99.   L. Eriksson, G. Fladda, and K. Bjorkqvist, Appl. Phys. Lett. 14, 195 (1969).

100. W. F. van der Weg, J. A. den Boer, F. W. Saris, and D. Onderdebinden, European Conference on Ion Implantation (Peregrinus, Stevenage, England, 1970), p. 198.

101. K. Masuda, K. Gamo, A. Imada, and S. Namba, Ion Implantation in Semiconductors, ed. by I. Ruge and J. Graul (Springer-Verlag, Berlin, 1971), p. 455.

102. S. Chou, Ph.D. Thesis, Stanford University (1971) (unpublished) and S. Chou, L. A. Davidson, and J. F. Gibbons, Appl. Phys. Lett. 17, 23 (1970).

103. F. Fujimoto, K. Komaki, M. Ishii, H. Nakayama, and K. Hisatake, Phys. Status Solidi A12, K7 (1972); F. Fujimoto, K. Komaki, K. Hisatake and H. Nakayama, Phys. Status Solidi A5, 737 (1971).

104. O. Meyer and J. W. Mayer, Solid State Electron. 13, 1357 (1970).

105. B. Domeij, G. Fladda, and N.G.E. Johansson, Radiation Effects 6, 155 (1970).

106. S. T. Picraux, N.G.E. Johansson, and J. W. Mayer, Semiconductor Silicon, ed. by R. R. Haberecht and E. L. Kern (Electrochem. Soc., N. Y., 1969), p. 422.

107. J. A. Davies, L. Eriksson, and J. W. Mayer, Appl. Phys. Lett. 12, 255 (1968).

108. Y. Akasaka, K. Horie, G. Nakamura, K. Tsukamoto, and Y. Yukimoto, Jap. J. Appl. Phys. 13, 1533 (1974).

109. G. Weyer, J. U. Andersen, B. I. Deutch, J. A. Golovchenko, and A. Nylandsted-Larsen (to be published); G. Weyer, B. I. Deutch, A. Nylandsted-Larsen, and J. U. Andersen, Hyperfine Interactions Studied in Nuclear Reactions and Decay, ed. by E. Karlsson and R. Wappling (to be published, Sweden, 1974).

110. L. Eriksson, J. A. Davies, J. Denhartog, J. W. Mayer, O. J. Marsh, and R. Mankarious, Appl. Phys. Lett. 10, 323 (1967).

111. J. A. Cairns and R. S. Nelson, Phys. Lett. 27A, 14 (1968).

112. E. Bøgh, <u>Interaction of Radiation with Solids</u>, ed. by
      A. Bishay (Plenum Press, N. Y., 1967), p. 361.

113. E. Uggerhøj and J. U. Andersen, Can. J. Phys. <u>46</u>, 543 (1968).

114. F. H. Eisen and E. Uggerhoj, Radiation Effects <u>12</u>, 233
      (1972).

115. F. H. Eisen, <u>Ion Implantation in Semiconductors</u>, ed. by
      I. Ruge and J. Graul (Springer-Verlag, Berlin, 1971), p. 287.

116. N. Iue, N. Matsunami, K. Morita, N. Itoh, M. Yoshida, and
      S. Hirota, Radiation Effects <u>14</u>, 191 (1971).

117. G. Della Mea, A. V. Drigo, P. Mazzoldi, G. Nardelli, and
      R. Zannoni, Radiation Effects <u>3</u>, 259 (1970).

118. G. Fladda, P. Mazzoldi, E. Rimini, D. Sigurd, and L. Eriksson,
      Radiation Effects <u>1</u>, 249 (1969).

119. L. Eriksson, G. R. Bellavance, and J. A. Davies, Radiation
      Effects <u>1</u>, 72 (1969).

120. J. P. Biersack and D. Fink, <u>Atomic Collisions in Solids</u>, ed.
      by S. Datz, B. R. Appleton, and C. D. Moak (Plenum Press,
      N. Y., 1974) in press.

120a. A. B. Campbell, J. B. Mitchell, J. Shewchun, D. A. Thompson
      and J. A. Davies, Can. J. Phys. <u>53</u> (1975)(in press).

121. J. F. Chemin, I. V. Mitchell, and F. W. Saris, J. Appl. Phys.
      <u>45</u>, 537 (1974).

122. K. Bjorkqvist, B. Domeij, L. Eriksson, G. Fladda, A. Fontell,
      and J. W. Mayer, Appl. Phys. Lett. <u>13</u>, 379 (1968).

123. K. Bjorkqvist, D. Sigurd, G. Fladda, and G. Bjarnholt,
      Radiation Effects <u>6</u>, 141 (1970).

124. A. W. Tinsley, K. C. Jones, P.R.C. Stevens, G. G. George,
      and E. M. Gunnerson, <u>European Conference on Ion Implantation</u>
      (Peregrinus, Stevenage, England, 1970), p. 187.

125. A. B. Campbell, J. B. Mitchell, J. Shewchun, D. A. Thompson,
      and J. A. Davies, <u>Ion Implantation: Science and Technology</u>,
      ed. by S. Namba (Plenum Press, N. Y., 1975) in press.

126. R. R. Hart. H. L. Dunlap, and O. J. Marsh, <u>Ion Implantation
      in Semiconductors</u>, ed. by I. Ruge and J. Graul (Springer-
      Verlag, Berlin, 1971), p. 134.

127. L. C. Feldman, W. M. Augustyniak, and J. L. Merz, Radiation Effects 6, 293 (1970).

128. G. Ilic, G. T. Ewan, and J. L. Whitton, Radiation Effects 18, 47 (1973).

129. K. Gamo, M. Takai, M. S. Lin, K. Masuda, and S. Namba, Ion Implantation: Science and Technology, ed. by S. Namba (Plenum Press, N. Y., 1975) in press.

130. T. G. Finstad, S. L. Anderson, and T. Olsen, Phys. Status Solidi A25, 515 (1974).

131. K. Gamo, M. Takai, K. Masuda, S. Namba, Proc. of the 4th Conf. on Solid State Devices, Supplement to the J. of the Japan Soc. of Appl. Phys. 42, 130 (1973).

132. I. V. Mitchell, J. W. Mayer, J. K. Kung, and W. G. Spitzer, J. Appl. Phys. 42, 3982 (1971).

133. F. H. Eisen, J. S. Harris, B. Welch, R. D. Pashley, D. Sigurd, and J. W. Mayer, Ion Implantation in Semiconductors and Other Materials, ed. by B. L. Crowder (Plenum Press, N. Y., 1973), p. 631.

134. G. Langguth, E. Lang, and O. Meyer, Ion Implantation in Semiconductors, ed. by I. Ruge and J. Graul (Springer-Verlag, Berlin, 1971), p. 228.

135. W. Akutagawa, D. Turnbull, W. K. Chu, and J. W. Mayer, J. Phys. Chem. Solids (to be published).

CHAPTER 5

ION IMPLANTATION IN METALS

G. Dearnaley

AERE, Harwell, England

INTRODUCTION

For several years now it has become obvious to all that ion implantation provides a versatile and controllable technique for the doping of semiconductors. Indeed, so important and successful has ion implantation proved in silicon device technology that the term has become almost synonymous with this field of semiconductor research and development. This point of view does less than justice to the scientific and economic importance of other branches of surface materials technology, where the ability to introduce controlled amounts of specific impurities into a solid can be valuable.

Already, in Chapter 6, Dr. Meyer has shown how the properties of superconductors have been modified by ion implantation. Many other materials properties, both physical and chemical, are susceptible to the presence of implanted impurities. The present chapter will focus attention on metals and alloys where the most important surface processes are probably wear and corrosion: each of these phenomena is responsible for a huge annual expenditure, exceeding the total sales of the semiconductor device industry. Other important effects are the modification of electrochemical behaviour or of the bonding ability of a metal surface. In many cases, particularly when one has exploited conventional materials technology to the full, the properties listed above are controlled by the composition within a depth of a few microns below the surface.

Why not choose the bulk composition to suit these surface requirements in the first place? After all, this is what is done

in fabricating articles of stainless steel. Consider however how
much of the world's limited resources of chromium and other
expensive metals are incorporated into alloys for reasons
connected solely with their influence on surface behaviour.
Quite often there is a conflict between surface and bulk properties,
and the cheapness, strength, thermal, electrical or nuclear
properties may determine the choice of material for a component.
Conflicting surface requirements are then usually met by a
coating applied, for example, by painting, electroplating,
diffusion, spray-coating, hot-dipping or cladding. For many
purposes these are satisfactory and well-established methods, but
in certain instances such techniques prove inadequate, usually
because of interfacial corrosion or bonding failure, and it is
not easy to detect potential trouble of this kind.

     There is a parallel here with semiconductor device technology,
where diffusion seemed perfectly adequate until ion implantation
arrived. In metals, ion implantation shares with diffusion the
ability to produce a surface which is coherent with the substrate,
with no interfacial weakness and negligible dimensional change.
Just as in semiconductors, there may be cases where implantation
is best coupled with diffusion in order to obtain an adequate
depth of protection.

     After reviewing the historical development of this relatively
new field, I shall summarise its present state particularly in
relation to wear and corrosion, and finally will discuss the
equipment necessary for ion implantation in metals, since the
requirements differ in some respects from those of semiconductors.

                          HISTORICAL PERSPECTIVE

     Metal targets have been bombarded with ion beams ever since
the first ion accelerators were developed over 40 years ago.
Usually, however, the ions available were of light, gaseous
species such as H^+ or He^+, the most obvious visible effect on the
material was an accummulation of cracked hydrocarbon due to poor
vacuum conditions, and the scientists concerned were far more
interested in the nuclear properties of their targets.

     Some chance observations were reported, however, and the
fact that ion bombardment even with inert gas ions can strongly
affect the oxidation of certain metals has been known since 1961,
when Trillat and Haymann[1] reported that argon ion bombardment
will inhibit the normally rapid tarnishing of a freshly-polished
uranium surface. In an important review, Thompson[2] suggested
that this effect is due to some damaging mechanism, not yet
fully understood, rather than any chemical influence of the
implanted argon. The importance of both mechanical and chemical

effects of ion implantation in metals is now apparent.

In a brief note, Crowder and Tan[3] reported that the implantation of boron ions, to doses of around $10^{16}/cm^2$, will reduce the atmospheric tarnishing of copper.

These isolated observations, while interesting, were not followed up. The reasons lay either in the prior claims of semi-conductor research, or in the lack of an adequate variety of ion beams in sufficient intensities to achieve the necessary doping levels. While semiconductor properties are readily altered by small concentrations of dopant species, so that a typical implanted dose, or fluence, may be 10^{14} ions/cm^2, in metals a typical dose may be 10^{16} ions/cm^2. Furthermore, one knows that in Si the most interesting implanted species are B^+, P^+ and As^+, while in metals one's 'palette' must be much larger, for the variety of useful additives is very wide.

Thus, further developments had to await the availability of high-intensity, versatile ion implantation machines, as well as the scientific incentive. What seems to have been the first systematic investigation of the field of ion implantation in metals and alloys began at Harwell about 1970.

This was because

(i)   suitable accelerators had been developed for the ion implantation project, and these had versatile ion sources;

(ii)   Harwell has a broad concern with materials, and particularly for the demands of reactor metallurgy. Expert knowledge of metallurgy and corrosion was available;

(iii) ion implantation in semiconductors was seen to be moving more and more into industrial laboratories (a healthy evolution) so that, while certainly not abandoning the semiconductor field, Harwell could take up some longer term developments.

The present Chapter will review much of the work carried out under this project, and also the studies made at the University of Salford, together with some published work from the U.S.A., France and Germany. The total world effort involved in the ion implantation of metals is still very small, compared with that in semiconductors, and the field is still a new one. Despite this, the results achieved seem promising and the interest of metallurgists and corrosion scientists is sufficiently aroused that jointly-organised conferences are planned. A new series of

conferences, commencing at Albuquerque in 1973 with the title
"Applications of Ion Beams to Metals"[4] has provided a useful
forum for discussion.

FRICTION AND WEAR

The classical theory of metallic friction[5] considers that
friction arises from the force required to shear numerous metallic
junctions or micro-welds which are set up at the contacting
asperities of two sliding surfaces. Friction, under non-
lubricated conditions, will depend, among other things, on the
strength and location of the shearing junctions, on whether an
oxide is formed due to localised heating in air, and on the
strength of the oxide-metal adhesion.

Wear, as discussed for example by Archard[6], involves
adhesion, which results in the tearing out of surface material.
It also involves abrasion, or the ploughing action of both hard
asperities and of loose debris. Under suitable conditions
corrosion may play a part since it may be the corrosion film that
wears (or alternatively, acts as a solid lubricant).

Thus, friction, wear and corrosion all play a part in the
field now often referred to as tribology[7], although the relationship between coefficient of friction and wear rate is never
simple. In general, we may say that wear is a more complex
phenomenon than friction: the role of previously-formed wear
debris makes this clear.

In 1971, Hartley began, in work at Harwell and the University
of Sussex, to investigate the effect of ion implantation on the
coefficient of friction of metals (mostly steels). Friction
measurements were obtained using the simple apparatus shown in
Figure 1 in which the implanted specimen is driven slowly beneath
a loaded tungsten carbide ball[8]. The horizontal frictional
force, registered by a transducer, is amplified and recorded by a
pen recorder.

Hartley found[8,9,10] that all ion implantations produced
significant changes in friction, except in the case of the inert
gas Kr^+. Large frictional changes occurred following the implantation of soft metals such as Ag^+, Sn^+, In^+ and Pb^+, the greatest
effect being obtained after about 3.10^{16} ions/cm^2 of Sn , at
380 keV, where the reduction was 60 per cent (see Table I).
Implantation did not always reduce the coefficient of friction
and 6.10^{16} ions/cm^2 of Pb^+ (at 175 keV) produced a succession of
very pronounced 'stick-slip' events, characteristic of local
adhesion followed by intermittent sliding: this rapid fluctuation
in friction causes a broadened trace in the pen recording

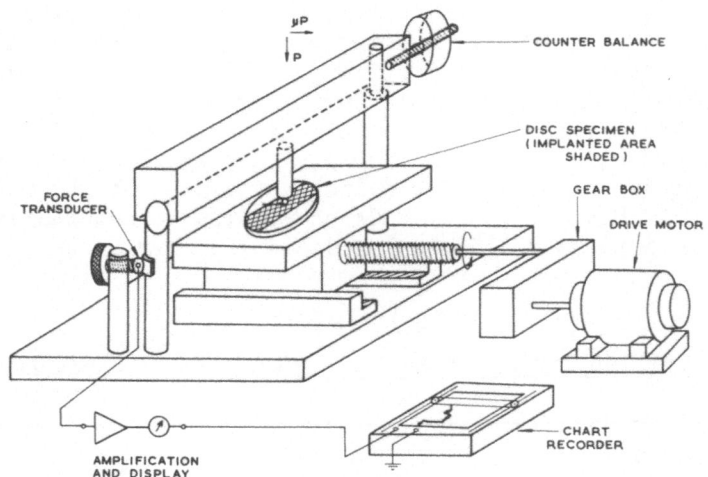

Figure 1.   Schematic diagram of apparatus used for measuring
            friction changes in ion-implanted metals. (from
            Hartley et al.(8))

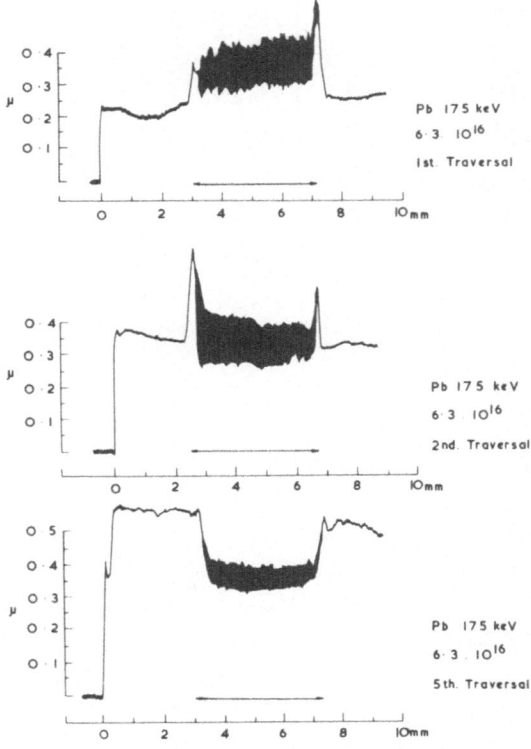

Figure 2.   Pen-recorder traces of the coefficient of friction,
            μ, in Pb+ implanted En 352 steel, showing the effect of
            repeated traversals of the same groove. Markers below
            the traces indicate the extent of the ion-implanted
            zone. (from Hartley et al.(10))

(Figure 2), absent in other instances (Figure 3). Retracking
several times over the same friction groove reduced the amplitude
of these oscillations, and the friction remains constant, while
rising considerably in the neighbouring areas of unimplanted steel
(Figure 2). This may be due to the formation of hard oxide
particles (Fe_3O_4) at hot spots during sliding, whereas the softer
lead oxides are known to act as a solid lubricant at high
temperatures[7]. Examination by ESCA showed that a substantial
fraction of the implanted lead was oxidised after friction
testing.

TABLE 1

Friction Data for Ion Implanted En 352 Tested in Air (from ref. [10])

Ion	Dose $(ions-cm^{-2})$	μ_{En352}	$\mu_{Impl.}$	Remarks
Kr	$2.8.10^{16}$	.24	.24	Friction Peak at Implantation Boundary
Sn	$2.8.10^{16}$	.24	.09	
In	$2.8.10^{16}$	.30	$.31 \pm .05$	Erratic; Transfer of ions during wear
Ag	$2.8.10^{16}$	.22	.26	As In^+; Adhesion
Pd	$6.3.10^{16}$	.23	$.33 \pm .08$	Stick-slip; Adhesion Friction Peak (as Kr)
Mo	$2.8.10^{16}$	.26	.24	
S	$6.1.10^{16}$	.20	.19	
Mo + 2S	$2.8.10^{16}$ + $5.6.10^{16}$	.26	.20	

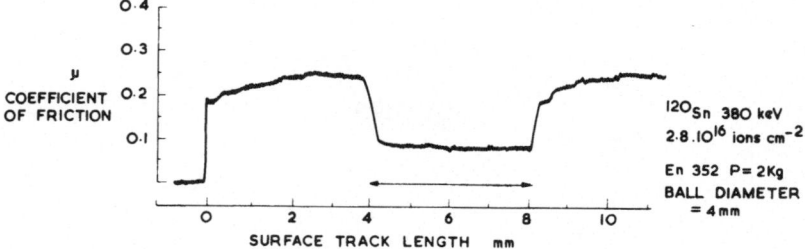

Figure 3.   Pen-recorder trace of the coefficient of friction, μ,
             in Sn^+ implanted En 352 steel. Markers below the trace
             indicate the extent of the ion-implanted zone.
             (from Hartley et al.[10])

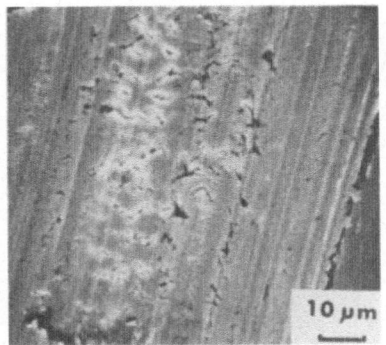

a

b

Figure 4.   (a) Scanning electron micrograph of the groove produced
            during friction tests in Pb+ - implanted En 352 steel.
            Transverse cracks are clearly visible. (from Hartley
            et al.(8))

            (b) Scanning electron micrograph of the groove produced
            during friction tests in Mo+ - implanted En 352 steel.
            No transverse cracks, and no stick-slip adhesion, are
            apparent in this case.

    Scanning electron microscope examination of the surface after
such a friction test reveals clear evidence for stick-slip adhesion
in a Pb+ implanted steel specimen (figure 4a). Transverse cracks
show how the surface layers have been sheared following local
adhesion. By contrast, a Mo+ implanted specimen of the same
steel shows no evidence of this behaviour, either in the friction
trace or in surface topography (figure 4b).

    The relative change in friction induced, during the first
traversal, is shown as a function of ion dose for three different
species in Figure 5. The ion energies were chosen to give a
similar mean range of about 300Å, and the highest doses employed

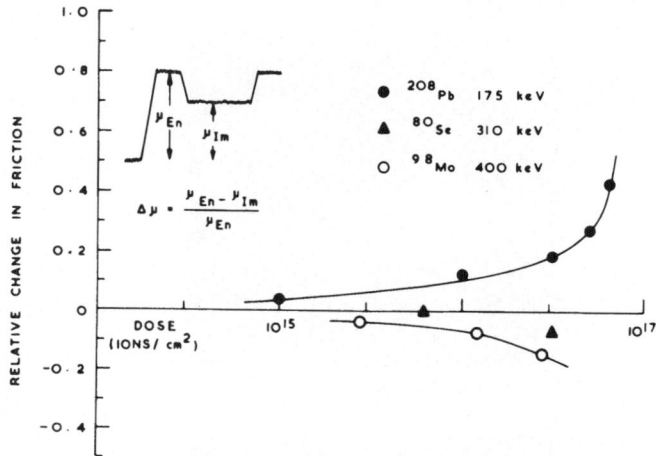

Figure 5.   The effect of ion dose on the relative change in
             friction, μ, produced by Pb⁺, Se⁺ and Mo⁺ ions in
             En 352 steel. (from Hartley et al.[10])

$(6.10^{16}$ ions/cm^2) give a peak concentration, at or close to the
metal surface, of about 4 per cent in the case of Pb⁺, as
measured by ion backscattering.

The differing behaviour of different species, the lack of any
effect produced by Kr⁺ ions and the reproducibility of the
phenomena dispel the suggestion that ion bombardment has merely
created a carbonaceous contaminant layer that modifies friction.
Obviously, however, clean hydrocarbon-free vacuum conditions
during implantation are necessary to avoid this.

It is interesting to note that Hartley observed a greater
effect following the implantation of Mo⁺ ions together with twice
the dose of S⁺ ions than in the case of either of these species
implanted alone. Although he was unable to prove that MoS$_2$ has
been formed within the steel, this possibility could not be ruled
out.

Ion backscattering, which has been reviewed thoroughly in
this volume, offers a most useful means of studying the behaviour
and transfer of implanted material as a result of friction tests.
For this purpose, Hartley used an ion microbeam 10 μm to 15 μm in
diameter developed by Cookson et al.[11]. This magnetically-
focused ion beam can be reduced to as little as 4 μm, but the
larger diameter was preferred for examination of 200 μm wide
friction groove, so as to minimize heating effects. By this
means, the concentration of implanted metal can be studied across
the profile of a groove created by the test ball (Figure 6), and
there is an obvious correlation with the profile of the surface
as measured by a 'Talysurf' instrument. Even though the groove
may be 3 μm below the original surface, much of the implanted

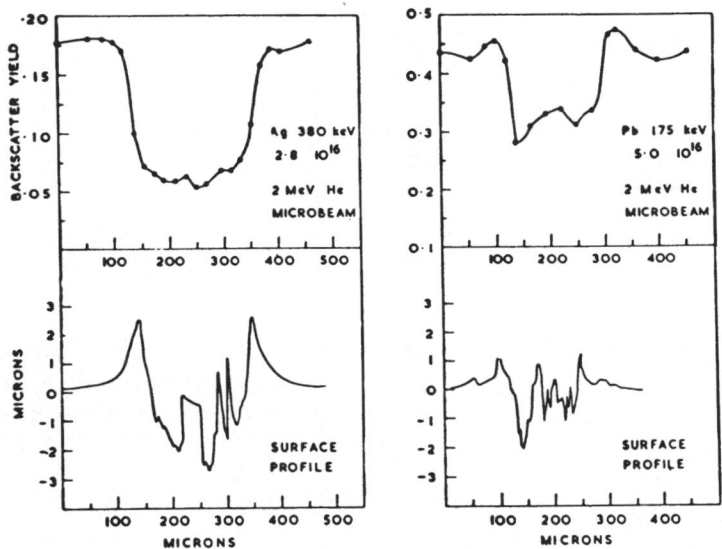

Figure 6.   Upper curves show the results of 2 MeV ^{4}He ion
             microbeam backscattering data for friction grooves
             in Ag$^+$ and Pb$^+$ implanted steel surfaces. Lower
             curves show the corresponding surface profiles as
             measured by a mechanical stylus. (from Hartley
             et al.[10])

material remains after 5 traversals. Hartley has measured the
presence of material 'smeared' into the originally unimplanted
area of the sample, at the base of the friction groove.
Presumably this is due to transfer from the steel to the tungsten
carbide ball, and from there back to the steel. For this kind
of study ion backscattering promises to be at least as useful as
earlier radiotracer experiments, because, using the ion microbeam,
it can be localised.

    Wear is a more important phenomenon than friction in most
practical circumstances, due to the effectiveness of modern
lubrication technology. It is important that wear tests are
carried out under standardized conditions, since otherwise the
results cannot be related at all easily to different testing
procedures. Hartley[10] therefore chose a standard design of
pin-and-disc machine for a series of experiments on ion implanted
metals. Although the disc was the implanted component, the wear
was measured in terms of the volume of material lost from the
tip of a conical pin: the two components (pin and disc) are to
be considered as a couple, however, and a low wear rate of one
implies a low wear for the other.

    The striking thing to emerge from these experiments is the
invariable decrease in wear parameter as a result of ion

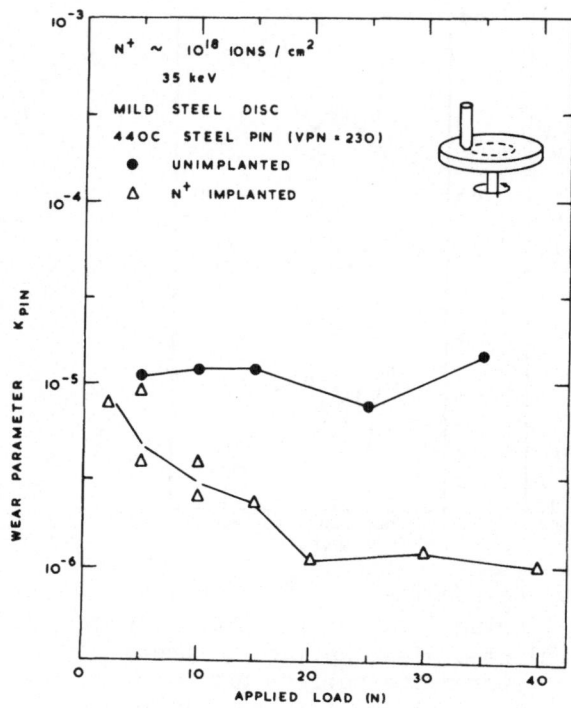

Figure 7. The decrease
in wear parameter, k,
measured in pin-and-disc
wear measurements for a
nitrogen implanted mild
steel disc, as a function
of the load applied to
the pin. (from Hartley
et al.(10)).

Figure 8. The decrease
in wear parameter, k,
measured in pin-and-disc
wear measurements for
an Mo+ implanted mild
steel disc, as a function
of the load applied to
the pin. (from Hartley
et al.(10))

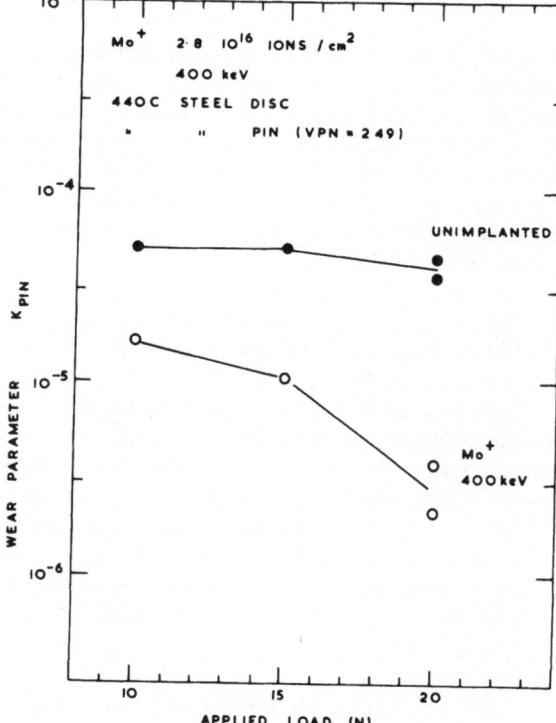

implantation (figures 7,8). The wear parameter, K, is here defined by the expression

$$K = \frac{3H}{Lx} \cdot V$$

in which V is the volume of material removed, H is its hardness, L is the applied load and x is the total distance moved. Care was taken by Hartley et al.[10] to make comparative measurements using the same pin, and to compare the front and back faces of the same disc. The improvement in wear parameter increases with increasing load, and remains constant over many hours of testing. That the metal specimens are indeed in contact at these loads was confirmed by electrical measurements of contact resistance.

There are several possible explanations for this effect. Doping of the surface with foreign atoms may inhibit adhesion, or may alter the strength of the micro-welds. Perhaps a more likely hypothesis is the creation of a considerable lateral compressive stress as a result of ion implantations at these high doses (1016-1017 ions/cm), simply by the non-equilibrium injection of additional atoms. The magnitude of the stresses induced has been measured by Hartley[12] using an adaptation of the technique employed by EerNisse[13] in which a reed-like specimen is canti-levered from one end, and its curvature due to an induced stress

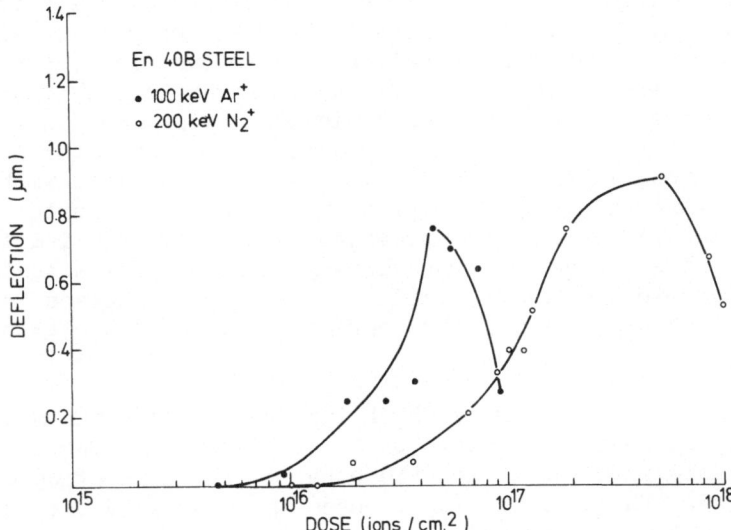

Figure 9.   Buildup of lateral compressive stress in annealed nitriding steel surfaces as a consequence of Ar+ and N+ ion implantation. (from Hartley[12])

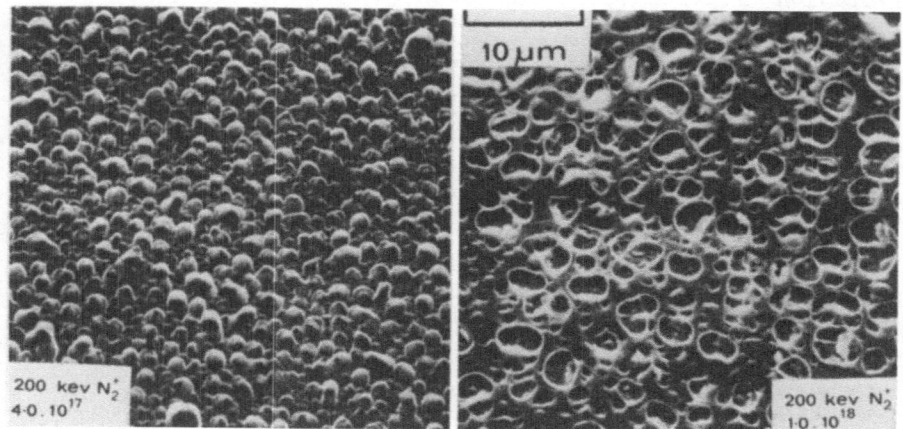

Figure 10.   Blisters and craters produced on the surface of N^+ -
             implanted annealed nitriding steel, examined by
             scanning electron microscopy. (from Hartley[12])

in the bombarded face is measured by a sensitive capacitative
probe. The stress builds up to a maximum, which appears to correspond to the yield stress of the material, and then diminishes
(figure 9) during nitrogen or argon bombardment. Scanning electron
microscope examination of the surface of annealed steel specimens
after N^+ bombardment shows the appearance (Figure 10) of some
remarkable bubbles and blisters. These are formed under an
inhomogeneous surface skin containing out-diffused and deposited
carbon, etc., but they serve to reveal the onset of gas emission
which, together with nitride formation, may account for the relief
of compressive stress. Preliminary results have shown that there
is an optimum dose of nitrogen for diminishing wear, and that this
dose cannot be far from that which produces a maximum lateral
compressive stress in the surface. We are therefore left with the
current view that the lateral compressive stress locks surface
grains of metal into place, and provides a means of reducing the
tendency for shear forces to drag these grains out of the surface
following micro-welding at asperities. The other processes
discussed at the beginning of this paragraph may also be taking
place.

     There are thus two ways by which ion implantation can modify
the friction and wear performance of a metal surface. One is the
chemical or alloying effect, and the other is the generation of a
surface stress. The implantation process is being used both to
provide new information regarding the physico-chemical mechanisms
involved during wear, and to improve the wear characteristics of
certain practical metallic components.

CORROSION

There are two distinct ways in which ion implantation, supplemented by ion beam and other, more conventional techniques of surface analysis, can aid corrosion science. Perhaps the more obvious approach is to consider ion implantation as a means of introducing into the surface layers of a metallic component those constituents (e.g. Cr) which are known to convey corrosion resistance, as an alternative to alloying them throughout.

A more far-reaching and informative approach is to consider ion implantation as a tool for studying and understanding corrosion mechanisms. It offers us the means of introducing essentially any doping species into a metal, in reasonably well-known concentrations and without altering the grain size. Subsequent corrosion tests allow one to determine which species are beneficial, and perhaps to understand the parameters which control the corrosion, under specified conditions. After this, one may consider how best to introduce the appropriate additives, and this may not necessarily be by ion implantation.

The usual method of developing a corrosion-resistant alloy is to prepare by conventional metallurgical techniques a series of alloys containing differing amounts of one or more additives. These are exposed to corrosive environments, resembling as closely as possible those to be encountered in the field, and the degree of attack is measured by the weight gain (a method introduced by Lavoisier, but useful nevertheless) and by optical inspection. More sophisticated studies may involve the application of a controlled stress during corrosion. A good example of the kind of work involved is provided by the development of the zircaloys under the general direction of the Naval Reactors Branch of the USAEC[14]. One difficulty in the comparison of specimens is that of achieving a consistent grain structure, since grain boundaries may play an important part in corrosion. Another problem is where to start in choosing alloying additions: some rule of thumb is required to supplement intuition. The only scientific guide is that provided by the so-called Wagner-Hauffe rules[15] which apply under conditions of parabolic corrosion (i.e. film thickness proportional to $(time)^{\frac{1}{2}}$) in which it is the rate of migration of charged species across the corrosion film that controls its further growth.

From now on we shall concentrate attention upon oxides and oxidative corrosion. For the statement of the Wagner-Hauffe rules it is useful to divide oxides into the following four categories:-

1. Oxides with anion defects

    (a) oxygen-deficient with anion vacancies (n-type semi-conductor) e.g. Nb_2O_5, ZrO_2,

(b) oxygen-excess, with interstitial anions (p-type semiconductor) e.g. UO_2.

2. Oxides with cation defects

(a) metal-deficient, with cation vacancies (p-type semiconductor) e.g. NiO, Cr_2O_3, FeO,

(b) metal-excess, with interstitial cations (n-type
semiconductor) e.g. ZnO.

In addition, one must know whether the rate-determining process is
the migration of ions or the counterflow of electrons (or holes).
The rules are then expressed in terms of the valence of the
additive species. For example, addition of trivalent ions such as
Cr^{3+} to a bivalent metal-deficient p-type oxide such as NiO will
increase the concentration of cation vacancies (due to the strict
maintenance of electrical neutrality), but will decrease the hole
concentration (due to the addition of donors). Since in this
instance it is the transport of cations that determines the oxidation rate, there is an enhancement of the oxidation as a result of
vacancy-assisted ion migration. If hole transport were rate-
determining, the oxidation would be inhibited. Conversely, additions
of monovalent ions, e.g. Li^+, have the opposite effect and reduce
the oxidation rate[15].

However, so many exceptions are observed in practice[15,25]
that these rules are of no great value. It has become recognized
that corrosion is often a highly complex phenomenon, and that
factors such as grain boundary diffusion, mechanical stresses and
layered scale formation lead to departures from rules based upon a
homogenous model of the oxide film. The versatility of the ion
implantation technique allows a test to be made of the Wagner-Hauffe
rules and, when they break down, the importance of other parameters
such as ionic size can be explored. Examples of this approach will
be given below.

ION BACKSCATTERING

A study of the effect of ion implantation on the corrosion of
a metal demands an accurate means of determining the oxygen uptake
in the implanted region, compared with the unimplanted area of metal.
The conventional method of weight gain determination is too cumbersome, requires long exposure times and large implanted areas. A
more sensitive differential technique is required.

Dearnaley et al.[17,18] developed for this purpose the ion
back-scattering technique which has been the subject of earlier
chapters of this book. It is therefore unnecessary to describe

the procedure in detail, except to point out that the elastic
scattering cross-section of O^{16} is normally quite low compared with
that of most metals. Since the oxygen peak is superimposed as a
large background from the metal atoms, it is difficult to achieve
the necessary accuracy. However, the $O^{16}(p,p)$ cross-section displays
a number of strong resonances (figure 11) and by using a proton
energy of 4.55 MeV and a scattering angle of about 165°, a cross-
section some 70-80 times the Rutherford value is achieved[16]. As
will be seen in spectra from oxidized zirconium (Z = 40) a very
useful increase in sensitivity is thus obtained.

Helium ion backscattering, at 1-2 MeV, is valuable for
establishing the stoichiometry of an oxide as a function of depth
below the surface. It also serves to reveal the distribution of
implanted material after oxidation. The ion transport processes
that occur during oxidation rarely leave the original depth
distribution of implanted atoms undisturbed; their redistribution
can be a guide to the corrosion mechanism, analogous to the use
of inert 'marker' wires embedded in a test specimen.

It is appropriate to stress at this point the value of both
ion backscattering analysis and nuclear reaction analysis in
corrosion science, for obtaining a quantitative and non-destructive
measurement of the composition of corrosion films up to several
microns in thickness. This is an important field which has been
pioneered over many years by Dr. G. Amsel and his colleagues at
the University of Paris[19], in both anodic and thermally-grown

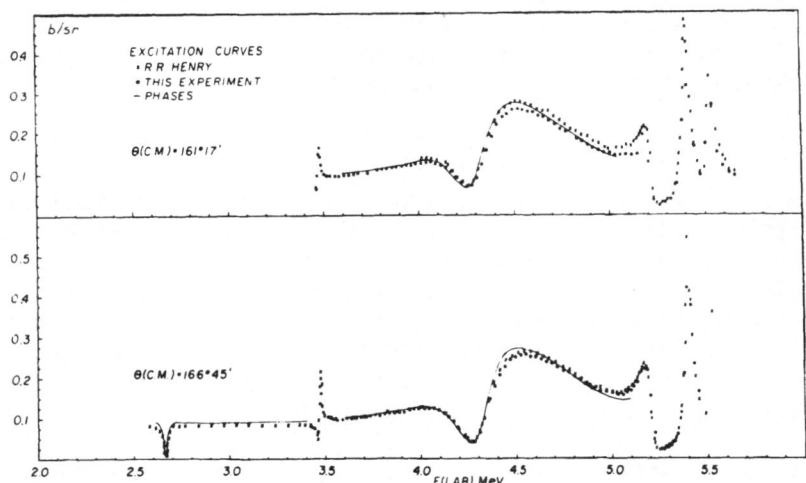

Figure 11. Differential elastic scattering cross-section, in barns
per unit solid angle, as a function of proton energy
for the $^{16}O(p,p)$ scattering process at two angles of
observation. (from Harris et al.[16])

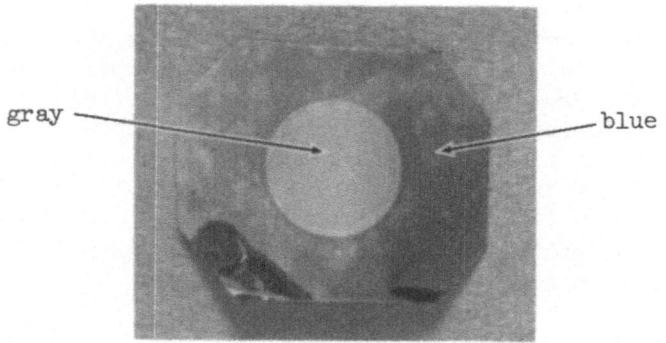

gray   blue

Figure 12.   Photograph of a thermally oxidized specimen of bismuth-
             implanted titanium foil, showing a contrast in optical
             interference color over the implanted (circular) area
             corresponding to an enhancement of the oxidation.
             (from Dearnaley[17])

oxides which have <u>not</u> been ion-implanted. An example of how these
ion beam techniques can be strengthened by combination with other
methods of surface inspection has been published recently by
Dearnaley et al.[20].

    We shall now move on to consider a series of examples of how
ion implantation, supplemented by ion backscattering analysis, has
been applied to study the initial stage of oxidation of the metals
Ti, Zr, Cu, Al and an 18% Cr, 8% Ni, 1% Ti (i.e. 18/8/1) stainless
steel. The results illustrate a variety of different oxidation
mechanisms and experimental techniques.

                     Titanium and Stainless Steel

    A survey of the effects of implantation on the thermal oxida-
tion of polycrystalline titanium and 18/8/1 stainless steel was
carried out by Dearnaley et al.[17,18]. These are both technologically important materials often employed because of their
combination of strength and corrosion resistance. The purpose was
to study the initial stage of oxidation during which the mechanism
may be expected to be relatively simple and free from the mechanical
complexities which accompany thick scale formation.

    Polished discs on 5 μm thick foil specimens of the two metals
were implanted with ten different ion species at energies chosen to
give an ion range approaching 1000Å. A 500 keV Cockcroft-Walton
accelerator and versatile ion source was used: the target chamber
pressure was about 3.10^{-7} torr. The specimens were then oxidised

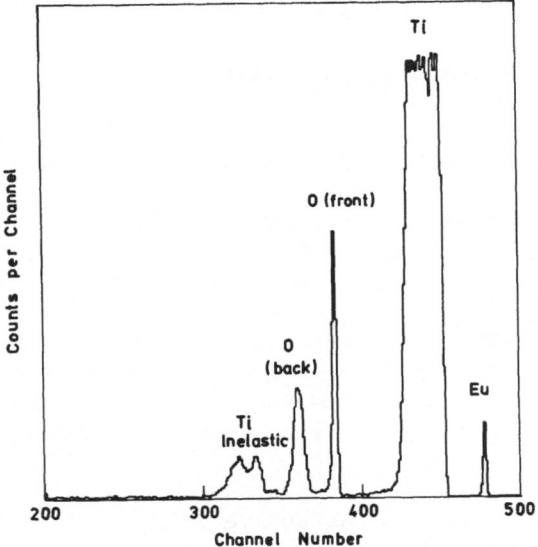

Figure 13. Energy spectrum of 4.0 MeV protons elastically scattered at an angle of 164° from a 5μm thick foil of titanium implanted with Eu⁺ ions and subsequently oxidized. (from Dearnaley(17)).

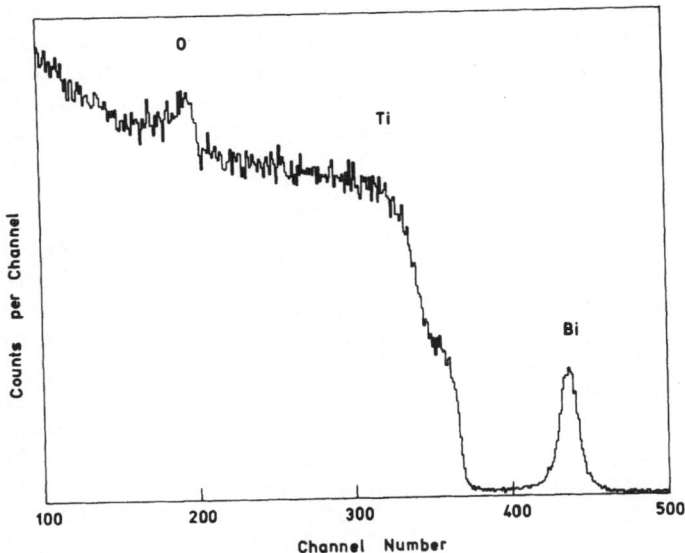

Figure 14. Energy spectrum of 1.5 MeV ⁴He ions elastically scattered at an angle of 164° from a 5 μm thick foil of titanium implanted with Bi⁺ ions and subsequently oxidized. (from Dearnaley(17)).

in dry O_2 at atmospheric pressure at $600^{o}C$ (for Ti) or $800^{o}C$ (for
steel) for times between 15 and 30 minutes. The optical interference colour gave an immediate qualitative indication of oxide
thickness (fig. 12), but it was shown that even if all the visible
oxide is removed from Ti (by vibratory polishing) there is still
an appreciable amount of oxygen dissolved in the underlying metal.

The oxygen take-up was determined quantitatively by proton
elastic scattering at an angle of 164°, the proton energy being
4.0 MeV. It was found preferable to use this method with thin foil
specimens as $Ti^{48}(p,p)$ scattering shows a strongly resonant
behaviour above 3.5 MeV. The peaks due to proton scattering from
oxygen and metal in a thin foil are then completely separated
(figure 13). Helium ion backscattering at 1.5 MeV shows that
there is an increase in the $Ti:O$ ratio as a function of depth
(figure 14).

Calcium ions were the most effective in inhibiting the oxidation of titanium, their effect increasing with fluence up to a
value of 5.10^{16} per cm^2. Europium ions similarly led to a reduction
in oxidation. All other ions investigated (Zn^+, In^+, Ce^+, Al^+, Y^+,
Ni^+ and Bi^+) enhanced oxidation, again to an extent which increased
with fluence. Argon ions produced only a minor effect, however,
suggesting that under the conditions of these experiments radiation
damage is of secondary importance.

In stainless steel the interesting result emerged that those
ions (Ca^+, Eu^+) which inhibit oxidation of Ti where the ones that
enhanced oxidation while, vice versa, those ions that enhance
oxidation in Ti were found to inhibit it in stainless steel.
Moreover, the relative order of their effectiveness appeared to be
the same.

It is clear that these results in Ti and steel do not agree
at all with the Wagner-Hauffe valency rules, although certainly
there is some trend from divalent Ca to pentavalent Bi (Note
however the position of Ni in Table II). However, if the ions are
arranged in the order of their relative effects on the oxidation
of the two metals (Table II) it is apparent that there is a
remarkably good correlation with the electronegativity of each
ion. In Pauling's terms[21] this is a measure of the power of an
ion to attract electrons, and hence it is reasonable that it
should be a more significant and sensitive parameter than the
valency in determining the distribution of charged defects in an
oxide lattice. This is particularly important in an oxide such as
TiO_2, which has high charge states and in which the Madelung energy
is high.

Since the parabolic oxidation law is obeyed for Ti at $600^{o}C$[15]
it is appropriate to consider the oxidation mechanism in terms of

TABLE II

Relative effectiveness of ion	Electro-negativity	Position in Electro-chemical Series	Ionization Potential			Ionic crystal Radius Å
			1st	2nd	3rd	
Calcium	1.0	−2.76V	6.1	11.87	51.2	0.99
Europium	1.1	−	5.67	11.24	n-q	0.95
Cerium	1.1	−2.34V	5.6	12.3	20.0	1.03
Yttrium	1.3	−2.37V	6.38	12.23	20.5	0.89
Zinc	1.5	−0.76V	9.39	17.96	37.7	0.74
Aluminium	1.5	−1.706V	5.98	18.82	28.4	0.51
Indium	1.7	−0.34V	5.78	18.86	28.03	0.81
Nickel	1.7	−0.23V	7.63	18.15	35.16	0.69
Bismuth	1.9	−	7.29	16.68	25.6	0.96

charge transport through the growing film, as did Wagner and Hauffe. TiO_2 is known to be an oxygen deficient n-type semiconductor[15]. Species which can accept electrons from Ti^{3+} donors will reduce the electronic conductivity and, if electron transport is rate-determining, the oxidation will be diminished. We cannot claim to understand why low electronegativity is correlated with this behaviour, but it is already established that calcium acts as a good electron trap in TiO_2.

These conclusions are moreover consistent with the measurements made by Suffield and Dearnaley[22] on the electronic conductivity of anodic TiO_2 films grown on ion-implanted Ti. Different impurities stabilised different impedance states of the oxide in a manner correlated with their electronegativity: calcium implantations produced the highest impedance Ti oxides that were observed.

In 18/8/1 stainless steel exactly the opposite effects were observed to those in Ti. Every ion which enhanced the oxidation of one metal was found to inhibit (the early stages of) oxidation in the other (Table III). The hypothesis was advanced[17] that the oxide film on this steel is a metal-deficient p-type semi-conductor, in which hole transport is rate-determining. This is consistent with the semiconducting behaviour of Cr_2O_3 which is known to be the main component of the initial oxide on 18% Cr steel[23].

TABLE III

The Effect of Ion Implantation on the Oxidation of 18/8/1
Stainless Steel

Spec. No.	Ion Type and Dose		Oxidation Effect	
		ions/sq cm		
S1	Ca	10^{15}	enhanced	20%
S2	Ca	10^{16}	enhanced	40%
S3	Ca	10^{16}	enhanced	51%
S4	Ca	5×10^{16}	enhanced	100%
S5	Y	10^{16}	reduced	30%
S6	Y	5×10^{16}	reduced	50%
S7	Eu	5×10^{16}	enhanced	50%
S8	Eu	5×10^{16}	enhanced	62%
S9	Bi	10^{16}	reduced	28%
S10	Bi	10^{16}	reduced	31%
S11	Bi	2×10^{16}	reduced	50%
S12	In	2×10^{16}	reduced	32%
S13*	Al	10^{17}	reduced	52%

It must be emphasized that these rather simple ideas are
bound to break down when the scale thickness increases and inhomogeneities such as those discussed by Wood[23] appear. Moreover,
they may well not apply in other temperature regimes, in which
parabolic oxidation does not occur.

Zirconium

There have been two studies of the effects of ion implantation
on the oxidation of zirconium and its alloys. In one[17,24], a
parallel survey was made in order to compare Zr and Ti, which are
chemically similar metals. In the other[32], the effects of ion
bombardment of the oxide were investigated, as part of a test as
to whether reactor irradiation will modify the corrosion rate of
the zircaloys.

Weidman and Dearnaley[17,24] measured the effect of ion
implantation on the oxidation of electropolished zirconium. Oxidation was performed in dry O_2 at 1 atm. for times of about 5 minutes

at temperatures between 380° and $400^\circ C$. Oxygen take-up was determined by proton scattering at 164° at an energy of 4.55 MeV (the peak of an O^{16} (p,p) resonance). In this case, since Zr itself shows no resonant behaviour, thick specimens could be used and the yield from O^{16} was readily distinguished from background (figure 15).

Table IV shows the ratio of oxygen take-up in the implanted area, compared with the unimplanted area of metal. It is obvious that the behaviour of Zr is totally different from that of Ti. Most ions enhance the oxidation of Zr, the exceptions being the transition metals (Fe, Ni, Nb, and Cr), over a certain fluence range. Zr^+ ions induce little or no effect, and so lattice disorder and strain in the metal appear not to be important. There is no correlation with electronegativity or valency: it was already known that the Wagner-Hauffe rules are violated in Zr[25].

TABLE IV

The Ratio of Oxygen Up-take in Implanted and Unimplanted Zr

Ion species	Fluence	Oxidation temperature $^\circ C$	Oxidation time, mins.	Oxygen Ratio
Electropolished specimens				
Ni	$5 \cdot 10^{15}$	400	5	0.53
Nb	$5 \cdot 10^{15}$	400	5	0.84
Fe	$5 \cdot 10^{15}$	400	5	0.86
Fe	10^{15}	400	5	0.93
Cr	$5 \cdot 10^{15}$	400	5	0.95
Ni	10^{15}	400	5	0.98
Cr	$5 \cdot 10^{16}$	400	5	1.03
Zr	10^{16}	400	5	1.04
P	$5 \cdot 10^{15}$	400	5	1.06
Ca	$5 \cdot 10^{15}$	400	5	1.10
Y	$5 \cdot 10^{16}$	390	5	1.13
Si	$5 \cdot 10^{15}$	400	5	1.18
Fe	10^{16}	400	5	1.47
B	$5 \cdot 10^{16}$	405	45	1.48
Nb	$5 \cdot 10^{16}$	380	7	1.56
B	$5 \cdot 10^{16}$	380	6	1.63
Fe	$2 \cdot 10^{16}$	400	5	1.80
Sn	$5 \cdot 10^{15}$	400	5	1.93
Fe	$5 \cdot 10^{16}$	400	5	2.18
Ca	$5 \cdot 10^{16}$	400	5	2.36
Eu	$5 \cdot 10^{16}$	410	3.5	5.75
Cu	$5 \cdot 10^{15}$	400	5	6.63

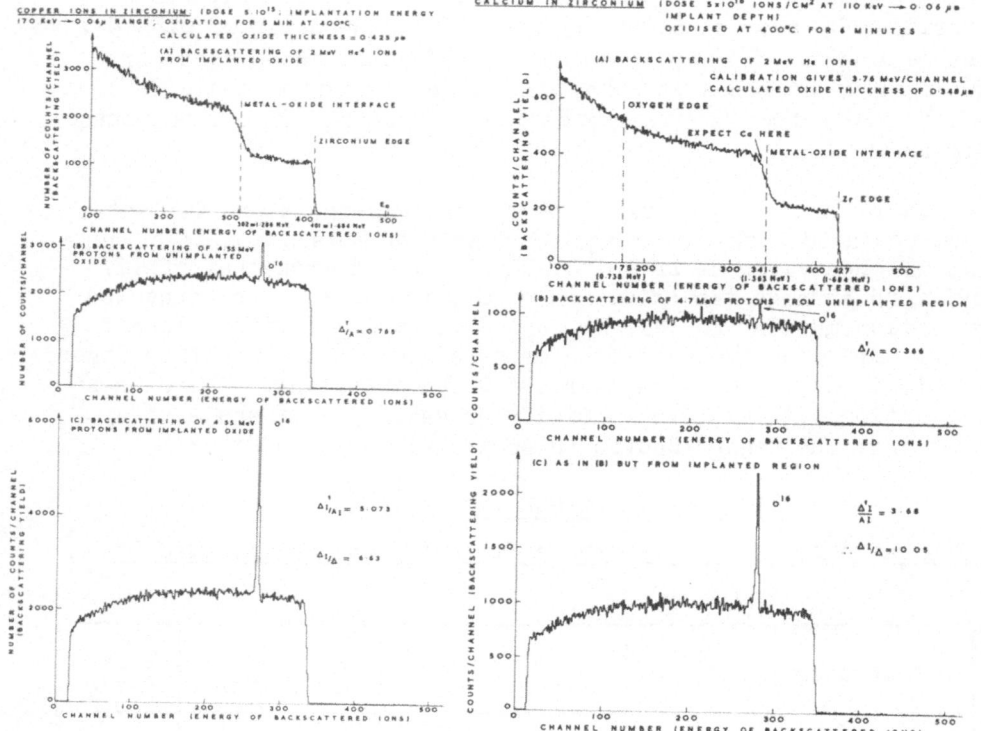

Figure 15   (a)   Ion backscattering spectra from Cu^+ implanted
            electropolished zirconium after thermal oxidation at
            $400°C$.   (from Dearnaley[17]).

            (b)   Ion backscattering spectra from Ca^+ implanted
            electropolished zirconium after thermal oxidation at
            $400°C$ (from Dearnaley[17]).

        Since the grain size of the electropolished specimens used by
Weidman et al.[17,24] was about 50 μm it was possible to see the
interference colours of individual grains (figure 16).   That
different grain orientations oxidise to different extents was
already known[26] but the colour changes show that ion implantation
affects the oxidation to an extent dependent upon grain orientation.
The proton and He+ ion beams used to probe these specimens were
about 500 μm in diameter and hence averaged the effects in about
100 grains.

        The effect of implanting different amounts of iron and nickel
into Zr is shown in figure 17.   There appears to be an optimum
fluence of about 3.10^{15} ions per cm^2 and it is interesting to note
that this corresponds to a mean concentration of transition metal
which is about the same as that found to give optimum oxidation
resistance in zircaloy 2[14].

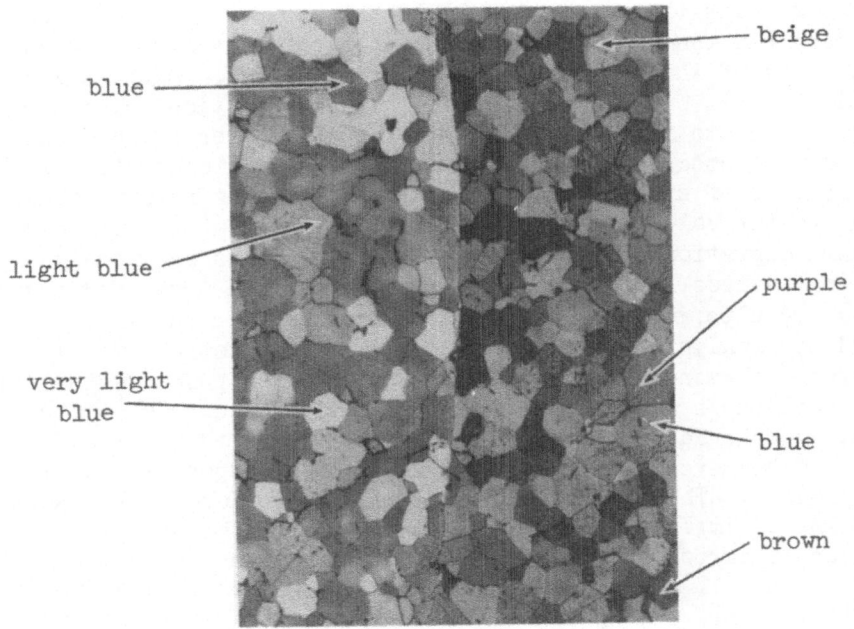

beige

blue

light blue

purple

very light
blue

blue

brown

Figure 16. Photograph of an oxidized ion-implanted specimen of
electropolished zirconium. The left half had been
implanted with 5.10^{16} calcium ions per cm^2 (from
(Dearnaley[17])).

Figure 17. The effect of
Fe^+ and Ni^+ implantation
upon the oxidation of
electropolished zirconium,
as a function of ion dose.
The minimum corresponds
to a surface concentration
of about 3000 ppm. (from
Dearnaley[17]).

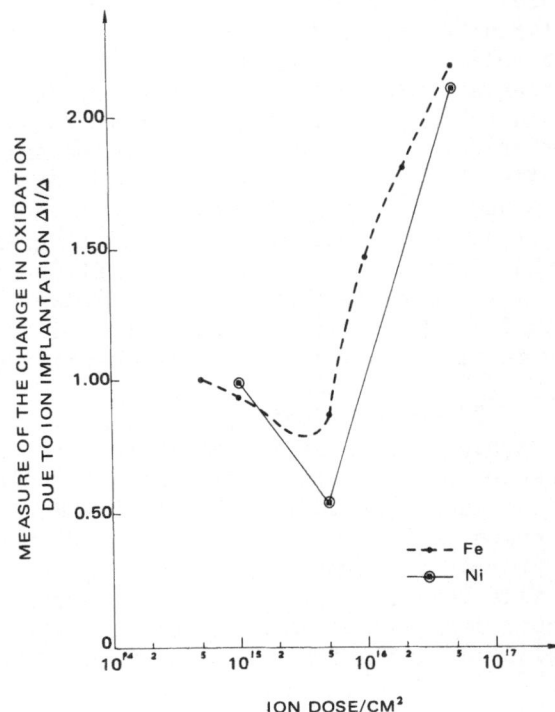

The species that inhibit oxidation in Zr have in common an ionic size (in the most likely valence state) that lies between 80% and 100% of the Zr^{4+} ion size. Other species, which enhance oxidation, would either not fit into the ZrO_2 lattice or, in doing so, would tend to increase its volume. This factor of ion size has been discussed also by Parfenov et al.[27] in connection with the oxidation of zirconium alloys. The oxide of zirconium occupies a much greater volume than that of the original metal, and grows by inward migration of oxygen ions. There is thus an enormous mechanical stress generated between the oxide and metal and the observed results in ion-implanted specimens favour the idea, put forward by Wanklyn[25] that it is the mechanical stress within the ZrO_2 that determines the formation of cracks and pores in the oxide, and so influences the rate of further oxide growth. Cox[28] has obtained good evidence of pore formation in even thin oxide films on Zr, and believes[29] that grain boundary transport is the most rapid process. Thus crystallite morphology, influenced by the ionic size of implanted impurities, can control the oxidation behaviour of zirconium. The reversal observed as a function of dose in figure 17 may be explained by a transition to a different crystal structure, or by an alteration in mechanical strength resulting in a change in crystallite size.

It may at first seem strange that the two Group IV A metals, Ti and Zr, should behave so differently after implantation with the same dopant species. The answer probably lies in the different crystal structure of their oxides and the remarkable ability of titanium oxides to eliminate oxygen vacancies by the mechanism of crystallographic shear[30]. Anion vacancies become ordered into planes to form so-called 'Wadsley defects': the lattice closes up and a lower stoichiometry exists locally (Figure 18). It is therefore erroneous to discuss the oxidation of titanium, as has often been done, in terms of the migration of oxygen ions via a vacancy-assisted diffusion. It is not yet established whether the diffusion is interstitial or by progressive shrinkage of Wadsley and other extended defects. What is clear, however, is that the crystal structure and, in particular, the defect behaviour of an oxide plays a very important part in determining the oxidation of a metal and the influence of impurities upon the oxidation rate.

Zirconium dioxide, by contrast, does not exhibit crystallographic shear. It therefore remains rich in anion vacancies and hence is one of the best ionic conductors, while this conductivity is enhanced by additions of calcium and yttrium[31]. For this reason, we can understand the uniform stoichiometry of the oxide apparent in Figure 15, compared with the non-uniformity of titanium oxide (Figure 14). The latter corresponds to an increasing concentration of Wadsley defects nearer to the metal, and these allow a continuous variation in stoichiometry.

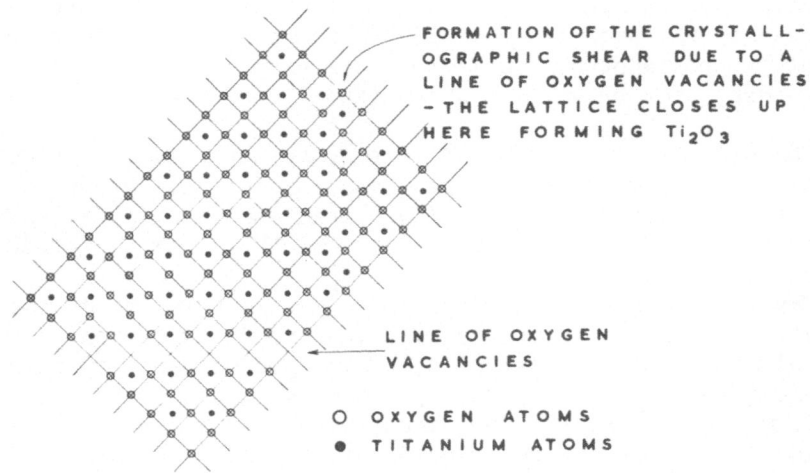

FORMATION OF THE CRYSTALL-
OGRAPHIC SHEAR DUE TO A
LINE OF OXYGEN VACANCIES
- THE LATTICE CLOSES UP
HERE FORMING Ti_2O_3

LINE OF OXYGEN
VACANCIES

O OXYGEN ATOMS
● TITANIUM ATOMS

Figure 18.   Schematic diagram illustrating (in two planes at right angles) the process of crystallographic shear in anion-deficient TiO_2.   Oxygen vacancies aggregate into sheets, after which the lattice closes up to create a locally modified stoichiometry.   (from Dearnaley[17]).

Spitznagel, Fleischer and Choyke[32] investigated the effect of O^+, Ar^+ and Xe^+ ion bombardment of oxide films on zircaloy 4 and 1% niobium-zirconium alloy on their corrosion in oxygenated water at $360°C$.   The bombardment diminished the subsequent weight gain of the specimens and invariably produced a more uniform oxidation than in the unbombarded control specimens.   Various explanations for the effect are discussed by Spitznagel et al.[32], but an interesting comment by EerNisse[33] on this work contains the suggestion that the ion bombardment leads to a relaxation of mechanical stresses.   This would correlate very well with the conclusions of Weidman et al.

Aluminium

Butcher[34] has investigated the effects of ion implantation on the thermal oxidation of pure Al and a 14% Cr, 4% Al steel alloy (Fecralloy), while Towler, Collins and Dearnaley[35] have measured the effect of similar implantations on the anodic oxidation of aluminium.

In both cases the behaviour was similar, at least up to a limiting ion dose, and followed that expected on the basis of the Wagner-Hauffe rules, for an n-type semiconductor with excess cations and electron transport as the rate-determining factor[36]. Thus low-valence species such as Au^+ and Ag^+ inhibited the

oxidation, while high-valence implants of Pb^+ and Bi^+ increased
the oxidation. High doses of Bi^+ ions produced the remarkable
effect of spontaneous room-temperature oxidation, leading to the
development of a brown film which finally exfoliated.

Once again, these results were found to be consistent with
electrical measurements made on anodic oxide films grown on ion-
implanted aluminium (Gleaves[37]). Lead and bismuth implanted
specimens showed a considerable increase in electronic conductivity.

Mackintosh and Brown[38] used He^4 backscattering to study the
migration of implanted ions during the anodic oxidation of
aluminium. Fifteen different metallic species were implanted at
low energies (20-40 keV) into specimens with and without a
pre-formed oxide layer. Some ions, such as Ag^+, were found to
retreat before the advancing oxide front, and dissolution of the
oxide revealed that most of the silver is located in the metal
rather than the oxide. Alkali metal ions moved outwards, however,
and were rapidly lost to the electrolyte. Inert gas ions remained
at the same relative position with respect to the original surface:
in effect, these served as immobile 'markers'. The transport
properties of the ions were correlated with their oxidation
potential, those with low values of oxidation potential migrating
outwards during anodic oxidation.

Towler, Collins and Dearnaley[35] made similar measurements
but during the growth of porous anodic oxide films on aluminium.
As is to be expected, the ion migration behaviour is different in
this case, because there is an essentially field-free region of
oxide due to conduction down many pores to a 'barrier layer' below.
Ion transport through the barrier layer depended upon the concentration of impurity ions left at the metal-oxide interface and
upon the valence of the ion. Pb^+ and Bi^+ remained in the outer
layers of oxide, while Au^+ was transported inwards.

Copper

Following the observation by Crowder and Tan[3] of an increase
in the corrosion-resistance of boron-implanted copper, Rickards and
Dearnaley[39] commenced a study of the effect of ion implantation
in single-crystal copper.

Aluminium ions implanted into the <110> face of electropolished Cu crystals reduced subsequent oxidation in dry O_2 at
$320°C$ by about 70 per cent. The dose in this case was 5.10^{16}
ions/cm^2 at 220 keV; lower doses produced a smaller effect.
Rickards also observed[39] that bombardment of the same specimens
with Cu^+ ions at 200 keV, to a dose of 0^{16} ions/cm^2, produced a
300 per cent increase in the take-up of oxygen. In this case it
appears that radiation damage, or strain, or surface sputtering

(exposing new crystal facets) result in an increase in oxidation rate: the 110 face of copper is known to resist oxidation strongly. By ion backscattering measurements under channelling conditions Rickards was able to measure the stoichiometry of the oxide, and to demonstrate its inhomogeneity.

Aqueous Corrosion

Ashworth et al.[40,41] have begun a series of investigations of the effect of ion implantation on the aqueous corrosion of copper, aluminium and iron. Polycrystalline specimens were implanted with Ar^+, Al^+, B^+, Fe^+ and Mo^+ ions at 40 keV. Salt-spray tests showed that high doses (10^{16} ions/cm^2) of Ar^+ and B^+ ions inhibit the tarnishing of copper and that similar doses of Ar^+ inhibit the rusting of iron.

Potentiostatic measurements of passivation and corrosion currents for implanted samples immersed in suitable electrolytes (e.g. 0.1M NaOH + Na_2SO_4) were also measured. High dose implantations in Cu gave rise to some variation in the polarisation curves, while for the Al samples a large increase in the passive region of the voltage: current diagram was obtained (figure 19). Here the anodic section of the polarization curve for the argon-implanted sample shows that passivity is retained up to the highest voltages used and no breakdown of the surface oxide film was observed. It is interesting that similar effects were observed with most of the implanted species used.

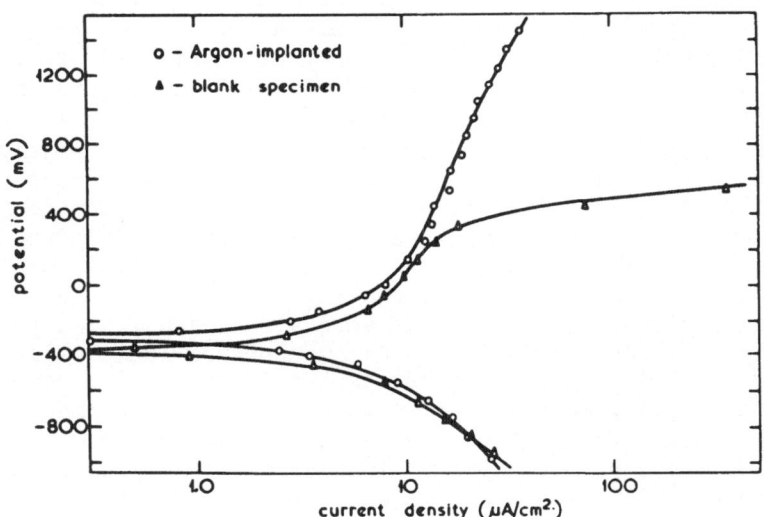

Figure 19. Potentiostatic polarization curves for argon-implanted aluminium (20 keV, 10^{16} ions/cm^2) in 0.01M sulphate solution at pH 6.6 (from Ashworth et al.[40]).

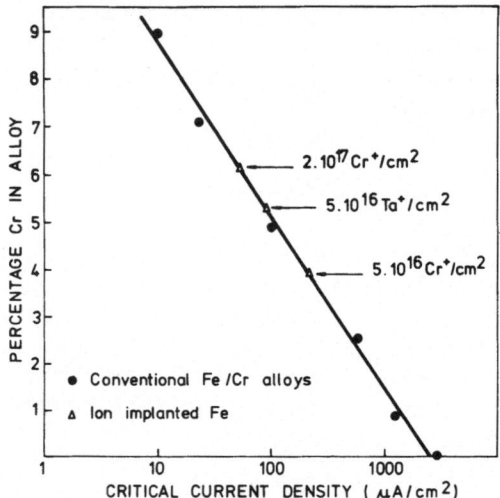

Figure 20. Critical current density for passivation as a function
            of chromium concentration in bulk alloys and ion-
            implanted specimens of iron. Also shown is the
            behaviour of a Ta^+-implanted sample. (after Ashworth
            et al.(41)).

In more recent work[41] Ashworth et al. have concentrated on
the corrosion of iron, implanted with Cr^+, Mo^+, Ta^+, W^+ or Fe^+
ions. Cathodic reduction in borate/HCl solutions was used to
obtain a measurement of the type and thickness of oxide present.
These tests showed that implanted Fe has an oxide approximately
60Å in thickness, i.e. twice that of normal air-formed oxide.
The film is a duplex of Fe_3O_4 and $\gamma-Fe_2O_3$, with the extra oxide
thickness present in the Fe_3O_4 layer. Potentiostatic polarization
tests on Cr^+ implanted iron showed a large increase in passivity
with both a lower critical current and average passive current
density, and the behaviour was compared with that of conventional
Cr-Fe alloys. Doses of 5.10^{16} and 2.10^{17} Cr^2ions/cm produced
surface alloys corresponding to conventional bulk alloys containing 4.2% and 6.2% chromium, respectively (figure 20). The iron
samples implanted with Fe^+ and Ta^+ were also investigated potentio-
statically. The Fe^+ implanted specimens behaved identically to
unimplanted samples once the first cathodic sweep had removed the
surface oxide film, which had been thickened by bombardment, as
mentioned above. The degree of passivation, as measured by the
critical current, produced by Ta^+ implantation was greater than
that produced by an equivalent dose of Cr^+.

Practical Applications in Corrosion

Here the objective is not so much to improve the understanding
of corrosion mechanisms as to make use of ion implantation as an

alternative means of introducing beneficial ion species into the
surface layers of an engineering component.

Alloyed additions of yttrium or rare earths are known to be
effective in inhibiting the high temperature oxidation of iron and
nickel based alloys which also contain chromium. The effects of
small concentrations of yttrium on the oxidation in CO_2 of an
austenitic stainless steel containing 20% Cr, 25% Ni and stabilised
with 1% Nb had previously been studied, within the temperature
range $800°C$-$1000°C$. A few parts per thousand of yttrium reduced
the overall attack and improved the oxide adherence considerably.

Both these are advantages in one potential application of the
material, namely to fuel cladding in nuclear reactors. However,
there are drawbacks to the use of yttrium-bearing alloys, the
chief of these being the reduction in ductility and tensile strength
of the material, probably due to grain boundary segregation of
yttrium oxide. The virtue of ion implantation is that it allows
the yttrium to be introduced into the surface of the steel
components, after fabrication, so that the surface corrosion
behaviour is improved without impairing the bulk properties of the
material (the wall thickness being about 0.38 mm).

Implanted yttrium, in concentrations of about 0.2 per cent
through a depth of 0.2 μm (obtained with a fluence of $3.5.10^{15}$ Y+
ions/cm^2 in a multi-energy implantation) was found to be just as
effective as alloyed yttrium in reducing oxidation[43] (Figure 21).
It also lessens the percentage of oxide that spalls. The implantation treatment is effective over a period of at least 6000 hours
(8 months) at $850°C$ during which time an amount of steel equivalent
to over 20 times the implanted depth has been converted to oxide.

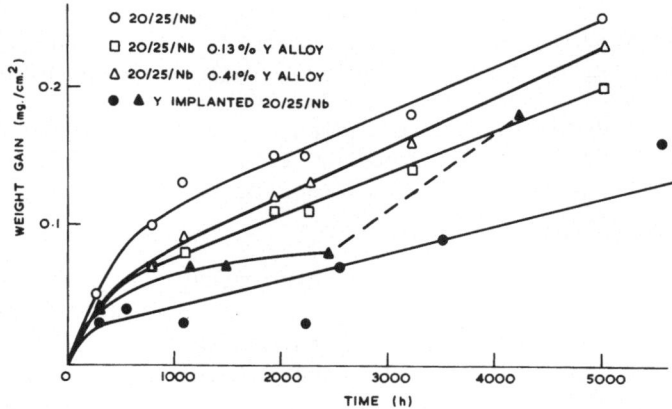

Figure 21.   Effect of yttrium implantation ($3.5.10^{15}$ ions/cm^2) on
             the oxidation of an austenitic stainless steel in CO_2
             at $700°C$, compared with the behaviour of alloyed yttrium
             specimens. (from Antill et al.[43]).

Ion microprobe analysis of a bevelled section of implanted steel, after oxidation, has shown the accumulation of yttrium at the metal-oxide interface, where it is believed to form an impermeable layer rich in the perovskite $YCrO_3$.   The improved adherence of the oxide may be due to a modification of the vacancy condensation mechanism at the interface where voids are formed that lead to spalling.   The effectiveness of an implanted layer rules out some other explanations of improved oxide adherence, such as one involving keying at grain boundaries.   Thus an initially shallow implantation may provide a long-lasting protection, and the evidence is that relatively little of the implanted material is lost by spalling.   It seems therefore that ion implantation offers a practicable means of improving corrosion resistance in these steel components, and attention has since been devoted to the achievement of a sufficiently intense Y^+ beam to meet the economic requirements of the process.

A second application is under assessment in the corrosion of steel burner tips for an oil-fired power station.   These components, fabricated of 13% Cr carbon steel, operate at $550^\circ C$ in a hostile environment rich in sulphur-bearing combustion products. Failure is often due to erosion, caused by the jet of fuel, of the burner orifice following its corrosion on the combustion face. Peplow, Dearnaley, Hartley and Poole[44] are investigating the effectiveness of ion implantation of various species for prolonging the life of these components.   Preliminary results appear to be encouraging.   In this instance also the objective is to use ion implantation as a means of introducing species which are known, from work on alloys, to be beneficial.   In addition, there is the prospect of modifying the adhesion of the oxide to the metal beneath, and so improving its resistance to erosion.

Ion implantation appears to be a viable means of providing a long-lasting inhibition of the corrosion of metal components which are important parts of costly systems, such as occur in the nuclear energy aerospace and military fields.   The use of small amounts of additives introduced by ion implantation into the surface of a metal rather than alloyed throughout is in keeping with the trend towards conservation of material resources and optimum control of material properties.

ELECTROCHEMISTRY AND CATALYSIS

It is obviously attractive to investigate whether small quantities of catalyst material, such as platinum, implanted into the surface layers of a metallic or other catalyst support, could be as effective as the much heavier surface coatings normally employed.   In the first published experiment of this kind Grenness, Thompson and Cahn[45] examined the electrocatalytic activity of platinum ions implanted into tungsten and anodic WO_3 electrodes.

The platinum ions were implanted at 200 keV energy at doses between 2.10^{15} and 2.10^{16} ions/cm^2: the mean ion range in WO_3 was calculated to be 350Å, but this may well be reduced at the higher doses as a result of surface sputtering.

Current-voltage measurements for the hydrogen evolution reaction were made in a conventional three-compartment cell with a vitreous carbon counter-electrode and a saturated calomel reference electrode. Nitrogen gas was introduced through a sintered glass disc in the bottom of the working electrode compartment.

In the case of tungsten samples which were implanted with platinum the initial performance (as judged by the Tafel slope, i.e. the relationship between the logarithm of the current and the potential) was even lower than that of untreated tungsten. However, anodization of the implanted tungsten brought about a striking improvement and, for example, three anodic sweeps to + 3V resulted in a change of Tafel slope from 120 mV/decade to 36 mV/decade in a sample implanted with 2.10^{16} Pt$^+$ ions/cm^2.

If the tungsten was anodized at 72V (corresponding to 1100Å of WO_3) before implantation it was found that 2.10^{15} Pt$^+$ ions/cm^2 gave (after four voltage sweeps) a Tafel slope of 30 mV per decade, which is very close to that of bright platinum itself. Moreover, the cyclic voltammograms observed in a sample implanted with 2.10^{16} Pt$^+$ ions/cm^2 and similarly activated displayed a striking resemblance (as regards the four peaks ascribed to hydrogen adsorption and desorption) to those of bright platinum tested under similar cell conditions. Over large regions of the diagram the two voltammograms coincided perfectly. Repeated cycling brought about a slight improvement in the Tafel slope, and no electrode deterioration was observed.

Grenness et al.[44] infer that a high concentration of Pt is present at the surface of the specimens after activation, but they do not claim to understand the role of the tungsten oxide layer or the migration of Pt$^+$ ions within it. The work of Macintosh and Brown[38] and of Towler Collins and Dearnaley[35] on anodized aluminium has shown that implanted species can, under suitable conditions, migrate to the outer surface of the anodic oxide. Possibly this is the case for Pt$^+$ in WO_3, but unfortunately the same technique of investigation (by ion back-scattering) could not be applied owing to the high atomic number of tungsten.

Wolf[46] has recently described how the catalytic hydration of unsaturated hydrocarbons was shown to be much faster when using ion bombarded Ni foils than when using untreated ones. It was inferred that this was a radiation damage phenomenon, rather than one of chemical doping, and that displaced Ni atoms occupying non-equilibrium positions on the surface of the foil possessed a higher degree of catalytic activity.

There can thus be two distinct ways in which ion implanted
metals are treated by ion bombardment so as to alter their electrochemical and catalytic properties. In view of the industrial
importance of this field, it is one in which many more investigations are likely to be made.

IMPLANTATION METALLURGY

Ion implantation can be used as a means of introducing a
chosen distribution of one metallic species in another, by what is
obviously a non-equilibrium process. Furthermore, the production
of numerous vacancies and interstitials during ion implantation
can significantly affect the final distribution and clustering of
the implanted species.

The behaviour of this 'exotic alloy' may then be studied
during its transition towards an equilibrium state. Some very
interesting phenomena have already been demonstrated in this way,
although this is a huge field which has scarcely begun to be
explored.

An early observation was made by Thackery and Nelson[47] who
implanted lead ions at 70 keV into aluminium. Lead is highly
insoluble in this metal, and on annealing 400°C the lead was seen
to precipitate as small oriented crystallites (figure 22). More
recent studies have concentrated upon the lattice site location of
implanted metal atoms, e.g. for hyperfine magnetic interaction
studies. A considerable amount of data is now available, and is
reviewed in the chapter by S.T. Picraux. The backscattering of
channelled ion beams has been of great value in these experiments,
and in metals (compared with semiconductors) there is less concern
that the ionization and other damaging effects of the probing beam
might lead to a displacement of the solute species. Poate et al.[48]
showed in this way that substitutional alloys can be formed by
high-dose implantations of gold into Ni, Cu, Pd and Ag with no
thermal annealing, and also tungsten implanted into Cu shows a
high substitutional occupancy despite its extremely small solubility
under conventional conditions.

The migration and precipitation of relatively insoluble species
at nucleation sites in a metal involves interesting phenomena that
can only rarely be studied by alternative means (e.g. quenching a
high-temperature equilibrium). The fact that ion bombardment
induces other effects besides doping must however not be overlooked:
dislocation networks and a lateral compressive stress will modify
the subsequent atomic migration. At very high fluence levels
(5.10^{16} ions/cm^2) the cooperative effect of additional atoms
in creating a compressive stress may cause dislocations to be
propagated far beyond the implanted zone, in materials such as
copper which have a low Peierls-Nabarro stress. Sood[49] has

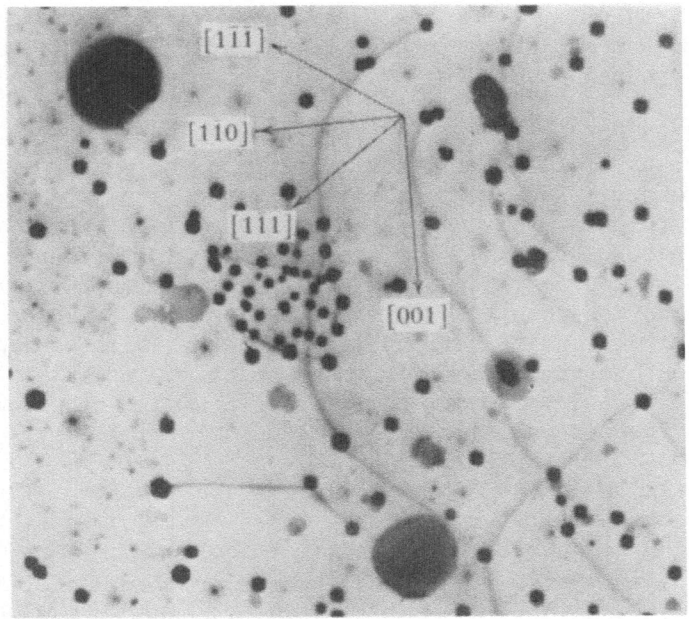

Figure 22.   Cubic precipitates of Pb in aluminium formed by ion
             implantation of 70 keV Pb$^+$ followed by thermal
             annealing. (from Thackery and Nelson[47]).

studied the disorder introduced and the lattice site occupancy
of impurities following the implantation of Cu and Al crystals
with a wide variety of ion species with a broad spectrum of
atomic sizes and electronegativities. He found that the conditions
for substitutional occupancy are more restricted in Al than in Cu.

It seems likely that ion implantation will yield a good deal
of new information regarding alloy formation, and the validity
of the Hume-Rothery rules. Some of the 'exotic alloys' which can
be produced by ion implantation may also be found to possess novel
properties, e.g. as superconductors or magnetic alloys.

EQUIPMENT FOR THE ION IMPLANTATION OF METALS

Research and practical exploitation of the ion implantation
technique in metals both raise different sets of requirements
from those of the semiconductor field. This has stimulated the
development of novel equipment aimed more specifically at this
relatively new area of technology.

Firstly, research studies e.g. of corrosion inhibition demand
a wide variety of ion species in order to explore the many

possibilities that result from the chief advantage of the technique, namely its versatility. On the other hand, factory utilisation of the process is more likely to call for a dedicated ion facility capable of producing only one or two ion species.

Secondly, while laboratory research will usually involve the implantation of small metal discs or coupons not unlike semi-conductor wafers, the practical application of ion implantation to metallic components will involve a great variety of workpiece shapes and sizes. Internal surfaces may have to be treated, and probe-like ion sources are under development for such purposes. Manipulation of objects through the ion beam will call for conveyor-belt designs of various kinds together with storage bays for processed components and for those awaiting implantation. Target chambers and therefore pumping systems will generally be larger than those required in the semiconductor industry.

Ion implantation in metals calls for ion doses, or fluences, which are substantially greater than those employed in semi-conductors. Doses between 10^{15} and 10^{17} ions per cm^2 are typical in metals applications. The trend towards greater use of high-dose implantations for pre-deposition in silicon has lessened the disparity to a level of about one order of magnitude in dose. This demands the use of rather massive ion sources, capable of ionizing metallic elements or compounds at high temperatures. The milliampere ion beams available from equipment of the type described by Freeman[50] are immediately applicable to work on metals, although some metallic species e.g. platinum are notoriously difficult to ionize. Improved techniques of operation of the Freeman source has resulted, however, in 0.5 mA beams of Pt^+ ions[51].

Ion sources capable of delivering tens or hundreds of milli-amperes of certain ion species are technologically feasible: such sources after all were developed thirty years ago as part of the Manhattan Project. However, the implantation of ions to an adequate depth in a target specimen, e.g. to penetrate air-formed oxide films, may well require ion energies of at least 100 keV. Then a 10 mA beam will dissipate 1 kW or more in the target, and the problems of heat dissipation in vacuo become severe. One solution is obviously to move the target objects through the ion beam so rapidly that the transient temperature rise is held within acceptable limits.

Since there are few a priori reasons for choosing a particular doping species and fluence in metals implantation than in the semi-conductors, each technological application will probably require a period of research aimed at optimisation of the ion implantation conditions. A good example of this is the work of Hartley et al.[10,12] on the wear resistance of ion implanted steels.

If an elemental species is found to be effective for the process (e.g. N^+ ions in steel), then it is feasible to consider a simple form of ion implantation equipment consisting only of an ion source, accelerating system and target chamber, without mass analysis. The efficiency and cheapness of the plant are thereby improved.

Control of the neutral component of an ion beam is often important in semiconductor applications, where precise knowledge of the implanted dose is necessary. In metals, however, the properties being modified are not critically dependent upon dose, and so this requirement of accurate dose and uniformity can be greatly relaxed. The feature of ion implantation which remains important is its versatility, i.e. its ability to introduce a given ion species into the target without diffusion or metallurgical processing.

It is interesting to consider the relationship between ion implantation and ion plating. In ion plating[52] metal atoms are vapourized and transmitted through a gas discharge (usually of argon) to the workpiece, which is maintained at a negative potential of a few kilovolts. A proportion, probably no more than 1 per cent, of the atoms arriving at the target are ionized and may have energies of a few keV. The specimen is also bombarded with argon ions, which may play an important part in clearing the surface (by sputtering) prior to ion plating.

The value of ion plating lies in the deposition of coatings, which may be many microns in thickness, possessing a high degree of adherence to the substrate. This adhesion has been attributed to the presence of an interface layer of graded composition, formed probably as a result of sputtering followed by re-deposition of the sputtered material owing to the high gas pressures employed (typically 10 Torr). Such removal and redeposition involves very low particle energies. Nevertheless, the total power dissipation can be large, the target specimen may be raised in temperature by several hundred degrees and diffusion may occur.

Ion implantation differs from this in several respects. It involves a beam which is almost wholly ionized; the particle energies are typically 100 keV or more, and the operation is carried out in a reasonably high vacuum. The bombarded surface will always retain a proportion of the substrate material, since sputtering will set a limit to the concentration of implanted species which can be retained. In metals which sputter readily, e.g. copper, this concentration may be rather low unless high ion energies are used, to increase penetration. A coating layer of distinct composition is never produced by ion implantation, and therefore the question of adhesion does not arise, since the implanted layer is an integral part of the original specimen.

Dimensional changes are also negligible.

Certain features of ion plating and ion implantation equipment could potentially be combined. Thus, sputter-cleaning in an argon discharge could be a necessary preliminary step before ion implant- ation (e.g. to remove oxide), and one which could be carried out in the same apparatus. A somewhat higher pressure than normal in the work-chamber could improve the 'throwing power' of ion implantation (i.e. its ability to treat edges and re-entrant configurations) at the expense of increasing the neutral component of the beam and therefore the control of dose. Calorimetric methods of dose control would then be preferable to electrical monitoring.

Since the penetration of the heavier ions that may be employed is limited, it is important to ensure that hydrocarbon contamination of targets is held to a minimum. This is parti- cularly important in studies of electrochemical charges, of aqueous corrosion, and of friction. Efficient cryogenic trapping and the use of appropriate construction materials and seals are therefore important parts of the design.

CONCLUSIONS

It is too early yet to predict what contribution ion implant- ation will make to the processing of metallic components. Despite the obvious physical and economic constraints of the process there are signs that it will prove useful for specialized applications in both high cost and low cost systems.

On the research side, ion implantation in metals has already provided a better understanding of the corrosion mechanisms of certain metals and alloys. The combination of chemical doping and the creation of a large compressive stress results in an inter- esting combination of surface properties, not achievable by other means. The friction and wear properties of ion implanted metal surfaces have proved most interesting, and the research has revealed that chemical effects and compressive stress induced can both affect the wear resistance. There is much still to be learned in this field. There is an equally large scope for research on the electrochemistry and catalytic properties of ion implanted surfaces of both metals and oxides.

The technological application of ion implantation provides a challenging set of problems for the equipment designer, but equally a new market for the equipment manufacturer. Requirements are so varied, however, that the need will be for custom-built machines and target chambers.

REFERENCES

(1) Trillat, J.J. and Haymann, P. (1961) in "Le Bombardement
     Ionique, Theories et applications", Ed. J.J. Trillot (Editions
     du C.R.N.S., Paris) p.25.

(2) Thompson, M.W. (1970) Proc. European Conf. on Ion Implantation,
     Reading (Peter Peregrines Ltd., Stevenage, England) p.109.

(3) Crowder, B.L. and Tan, S.I. (1971) IBM Technical Disclosure
     Bulletin 14, No. 1, 198.

(4) International Conference on Applications of Ion Beams to
     Metals, Albuquerque, N.M., U.S.A., October 1973, edited by
     Picraux, S.T. EerNisse, E.P. and Vook, F. (Plenum Press,
     N.Y., 1974).

(5) Bowden, F.P. and Tabor, D. "The Friction and Lubrication of
     Solids", Oxford Univ. Press Part I (1950), Part II (1964).

(6) Archard, J.F. J. Appl. Phys. 24, 981 (1953).

(7) Wakelin, R.J., Am. Rev. Mat. Sci. 4, 221 (1974).

(8) Hartley, N.E.W., Dearnaley, G. and Turner, J.F., Proc. 3rd
     International Conf. on Ion Implantation, Yorktown Heights,
     N.Y., U.S.A., December 1972, edited by Crowder, B.L. (Plenum
     Press, N.Y. 1974).

(9) Hartley, N.E.W., Swindlehurst, W.E., Dearnaley, G. and
     Turner, J.F. J. Mat. Sci. 8, 900 (1973).

(10) Hartley, N.E.W., Dearnaley, G., Turner, J.F. and Saunders, J.
     Proc. Int. Conf. on Applications of Ion Beams to Metals,
     Albuquerque N.M., October 1973 (Plenum Press, N.Y. 1974) p.123.

(11) Cookson, J.A., Ferguson, A.T.G. and Pilling, F.D. J. Radio-
     analytical Chem. 12, 39 (1972).

(12) Hartley, N.E.W. J. Vac. Sci. Tech. 1975 (to be published).

(13) EerNisse, E.P. Appl. Phys. Lett. 18, 581 (1971).

(14) Kass, S. "Corrosion of Zirconium Alloys"
     ASTM Publication No. 368, p.3 (1964).

(15) Hauffe, K. "Oxidation of Metals" 1965 (Plenum Press, N.Y.)

(16) Harris, R.W., Phillips, G.C. and C. Miller Jones, Nucl. Phys.
     38, 259 (1962).

(17) Dearnaley, G., Proc. Int. Conf. on Applications of Ion Beams to Metals, Albuquerque N.M., October 1973 (Plenum Press, N.Y. 1974) p.63.

(18) Dearnaley, G., Goode, P.D. Miller, W.S. and Turner, J.F. Proc. Int. Conf. on Ion Implantation in Semiconductors and Other Materials, Yorktown Heights, N.Y. December 1972 (Plenum Press, N.Y. 1973) p.405.

(19) Amsel, G. et al. Nucl. Instr. & Methods, 92, 481 (1971).

(20) Dearnaley, G., Garnsey, R., Hartley, N.E.W., Turner, J.F. and Woolsey, I.S., J. Vac. Sci. & Tech. 12, Jan/Feb. 1975 (to be published).

(21) Pauling, L. "The Nature of the Chemical Bond" 1945 (Cornell Univ. Press) p.58.

(22) Suffield, N.W. and Dearnaley, G. Proc. Conf. on Ion Implantation in Semiconductors & Other Materials, Yorktown Heights, N.Y., December 1972 (Plenum Press, N.Y. 1973) p.541.

(23) Wood, G.C., Oxidation of Metals, 2, No. 1, 11 (1970).

(24) Weidman, L., M.Sc. Thesis, Brighton Polytechnic (1973).

(25) Wanklyn, J. "Corrosion of Zirconium Alloys" ASTM Publication No. 368, p.58 (1964).

(26) Pemsler, J.P., J. Electrochem. Soc. 105, 315 (1958).

(27) Parfenov, B.G., Gerasimov, V.V., Venediktova, G.I. - "Corrosion of Zirconium and Zic. Alloys" - Israel program for scientific translations (1969), U.S. Dept. of Commerce.

(28) Cox, B., J. Nucl. Mat. 29, 50 (1968).

(29) Cox, B. and Pemsler, J.P., J. Nucl. Mat. 28, 73 (1968).

(30) Andersson, S. and Wadsley, A.D. Nature 211, 381 (1966).

(31) Steele, B.C.H. "Solid State Chemistry" Ed. Roberts, L.E.J. (Butterworths, London) p.117 (1972).

(32) Spitznagel, J.A., Fleischer, L.R. and Choyke, W.J. Proc. Int. Conf. on Applications of Ion Beams to Metals, Albuquerque, N.M., October 1973 (Plenum Press, N.Y. 1974) p.87.

(33) EerNisse, E.P. see comment during discussion following ref. 32.

(34) Butcher, D.N., M.Sc. Thesis, Brighton Polytechnic (1974).

(35) Towler, C., Collins, R.A. and Dearnaley, G. J. Vac. Sci. & Tech. (to be published) 1975.

(36) Kubaschewski, O. and Hopkins, B.E. "Oxidation of Metals and Alloys" (Butterworths, London) 1953.

(37) Gleaves, G.L. M.Sc. Thesis, University of Lancaster, (1973).

(38) Mackintosh, W.D. and Brown, F. Proc. Int. Conf. on Applications of Ion Beams to Metals, Albuquerque, N.M. October 1973 (Plenum Press, N.Y. 1974) p.111.

(39) Rickards, J. and Dearnaley, G. Proc. Int. Conf. on Applications of Ion Beams to Metals. Albuquerque, N.M. October 1973 (Plenum Press, N.Y. 1974) p.101.

(40) Ashworth, V., Carter, G., Grant, W.A., Jones, P.D., Proctor, R.P.M., Sayegh, N.N. and Street, A.D. Proc. Int. Conf. on Ion Implantation in Semiconductors and Other Materials, Yorktown Heights, N.Y. December 1972 (Plenum Press, N.Y., 1973) p.443.

(41) Ashworth, V., Proctor, R.P.M., Wellington, T.C., Baxter, D. and Grant, W.A. Proc. Int. Conf. on Ion Implantation in Semiconductors & Other Materials, Osaka, August 1974 (to be published).

(42) Antill, J.E. and Peakall, K.A., J. Iron & Steel Inst. 205, 1136 (1967).

(43) Antill, J.E., Bennett, M.J., Dearnaley, G., Fern, F.H., Goode, P.D. and Turner, J.F., Proc. Int. Conf. on Ion Implantation of Semiconductors & Other Materials, Yorktown Heights, N.Y. December 1972 (Plenum Press, N.Y. 1973) p.415.

(44) Peplow, D., Dearnaley, G., Hartley, N.E.W. and Poole, M.J. (to be published) 1975.

(45) Grenness, M., Thompson, M.W. and Cahn, R.W. J. Appl. Electrochem. 4, 211 (1974).

(46) Wolf, G. Proc. Int. Conf. on Ion Implantation of Semiconductors and Other Materials, Osaka, August 1974 (to be published).

(47) Thackery, P.A. and Nelson, R.S. Phil. Mag. 19, 169 (1969).

(48) Poate, J.M., DeBonte, W. J., Augustyniak, W.M. and Borders, J.A., Appl. Phys. Lett. 25, 698 (1974)

(49) Sood, D.K. and Dearnaley, G., J. Vac. Sci & Tech. 12, no. 1.
     (1975) (to be published).

(50) Freeman, J.H. et al., Proc. Int. Conf. on Ion Implantation in
     Semiconductors & Other Materials, Osaka, August 1974 (to be
     published).

(51) Freeman. J.H. (priv. comm.).

(52) Mattox, D. Proc. 6th International Vacuum Symposium, Kyoto
     1974 (to be published in Jap. J. of Appl. Phys., Supplement
     2, 1974).

CHAPTER 6

ION IMPLANTATION IN SUPERCONDUCTORS

O. Meyer

Institut für Angewandte Kernphysik

Kernforschungszentrum Karlsruhe

INTRODUCTION

The possibility to introduce controlled amounts of impurities into material with the additional advantage to overcome restrictions due to solubility and diffusibility offers an attractive research tool to study near surface properties of material in a systematic manner. Numerous work has been performed in order to study the influence of implanted ions on the physical properties of material and it is well known that the technique of ion implantation is widely used in semiconductor integrated circuit industry. The main topics of a first international conference on this subject [4,5,11,15 etc.] such as implantation modification of superconducting properties, ion induced surface reactions (corrosion science),[60] alloying and migration in high fluence implants, implanted atom location, ion lattice damage together with ion implanted gas build up and implantation simulation of neutron damage reflect the broad spectrum of applicability of ion implantation and promising progress has been obtained in all these fields.

There has been also considerable interest in the effects of irradiations of superconductors since 1960 and many studies have been concerned with the influence of radiation damage on the superconducting critical current and critical field. Irradiations were performed using neutrons, protons, deuterons and alpha-particles and striking effects were reported for the critical current density, however little influence of radiation damage on the superconducting transition temperature has been observed. A summary of this work up to 1968 is given by Cullen[8].

In the last few years heavy ions have been implanted into

superconductors and similar to the experience from ion implantation
in semiconductors the influence of radiation damage effects on the
superconducting properties has to be separated from the influence
due to doping or alloying effects. As the maximum obtainable concentrations by ion implantation is not controlled by equilibrium
processes there is a possibility to produce new stable structures
as well as metastable phases which may have unexpected superconducting properties. A further possible application of ion implantation is the controlled change of composition in compounds, where
it has been demonstrated that the superconducting transition temperature is strongly dependent on composition and density, and
these parameters however can not be optimized by conventional techniques.

        The purpose of the following contributions is to discuss prob-
lems which are connected with the influence of implanted atoms on
the superconducting properties and to summarize some work which
has been performed in the last few years. For convenience a short
introduction to superconductivity is given mainly in order to define
the superconducting parameters used in the following sections.
The results of damage effects on the superconducting properties are
summarized in the next paragraph with emphasize mainly on more recent work performed by heavy ion irradiation. In the last section
the influence of metallurgical effects produced by implanted ions
on the superconducting properties are discussed. As the superconducting properties strongly depend on the crystal structure of the
material it seems reasonable to organize the subsections in this
respect.

DEFINITION OF THE SUPERCONDUCTING PARAMETERS

        The most obvious property of a superconductor is the total ab-
sence of resistance. The transition temperature, T_c, usually defined as the temperature where the resistance R decreases to half
of its normal value (Fig.1) is a very specific property of a given
element, alloy or compound and is af-
fected by the metallurgical condition of the sample. T_c measurements
can therefore provide information
about what has happened to the
sample and have been used to determine the purity of the sample, the
solid solubility level, the composition and phases which are present and the appearance of new
phases. A sharp transition width
indicates a very pure element or a
well annealed alloy, a broad transition width on the other hand reflects inhomogeneities or unusual
strain

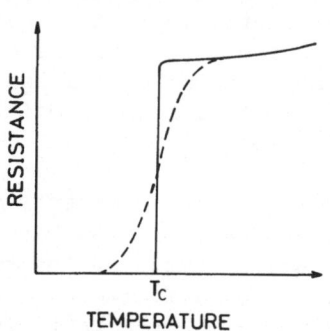

Fig. 1 Schematic diagram showing the decrease of resistance
at the superconducting transition temperature T_c.

distributions. Our fundamental understanding of what crystalline properties determine T_c in a particular superconductor is still limited especially as the high T_c-superconductors have complex electronic band structures and even their normal state properties are not well understood. The theory of Bardeen, Cooper and Schrieffer[1] (BCS) shows, that T_c depends sensitively on the product $N(o) \cdot V$, with $T_c = 0.85\ \theta_D\ \exp - |N(o)\ V|^{-\frac{1}{2}}$, where $N(o)$ is the density of electronic states at the Fermi surface and V is the electron - electron interaction parameter. This is the difference of two large interactions ($V = V_{ph} - V_{Coul.}$), the attractive electron-phonon interaction ($V_{ph} \simeq 1$ eV) and the repulsive Coulomb interaction ($V_{Coul.} \simeq 1$ eV). The Debeye temperature, θ_D, indicates the influence of the phonons and explains the experimentally observed isotope effect (i.e. T_c has been found to be inversely proportional to the square root of the mass of the isotopes for some soft metals). In extension of this theory, Mc Millan[35] derived the following equation, which has found a wide spread use:

$$T_c = 0.69\ \theta_D\ \exp - (1+\lambda)/(\lambda-\mu^+) \qquad \mu^+ < \lambda \text{ and } \lambda = N(o)\ I^2/M\theta_D^2$$

where I^2 is the electron-phonon interaction constant and M is the mass of the lattice atoms, μ^+ corresponds to the Coulomb interaction and is about 0.1.

Pure soft metals exhibit perfect diamagnetism (Meissner - Ochsenfeld effect) in the superconducting state and exclude an externally applied magnetic field up to some critical field H_C: for higher applied fields the superconductor reverts to the normal conducting state. The dependence of H_C on the temperature is nearly parabolic and can be approximately described by $H_C(T) = H_C(o) \times |1 - T^2/T_c^2|$. For a long thin wire parallel to the magnetic field, the magnetization M is proportional to the applied magnetic field until H_C is reached (Fig. 2), where the superconducting state switches to normal. Superconducting materials showing this magnetization behaviour are called type I superconductors. If a current is sent through a type I superconductor, the current flow is restricted to the surface. The maximum current density I_C will be reached if the generated field becomes equal to H_C. (For the above mentioned geometrical conditions I_C is given by $I_C = 2\pi\ RH_C/\mu_o$ with R = radius of the wire and $\mu_o = 4\pi \cdot 10^{-7}$ Vs/Am.)

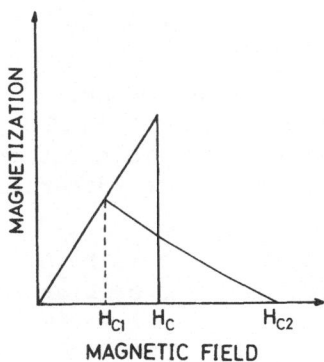

MAGNETIZATION

H_{C1} H_C H_{C2}

MAGNETIC FIELD

Fig. 2 Schematic drawing of magnetization curves for type I and type II superconductors.

In a type II superconductor the externally applied magnetic field can partly penetrate the superconductor so that in this "mixed state" superconducting and non superconducting regions exist at the same time. This state is also called the Shubnikov-phase. The magnetization curve for a type II superconductor is included in Fig. 2. At H_{c1} the applied magnetic field starts penetrating into the superconductor and forms a regular arrangement of filaments, the so-called flux thread lines. These fluxoids consist of a normal conducting core and a surrounding supercurrent vortex. If a current is sent through a type II superconductor an interaction will occur between the magnetic field of the flux threads and the current (Lorentz-force). The flux threads will move in response to this force, thus producing electrical resistance and losses. Therefore type II superconductors are not appropriate for technical applications in magnets. The flux threads have to be "pinned" and materials which contain pinning centers are called type III superconductors. A great variety of microstructural features can act as pinning centers such as for example dislocations, grain boundaries, second phase precipitations as well as voids and defect clusters produced by particle irradiation. Both the lower and the upper critical field, H_{c1} and H_{c2} depend on H_c. The BCS theory connect T_c and H_c, and for $T \rightarrow 0$ it is: $H_c(0) = T_c \gamma^{0.5}/0.17$, where γ is the coefficient of the specific heat, which is proportional to $N(E)$, the density of electronic states.

There are mainly three structural classes where alloy and compound superconductors with high critical fields have been found and used:

1. Most superconducting magnets have been produced with alloys having the bcc-structure such as NbZr ($T_c = 11$ K) and NbTi ($T_c = 10$ K). NbTi is slightly superior to NbZr because of easier fabrication and higher critical fields.

2. Interstitial compounds with NaCl(B) crystal structure are difficult to handle and have not been used. However thin NbN-films (17 K) produced by reactive sputtering have shown rather high current carrying capacities and are found to be rather resistant to radiation as will be shown later.

3. Compounds with the A-15 (ß-tungsten) structure such as Nb_3Sn (18.2 K) V_3Ga (17 K) reveal the best performance in as high field magnets known up to now.

INFLUENCE OF RADIATION DAMAGE ON THE SUPERCONDUCTING PROPERTIES

There are three main reasons which have led to the study of irradiation effects in superconductors:

1. In order to select suitable superconducting material for mag-
    nets used in accelerators and possibly in fusion reactors one
    has to know the influence of radiation damage on I_c, H_c and T_c.

2. The question as to whether irradiation or implantation can pro-
    vide more effective pinning centers than can be obtained by
    conventional techniques is still not answered, even if there
    are some indications in recent work that heavy ion implantation
    may have some advantage.

3. The study of metallurgical or chemical effects of implanted
    ions requires a knowledge of possible damage effects.

In the following sections more recent work mainly performed by heavy ion bombardment is discussed.

a) Non Transition Metals

      Only small effects of irradiation at 4.2 K on the properties
of soft superconductors have been observed, far larger effects have
been realized by cold working or in films evaporated on a liquid
He cooled substrates (quenched condensed films). Radiation damage
effects generally anneal below or at room temperature. This is in
agreement with results obtained by Kübler and Meyer[29] for Sn and
Pb films bombarded with $5 \cdot 10^{16}$ Ar$^+$/cm^2 at room temperature, where
no influence on T_c has been observed.

      A slight increase in T_c from 1.20 → 1.21 K of Al films however
has been found after implanting high fluences of Ar ions (10^{17} Ar$^+$
/cm^2). In these experiments, the range of the Ar ions was chosen
equal to a larger than the film thickness, so that the observed
effects are probably due to intrinsic damage. For Sn layers evaporated at room temperature and homogeneously implanted with Cu at
4 K, Buckel and Heim[4] found that the T_c increased by about 0.1 K
and that the residual resistance as well as T_c saturated at a concentration of 50 ppm of Cu. In addition it has been found that the
resistance of a quenched condensed Sn film (evaporated on a 4 K
cold substrate) decreased during implantation of Mn ions in the
same concentration region to about the same saturation level and
therefore Buckel and Heim[4] supposed that an equilibrium state of
radiation damage production and radiation annealing had been reached.
For Pb films a small decrease in T_c of about 0.1 K has been observed
after implanting Zn$^+$ ions. These results are in agreement with changes in T_c caused by cold working or by rapid quenching at liquid
He temperatures. In contrast to this, earlier measurements by Van
Itterbeck et al.[53] on Tl, In and Sn foils irradiated with 5.3
MeV α-particles at 4 K, showed a decrease in T_c of a few tenth of
a percent. As the residual resistance ρ had not yet saturated in
these experiments and a positive curvature dT_c/dC was indicated,
further irradiation may have led to an increase in T_c.

b)   Transition Metals

Neutron irradiation of Nb at room temperature have been found
to alter the superconducting behaviour (and thus the effects observed are stable at this temperature). Neutron induced flux pinning centers are more effective in Nb than inhomogeneities introduced by mechanical deformation. Tsypkin and Chudnova[53] have bombarded Nb foils of different oxygen content (<2 at %) with 10 - 40
MeV N and O ions at fluences between 10^{12} and 3.10^{15} ions/cm^2
and have observed a significant increase in I_C and a slight increase
in H_{C2}. Freyhardt, Loomis and Taylor[15], implanted 3.5 MeV Ni ions
in Nb foils at 900°C and found a substantial increase in both I_C
and H_{C2}. In both studies it is assumed that the regular arrangement
of voids, which has been observed with transmission electron microscopy may be responsible for the observed results, as the super-
lattice of the voids advantageously fits the flux thread lattice.
These results seem to be a first indication that heavy ion irradiation may produce a very effective pinning center arrangement which
cannot be obtained by conventional techniques. In the above mentioned experiments T_C is only little affected at fluences which
already have large effects on the current carrying properties. Using
heavy ions and high fluences and thereby increasing, the energy
density deposited in nuclear collisions, Crozat[11] and coworkers
found a substantial decrease in T_C for Nb films, and Meyer, Mann
and Phrilingos[36] for V, Nb and Ta films. T_C reductions in V layers have been observed by Kübler[28] in dependence on the mass,
energy and fluence of the bombarding ions; the results showed that
the reductions could be assigned to defects produced in the layers.
Kübler[28] further investigated possible impurity effects due to
the implanted ions by implanting Ga ions in V layers. By comparing
the amount of T_C reductions observed for Ga$^+$ implantations with
those for Kr$^+$ implantation, no additional effect could be detected
up to about 20 at % of Ga atoms homogeneously distributed in the
V films.

In a recent more quantitative study by Linker and Meyer[31]
the damage production and distribution from Ne ions has been calculated and its influence on the T_C of V, Nb and Ta thin films has
been measured. The energy of the Ne ions was chosen such that they
penetrated the layers and came to rest in the substrates. In order
to obtain a quantitative measure for the distribution and the total energy lost in nuclear collisions by the Ne ions in different
layers, the primary energy deposition profile for the projectiles
has been calculated from universal $(d\varepsilon/d\rho)_n$ data (ε and ρ are di-
mensionless range and energy parameters introduced by Lindhard,
Scharff and Schiott (LSS)[32] using in the actual computations an
analytical expression for $(d\varepsilon/d\rho)_n$ from Winterbon, Sigmund and
Sanders[55] which is a good fit to Schiott's data and electronic
stopping values from the LSS theory.

Some calculated primary energy deposition profiles for an Ne
ion in V, Nb and Ta are given in Fig. 3; the R_p and ΔR_p values
from LSS theory have also been included. These profiles are rather
homogeneous in layers of about 1000 Å thickness as compared to
profiles resulting when particles come to rest in the layer.
It should be emphasized however that these are profiles for an
average particle, as fluctuations in the range were neglected in
the calculations. The influence of fluctuations on primary deposition profiles is evident mainly in the region where the particle
come to rest. Similar densities of energies deposited in different
materials were obtained by choosing appropriate fluences for the
Ne ions.

Relative T_c reductions $\Delta T_c/T_c$ from the work of Linker and
Meyer[31] are shown in Fig. 4 as a function of the density of energy deposited in the layers. For energy densities below 10^{18} eV/Åcm^2
the relative reduction increases slightly with increasing energy
density and similar values are obtained for V, Nb and Ta. This result is not unreasonable for these groups VB transition metals as
they all have the same crystal structure (bcc) and similar electronic density distributions. It is assumed that the heavy ion radiation damage reduces the density of states at the fermi level and/or

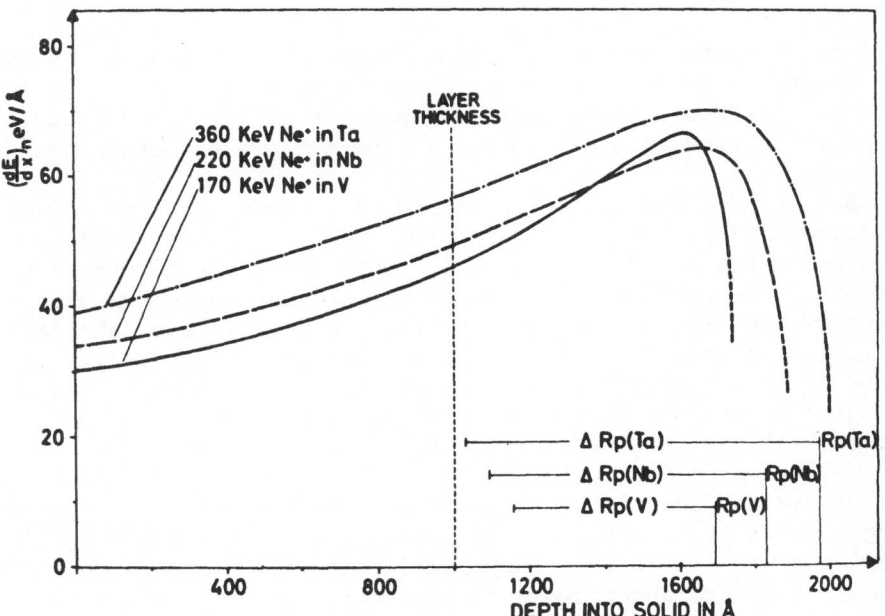

Fig. 3   Calculated primary energy deposition profiles for an
         average Ne ion in V, Nb and Ta; R_p and ΔR_p values from
         LSS theory are included in the figure.

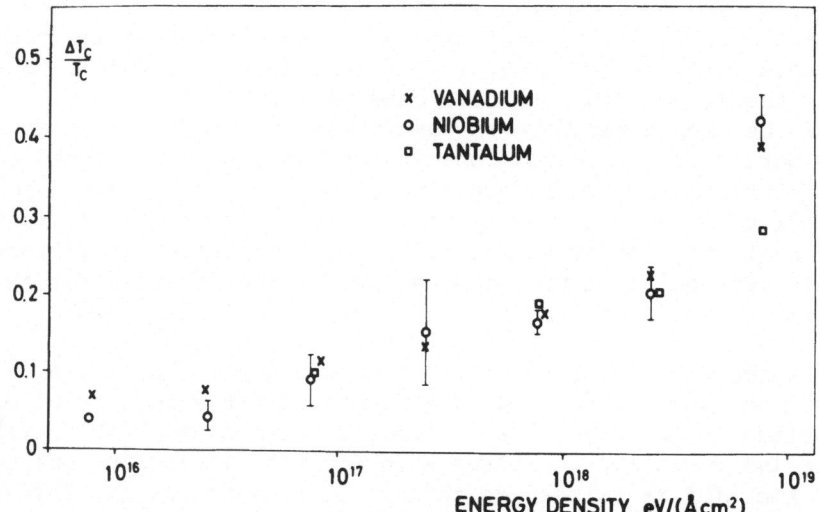

Fig. 4   Relative decrease of the superconducting transition tempe-
         rature $\Delta T_c/T_c$ as a function of energy density deposited
         in nuclear collisions (ded).

increases the average phonon frequency as radiation hardening is
known to occur in bcc metals. Of course, these are only two possible factors for a microscopic explanation of the T_c-reductions
and no solid arguments can be given from the present experiments.

Beyond 10^{18} eV/$\overset{\circ}{A}$ cm^2, corresponding roughly to 50 to 100 dpa
(displacements per atom), a pronounced change in slope is observed
which indicates a different process influencing the superconducting
transition temperature. Backscattering analyses indicated that the
layers became slightly porous suggesting discontinuous material removal. In addition, from thin film x-ray analysis of heavily dis-
torted   layers only the two strongest reflections were recorded
and these showed distinct broadening as compared with the equivalent reflections from unbombarded layers.

Deprez and coworkers[12] determined the low frequency (30 Hz)
energy losses in superconducting V before and after implantation
of 10^{16} Ga$^+$/cm^2 at 75 keV. They found that the losses for the lower
ac amplitudes are about two orders of magnitude smaller in the
sample implanted with Ga due to a pronounced increase of the surface
shielding capacity.

c)   Transition Metal Alloys

For NbZr and NbTi alloys large changes in the superconducting

properties have been obtained by mechanical deformation or by introduction of impurities forming a second phase precipitation. Alloys
that have been optimized in this way are found to be rather resistant against irradiation with neutrons and light ions. Meyer, Mann
and Phrilingos[36] implanted Mo films with C ions up to concentration of about 10 at % and were able to increase the T_C from 1.2
to 7 K. In a second irradiation these layers have been bombarded
with Ar ions up to $5 \cdot 10^{16}$ ions/cm^2 at room temperature. Only a
small decrease in Tc from 7 to 6.5 K has been observed again indicating that transition metal alloys are rather stable against irradiation.

Chang and Rose-Innes[6], altered the composition of a cylin-
drical $Nb_{77}Mo_{23}$ rod by implanting extra Mo ions into a narrow surface strip parallel to the cylinder axis. This leads to a decrease
of T_C where it is not certain if this decrease in T_C is due to a
damage or an alloying effect. As T_c decreases rapidly with Mo concentration a local reduction of the surface pinning is observed in
the case where the externally applied magnetic field is tangential
to the surface strip in one of two possible directions. A reduction
of the critical current is observed when the flux threads move towards the implanted surface and are less effectively pinned because superconductivity has been depressed in that region. These
results confirmed that the current of a type II superconductor is
determined by pinning of flux threads at the surface.

d)   Superconductors with A-15 and NaCl-Structure

Numerous studies have been performed in order to investigate
the influence of irradiation on the superconducting properties of
compounds with the A-15 structure. The results obtained with neutron irradiation of Nb_3Sn have been well documented by Cullen[8].
Studies by Wohlleben[56] with protons and deuterons and by Ischenko
and coworkers[22] with oxygen ions agree with earlier observations
that differences in the effects of irradiation are mainly due to
the different microstructures obtained for material produced by
different material preparation techniques. In several compounds I_c
can be increased by more than an order of magnitude and this enhancement is stable up to annealing temperatures well above room
temperature.
Other superconducting properties are less affected; T_c for example
was found to decrease only by about 1 % in these experiments. In
a more recent work Seibt[45] has studied radiation effects on commercial NbTi and V_3Ga multifilament wires homgeneously irradiated
with 50 MeV deuterons up to a fluence of $2.6 \cdot 10^{17}$ ions/cm^2 near liquid He temperature. While irradiation effects on I_c and H_c were
found to be negligible for NbTi in agreement with earlier observations, the V_3Ga wires showed a reduction in I_c by about 50 % and T_c
decreased from 14.7 to 12.3 K. The influence of the deuteron

fluence on I_c is presented in Fig. 5, I_{co} is the critical current before irradiation. Annealing experiments up to 100°C resulted in a small (15 %) recovery of the superconducting properties of the V_3Ga wire.

Large reductions in T_c of A-15 superconductors have been observed recently in both neutron irradiated and heavy ion irradiated material. Schweitzer and Parkin[44] performed neutron irradiations at ambient temperatures with fluences between 10^{18} and 10^{20} fast neutrons/cm^2 . The A-15 materials Nb$_3$X, with X = Sn, Al, Ge, Ga and V$_3$Si showed a rapid monotonic reduction in T_c from about 0.95 T_c to 0.15 T_c of the unirradiated value in the fluence range indicated above. T_c changes for Nb, Nb-Ti and NbN are reported to be about 1 K after exposures of about 10^{20} n/cm^2. Large T_c reductions by thermal neutron irradiation have been observed by Bauer and Saur[2] for Nb$_3$Sn doped with boron and uranium, due to a large energy density deposited by the reaction products.

The influence of radiation damage produced by Ar ion bombardment with fluences between 10^{13} to 10^{17} ions/cm^2 on the T_c of thin films of various superconducting material observed by Meyer, Mann and Phrilingos[36] is summarized in Fig. 6. The relative decrease

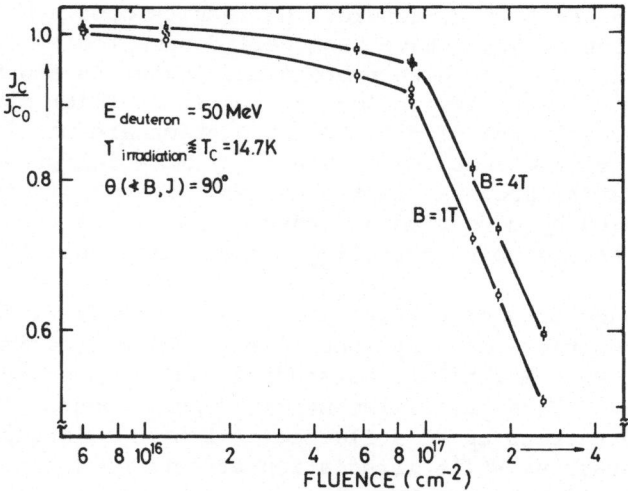

Fig. 5 Critical current ratio I_c/I_{co} versus deuteron fluence for a V$_3$Ga multicore wire (360 filaments, fil, diam. 10 μm). (from Seibt[45])

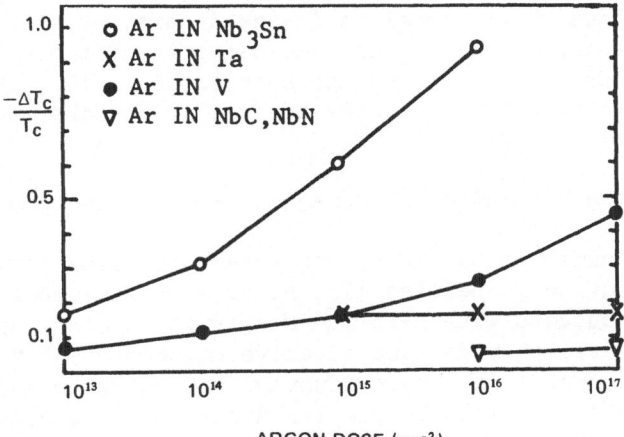

Fig. 6 Decrease of T_c, ΔT_c, normalized to the T_c of the unirradi-
        ated layer for V, Ta, NbN, NbC and Nb_3Sn in dependence on
        the Ar fluence.

in T_c, $\Delta T_c/T_c$, normalized to the corresponding T_c values of the unirradiated films is found to be quite different for Nb_3Sn from what
it is for NbN and NbC. In Nb_3Sn layers, for example, the initial
T_c of 18.2 K has been reduced to 2 K after implantation of 10^{16}
Ar^+/cm^2 whereas for NbN and NbC with the NaCl structure only a
slight influence has been found under similar irradiation conditions. In a preliminary experiment Linker and Meyer[31] have reduced the T_c of a Nb_3Ge layer from 20.7 to 7.1 K after irradiation with 10^{17} He^+/cm^2 at 300 keV.

e) Transition Metal Layer Compounds

        Tsang and coworkers[57] studied the structural disorder in-
duced by Ar ion bombardment of TaS_2 and $TaSe_2$ at room temperature
and observed an increase of T_c to 4,2 K and 2,5 K respectively.
Before irradiation the T_c in both cases was below 1,2 K. In NbSe
however, T_c did not increase after heavy ion bombardment. This observed increase of T_c, due to inert ion bombardment is far more
dramatic than that observed for non transition metals (see chapter Ia).As the crystal bonding in these compounds is predominantly
ionic and/or covalent the T_c enhancement is probably due to the
fact that in contrast to non transition metals the radiation defects
are stable at room temperature. For TaS_2 and $TaSe_2$ as well as for
$NbSe_2$ it is thought that the density of states is very high and
has a maximum at the Fermi surface. As an increase of T_c has been

observed for TaS_2 and $TaSe_2$, a model stating that radiation damage smoothes out sharp anomalies in the density of states which
will lead to a decrease in T_c in some cases is probably not appropriate. Tsang et al. associated the changes in T_c in these compounds with the quenching of phase transitions by the introduction of crystal disorder.

f) Quantitative Estimation of Damage in Superconductors

Although numerous papers on the effects of irradiations on
the superconducting properties (I_c, H_c, T_c) of elemental and compound superconductors with A-15 and B1-structures have appeared
in the last years, only few quantitative connections have been
drawn between changes in superconducting parameters and damage
generated in the materials by the irradiations. This situation
makes it difficult to compare the results obtained by different
authors, especially when different particles like n, p, d or heavy
ions are used in the irradiation procedures.

Some attempts however were made in this regard and Wohlleben
[56], using a Rutherford scattering cross section for the estimation of the density of primary knock on atoms (PKA) produced with
p and d irradiations in the MeV range, found that the maximum
in the changes of the critical current ΔI_c for Nb_3Sn appeared
at the same position on a fluence scale, when a normalized fluence

$$\Phi_n = \Phi \frac{M_1 Z_1^2}{E}$$

(Φ=ions/cm^2, M_1, Z_1,E: mass, charge number, and energy of the incident particles) was used. With this same procedure Ischenko
et al.[22] were able to compare their results on the influence of
24 MeV O^{5+} irradiations on the critical current of Nb_3Sn with the
findings obtained with proton and deuteron irradiations. Further
Freyhart, Loomis and Taylor[15] converted the irradiation fluence
of 3.5 MeV Ni^+ ions in Nb to a number of atom displacements per
atom (dpa) whereas Crozat et al.[11] set the number of defects
produced in Nb by Nb^{++}, oxygen$^+$and Er^+bombardment with energies of
130 keV, 15 and 40 keV and 50, 110 and 260 keV respectively, proportional to the ions energy. Linker and Meyer[31] recently compared the influence of heavy ion radiation damage produced with
Ne^+ ions on the superconducting transition temperature T_c of V, Nb
and Ta layers with the total energy density deposited in nuclear
collisions (s. chapter I_b).

In principle,just as damage production rates and deposition
profiles have been calculated by several authors for neutron and
ion bombardment of solids (especially reactor materials) these procedures can also be applied to damage produced in superconductors.
In the slowing down process of a particle, damage is generated

mainly by nuclear stopping and, for a quantitative estimation, the
appropriate cross section σ for nuclear stopping, for the ion-
target systems and the ion energies under consideration, must be
known for a calculation of the amount of energy deposited into
nuclear collisions, and a reasonable displacement production function ν must be used to estimate the number of displaced atoms.
If a damage density as a function of depth is desired the depth
dependence of these functions must also be known.

The atomic concentration of displaced atoms C_d for monoenergetic ions may then be expressed by the following formula according to e.g. Thompson[51]:

$$C_d = \Phi \int_{T_{min}}^{T_{max}} \nu(T) \; \frac{d\sigma}{dT} \; (E,T) \; dT$$

The symbols in this formula are:

Φ   total fluence
T   energy transferred to a target atom in a recoil
         with $T = 4 \; M_1 M_2/(M_1+M_2)^2 \; E \; \sin^2 \Theta/2 = \Lambda \; \cdot E \cdot \sin^2 \Theta/2$
E   energy of the incident particles
Θ   scattering angle in the center of mass system
M_2 mass of the target atoms
T_{max} ΛE, maximum possible energy transfer to a struck atom
T_{min} threshold energy for displacement production

For neutrons a similar general formula may be used e.g.
Köhler and Schilling[24], Kulcinski et al.[27] and the concentration of displaced atoms is given by:

$$C_d = \Phi \cdot N \iint \rho(E) \sigma(E,T) \; \nu(T) dE \; dT$$

with
N   number of scattering centers per cm^3
ρ(E)dE   neutron fluence in energy range, E, E + dE
σ(E,T)dE   cross section for a neutron with energy E to produce
            a PKA in energy range T, T + dT

In these general formulas for the displacement production
function ν(T), a simple KINCHIN-PEASE[23] model with $\nu(T)=T/2E_d$
(E_d-displacement threshold energy) may be used whereas for the
differential cross section dσ/dT a form suitable for the particle,
energy and target-system under consideration depending on an appropriate interaction potential must be applied.
For light ions like p or d with energies in the MeV range a Rutherford scattering cross section (dσ/dT ≈ $1/ET^2$) can be used; for
heavy ions a differential cross section given by Lindhard,

Nielsen and Scharff[59] based on a Thomas-Fermi interaction potential which also covers the Rutherford cross section range is most
suitable.
This scattering cross section is given by:

$$d\sigma = \pi a^2 f(t^{1/2}) dt/2t^{2/3}$$

where
$t = \varepsilon^2 \sin^2 \Theta/2$

$a = 0.8853 a_o (z_1^{2/3} + z_2^{2/3})^{-1/2}$

a_o = Bohr radius

ε = reduced energy parameter

Θ = scattering angle in the center of mass system.
$f(t^{1/2})$ is a function given in numerical form. A good analytical
approximation for this function has been published by Winterbon,
Sigmund and Sanders [55]. Sophisticated calculations on damage distributions have been performed by Brice[3], who was able to include the energy transport of recoiling target atoms into the
energy deposition distribution calculations.

INFLUENCE OF IMPLANTED IONS ON THE SUPERCONDUCTING
TRANSITION TEMPERATURE

A continuous aim in the field of superconductivity is the
search for material with large T_C values and this search has led to
the observation, that material showing structured instabilities
exhibit the best results in this respect. Instabilities in compounds
or alloys occur at compositions close to phase transition regions
and also in metastable compounds produced by non equilibrium processes such as splat cooling, evaporation and sputtering. The most
obvious example for this is Nb_3Ge which has the highest known T_C
of 23 K and which has been produced by all these techniques.

Ion implantation as a nonequilibrium technique is thought to
be of interest in this respect because of its inherent possibility
to increase solubility levels above thermal equilibrium values and
to produce metastable alloys. A further possible application of this
technique is the use of ion implantation in order to compensate deviations from stoichiometry and also to fill vacancies in sublattices
of compounds which cannot correctly be formed by conventional techniques. A first indication that this might be possible has been obtained by Meyer and Hofmann[38], who implanted N ions in $NbC_{0.95}$
at room temperature and observed an increase in T_C from 9 to 11 K.
The metallurgical nature of an ion implanted layer is known to be
very complex as radiation damage and associated effects such as radiation enhanced diffusion and precipitate growing during implanta-

tion can significantly modify the metallurgical conditions. Therefore T_C measurements besides X-ray diffraction technique and
transmission electron microscopy etc. can be used as an additional
simple means to obtain information on the fate of the implanted
ion. In the last few years ion implantation has been used by several groups in order to study the effect of implanted ions on T_c
and their results will be summarized in the following chapters.

a)   Magnetic Impurities in Non Transition Metals

The interaction of paramagnetic impurities in a metal with
electrons in the normal as well as in the superconducting state
will change the resistance and will lead to a decrease in T_C.
With conventional techniques the preparation of interesting alloys
is often limited because of very low solubility levels. Geerk,
Heim and Kessler[18], and also Buckel and Heim[4] have produced
such alloys by ion implantation near liquid He-temperature in order to prevent diffusion and precipitation of the implanted ions
and to obtain metastable alloys beyond the solubility limit. On
the other hand, very small concentrations were needed in order to
avoid interaction among impurities. Ion implantation is known
to provide small concentrations in the ppm range with the same relative accuracy as at very high concentrations. Results of this
group[4] are presented in Fig. 7 where it can be seen that small
concentrations of Mn cause a strong depression of the T_c of homogeneously doped Hg, Pb and Sn-films. The slope of the curves is
approximately constant at concentrations above 100 ppm in agreement with theoretical calculations. At concentrations below 100
ppm a curvature in the T_c curves can be seen, being positive for
the system Mn in Sn and negative for Mn in Pb and in Hg. These
deviations from the linear dependence are thought to be due to
lattice defects. In order to verify this assumption Buckel and
Heim[4] implanted the nonmagnetic ions Cu and Zn which are close
to Mn in the periodic system, into Sn and Pb layers respectively.
The observed changes in T_c are also included in Fig. 7 and they
explain the observed deviations at concentrations below 100 ppm.
In several implanted dilute magnetic alloys a resistance minimum
had been found at low temperatures. The observed logarithmic increase of the resistivity towards low temperatures indicates that
the Kondo[26] effect is present in these alloys. In the theoretical
work of Kondo this behaviour is explained by a negative exchange
interaction between the spin of the conduction electrons and the
localized spins of the magnetic impurities.

b)   Pd-, Pd-Noble Metal Alloy, -Hydrogen System

A further excellent example of the usefulness of ion implantation has been demonstrated by Buckel and Stritzker[5], who succeeded in preparing superconducting Pd-H alloys by means of H im-

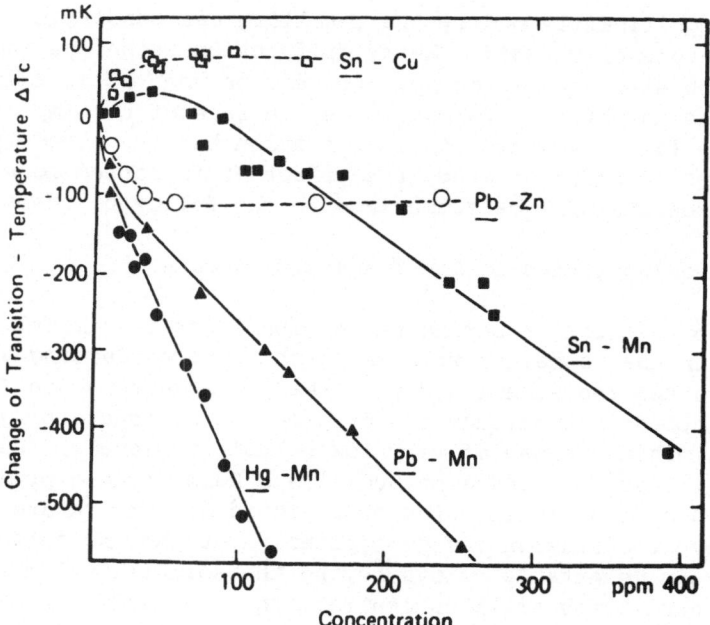

Fig. 7   Change of transition temperature ΔT_c for superconducting
         films of Pb, Sn, and Hg, as caused by implantation of Mn,
         Cu and Zn ions. (from Buckel[4])

plantation at liquid He temperature. They found a maximum in T_c with
increasing fluence of H and D ions of almost 9 K for the Pd-H system and of 11 K for the Pd-D system. These maximum T_c values were
estimated for a H/Pd ratio of about 1, which is not easily obtainable by other techniques. One of the main reasons for this T_c increase is thought to be due to the supression of the spin-fluctu-
ations of the Pd as it is known, that the system becomes diamagnetic at a H/Pd ratio of about 0.7. In an extension of this work
Stritzker[17] implanted H in Pd-noble metal alloys which are known
to become diamagnetic if the noble metal concentration exceeds
60 at %. These Pd-noble metal alloys however do not show superconductivity above 10 mK. After implantation of H or D high T_c values
have been measured which are probably due to a different T_c-enhance-
ment mechanism. Foils with different noble metal concentrations for
the three systems Pd-Cu, Pd-Ag and Pd-Au have been implanted with
H ions and the fluence has been increased until T_c passed the maxi-

mum. These maximum T_C values are shown in Fig. 8 in dependence on
the noble metal concentration. In the system Pd-Cu-H the highest
T_C of 16.6 K has been observed.
With increasing noble metal concentration a decreasing H content is
necessary to obtain the optimal T_C values. By further increasing
the H concentration Buckel and Stritzker observed an irreversible
change to the normal state. The resistance of the layers dropped
and no superconductivity was found above 1 K. It is believed that
this result is due to a phase transition from a lattice instabili-
ty to a non superconducting phase with lower resistivity. Therefore
it is assumed that the main precondition for the superconductivity
in these systems is the suppression of the spin fluctuations, but
that in addition, a peculiar phase transition indicates that weak
phonon modes might be responsible for the high T_C values.

c) Ion Implanted Transition Metal Systems

    Enhancements of T_C in various ion implanted transition metal
systems based on Mo, W, Re and Ti have been reported[36]. In a
systematic study Phrilingos[42] tried to find out the physical reason for this T_C enhancement of implanted system produced at room
temperature in comparison with the dependence of T_C on impurity concentration observed for solid solutions produced by conventional
melting techniques.

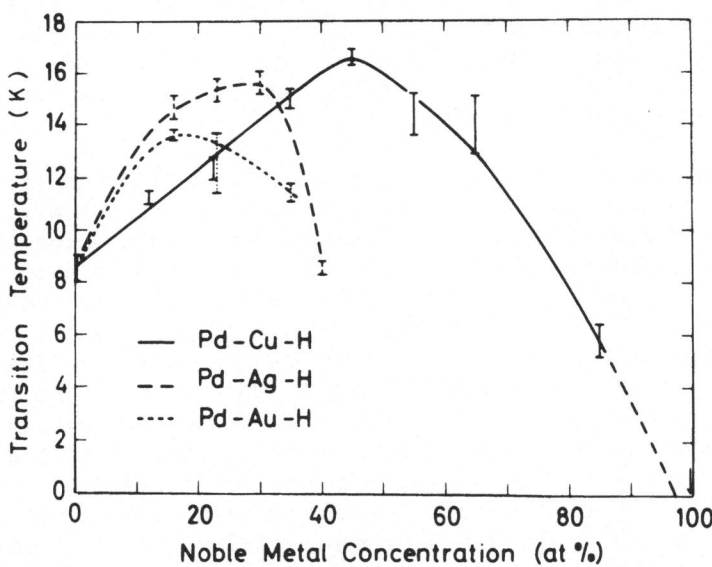

Fig. 8   Maximum T_C values after H implantation versus noble metal
         concentration in the systems Pd(Cu, Ag, Au)H. (Stritzker[47])

     The superconducting transition temperature, T_C of transition
metal alloys is known to vary strongly with the variation of the
number of electrons per atom, e/a. Matthias[34] has formulated
useful empirical rules for the T_C behaviour in transition metal
alloys, suggesting that T_C depends on the e/a-ratio and maxima
in T_C should occur at an average e/a near 4.8 and 6.8 (the number
of e is set equal to the group number). In a theoretical work by
McMillan[35] on transition metal alloys the rigid band model is
assumed to hold for alloys of metals which are neighbours in the
periodic table. Miedema[40] assumed that the two metals in the solution contribute individually to the density of states at the
Fermi energy and he predicted a linear relation between the change
of T_C with concentration, dT/dC, and the electron transfer which is
proportional to the electronegativity difference.

     Numerous work has been performed in order to study the influ-
ence of alloying on T_C of transition metals. For Mo it has been
reported[19] that T_C is raised when the elements Tc, Re, Ru, Os,
Rh, Ir, Pd and Pt, which have larger e/a ratios than Mo are dissolved in it. On the other hand, transition metals with smaller
e/a ratios as for example V, Nb, Ta, Ti, Zr and Hf are known to
decrease T_C upon alloying with Mo. The increase in T_C of Mo based
alloys have been explained by Miedema [40] by a charge transfer
from Mo to the more electronegative metals.

     Ion implantation at room temperature has been used in order to
introduce ions from all groups of the periodic system over a wide
range of concentrations into the near surface region of molybdenum. This method has the advantage to overcome restrictions due
to the solubility limit, as the solubility in a solid solution of
elements with large differences in electronegativity are strongly
restricted by the formation of stable intermediate compounds during recrystallization from melt.

     In a systematic study Phrilingos tried to find out if the con-
cept of charge transfer is a more general concept that will also
hold for non transition metals in molybdenum. Ion species with different electronegativities and with concentrations of about 5, 10
and 15 at % in the peak maximum of the distribution of the implanted ions for each system have been used in this systematic study.
The T_C-values of the implanted systems have been recorded and the
results have been summarized in Fig. 9. This plot, as proposed
by Darken and Gurry[14], combines the electronegativity and the
atomic radii, the most important parameters involved in the Hume-
Rothery rules[21] for the alloying behaviour of metals. In this plot
Goldschmidt's[21] atomic radii for the coordination number 12 and
Pauling's[41] values for the electronegativity have been used.
The square in Fig. 9. around Mo, indicates the region where, concerning to the Hume-Rothery rules, elements have a high probability

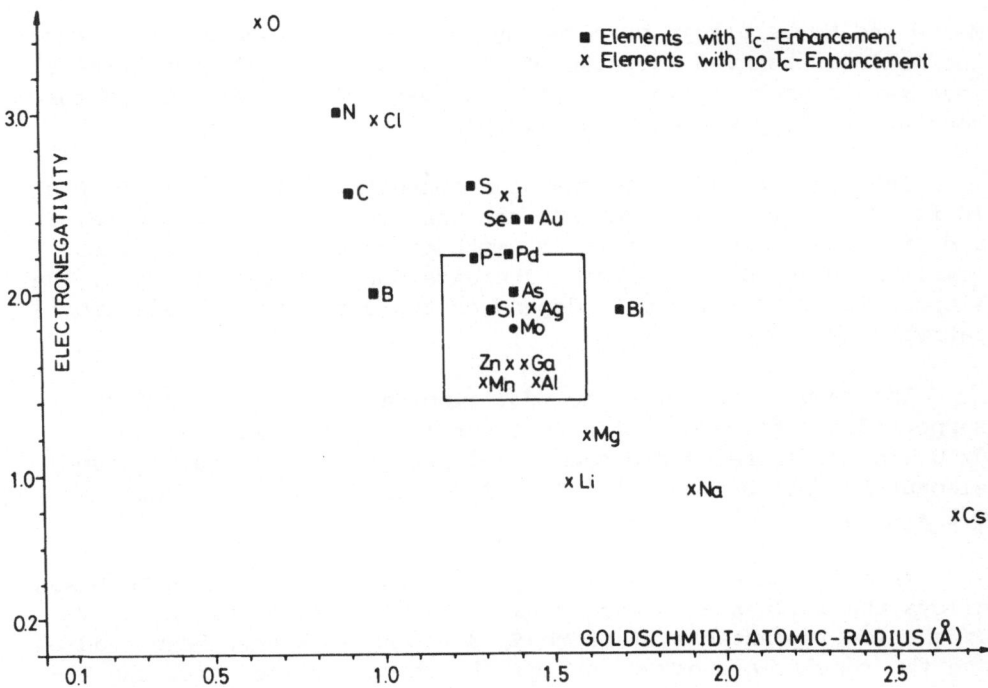

Fig. 9   Darken-Gurry Plot of ion implanted Mo-based systems sum-
marizing the results of T_c measurements.

of forming a substitutional alloy with Mo. For the elements located in the square Ag, Zn, Ga, Mn and Al do not show any T_c enhancement, whereas for Pd (in agreement with conventionally prepared alloys), P, As and Si a T_c increase above 1.2 K has been
observed. T_c enhancement is also found for elements located outside the square and therefore it is concluded that the formation
of substitutional alloys is not of primary importance for the T_c
enhancement mechanism. The results indicate that electronegativity is more important as for elements being more electronegative
than Mo, an increase of T_c usually is observed.

The increase of T_c in dependence of the ion fluence has been
measured for the systems Mo-C, Mo-N, Mo-S, Mo-Si, Mo-Au, Mo-As,
Mo-B, Mo-Se, Mo-P, Mo-Pd and Mo-Re. The slopes dT_c/dC of the last
two transition metal systems were found to agree with those of

Mo-Pd and Mo-Re alloys prepared by conventional techniques. However the maximum observed T_C value of 10.5 K for ion implanted Mo-Re alloys was found to be smaller than T_C (max) of 12 K determined for conventionally prepared Mo-Re alloys.

The increase of T_C versus atom concentration is shown in Fig. 10 for Mo layers implanted with the non transition metals C, N, S and Si. After a sharp rise in T_C with concentration a saturation level is observed with a small decrease for high ion fluences. The slopes dT_c/dc were found to depend on the difference in electronegativity $\Delta\chi$.

The result that $\Delta\chi$ is the most important factor is further supported by the observation that for the Mo-Ag system with $\Delta\chi\approx0.1$ no T_C increase has been found whereas for the same group element Au with $\Delta\chi$ (Mo-Au) = 0.6 dT_c/dc is found to be about 0.2 K/at %.

In order to study the stability of the ion implanted Mo-based alloys the influence of isochronal annealing processes on the T_C of the ion implanted systems Mo-N, Mo-C and Mo-S have been studied and the results are given in Fig. 11. It can be seen that the T_C-

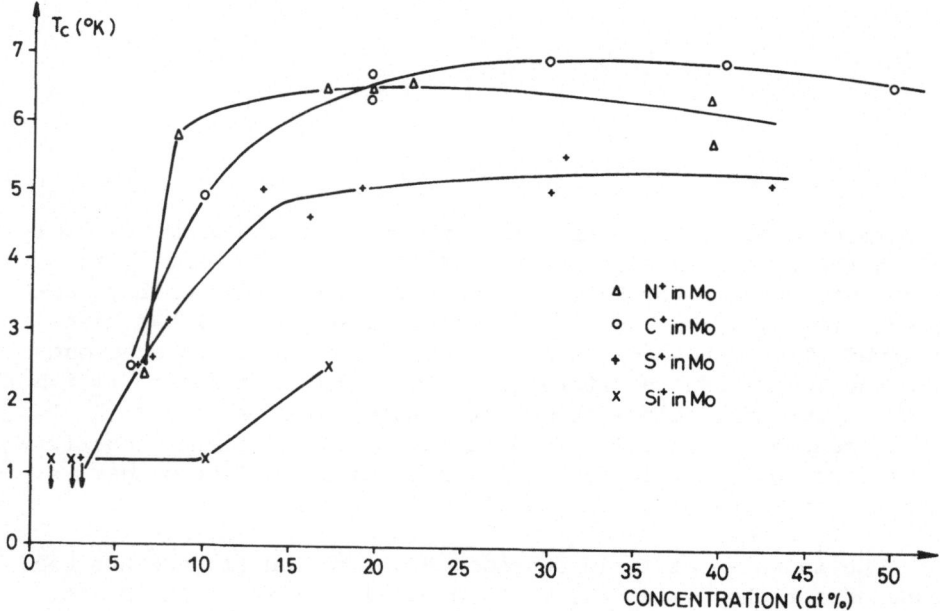

Fig. 10   Increase of the T_C of Mo layers after implantation of Si, S,C and N ions in dependence on impurity concentration.

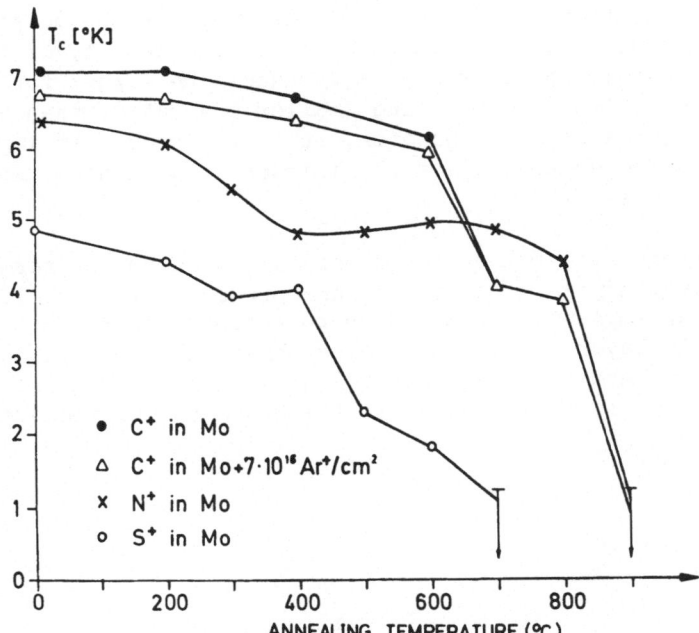

Fig. 11   Influence of an isochronal anneal process on the T_c of
implanted Mo layers.

enhancement of this alloys is stable for annealing temperatures up
to 400°C in the system Mo-S and up to 800°C in the systems Mo-N
and Mo-C.

The influence of lattice disorder as discussed in chapter Ic
is included in Fig. 11. Backscattering spectra from N^+ implanted
Mo layers have shown that at 800°C the N has moved throughout the
Mo layer and that at 900°C annealing temperature, additional oxygen had been incorporated. The sulphur profile however had not
been found to change in the temperature range between 400 and 600°C.

Chu, McMillan and Luo[7] observed that the T_c of Re always increased by adding small amounts of transition metal impurities. As
the density of states for Re sharply increases just below the Fermi
level this authors believed that the observed increase of T_c is due

to impurity scattering of electrons smearing out the density of
states and thus leading to an increase of the density of states at
the Fermi energy. The increase in T_C should be larger for Re samples doped with impurities having smaller e/a ratios than Re. In
terms of Miedema's theory however, it is expected that impurities
with an electronegativity smaller (larger) than that of Re will increase (decrease) T_C.

In order to test these predictions the author performed a
preliminary study and the results are presented in Fig. 12 for Zr,
N and S ions implanted in Re. A sharp increase of T_C for the Zr^+
implanted Re layers is found and the observed slope follows the results of Chu, McMillan and Luo for the W-Re-system. A relatively
small increase in T_C is found for the N and S implanted Re layers

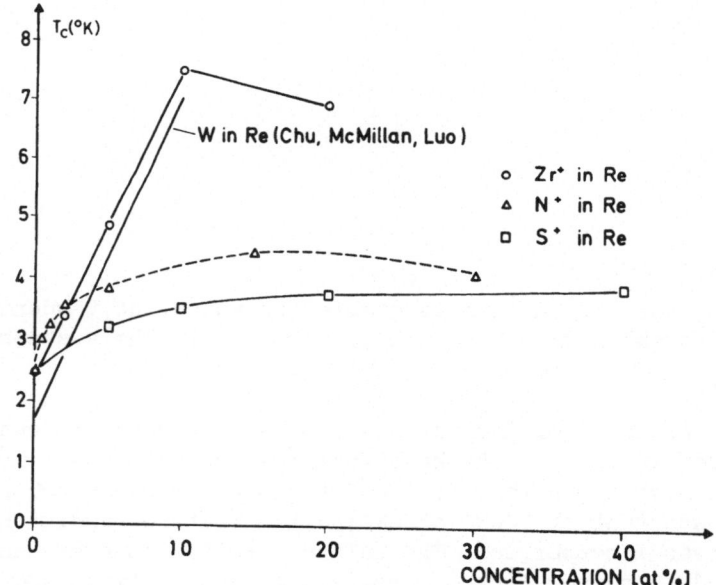

Fig. 12   Increase of the T_C of Re layers after implantation of
          Zr, N and S ions in dependence on impurity concentration;
          the T_C values of the W-Re alloys versus W concentration
          produced by Chu, McMillan and Luo are included for com-
          parison.

which is consistent with the impurity scattering model discussed above. Further experiments are in progress in order to test if the charge transfer mechanism is also a useful concept for Re as it was for Mo.

d) Aluminum Based Ion Implanted Systems

In contrast to transition metals, alloying of the simple (sp) metal elements is nearly ineffective in enhancing T_C (alloys with transition metals are excluded) and usually the solid solubility level is very limited. In a preliminary study the author[37] used the simple metal Al as a target material mainly because numerous studies had been performed on the influence of impurities, pressure, strain and damage on T_C. In addition Thackery and Nelson [50] as well as Thomas and Picraux[51] have already performed studies on ion implantation in Al and information on radiation damage and precipitate growing during bombardment is available. Therefore it was hoped that T_C measurements would provide new information about the metallurgical nature of ion implanted Al layers.

In order to compare T_c-values measured for Al based alloys produced by conventional alloying and produced by ion implantation at room temperature, the elements Ge, Zn and Mg have been implanted. The solid solubility levels for these elements in Al are rather high (Ge 2 at %, Zn 25 at %, Mg 15 at %) and the influence of these impurities in Al samples produced by conventional alloying techniques on T_c has been studied by Chanin, Synton and Serin [19]. They found that at low concentration (<0.1 at %) these impurities produced a decrease of a few percent in the T_C of Al, independent of the nature (electronegativity, valence, atomic size) of the dopant. This decrease in T_C is explained by Markowitz and Kadanoff[33] to be due to the removal of the anisotropy effect. A further increase in concentration C, yields a positive curvature in dT_C/dC, and the magnitude of this curvature seems to depend on the valence of the solute.

Seraphim, Chiou and Quinn[46] extended the experiments to higher (>1 at %) dopant concentrations and described the T_C dependence on the concentration by a single equation: $T_c = C(k_1+k_2 \ln C)$ where the parameters k_1 and k_2 seem to depend on the valence of the dopant and also on the mean free path of the electrons. They believe that clustering will effectively remove the dopant from the solution.

In Fig. 13 the results on T_C of Chanin et al. and Seraphim et al. are presented together with results observed for ion-implanted systems (Ge, Zn, Mg in Al) produced at room temperature. For Ge and Zn implants an increase ΔT_C is observed in the concentration region between 0.001 and 0.01 at %. The slope dT_C/dC decrea-

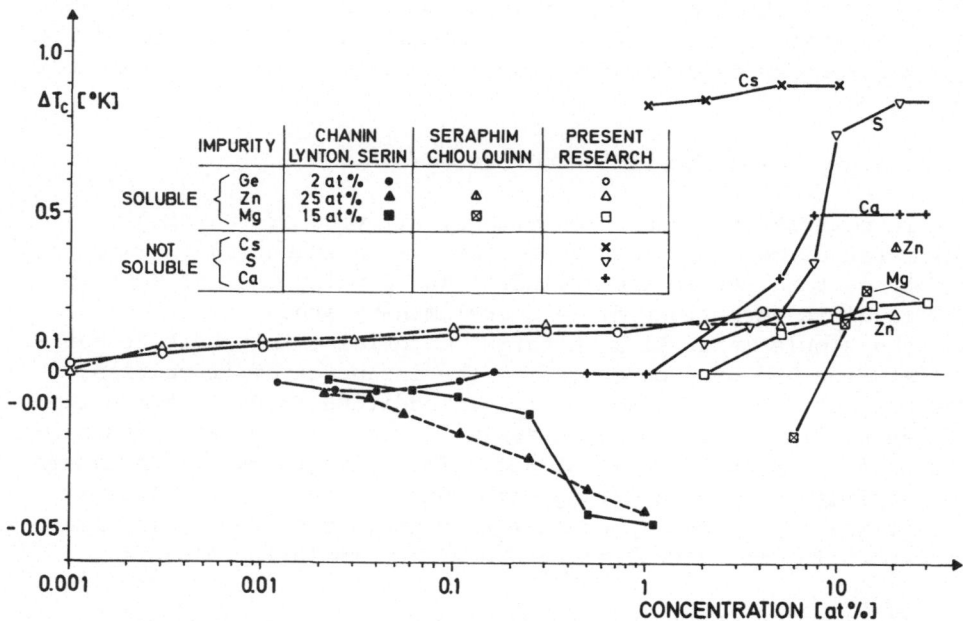

Fig. 13   Changes of the superconducting transition temperature ΔT_C
          in Al layers after implantation of different ions as a
          function of concentration compared with results obtained
          with conventional alloying techniques from other authors.

ses and is found to be nearly constant between 0.1 and 20 at %. The
T_C value given for the system Al-Zn (20 at %) is found to be larger than the corresponding value for an implanted system. For Mg
on the other hand an increase in T_C is observed for concentrations
above 2 at % and a saturation level in T_C, T_C(max) is found for
concentrations above 20 at %. In this case the behaviour of the
ion implanted alloy is not too different from results obtained for
conventional alloys although the different slope and the increase
in T_C at smaller concentrations indicate that additional effects
may influence the increase in T_C in implanted alloys.

It is obvious from the figure that the results obtained for
as implanted samples cannot be described by the equation given by
Seraphim, Chiou and Quinn[46] and one may conclude, that valence
effects and the influence of impurities as scattering centers for
electrons on the mean free path are not important. The results on

the influence of pressure and strain as mentioned in the introduction lead to the hypotheses that localized strain fields associated with the impurity ions may be responsible for the observed increase in T_c. The possibility that local strain fields of the impurity atoms could influence the energy gap of superconducting Al-based alloys has already been treated by Claiborne[10].

In order to test the assumption that strain fields connected to the impurity atom or to an impurity-defect complex may cause the T_c-enhancement, ions, known to exhibit large size effects in Al have been implanted. The size mismatch between solvent and solute atoms results in a deviation from Vegard's law[21] and is proportional to the elastic strain energy as has been shown for example by Friedel[16]. The size factor Ω_{sf} is defined by $(\Omega^x-\Omega_{Al})/\Omega_{Al}$, where Ω_{Al} is the atomic volume of Al and Ω^x is the value of the volume for a 100 at % impurity concentration extrapolated from the slope of the lattice parameter versus concentration obtained at low impurity concentrations. Table 1 shows the results for such impurities in Al, where the size factors are known and have been summarized by King[25]:

Dopant	Ca	Mg	Ge	Ga	Ag	Zn	Cu	Pb
Ω_{sf}	+177	+ 41	+ 13	+ 5	+ 0.12	-5.7	-38	-53
T_c (max)	+ 0.5	+0.23	+ 0.2	+0.22	+ 0.1	+0.13	0	+0.2

Table 1: Size factor and T_c(max) in ion implanted Al based systems.

In the limited number of Al based alloys for which size factors are available, Ca undoubtely has the biggest value and this value correlates with the largest value of T_c(max) in this special selection of elements. As an example the dependence of T_c for Ca on the concentration is also included in Fig. 13. For the other elements the correlation is not obvious and Cu, for example, which has a negative size factor does not show any increase in T_c at all. This observation may indicate that Cu completely precipitates during implantation at room temperature. The strain fields connected to the precipitates are believed to have no influence on T_c otherwise such influence should have been observed in conventional alloys with high impurity concentration levels.

In a further study, ions of different electronegativity, X have been selected in order to see if the chemical nature of the element is important for T_c enhancement.

In Table 2 the results from this systematic study are summarized:

Dopant	Cs	K	Ba	La	In	Pb	Sb	B	Au	C	S	N
χ	0.7	0.8	0.9	1.2	1.7	1.8	1.9	2	2.4	2.55	2.58	3
T_c(max)	+0.9	0	+0.25	+0.1	+0.2	+0.2	+0.25	+1.1	0	+0.6	+0.9	+0.65

Table 2:   Electronegativity and ΔT_c(max) in ion implanted Al based
           systems.

The T_c-enhancements seem to be pronounced for elements with large electronegativity values but there are two important exceptions: Au has a large electronegativity value but it does not show any T_c enhancement; this is similar to the result obtained for Cu. Cs has a very low electronegativity value but a large increase in T_c is observed. Here it is believed that, similar to the results obtained for Ca, a large size mismatch is responsible for this observation. From this study on electronegativity one may conclude that the chemical behaviour of the implanted ion itself does not play a major role for the T_c-enhancements observed in ion implanted Al base alloys produced at room temperature.

Large electronegativity values on the other hand may reduce the migration probability and prevent precipitation. The T_c dependence on the concentration of implanted S and Cs ions are also included in Fig. 13. It is assumed that a density effect is responsible for the observed results. A lowering of the density can lead to a reduction of the bonding forces between the Al atoms in the disturbed lattice and this in turn will reduce the average frequency of the lattice vibration and will increase the electron-phonon coupling strength in Al. This explanation is supported by several experiments which are summarized as follows: Hauser[20], studied mixtures of Al and 3.6 at % Ge sputtered at room temperature and observed a maximum T_c of 2.5 K. He explained his results by an observed 5-10 % increase in volume. This explanation is in accordance with a possible negative pressure effect, as Levy and Olsen[30] found that hydrostatic pressure p applied to Al decreased T_c. At low pressures dT_c/dp shows a linear variation for $p \to 0$, indicating that a negative pressure effect (decrease in density) may result in an increase in T_c. This assumption is supported by Notary's[39] observation of T_c-enhancement obtained by applying tensile strain on Al films evaporated on mylar.

APPLICATION TO SUPERCONDUCTING DEVICES

One of the first attempts to use ion implantation as a tool in the production of superconducting devices and integrated circuits has been performed by Harris[58], who implanted patterns of N and S ions into Mo films making use of the results already described in chapter III b. Harris successfully produced Josephson weak link structures in which two heavily doped Mo regions with relatively high transition temperature (T_c=4 K) are separated by a

short (1 μm long) lightly doped Mo region with much lower transition temperature (T_c'=1.5 to 2 K). Near T_c' the lightly doped region
acts as a weak link between the heavily doped regions and passes
a supercurrent having an oscillatory diffraction pattern dependence
on the magnetic field similar to that of Josephson tunnel junctions. A common feature of all weak link structures is the abrupt
spatial variation of the superconducting properties and ion implantation may provide a technique to control the superconducting properties on a microdimensional scale and to produce very abrupt
boundaries between implanted and unimplanted regions avoiding unwanted edge effects.

CONCLUSIONS

Numerous work has already been performed in order to study the
influence of radiation damage on the superconducting properties as
material has to be selected for superconducting magnets used in
radiation environment and similar to reactor material, neutron radiation damage can quickly be simulated by heavy ion bombardment.
The results of such work indicate whether the current carrying capacity of the material has been optimized or not and the irradiation
equipment can therefore be used as a material test facility. It is
not certain however if pinning centers or special arrangements of
these centers produced by ion implantation will have some principle
advantage as compared to pinning centers produced by metallurgical
techniques; more work has to be done with heavy ion implantation.

Ion implanted systems have shown promising effects as far as
the superconducting transition temperature is concerned and these
results are mainly based on the fact that ion implantation is a
versatile nonequilibrium technique. Figures of merit in this respect are the possibility to form metastable structures and the
enhancement of solubility as has been shown for the Pd-H system
and for non transition metals implanted into Mo and Re films, further advantage is the exact control of composition and stoichiometry. As the metallurgical state of an ion implanted system is difficult to analyze a detailed understanding requires more intensive
work and it is believed that the determination of the superconducting properties of implanted systems together with the large experience from metallurgical superconductivity will provide further
insight in these problems. The implantation technique itself has
the advantage that systematic studies can be performed rather
quickly and the optimized conditions for various parameters of a
promising system under consideration such as atomic species, concentration, reaction temperature etc. can be determined in a
reasonable time. Based on these data a special production procedure
may be found which is not necessarily ion implantation.

REFERENCES

1.)   Bardeen, J., Cooper, L.N., and Schrieffer, J.R., Phys. Rev.
      108 (1957) 1175.

2.)   Bauer, H., Saur, E., to be published in the Proc. of the
      5th Int. Cryogenic Engineering Conf. Kyoto May 1974, Japan.

3.)   Brice, D.K., Rad. Eff. 6 (1970) 77, and Proc. of the 4th
      Internatl. Conf. on Ion Impl. in Semiconductors and Other
      Materials, Osaka 1974.

4.)   Buckel. W., Heim, G., Application of Ion Beams to Metals,
      S.T. Picraux et al., Eds., Plenum Press (1974).

5.)   Buckel, W., Stritzker, B., Application of Ion Beams to
      Metals, S.T. Picraux et al. Eds., Plenum Press (1974).

6.)   Chang, L.L., Rose-Jnnes, A.C., Proc. 12th Conf.on Low
      Temperature Physics, Kyoto (1970) 381.

7.)   Chu, C.W., McMillan, W.L., Luo, H.L., Phys. Rev. B 3 (1971)
      3757.

8.)   Cullen, G.W., 1968 Proc. Summer Study on Superconducting
      Devices and Accelerators, Brookhaven National Laboratory,
      p. 437.

9.)   Chanin, G., Lynton, E.A., and Seri, B., Phys. Rev. 114
      (1958) 719.

10.) Claiborne, L.T., J. Phys. Chem. Solids 24 (1963) 1363.

11.) Crozat, P., Adde, R., Chaumont, J., Bernas, H., Zenatti, D.,
      Appl. of Ion Beams to Metals, S.T. Picraux et al., Eds.
      Plenum Press, 1974.

12.) Deprez, E., Brynserade, Y., Thys, W., and Reynders, L.,
      Phys. Lett. 46 A (1974) 458.

13.) Doran, D.G., Kulcinski, G.L., Rad. Eff., 7 (1971) 283.

14.) Darken and Gurry, Phys. Chem. of Metals, Mc-Graw Hill Book
      N.Y., 1953.

15.) Freyhart, H.C., Taylor, A., Loomis, B.A., Application of Ion
      Beams to Metals, S.T. Picraux et al. Eds. Plenum Press
      N.Y. London, 1974.

16.) Friedel, J., Phil. Mag. <u>46</u> (1955) 514.

17.) Gavaler, J.R., Appl. Phys. Lett. 23 (1973) 480.

18.) Geerk, J., Heim, G., and Kessler, J., Z. Physik <u>245</u> (1963) 14.

19.) Matthias, B.T., Geballe, T.H., and Compton, V.B., Rev. of Modern Physics 35 (1963) 1.

20.) Hauser, J.J., Phys. Rev. <u>B3</u> (1971) 1611.

21.) Hume-Rothery, W., Electrons,Atoms, Metals and Alloys Dover (1963) and G.V. Raynor, The Inst. of Metals, London (1962).

22.) Ischenko, G., Mayer, H., Voit, H., Besslein, B., Haindl, E., Z. Physik <u>256</u> (1972), 176.

23.) Kinchin, G.H., Pease R.S., Rept. Progr. Phys. <u>18</u> (1955) 1.

24.) Köhler, W., Schilling W., Nukleonik, <u>7</u> (1965), 389.

25.) King, H.W., J. of Mat. Science <u>1</u> (1966) 79.

26.) Kondo, J., Progr. Theor. Phys. <u>32</u> (1964) 37.

27.) Kulcinski, G.L., Laidler, J.J., Doran D.G., Rad. Eff., <u>7</u> (1971) 195.

28.) Kübler, G., Diplomarbeit, Karlsruhe, 1973.

29.) Kübler, G. and Meyer, O., unpublished work.

30.) Levy, M., and Olsen, J.L., Solid State Communications <u>2</u> 137 (1964).

31.) Linker, G., Meyer, O., Proc. of the 4th International Conf. on Ion Implantation into Semiconductors and other Materials, Osaka, 1974, Japan.

32.) Lindhard, J., Scharff, M., M. Schiott, H.E., Mat.-Fys. Medd. <u>33</u> (1963), Nr. 14.

33.) Markowitz, D., and Kadanoff, L.P., Phys. Rev. <u>131</u> (1963) 563.

34.) Matthias, B.T., Phys. Rev. 97, (1955) 74.

35.) McMillan, W.L., Phys. Rev. 167 (1968) 331.

36.) Meyer, O., Mann, H., Phrilingos, E., Application of Ion Beams
     to Metals, S.T. Picraux et al., Eds. Plenum Press, 1974.

37.) Meyer, O., Proc. of the 4th Internatl. Conf. on Ion Impl. in
     Semiconductors and Other Materials, Osaka Ana. 1974, Japan.

38.) Meyer O., and Hofmann, B., unpublished work.

39.) Notary, H.A., Appl. Phys. Lett. 4, (1964) 79.

40.) Miedema, A.R., J. Phys. F. Metal Phys. 4 (1974) 120.

41.) Pauling, L., The Nature of the Chemical Band Oxford University
     Press London (1960).

42.) Phrilingos, E., Diplomarbeit Karlsruhe (1974).

43.) Schiott, H.E., Mat.-Fys. Medd. 35 (1966), Nr. 9.

44.) Schweitzer, D.G., Parkin, D.M., Appl. Phys. Lett. 24 (1974)333.

45.) Seibt, E., 1974 Appl. Supercond. Conf. Oakbrook, Illinois,
     USA, Sept./Oct. 1974.

46.) Seraphim, D.P., Chiou, C., and Quinn, D.J., Acta Metallurgica
     9 (1961) 861.

47.) Stritzker, B., Z. Physik 268 (1974) 261.

48.) Tarutani, Y., Kudo, M., Taguchi, S., to be published in the
     Proc. of the 5th Int. Cryogenics Engineering Conf. Kyoto, May
     1974, Japan.

49.) Testardi, W.R., Wernick, J.H., Royer, W.A., Solid State Comm.
     15 (1974) 1.

50.) Thackery, P.A., and Nelson, R.S., Phil. Mag. 19 (1969) 169.

51.) Thomas, G.J., and Picraux, S.T., Proc. Appl. of Ion Beams to
     Metals, S.T. Picraux et al. Eds. Plenum Press, 1974.

52.) Thompson, M.W., Defects and Radiation Damage in Metals,
     Cambridge, at the University Press, 1969.

53.) Tsypkin, S.I. and Chudnova, R.S. Soviet Physics, Solid State
     13 2588 (1972).

54.) Van Itterbeek, A. Van Poucke, L. Bruynseraede, Y., Physica
     34 (1967) 361.

55.) Winterbon, K.B., Sigmund, P., Sanders, J.B., Mat. Fys. Medd.
     37 (1968/70), Nr. 14.

56.) Wohlleben, K., Z. angew. Phys., 27 (1969) 92.

57.) Tsang, J.C., Shafer, M.W., and B.L. Crowder, IBM Report
     RC 4870, to be published in Physical Review.

58.) Harris, E.P., IBM Report RC 5055 and Applied Superconductivity
     Conference Oakbrook, Ill., Sept. 1974.

59.) Lindhard, J., Nielsen, V., Scharff, M., Mat. Fys. Medd. 36
     (1968) Nr. 10.

60.) Dearnaley, G., this book.

ION-INDUCED X-RAYS FROM GAS COLLISIONS

W. W. Smith[*]

Department of Physics

and Q. C. Kessel[†]

Department of Physics and The Institute of Materials

Science, The University of Connecticut

Storrs, Connecticut   06268

1. INTRODUCTION

With the increasing availability of low energy ion accelerators, sophisticated nuclear instrumentation and detection
systems has come a serious interest in using x rays to study the
basic physics of ion-atom collision processes. A many-electron
ion colliding with an atom can produce a large number of atomic
states, singly and multiply excited, that are not seen when the
atoms are excited by photons or electrons. Spectroscopic and
lifetime measurements on these states are of fundamental interest.
A very effective way to make such studies is by looking at single
collisions of ions with atoms or molecules in a gas. Single-
collision excitation of x rays by heavy ions is also important
because of its applications in such diverse fields as atmospheric
physics, controlled thermonuclear fusion, nuclear physics, chemistry, solid-state physics, and materials research. It is known
that x ray yields resulting from inner-shell vacancy production
in solid targets depend, in some cases, on the configuration of
the outer-shell electrons. For this reason investigations of
isolated atomic and molecular targets can be helpful in elucidating
the excitation mechanisms and selection rules involved not only

* Supported by the U.S. Army Research Office - Durham.
† Supported by the National Science Foundation.

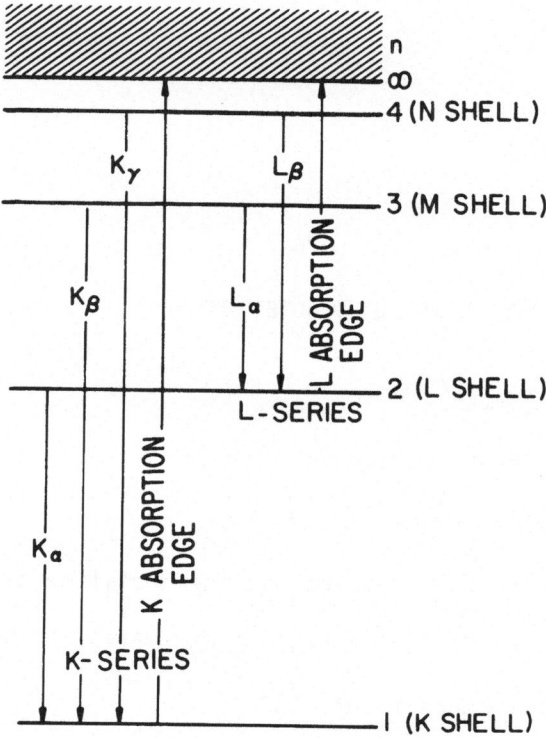

Figure 1.   One-electron energy levels in an atom with diagram
x-ray lines (transitions) indicated.

in single collisions but also in collisions of ions with solid
targets. Ordinarily, x-ray spectra from atoms excited with x rays,
electrons or protons are dominated by the so-called "diagram" lines
or transitions illustrated in fig. 1. A single atomic inner-shell
vacancy (e.g., K or L) is filled by an outer-shell electron, thus
transferring the vacancy to an outer shell. With heavy-ion
excitation of inner-shell vacancies, there is a strong tendency
for these diagram lines to be suppressed in favor of "satellite"
transitions (involving initial states with multiple outer-shell
vacancies,) and even "hypersatellite" transitions involving
initial states with multiple inner-shell and outer-shell vacancy
configurations.

The dramatic difference between excitation by simple charged particles, e.g. a proton beam, and many-electron ions is clear from fig. 2. With N⁺ excitations of aluminum K vacancies (solid target) at 5 MeV, the satellites completely dominate the spectrum

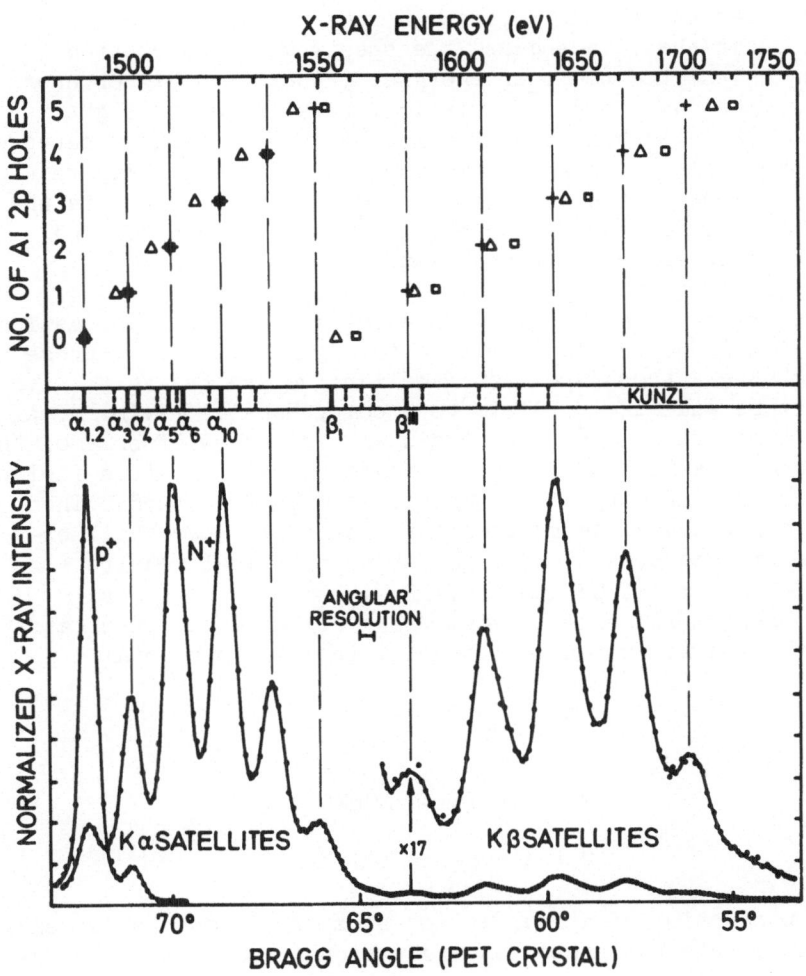

Figure 2.   The Al K x-ray spectra excited by 5 MeV protons and by nitrogen ions (Knudson et al. (1)).

(1). There is also a shift of the lines to higher energies as additional outer-shell electrons are removed. Similar phenomena have been seen with gas targets (2). These data show that with heavy particle excitation, the production of multiple outer-shell vacancies (Al 2p holes in the case of fig. 2) is the rule, not the exception. Auger electrons emitted following inner-shell ionization by heavy particles also have a spectrum dominated by satellites; an example, taken from the work of Volz and Rudd (3), is shown in fig. 3. In the Auger effect, the energy released in the transition of an outer-shell electron into an inner-shell vacancy goes into ionization of a second outer-shell electron. When one looks at the Auger-electron spectrum in the presence of multiple outer-shell vacancies, the spectrum of the emitted electrons is shifted to lower energies rather than higher ener- gies as in the case of x-ray emission; this is due to tighter binding of the electron to be ionized because of reduced screen- ing when multiple outer-shell vacancies are present. Figure 4 illustrates some of the principal processes which result in the filling of an inner-shell atomic vacancy: in low Z atoms the Auger processes compete strongly with x-ray emission.

This article is intended to present some of the systematics of heavy-ion induced x rays emitted from gas collisions: those with which the authors are most familiar; we are trying to give a broad picture and make no claim to a comprehensive review of all of the literature (4,5,6). We begin with a review of some of the collision models that have been found most useful in this field. The regions of validity of the models will be discussed. Then specific experimental techniques for observing inner-shell ex- citations will be presented, together with typical results from the work of several laboratories. Examples of situations in which the x-ray fluorescence yield is a strong function of atomic outer-shell configuration will be explained in some detail.

2. COLLISION MODELS

2.1 Survey of Models

The production of x rays in ion-atom collisions can usually be separated into two steps: a) excitation of an initial inner- shell vacancy configuration, followed by b) decay of that vacancy configuration by x-ray emission or a competing radiationless process. Most treatments of x-ray yields (7) assume that the process of inner-shell vacancy production occurs on a much shorter time scale than vacancy decay by x-ray or Auger emission. As an example, the velocity of a 100 keV He^+ ion is $\sim 2 \times 10^8$ cm/sec,

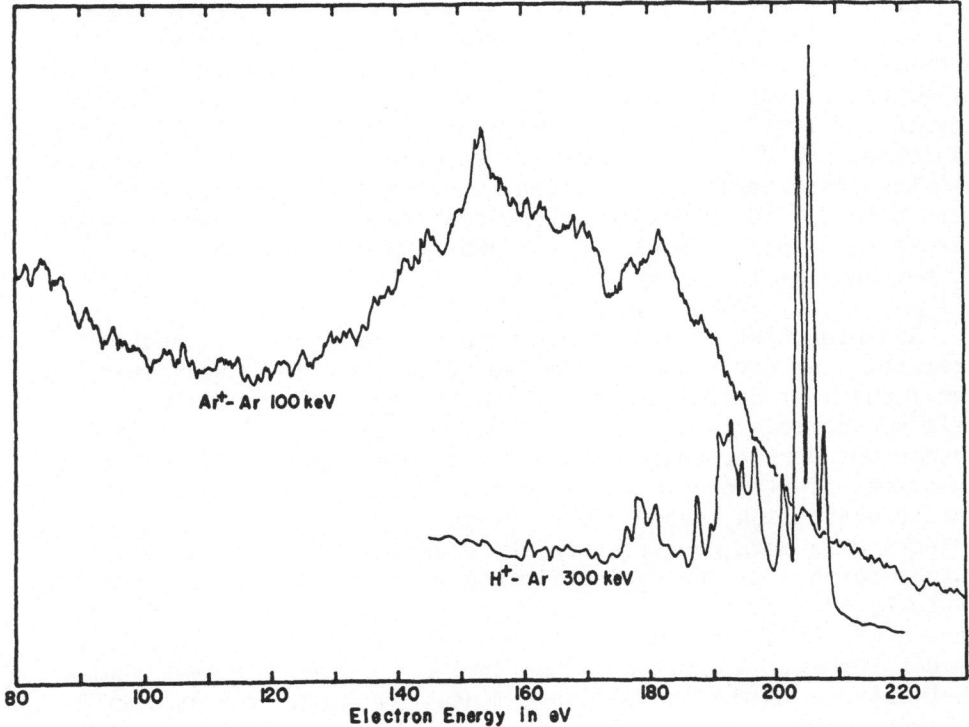

Figure 3. Spectra of electrons from argon bombarded by protons
and by Ar⧧ ions (Rudd and co-workers (3)).

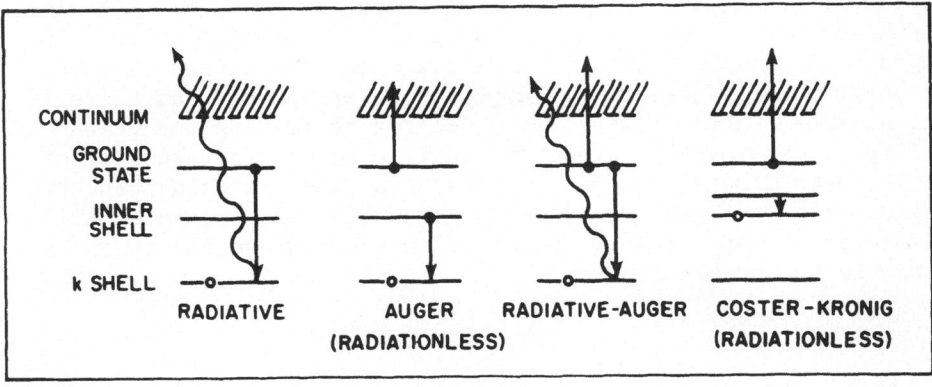

Figure 4. Some of the competing types of transitions known to occur
after the creation of a K-shell vacancy in an atom. The
Coster-Kronig transitions are from one <u>subshell</u> to another.

so that the ion can cross the diameter of an atomic K shell ($\sim$ 10^{-8} cm) in about 5 x 10^{-17} sec. We take this as a rough estimate of the time scale for inner-shell vacancy production. Vacancy-decay lifetimes obtained from linewidths observed in K x-ray emission (nuclear charge Z in the range Z = 20 - 50) (8) lie between 4 x 10^{-15} sec. and 4 10^{-16} sec. (widths 1 - 10 eV). As a rule then, the decay lifetimes are an order of magnitude longer than the characteristic collision time for K-vacancy production, even up to Z = 80, provided we are dealing with ion-beam energies greater than approximately 25 keV/AMU (AMU = atomic mass unit). For L-vacancies the decay times are usually longer.

Assuming that vacancy production and decay can be treated separately, we now consider the two collision models that have been found most useful in describing the <u>production</u> of inner-shell vacancies. These are the "Coulomb excitation" model in first-order Born approximation and the molecular-orbital electron promotion model (sometimes called the Fano-Lichten model). The Coulomb excitation model is most useful when the velocity of the projectile is much greater than the classical Bohr velocity of bound electrons in the <u>projectile</u> and where the projectile nuclear charge Z_1 is much less than the target nuclear charge Z_2; under these conditions the inner-shell electron orbitals of the target are not strongly polarized by the incoming projectile.--At the other extreme is the Fano-Lichten model (9) which is most useful at lower collision velocities, at which at least the <u>inner-shell</u> atomic electrons can "follow" the nuclear motion during the collision, momentarily forming a diatomic molecular state or "quasi-molecule." The Fano-Lichten model was originally conceived and applied to symmetric collisions in which the nuclear charges of projectile and target are equal; it was later extended to near-symmetric collisions by Barat and Lichten (10). The ranges of validity of the two models are not completely obvious; but unfortunately they do not appear to overlap, so that there is not, at present, a satisfactory model for estimating the probability of inner-shell vacancy production or of the intensity of collision-induced x rays in the "no-man's land" between these two models. Figure 5, from a diagram by Madison and Merzbacher (11), delineates the approximate ranges of nuclear charge and velocity where the two models are valid.

2.2 Coulomb Ionization

We now discuss the Coulomb ionization of inner-shell electrons in more detail within the framework of the plane-wave Born approximation (PWBA). Figure 5 indicates that this model works well for moderate and high velocities, when the ratio Z_1/Z_2 is small, in other words: for proton and alpha-particle excitation of inner-

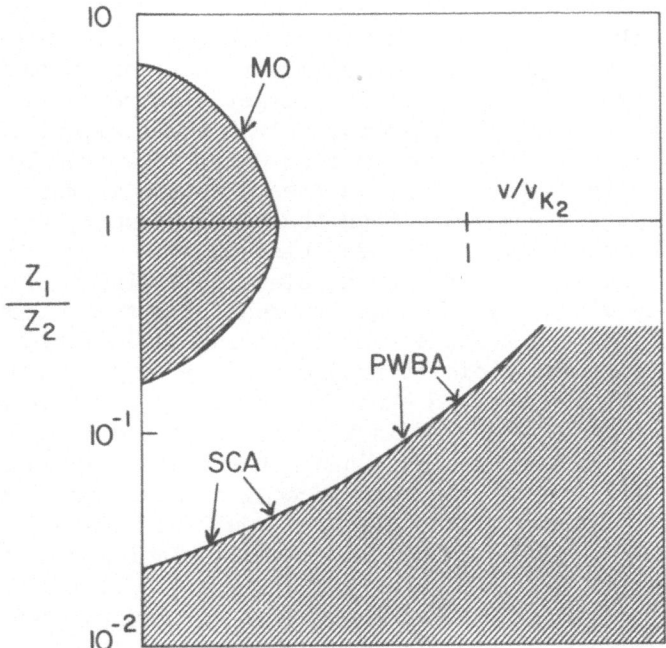

Figure 5.   Approximate regions of validity for various collision
            models as a function of Z_1/Z_2 and collision velocity v
            (in units of the target K-shell Bohr orbital velocity v_{K_2}).
            PWBA = plane wave Born approximation; SCA = semiclassical
            approximation; MO = quasimolecular electron promotion
            model. (Adapted from Madison and Merzbacher (11)).

shell vacancies in heavy targets. Much of the theory stems from
early work in the 1930's by Bethe and his students and has been
reviewed in detail by Merzbacher and co-workers (11,12). For the
PWBA to be valid, the projectile velocity v must be much greater
than v_{K_1}, the Bohr-orbital velocity in the projectile's K shell.
Since we also require $Z_1 \ll Z_2$, a simpler expression of this

condition is $v/v_{K_2} \ll Z_1/Z_2$, where we have used the fact that v_K
$\propto Z$.

Even when the energy of an incident heavy ion is greater
than the binding energy of the target inner-shell electron to be
ionized, is not possible to treat the target electron as if it
were free in formulating the theory. Unless the incident ion
kinetic energy is very high, the maximum energy that can be
transferred from the projectile to a free electron at rest, in an
elastic collision, is much less than the binding energy of an
inner-shell atomic electron. To see this, one can show, from
conservation of momentum, that for an elastic collision of a
heavy projectile with a light target at rest (the free electron),
the velocity of the ejected electron in a head-on collision is
approximately twice the incident projectile velocity. This means
that the energy of the ejected electron (assuming it to be a free
electron) is $4(m/M)E_P$, where E_P is the energy of the incident
projectile, m is the electron mass and M the projectile mass.
Thus, for a 10 keV proton in a head-on collision with an electron,
the final kinetic energy of the electron would be only $\sim$ 20eV,
which is much less than a typical inner-shell binding energy.
Nevertheless, we know experimentally that inner-shell ionizations
can be produced easily by either low-energy protons or even
heavier projectiles. If the target electron is considered bound
by Coulomb forces to a heavy nucleus, then all of the available
energy in the center-of-mass system can become potential energy
in the collision, and a much greater fraction can be transferred
to a single electron. The classical prediction is that the
inner-shell ionization cross section will be a maximum when the
projectile velocity is comparable to the classical Bohr orbital
velocity of the electron to be ionized, i.e. for a K-shell ionization, the maximum occurs near $v = v_{K_2} = Z_2\alpha c$, or when $mE_p/M =$
Z_2^2 Rydberg (1 Rydberg = 13.6 eV).

The PWBA (first order) gives satisfactory values for the
total cross section for K- and L-shell Coulomb ionization (11,12).
In the PWBA, the initial and final wave functions for the colliding
ion-atom system are taken as product functions of plane waves
representing the nuclear motion and hydrogenic target electron
wave functions. The differential inner-shell excitation cross

section is then given by (11):

$$\frac{d\sigma}{d\Omega} = \frac{M^2}{4\pi^2\hbar^4} \frac{v_f}{v_i} \left| \int \phi_f^*(\vec{r}) e^{-i\,\vec{p}_f\cdot\vec{R}/\hbar} Z_1 \frac{e^2}{|\vec{R}-\vec{r}|} e^{i\vec{p}_i\cdot\vec{R}/\hbar} \phi_i(\vec{r})\, d^3r\, d^3R \right|^2 \tag{1}$$

Here $d\Omega$ is the element of solid angle into which the incident particle is scattered, M the reduced mass of the system, $\vec{R}$ the position vector from the target to the projectile; v_i, $\vec{p}_i$ and v_f, $\vec{p}_f$ are, respectively, the relative speeds and momenta before and after collision; $\vec{r}$ is the position coordinate of the electron that is excited (relative to the target nucleus); and ϕ_i and ϕ_f are the initial and final wave functions of this electron. The matrix element in this equation for the cross section can be re-written in terms of the momentum transfer vector $\vec{q} = (\vec{p}_f - \vec{p}_i)/\hbar$.

It is proportional to the quantity $\int F_{fi}(\vec{q}) \frac{e^{i\vec{q}\cdot\vec{R}}}{q^2} d^3q\, d^3R$, where $F_{fi}(\vec{q}) \equiv \int \phi_f(\vec{r}) e^{i\vec{q}\cdot\vec{r}} \phi_i(\vec{r}) d^3r$. We see at once from (1) that the cross section scales as Z_1^2 and that it depends on the target-electron wave function only through the "form factor" F(q). Note that the largest contributions to the cross section come from small values of the momentum transfer $\vec{q}$, that is from nearly forward scattering. One can proceed to approximate the form factor by expanding the exponential in it as $e^{i\vec{q}\cdot\vec{r}} \approx 1 + i\vec{q}\cdot\vec{r}$. Then the matrix element of the first term in the expansion will be zero by the orthogonality of the unperturbed bound-state one-electron wave functions, so that the leading term in the matrix element will be proportional to the matrix element of the vector r sand-wiched between initial and final target wave functions. Hence the same matrix element comes in as appears in the oscillator strength for optical excitation by a beam of photons. Those ex-citations predominate that are connected to the ground state by allowed electric dipole (E1) transitions. This leads to the con-cept of "generalized oscillator strength" in collisional excita-tion. Generalized oscillator strengths in the context of elec-tron impact excitation have been reviewed by M. Inokuti (13).

Equation (1) is general enough to include excitation of a target electron to either a bound state or to a continuum state, i.e. ionization. In the case of inner-shell ionization, the ex-cited electron will have some kinetic energy ε when it leaves the atom. One must then integrate over all allowed values of ε to obtain the total cross section for ionization from a given initial state i. Merzbacher and Lewis (12) present numerical calculations

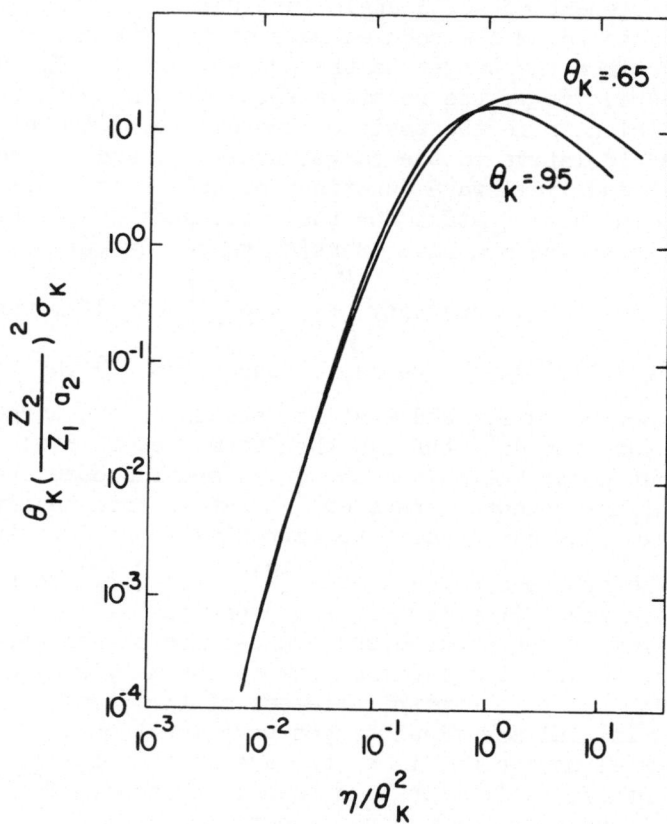

Figure 6.   Born approximation (PWBA) K-shell ionization cross section σ_K as a function of the reduced energy parameter
η/θ_K^2.   The reduced cross section is in units of a_2^2,
where a_2 is the target Bohr radius a_o/Z_2 (From Madison
and Merzbacher (11)).

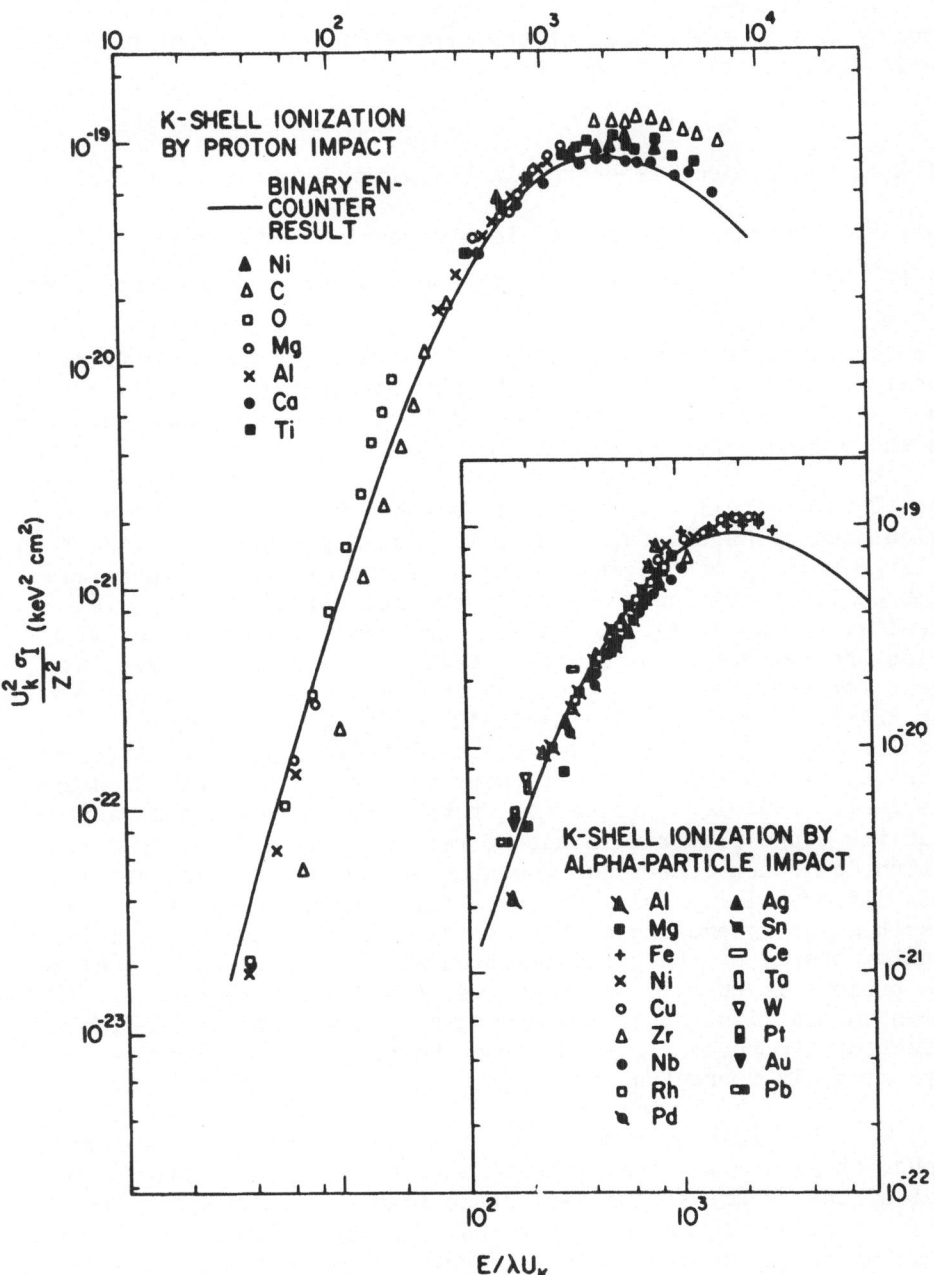

Figure 7.   Comparison of BEA results of Garcia with experimental
ionization cross sections. U_K is the K-shell electron
binding energy and $\lambda = M_1/M_P$ (M_p = proton mass)
(From Garcia (14)).

of the total cross section for K-shell ionization by low Z_1
projectiles as a function of two parameters. The first parameter
η_K is a scaled projectile energy:

$$\eta_K = \frac{m}{M_1} (E/Z_K^2 \text{ Ry}) \rightarrow \frac{37E}{M_1 Z_2^2},$$

if E is the projectile energy in MeV with the projectile mass M_1

in AMU. The second parameter is a screening constant $\theta_K \equiv$
$I_K/(Z_K^2 \text{ Ry})$ which, in practice, is constrained to lie in the range
$0.7 < \theta_K < 0.9$, increasing slowly with the Z of the target.
The screening parameter θ_K affects the value of the PWBA total K-
shell ionization cross section hardly at all when properly scaled
as in fig. 6, and by less than a factor 2 if the scaling factor θ_K
in the ordinate is ignored.

Total K-shell ionization cross sections have also been
calculated by Garcia for both proton and alpha-particle impact on
a large variety of targets, using a classical impulse approximation called the "binary encounter approximation" or BEA. The
results, shown in fig. 7, indicate the remarkable accuracy with
which the K-shell cross sections fall on a universal curve, at
least for energies up to the peak in the curve. The essence of
the BEA is to calculate, classically or quantum-mechanically (the
results are the same for the Coulomb interaction), the differential
cross section for exchange of energy between a free electron and
the incident Coulomb projectile, taking into account the fact
that the bound electron is not at rest but has a velocity distribution, and then conserving momentum and energy in each collision.
This differential cross section is first integrated over all
possible energy exchanges (between a lower limit equal to the
binding energy of the bound electron and an upper limit equal to
the projectile energy) and then over the quantum-mechanical
momentum distribution of the bound, hydrogenic electron. The
effect of the nuclear repulsion on the projectile motion is
approximately corrected for.

The BEA cross section, without the nuclear repulsion corrections, obeys the scaling law used in fig. 7, namely that the product of the square of the K-electron binding energy U and the
total cross section σ is a universal function of the ratio of the
incident energy to the binding energy: $U^2\sigma = f(E/U)$. Comparison
with fig. 6 shows that this scaling law differs from the PWBA
scaling by a factor θ_K in the ordinate, yet this means that the
PWBA and BEA calculations obey the same scaling low to within
better than a factor 2 in the cross section because of the limits
$0.7 < \theta_K < 1$. The universal function $f(E/U)$ must be evaluated
numerically; it can be obtained approximately from the graphs of

figs. 6 and 7, and has been given in tabular form recently by McGuire and Richard (15).

To summarize the situation for small Z_1/Z_2, both the plane-wave Born approximation (PWBA) and the binary-encounter approximation (BEA) calculations of total K-shell ionization cross sections can be used to fit the experimental x-ray emission cross sections to considerably better than a factor 2 in the velocity range from $v/v_{K_2} \sim 1$ down to $v/v_{K_2} \sim 0.01$, provided that recent theoretical estimates of the K-shell fluorescence yields at each Z_2 are included (16). At low velocities the cross sections begin to depart from the Z_1^2 scaling (17) predicted by these simple theories due to the breakdown of first-order theory and the onset of polarization effects. Bang and Hansteen (18) have obtained better agreement with the data by using time-dependent perturbation theory with a <u>classical path</u> for the atomic nuclear motion: this is the semi-classical approximation: or SCA. A hyperbolic Coulomb trajectory is used at low velocities where there is considerable projectile deflection. This deflection from a straight path reduces the ionization at low energies (19).

The same perturbation calculations work for L-shell x rays as for K (20), but there are some complications: subshell differences in vacancy populations have been observed experimentally (21) (e.g. between 2s, $2p_{1/2}$ and $2p_{3/2}$ vacancies). Also, because of weaker nuclear binding, the L-shell cross sections are more sensitive to the exact kind of target wave functions used. Finally, as previously noted, L-shell fluorescence yields can vary much more than is the case for the K-shell yields due to "chemical effects," such as the degree of outer-shell (M and higher) ionization and the specific outer-shell electron configuration involved. (See the discussion in Section 4.4). For high-Z targets and for projectile velocities approaching that of light, relativistic wave functions are presumably needed for accurate results, although we are aware of little detailed work in this area (22).

2.3 The Molecular-Orbital Model

Figure 5 indicates that the molecular-orbital model is appropriate for describing the production of inner-shell vacancies in a regime where the PWBA and the BEA calculations are known to fail: in the limit of low-energy collisions with $Z_1/Z_2 \sim 1$. The range of Z_1/Z_2 over which the MO model works is actually not restricted to values close to 1. Historically, the importance of using <u>molecular</u> wave functions rather than an atomic description in low energy, heavy particle collisions goes back at least as

far as the work of Weizel and Beek in the 1930's (23). They
applied the newly created molecular-orbital (MO) model of Hund
and Mulliken to the analysis of the ionization thresholds of
noble gases in collisions with alkali-metal ions. Electronic
states of diatomic molecules were used by Landau, Zener and
Steuckelberg (24) to describe both predissociation (25) from
bound states of such molecules and also low-energy inelastic
collisions involving near crossings of two adiabatic molecular
states. Further examples can be found in Mott and Massey's
classic, The Theory of Atomic Collisions (26).

   As differential inelastic scattering data involving inner-
shell excitations became available through the pioneering work of
Federenko and associates (27) in the Soviet Union and Everhart's
group at the University of Connecticut (28), it became necessary
to pay close attention to the molecular states to be used in
describing the collision process.--An instructive example is the
differential cross section for elastic scattering of He^+ ions on
helium gas (29). Resonant electron exchange between the two core
He^+ ions of the He_2^+ diatomic system can take place, leading to
oscillations in the differential charge-exchange cross section as
a function of incident energy that are out of phase with similar
oscillations in the differential elastic cross section. In the
conventional Born-Oppenheimer (30) treatment of diatomic molecular
states, the nuclear and electronic motions are separated by first
solving the Schrodinger equation for the stationary states of a
diatomic system with the nuclei held fixed. The total energy of
the electrons plus the Coulomb repulsion as a function of the
inter-nuclear distance, then represents the potential energy
function for the nuclear motion. This separation is possible
because of the large nuclear/electron mass ratio and the fact
that the nuclei are usually moving slowly compared to characteristic electronic velocities. In the He_2^+ case, the most loosely
bound electron of the three occupies a molecular-orbital in which
it is shared equally by both nuclei. When the nuclei are held
fixed, there are two degenerate states of the molecular system at
large separation, an even parity state (gerade or "g") symmetric
upon reflection about the center of the molecule, and an odd
parity state (ungerade = "u"), anti-symmetric upon reflection
about the molecular center. The initial state in a He^+ -He
collision is an equal linear combination of the g and u MO states
with a relative phase factor that depends on time. The oscillations in the resonant charge-exchange cross section can then be
viewed as quantal interference effects between symmetric and
antisymmetric molecular states (31). In general, the MO description is most appropriate when each electron is strongly bound by
both nuclei (the He^+ on He case), i.e. where $Z_1 \sim Z_2$. Because

the MO's are usually based on states of the Born-Oppenheimer
approximation using fixed nuclei, the MO model requires collision
velocities that are sufficiently low that the electrons can "follow"
the nuclear motion and adjust continuously (or with a finite number
of discontinuities near crossing points) to the changing internuclear distance in the collision. Thus the collision velocity
must be $\leq$ the characteristic Bohr electron velocities in the
orbitals involved. For high Z massive particles, "low velocity"
need not necessarily mean low energy: for large Z-values the MO
model can be useful for inner shells at collision energies of
several MeV/nucleon.

The application of the MO model to the theory of ionization
(or excitation) of inner-shell electrons, specifically the case
of symmetric Ar^+-Ar collisions (32), was first accomplished by
Fano and Lichten in 1965. The model was elaborated upon further
by Lichten (9) and extended to include asymmetric collisions by
Barat and Lichten (10). This model utilizes molecular-orbital
energy level diagrams, an example of which, taken from Fano and
Lichten's original paper, is shown in fig. 8. The molecular
states used in the model for describing medium energy collisions
have been called "diabatic states" by Lichten, in contrast to the
usual Born-Oppenheimer or "adiabatic" molecular states used in
conventional molecular spectroscopy. These states are sometimes
assumed (in a useful but not rigorous approximation) to be represented by a wave function that is written as an antisymmetrized
product of single-electron MO wave functions. The diagram of
fig. 8 shows a set of one-electron energy levels (analogous to
the energy levels of the H_2^+ molecule) plotted versus internuclear distance R, where each electron is assumed to move in the
combined field of the two colliding nuclei screened by all of the
other electrons in the quasi-molecule. In the independent electron approximation, the total energy of a many-electron diabatic
state is simply the sum of the one-electron energies (from fig.
8) of the individual MO's that go into it. Such diabatic state
energies become degenerate with the energy of other diabatic
states at certain critical internuclear distances (crossing
points), even when the two diabatic states have the same molecular symmetry quantum numbers (like two Σ molecular states, for
example). The molecular orbitals in the diabatic states are
occupied only to the extent allowed by the Pauli exclusion principle, by virtue of the antisymmetrization of the wave function.
Although a complete set of adiabatic Born-Oppenheimer states can
be rigorously defined, a rigorous and unique definition for the
diabatic states has not yet been found (33). Nevertheless the
usefulness and physical reality of the diabatic states as an
approximation is strongly suggested by calculated adiabatic
energy curves, an example of which is shown in fig. 9.

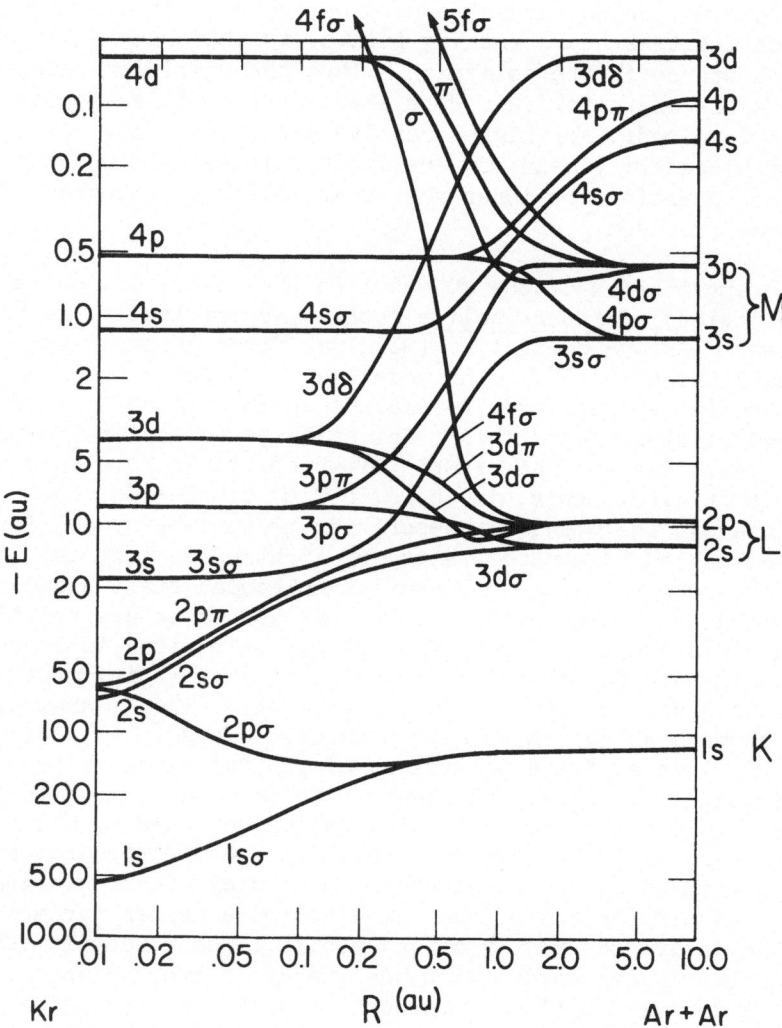

Figure 8. Energies of diabatic molecular orbitals suitable for
           analyzing fast collisions between Ar^+ and Ar. The
           energies at $R = \infty$ are one-electron energies for Ar
           atoms, and the energies at $R = 0$ are one-electron
           energies for the united Kr atom. For details, see
           Fano and Lichten (9).

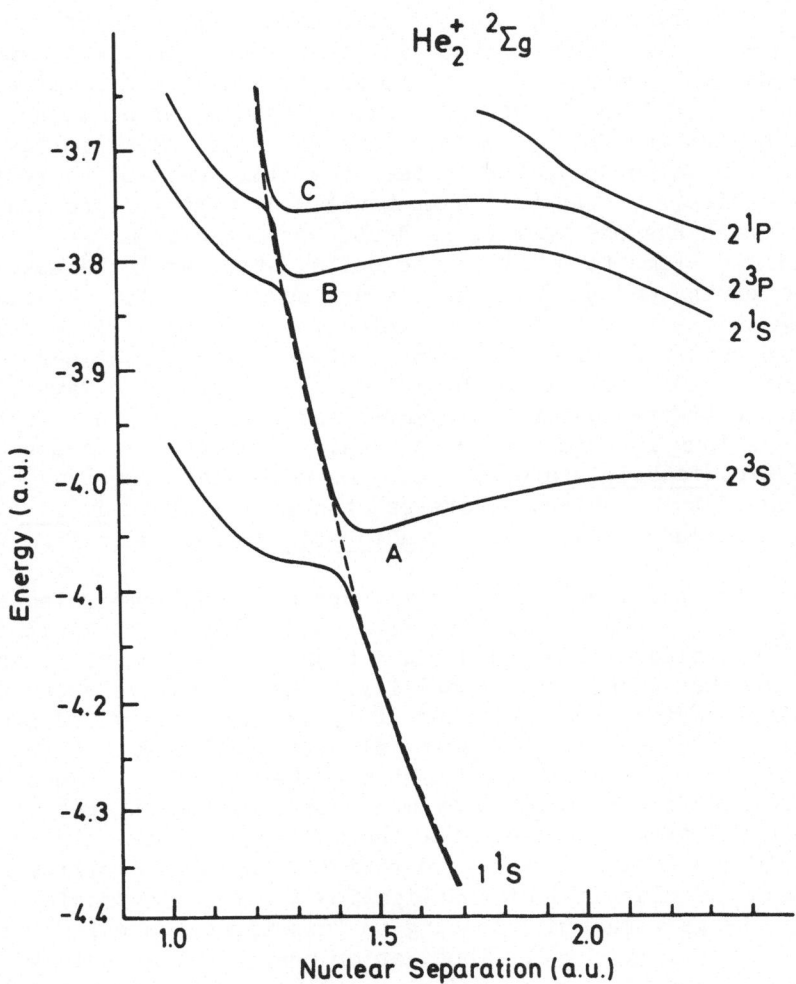

Figure 9.   Adiabatic excited state curves for $^2\Sigma_g$ states of the He$_2^+$ molecule as calculated by Bardsley (34) shown as solid lines. The dashed curve interpolated between the avoided crossings at A, B, and C indicates the molecular potential curve for a diabatic state (from (4)).

The term <u>electron promotion</u> arises out of the correlation of
one-electron H_2^+ - like states from the separated atom to the
united atom limits. A MO that correlates with a certain principal
quantum number n of the separated atoms may be "promoted" to a
higher principal quantum number n in the united atom limit. In
fig. 8, the σ -symmetry MO that correlates the 2p state of the
separated atoms with the 3d state in the united atom limit, designated 3dσ, is promoted from n=2 to n=3 because there is already
a full n = 2 shell in the united atom. On the other hand, the σ
orbital, designated 2pσ, because of its 2p united-atom limit, is
promoted from 1s to 2p in the united-atom limit because there is
no n = 1 orbital for the united atom with odd parity. The 4fσ
orbital, which has odd parity, is doubly promoted from n = 2 in
the separated atoms to n = 4 in the united atom, again because
the lower energy odd parity orbitals are normally fully occupied
according to the Pauli principle. Electron promotion gives rise
to a large number of crossing points between diabatic states
(taken here to be represented by a single configuration wave
function constructed as an antisymmetrized product of one-electron
MO's) at various internuclear distances. At every such crossing
point of two <u>diabatic</u> states of the same molecular symmetry (e.g.
a crossing of two Σ molecular states) there will be an <u>avoided</u>
crossing or pseudocrossing of two <u>adiabatic</u> states as in fig. 9.

Near a crossing of two diabatic curves of the same molecular
symmetry, certain residual interactions and dynamic couplings
(due to the nuclear motion) can cause transitions from one diabatic
state to another with high probability. The residual electrostatic
interactions between two electrons (they are not really independent)
can cause "radial coupling" between diabatic states with the
selection rule $\Delta \Lambda = 0$ (Λ is the quantum number for the projection
of the electronic orbital angular momentum onto the internuclear
axis). In addition, the fact that the internuclear axis is
rotating in any collision except a strictly "head-on" collision
causes the molecular axis of quantization for the electronic
states to change with time, leading to "rotational coupling" with
the selection rule $\Delta \Lambda = \pm 1$. Thus, at or near curve crossings
between two diabatic states, radial coupling can cause transitions
between two states of the same Λ -symmetry, like Σ - Σ or Π - Π
transitions, etc. Rotational coupling can cause Σ-Π or Π- Σ
or Π - Δ state transitions, etc. The radial coupling case was
treated by Landau and Zener (24). The case of rotationally
induced transitions (35) has been treated numerically in the case
of a simple crossing between two interacting states by Russek
(36). Thus, in a heavy-particle collision, electrons in inner-
shell MO's which are promoted can make transitions at or near a
crossing with an outer-shell MO, always <u>provided the outer-
shell MO is not already fully occupied within the limits
imposed by the Pauli principle</u>. By means of such electronic

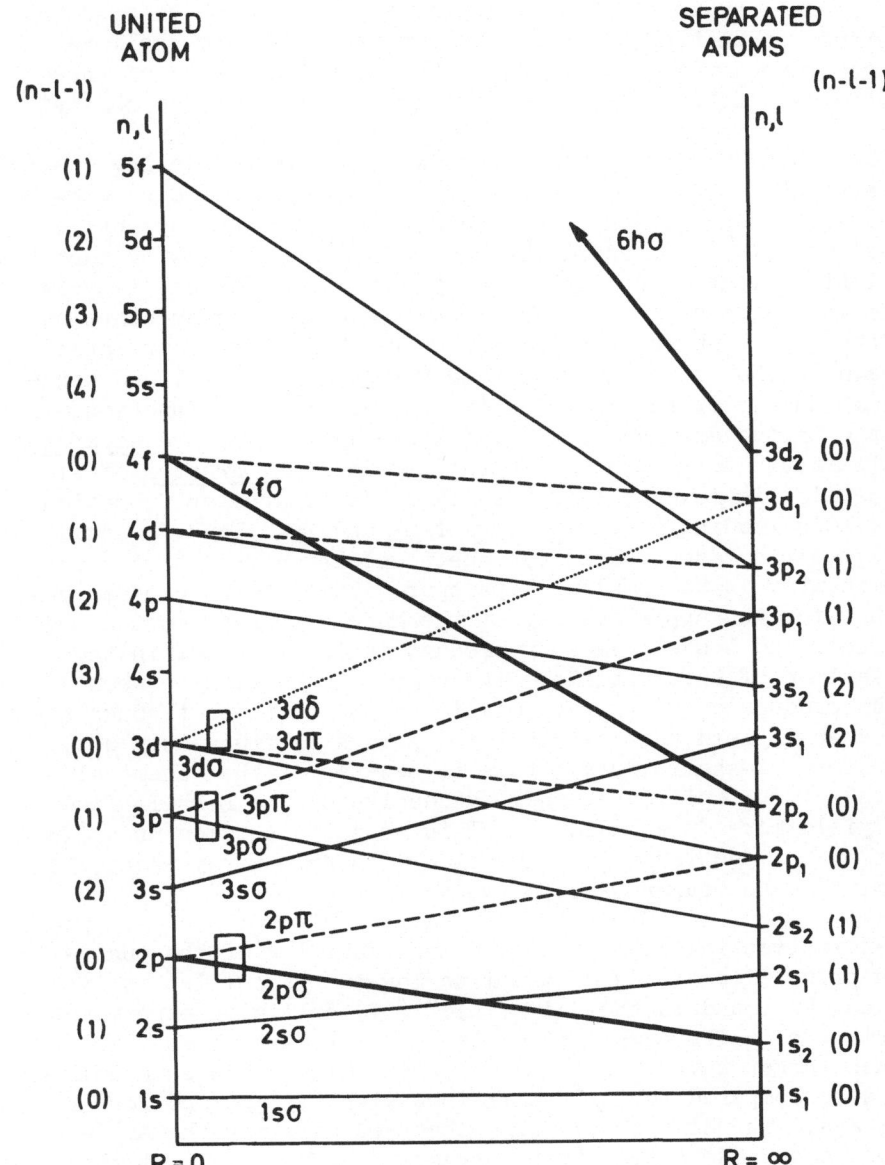

Figure 10. MO correlation diagram constructed using the Barat-Lichten rule. This diagram is for the case of Z_1 slightly larger than Z_2. The solid lines correspond to σ states, the dashed lines to π states, and the dotted lines to δ states. The prominent MOs: 2pσ, and 4fσ, and 6hσ are responsible for promotion of K, $L_{2,3}$, and $M_{4,5}$ electrons, respectively. The rectangles show interactions for which rotational coupling between two MOs may be expected (From (4)).

transitions from initially unoccupied outer-shell MO's, promoted electrons can be trapped in excited states and inner-shell vacancies created with high probability. Autoionization of these promoted and excited electrons usually occurs after the collision.

Diabatic MO correlation diagrams useful in predicting qualitatively the outcome of a collision as a function of distance-of-closest-approach R_0, can also be drawn (10) for near-symmetric collisions ($Z_1 \sim Z_2$), as in fig. 10. The rule for making such a correlation diagram, as explained by Barat and Lichten, is very simple in most cases: one adds orbitals to the diagram starting with the lowest energy orbital and working up. The correlation is drawn so that an MO has the same values of (n - ℓ), and hence the same number of radial nodes (n-ℓ-1) in the wave function, in both the united atom and separated atom limits. In the separated atom limit the corresponding energy level of the greater of Z_1 and Z_2 will lie lowest; this usually leads to greater promotion of the MO's leading to the lesser of Z_1 and Z_2. Thus for the near-symmetric cases, vacancies due to electron promotion occur predominantly in the smaller Z partner. Some of the experimental evidence for the applicability of the MO electron-promotion model to K-shell and L-shell vacancy production is discussed in the next section of this article. The simple connection between vacancy production and x-ray emission cross sections through the fluorescence yield that exists for K x rays is frequently not possible for L-shell x rays; this is because of the "chemical effects" of multiple outer-shell vacancies on the L-shell fluorescence yield already alluded to. This subject is taken up again at the end of Section 4 in connection with Ar L x-ray--scattered-ion coincidence measurements.

An interesting example of the applicability of the qualitative principles of the MO model to low-energy collisions in gases can be found in the "carambole" (37) collision experiments of Saris (38). The number of emitted Cl K x rays, normalized to the much larger number of Ar L x rays emitted in the same collision, were compared for Ar^+ collisions with Cl_2 gas and for Ar^+ on HCl gas. The Cl K x rays were observed only for the Ar^+ - Cl_2 collision. Recalling that corresponding MO levels of the higher Z partner lie lower in the separated-atom limit (fig. 10), one can deduce that in a binary Ar- Cl-atom collision a Cl K vacancy can be created only if the Cl 1s electron can be excited to the 2p state of the Ar. For this to happen, an Ar 2p vacancy must already exist prior to the collision. When Ar collides with a Cl_2 molecule, there can be, in effect, two successive collisions: the first Ar -Cl collision produces a vacancy in the argon 2p state; this is then transferred (via 1sσ -- 2pπ rotational coupling near R = 0) to the Cl K-shell when the Ar collides with the second Cl atom. This ultimate Cl 1s vacancy then gives rise to a Cl K x ray.

An important exception to the generalization that vacancy production and decay can be treated independently is the so-called "molecular orbital x-ray" that has been reported by several laboratories (39). These are x rays, usually observed with rather low intensity, that have a broad spectrum not characteristic of either isolated collision partner or of the united-atom limit but rather with an energy range intermediate between the separated and united atom limits. It is obviously not valid to separate the production of inner-shell vacancies from their decay in describing these MO x rays; rather, one must view the collisional x-ray emission in this case as a single quantum-mechanical process. MO x rays have been explained by Lichten (40) as analogous to the process of optical-line broadening by collision in the visible region of the spectrum. Lichten shows that the <u>maximum</u> emission cross section for photons emitted by an ion-atom system during collision is $\sim 10^{-5}\pi a_o^2$, where a_o is the Bohr radius; this result is independent of atomic number Z. The intensity of these x rays becomes significant compared with the characteristic x rays at large Z, when the maximum cross section for characteristic K x rays (geometric) becomes small due to the small K-shell radius. For the actual MO x-ray cross section to be appreciable, there must be a pre-existing K vacancy before the collision, as there could be for a collision in a solid. For details of this process, see the succeeding article by Cairns and Feldman.

3. MEASUREMENTS OF INNER-SHELL EXCITATIONS

3.1 Introduction

The measurements of inner-shell excitations fall into two complementary categories. To understand why, we must recall our consideration of the collision as a step-by-step process. The initial energy of the collision is that of the incident projectile. As the collision proceeds, the electron clouds surrounding the two nuclei interact, and through this interaction, a portion of the projectile's original kinetic energy is transferred to the electron clouds and thus becomes energy of electronic excitation. Both during and following the actual collision, the electron shells relax, resulting in removal of this excess energy. This de-excitation can evolve only through combinations of two processes: photon emission and non-radiative processes such as Auger-electron emission. These de-excitation processes and the experiments measuring the resulting photons and electrons constitute one of the experimental categories in question. The second category of experiments measure the original excitation energy. This energy, the inelastic energy loss Q, represents the total energy lost from the translational motion of the nuclei and

must equal the sum of the energies of the emitted photons and
electrons plus the ionization energies of the removed electrons.
This excitation energy is customarily determined through a knowledge of the kinetic energies of the nuclei before and after
collision.

By themselves, neither of these techniques provide a full
picture of the collision interaction. A measurement of Q, by
itself, gives no indication of how the inelastic energy is divided
between the two atoms, and says nothing about the energy received
by individual electrons or their subsequent behavior. On the
other hand, spectroscopic measurements of the photon and electron
energies provide little information about the initial excitation
process. It is through a correlation of the results of these two
types of experiments that major gains in our understanding of
these collisions have been made. In order to make this correlation sensibly, the limitations of each type of measurement must
be understood.

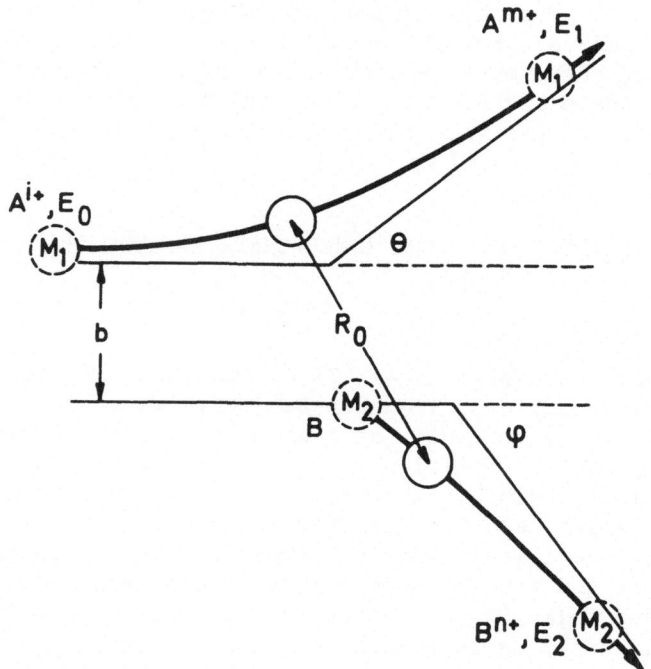

Figure 11. Schematic of the collision $A^+ + B \rightarrow A^{m+} + B^{n+} + (m+n-1)e$
in laboratory coordinates. The ion energies E, the
scattering angles θ and ϕ, and the impact parameter b,
are indicated. The solid circles show the positions of
the ions at the distance of closest approach R_O (From (4)).

Figure 11 shows a schematic diagram of an atomic collision. The inelastic energy loss Q is equal to the incident ion energy E_0 minus the energies E_1 and E_2, of the scattered incident ion and the recoiling target atom. A value for Q can be determined if E_1 and E_2 are measured from the same collision, or by kinematic calculation, if the θ and ϕ are measured in a coincidence experiment. If one assumes a reasonable form for the interatomic potential, one can also calculate using the angle θ, the impact parameter of the collision b, and the distance of closest-approach of the collision R_0 (41,42). Because the production of inner-shell excitations has been found to depend primarily on the quantity R_0, the ability to determine this quantity is an important strength of the differential measurements. A discouraging aspect of differential measurements is their time-consuming nature and their relatively low energy resolution.

Spectroscopic measurements, on the other hand, can provide exceedingly high energy resolution. However, spectroscopic measurements provide little information about R_0. In fact, since most spectroscopic measurements have no way of discriminating between collisions having different values of R_0, they provide information which is inherently an average from collisions having a continuous range of R_0 values. It is for this reason that such measurements are considered "total" in nature, as opposed to the differential measurements already considered. While total cross section measurements shed little light on primary excitation processes, they do provide detailed information about the behavior of individual electrons during the de-excitation process.

3.2 Theory of Energy-Loss Measurements

The schematic outline of a versatile scattering apparatus is shown in fig. 12 (43). It is one which is capable of measuring the energy E_1 of a projectile which has been scattered through an angle θ, as well as the corresponding parameters E_2 and ϕ, for the recoiling target particle. The analyzers A_1 and A_2 may be used not only for the measurement of E_1 and E_2, but also to determine the ionization states of the ions after the collision. Details of this apparatus will be presented later; for now it is sufficient to note that the quantities E_0, E_1, E_2, θ, and ϕ as well as the ionization states m and n described in fig. 11 may be determined experimentally.

The measurement of the final <u>charge states</u> of the scattered or recoil ions constitutes a differential measurement which can provide information about the impact-parameter dependence of inner-shell excitation. Because it is one of the simplest differential measurements to make, we will discuss it first. In an apparatus such as that shown in fig. 12, the scattered-ion

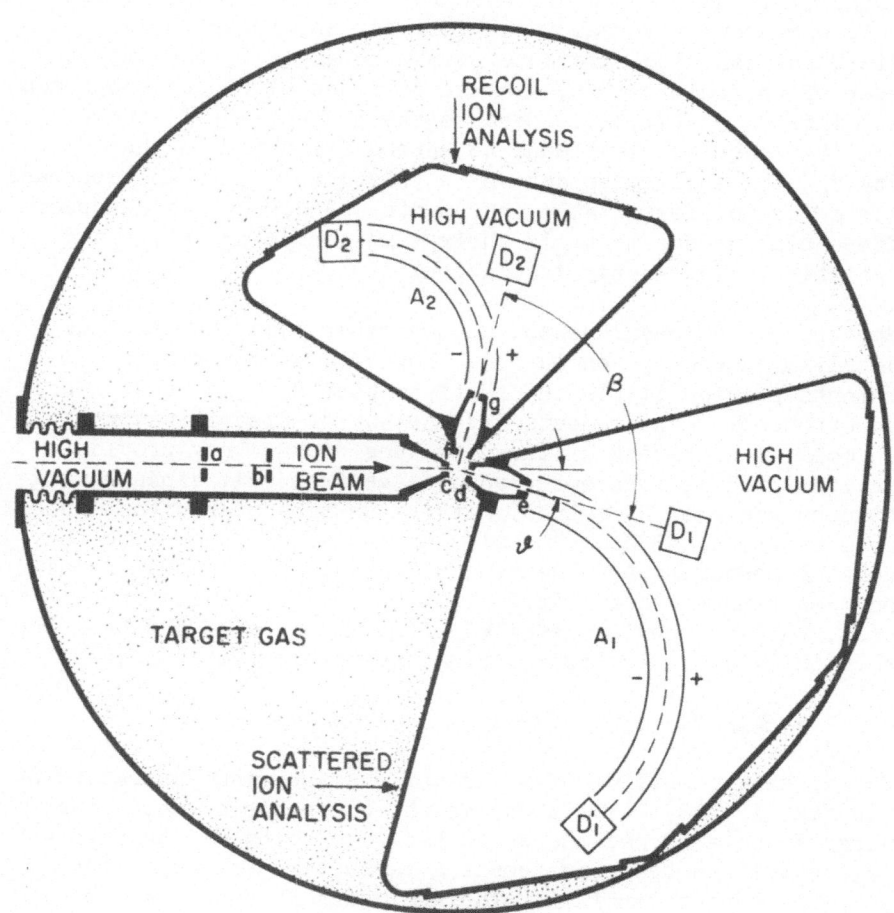

Figure 12. Schematic of coincidence scattering chamber. A_1-scattered ion charge analyzer; D_1, D_1'-surface barrier ion detectors; A_2-recoil ion charge analyzer; D_2 D_2'-secondary electron multipliers; a, b, c, d, e, f, g-collimation apertures. (Kessel (43)).

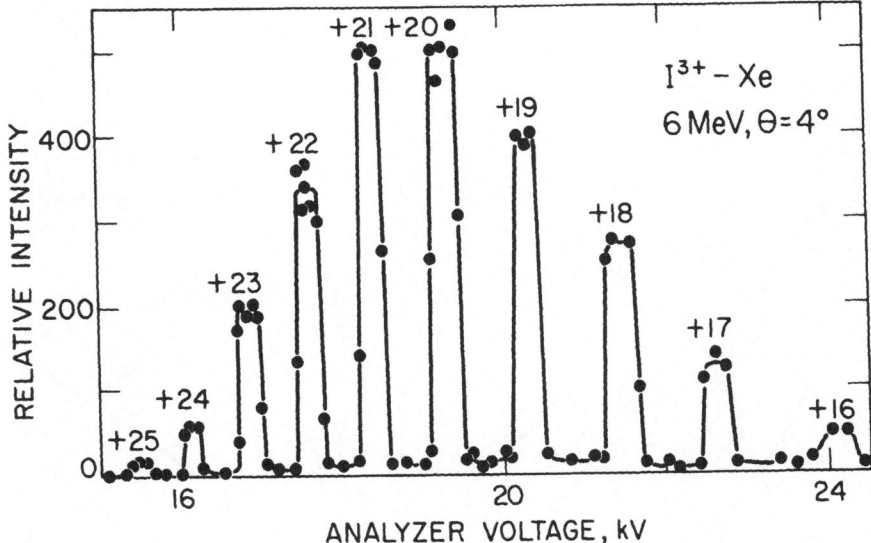

Figure 13. Final charge state distribution for I^{3+} ions scattered through four degrees by xenon atoms (Kessel (44)).

collimator is set at the angle θ and ions passing through it are deflected by an electrostatic analyzer. By varying the voltage on this analyzer, different charge states are deflected into the detector; through a knowledge of E_0, θ and the analyzer voltages, the charge states of the ions may be determined. Figure 13 shows a typical distribution of charge states that may be obtained in this manner (44). Although the data are for 6 MeV, 4 degree I^{3+} - Xe collisions, similar spectra have been obtained for collisions having energies from a few keV to tens of MeV. From the relative number of counts N_m represented by each peak, the probabilities P_m, of ions being scattered through the angle θ having charge state m, is given by

$$P_m = N_m/(\Sigma_m N_m)$$

The average charge state of all the ions scattered through θ at that energy is then given by $\bar{m} = \Sigma_m m P_m$. In a typical experiment E_0 and θ will be varied, thus giving rise to families of curves, such as those shown in fig. 14.

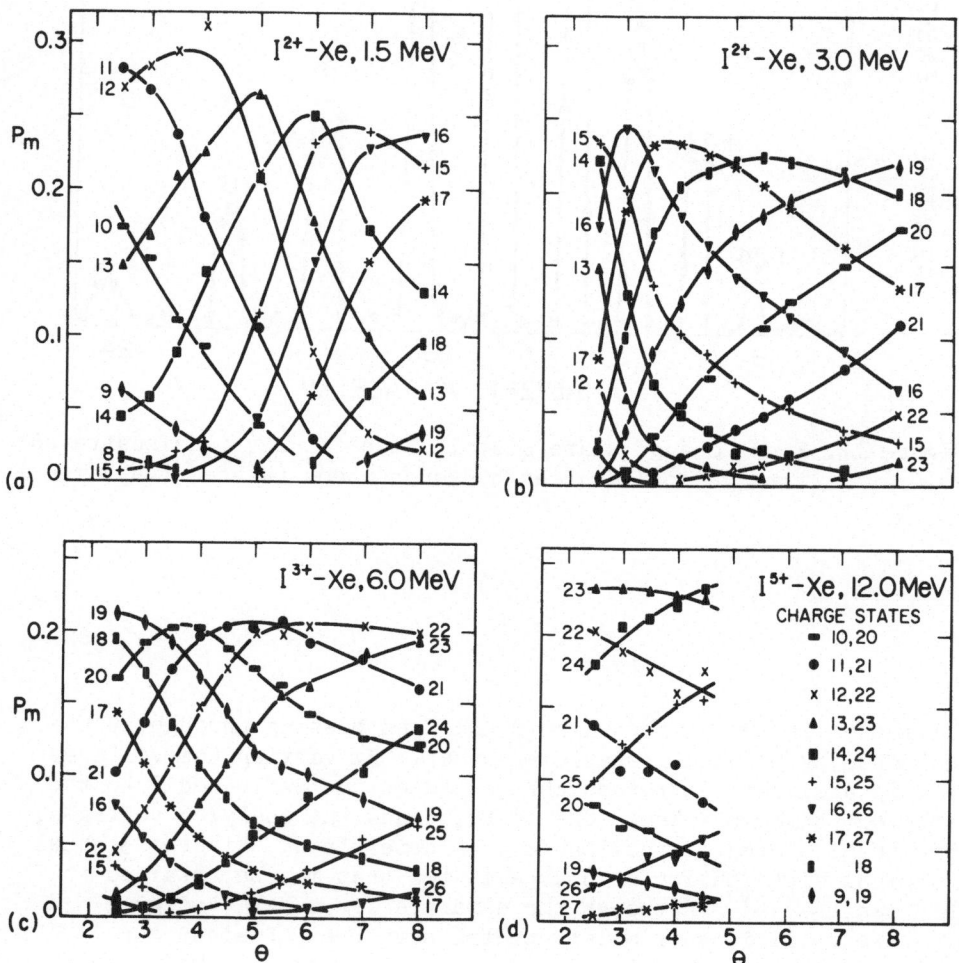

Figure 14. Probabilities P_m for the scattered iodine ions having charge m after collision, as a function of the scattering angle θ: (a) I^{2+} - Xe, 1.5 MeV; (b) I^{2+} - Xe, 3.0 MeV; (c) I^{3+} - Xe, 6.0 MeV; and (d) I^{5+} - Xe, 12.0 MeV. (Kessel (44)).

The differential measurements which have contributed most toward our present-day understanding of x-ray and Auger-electron emission, in the regime where the vacancies are produced by electron promotion, are those which measure the inelastic energy loss Q. This quantity, as defined earlier, is

$$Q = E_o - E_1 - E_2$$

and represents all the kinetic energy removed from the translational motion of the system. It is also, by necessity, equal to the sum of all of the energies absorbed by the various electron shells of the atoms. The quantity Q may be calculated classically for the collisions under consideration here. If any three of the parameters E_o, E_1, E_2, θ and ϕ are measured independently, the conservation of energy and momentum may be used to determine Q. This divides the experiments into three categories. First, the scattered-particle experiment where E_o, E_1 and θ are measured. For this case

$$Q = 2\gamma(E_o E_1)^{\frac{1}{2}} \cos\theta + (1-\gamma)E_o - (1+\gamma)E_1 \qquad (2)$$

where $\gamma = M_1/M_2$, is the mass ratio of the incident particle to the target particle. Second, is the recoil-particle experiment for which E_o, E_2 and ϕ are measured. Here,

$$Q = (E_o E_2/\gamma)^{\frac{1}{2}} \cos\phi - (1+1/\gamma)E_2 \qquad (3)$$

Finally, there is the delayed coincidence experiment wherein the quantities E_θ, θ and β ($\beta \equiv \theta + \phi$) are measured and for which

$$Q = E_o (1 - \sin^2 (\beta-\theta)/\sin^2 \beta - \gamma\sin^2\theta/\sin^2\beta). \qquad (4)$$

It is not immediately obvious which of these three methods is best for the investigation of inner-shell excitations. Morgan and Everhart (45) used the recoil-particle method and were the first to resolve energy losses due to inner-shell excitations. However, it turns out that both their method and the coincidence method lack the resolution available to the scattered particle method. This is due to the thermal motion of the target atoms.

The thermal velocities, superimposed upon the velocity given to the target atom by the collision itself, contribute in a major way to the observed line widths. With all three of these methods for measuring Q, the observed line widths δQ_{obs} have three important contributions: the natural line width δQ_{nat}, the instrumental line width δQ_{inst} and the thermal width δQ_T. As these contributions are independent of one another

$$(\delta Q_{obs})^2 = (\delta Q_{nat})^2 + (\delta Q_{inst})^2 + (\delta Q_T)^2.$$

Fastrup and co-workers (46) have derived expressions for the
thermal width δQ_T (at 1/e height) of each of the above experiments.
Their results, with δQ_T given in eV, are given below:

Scattered-particle: $\delta Q_T = 2(kTE_2)^{\frac{1}{2}} = 10(E_2)^{\frac{1}{2}}$ eV

recoil-particle $\delta Q_T = 2(kTE_1/\gamma)^{\frac{1}{2}} = 10(E_1/\gamma)^{\frac{1}{2}}$ eV

coincidence: $\delta Q_T = 2(kTE_0/\gamma)^{\frac{1}{2}} = 10(E_0/\gamma)^{\frac{1}{2}}$ eV

In these equations, k is the Boltzman constant, T the absolute
temperature, and the energies E_0, E_1, and E_2 are expressed in
keV. In most experiments the angle θ is less than 20° and thus
E_2 is much smaller than either E_0 or E_1. Becuase of this, the
thermal contribution to the measured widths is the smallest in
the scattered-particle experiments. A typical example is that of
30 keV Al ions being scattered through 4.5° by argon atoms. In
this case E_0 = 30.00 KeV, E_1 = 29.75 keV and E_2 = 0.13 keV. For
these values the above equations for the thermal width yield
values of 3.6 eV, 67 eV and 67 eV, respectively. Under most
conditions, instrumental resolution is what limits the resolving
power of scattered-particle experiments while the thermal broadening is the limiting factor in recoil-particle and coincidence
experiments. Afrosimov and co-workers (47) have performed experiments where they measured not only E_0, θ, and ϕ (a coincidence
measurement) but E_1 and E_2 as well. By doing this they combined
the advantages of coincidence measurements with the superior
resolution of the scattered-particle method.

    A typical scattered-particle energy spectrum is shown in
fig. 15. These data were taken by Fastrup and co-workers (48)
and show the result of 10° scattering of Na^+ ions by Ne targets.
The ions scattered through 10° are found to have a range of
charge states; the energy loss for each charge state must be
determined. In fig. 15 only the energy spectrum for the Na^{+4}
ions is shown. Below the energy of the scattered ion is shown
the calculated value of the inelastic energy loss Q_I for the
event. These data were complicated by not using isotopically-
pure Ne^{20} gas for a target. Naturally occurring neon includes
approximately nine percent Ne^{22}. Since this results in a different value of γ in eq. (2), the Na^+ ions scattered through 10°
by the heavier isotope will have a slightly higher energy. The
calculated value of Q for the Ne^{22} peak in fig. 15 is the same as
that for the Ne^{20} peak.

    Inner-shell excitations occur only for certain ion-atom
collisions, and then only for certain combinations of E_0 and θ.

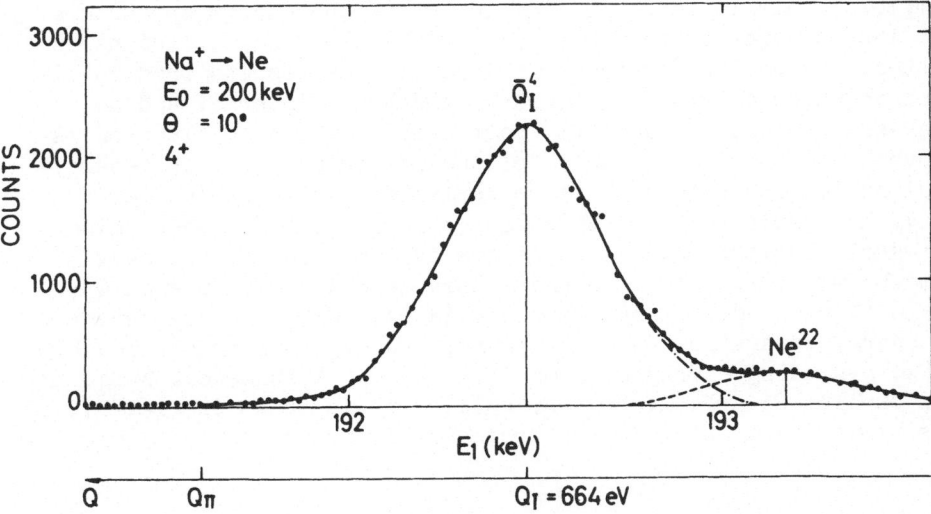

Figure 15. Energy spectrum of the scattered 4+ sodium ions from Na[+]
            -Ne collisions with incident energy E_O = 200 keV and
            scattering angle θ = 10°. The estimated position of a
            possible Q_{II} peak is indicated. (Fastrup et al.(48)).

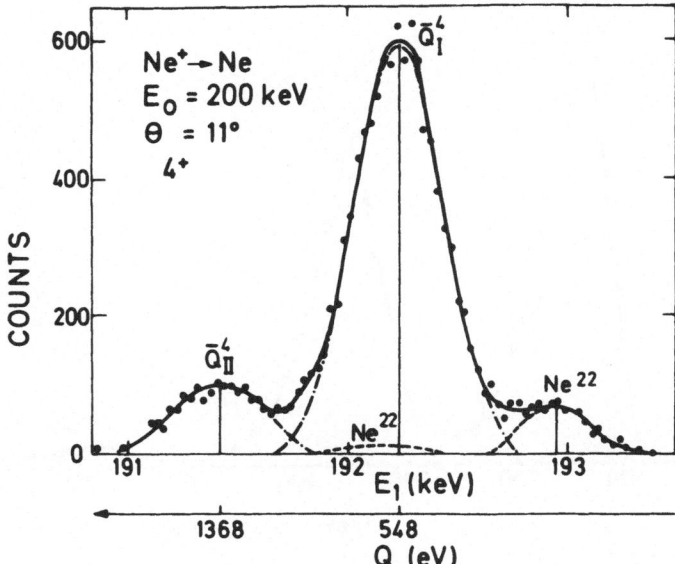

Figure 16. Energy spectrum of the scattered 4+ neon ions from Ne[+]
            -Ne collisions with incident energy E_O = 200 keV and
            scattering angle θ = 11°. The contributions from
            [22]Ne target atoms are indicated (Fastrup et al (48)).

When inner-shell excitations can be resolved, they often appear in the energy loss spectrum as double or even triple peaks. An example of this is shown in fig. 16, where the energy spectrum of Ne^{+4} ions scattered through 11° in 200 keV Ne^+- Ne collisions is plotted. Here, the lower energy-loss Q_I peak results from collisions in which only the outer-shell electrons interact and no inner-shell electrons actively participate. When an electron is removed from the inner K shell of one of the Ne ions, the energy required for its removal must be reflected in the inelastic energy loss spectrum. In this case, those events corresponding to K-shell excitation collisions are shifted, in energy, away from the Q_I peak. The area under this second peak, labeled Q_{II} in the figure, corresponds to the relative number of such events. The energy separation of these peaks, 820 eV, corresponds roughly to the K-binding energy of neon (870 eV). A triple-peak spectrum, this one for 25 keV, 16°Ar^+- Ar collisions is shown in fig. 17. This shows the relative cross sections for producing zero, one,

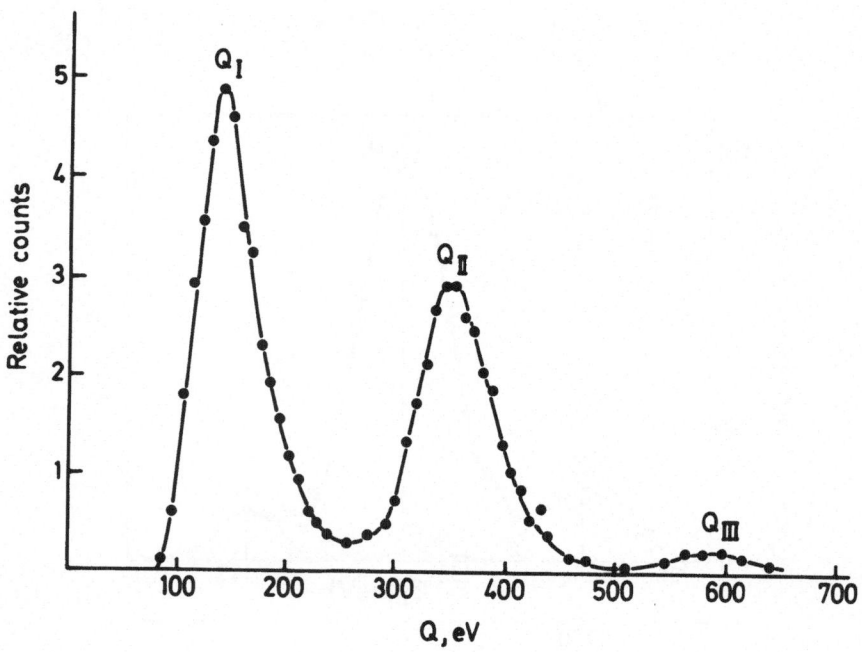

Figure 17. The relative cross sections for producing zero, one, or two (Q_I, Q_{II}, or Q_{III}) $L_{2,3}$ vacancies in Ar^+-Ar → Ar^{2+} + Ar^{2+} + 3e collisions are plotted versus Q for 25 keV, θ = 16° collisions. These high resolution data were obtained by Afrosimov et al. (47) with an improved coincidence technique.

or two (peaks are labeled Q_I, Q_{II}, or Q_{III} respectively) $L_{2,3}$ vacancies in $Ar^+ + Ar - Ar^{+2} + Ar^{+2} + 3e$ collisions. These high-resolution spectra, obtained by Afrosimov and co-workers (47), were taken by the coincidence technique with the additional resolution provided through the determination of E_1 and E_2. Such a measurement specifies not only Q, but also the final charge states of both the recoil and the scattered ions.

3.3 X-Ray and Electron Emission

No attempt will be made to go into the details of Auger electron and characteristic x-ray measurements here. Rudd and Macek (49) have written an excellent review on electron produc-tion in ion-atom collisions and the following article by Cairns and Feldman describes x-ray emission measurements. However, to show how such measurements relate to inelastic energy loss measure-ments it is necessary to discuss them briefly.

The inelastic energy loss values derived from the previous figures represent the sum of many events whose losses may be considered separately. The possible processes include electron emission, photon emission and the production of metastable states. Of these processes, electron emission dominates for most of the collisions under consideration here. This is because the fluor-escence yields are generally small for the inner-shell vacancies which are created in low Z ion-atom collisions. For this reason non-radiative processes, such as the Auger process, are the important de-excitation processes. The detailed information from coincidence measurements in fig. 17, still give only an incomplete picture of an inelastic collision. The final states may be known, and the spectroscopic ionization energies may be used to calculate the change in potential energy, but no information is available concerning the <u>distribution</u> of the remaining inelastic energy among the free electrons. This latter information is better obtained by measuring the energy spectrum of the emitted electrons directly. Rudd has pioneered such measurements: a portion of one of his early spectra was shown in fig. 3, and the complete spectra are now shown in figure 18. These data, obtained from H^+ - and Ar^+ - Ar collisions, show that electrons having energies from zero to about 300 eV are observed. Most of these are represented by the broad continuum; however, superimposed upon this continuum are discrete peaks. The peaks in the 5-20 eV range are due to outer-shell ionization processes, while those between 100 and 300 eV are due to the same Argon $L_{2,3}$ inner-shell processes as are the inelastic peaks in fig. 17.

Measurements of the de-excitation products fall into two categories: those experiments which concentrate on measuring the <u>energies</u> of the secondary products (spectroscopic measurements)

and those which place primary importance on the <u>intensities</u> of
these products (cross section measurements). For example, to
determine the cross section for production of $L_{2,3}$-MM Auger electrons from the curve in fig. 18, the area between the continuum
(estimated by the dashed line) and the solid curve must be determined between 100 and 300 eV. Similar measurements at other
energies allow the generation of cross-section versus energy
curves, and the resulting curves for Ar^+ - Ar collisions will be
shown later in fig. 30.

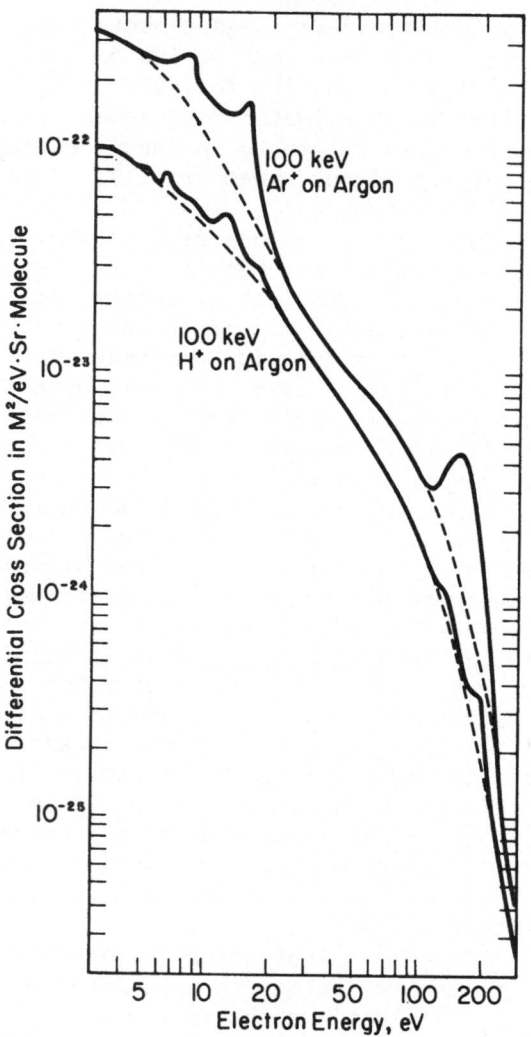

Figure 18. Doubly-differential cross sections for the ejection of
electrons from argon by protons and argon ions at 100 keV.
Electrons are detected at 160° from the beam direction.
(Rudd et al. (50)).

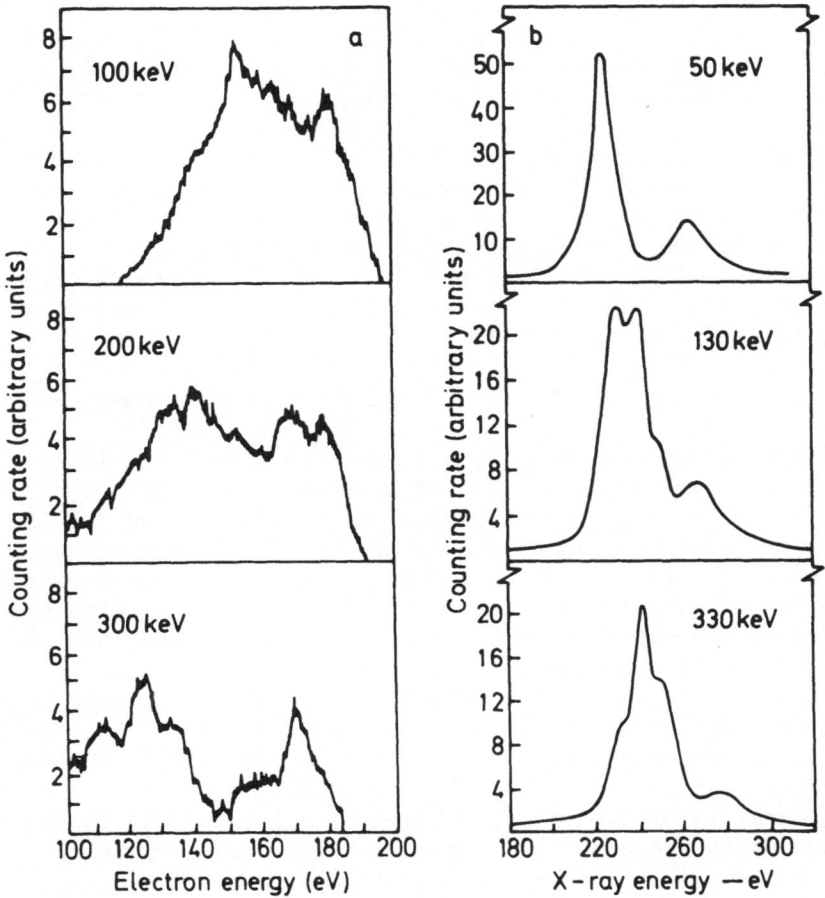

Figure 19. (a) Energy spectra of $L_{2,3}$-MM Auger electrons from Ar^+-
           Ar collisions (Rudd et al. (50)).
           (b) Energy spectra of L X-rays from Ar^+ -Ar collisions
           (Cunningham et al. (51)).

When the peaks in fig. 18 are observed with higher resolution,
structure is evident. Examples of this are shown in fig. 19,
where the electron data are compared with the corresponding x-ray
data of Cunningham and co-workers (51). This comparison illustrates certain features of the two types of measurements. The
electron spectra in fig. 19a are quite broad, much of this being
due to the kinematic "Doppler" shifting of the electron energies,
but the Doppler shifts of the x-ray peaks in fig. 19b are negligible and hence greater resolution is possible. On the other
hand, the rather soft x rays and the low fluorescence yields for
Ar L shells makes this a very difficult experiment. Another

complicating aspect of such collisions may be seen by comparing
the energy shifts of the spectra in figs. 19a and 19b. Although
both the x rays and electrons result from the filling of $L_{2,3}$
vacancies in argon, new x-ray peaks appear at higher energies
when the collision energy is increased while the new electron
peaks appear at lower energies. This is due to the increased
ionization of the outer M shell which occurs at the higher energies. The effect of having fewer electrons in the outer shells
is to increase the binding energy of the inner shells. Larkins,
through Hartree-Fock calculations of the binding energies (52)
has confirmed that this increased binding energy is responsible
for the shifts shown in fig. 19.

3.4 Typical Apparatus-Ionization and Inelastic Energy Loss

Before discussing these data in greater detail, it is appropriate to describe typical types of apparatus that have been used
for the measurement of the parameters E_o, E_1, E_2, θ and ϕ. The
incident ion energy E_o is generally known: if not from the voltage
of the accelerating source, then from an analyzing magnet placed
between the accelerator and the scattering apparatus. With the
exception of the experiments reported by Afrosimov and co-workers
(47), no attempts have been made to measure E_1, E_2, θ, and ϕ
simultaneously because of the time-consuming nature of such an
experiment. Apparatus suitable for the measurement of E_1, and θ
or E_2, and ϕ are relatively simple. Such a scattering chamber,
used by Everhart and co-workers (53) in 1956, is outlined in fig.
20. It consists of two parts, bridged by a flexible bellows.
These two parts are contained within a large vacuum system and
the bellows serve to contain the target gas. The first part,
containing the entrance aperture a is rigidly fixed and provides
access to the target cell for the target gas and pressure-measur-
ing devices. The flexible bellows allows the second part to
rotate through any angle of scattering up to 38° about the
scattering center. This part rotates on a shaft which goes
through the outer vacuum wall by means of an O-ring seal. This
way the angle of scattering may be changed and determined from
outside the vacuum chamber.

Unlike target gas cells used in nuclear physics, no foils
may by used at apertures a and c to contain the target gas.
Instead, a pressure gradient must be maintained across open holes
by differential pumping. Typically holes of one mm or less allow
target gas pressures of 10^{-3} torr while maintaining vacuums of
10^{-6} torr in neighboring parts of the apparatus. In the above
apparatus, apertures c and d serve to collimate the scattered
particles. Following this collimation the scattered ions are
passed between the plates of an electrostatic analyzer and separated
according to their charges. A simple analyzer such as this can

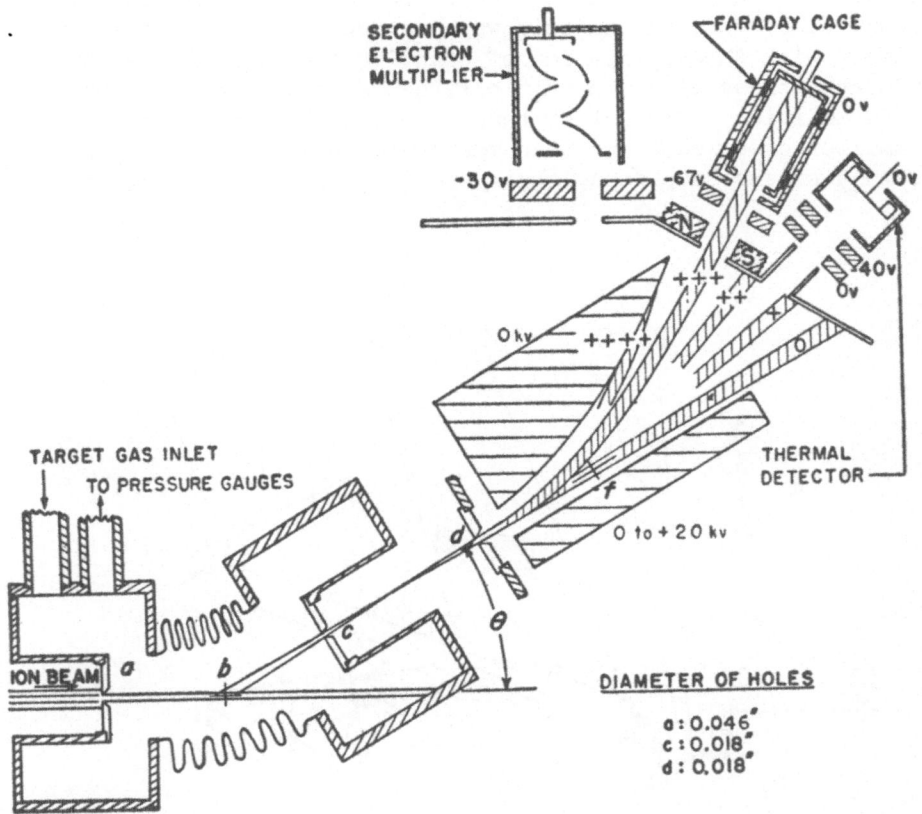

Figure 20. Schematic of scattering apparatus used by Everhart and
          co-workers (53) for the measurement of E_1 and θ of
          scattered ions.

resolve the charge states because all the ions scattered to a
given angle have approximately the same kinetic energy. If a
more exact determination of the energy of the scattered particle
is required, as is the case for inelastic energy loss measurements,
an analyzer of the type shown in fig. 20 is not sufficient but
must be replaced with one having greater energy resolution. The
three detectors in fig. 20 can be rotated about the analyzer axis
so that any of the detectors can be used to examine an individual
component of the scattered beam. The detectors used in these
early experiments were usually thermal detectors or Faraday
cages. Everhart also used a photomultiplier from which the
photosensitive cathode and the glass envelope had been removed.

Particles striking the first dynode would cause the emission of
secondary electrons, and in this way, the neutral atoms were
detected. With the application of counting techniques to such
experiments, these electron multipliers became widely used for
counting individual particles, thus allowing the measurement of
much smaller cross sections than was previously possible. Now
these multipliers have been largely replaced with curved glass

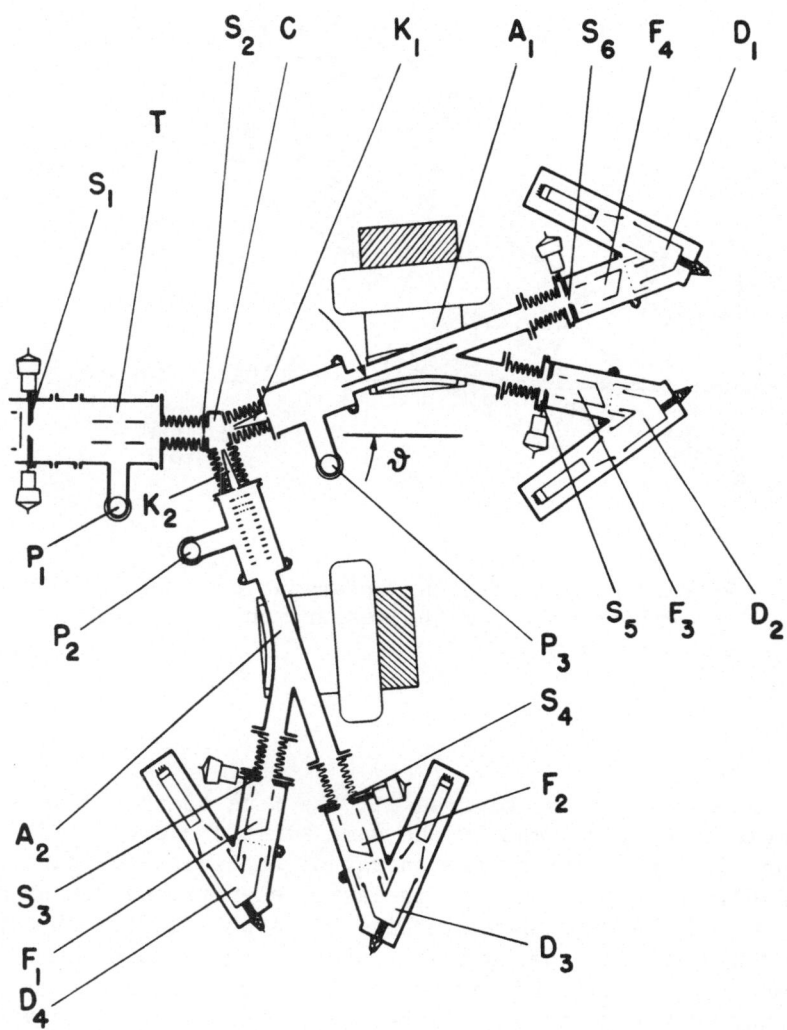

Figure 21. Coincidence scattering apparatus constructed by Afrosimov
          and co-workers (54).

channel multipliers with semiconducting walls, such as the Bendix Channeltron multiplier. The apparatus of fig. 20 was later altered by Morgan and Everhart and used for the recoil-particle inelastic energy loss measurements which were the first to show evidence of inner-shell excitation in heavy ion-atom collisions (45).

A more sophisticated apparatus is shown in fig. 21. This is the coincidence apparatus designed by Afrosimov and co-workers (54). The incident ion beam enters the target gas chamber C, after being collimated by slits S_1 and S_2. A high vacuum is maintained in this collimating region by pump P_1. A fast ion scattered through the angle θ passes through collimator K_1 and into the magnetic analyzer A_1. By varying the magnetic field A_1, the different scattered charge states may be identified and deflected into detector D_2, where they are counted. Neutral particles are not deflected and enter detector D_1. The charge states of the slower ions passing through the collimator K_2 are identified in the same way by analyzer A_2. Detectors D_1 - D_4 are of an early design, now superceded by the electron multipliers. (In these early detectors the scattered ions were post accelerated through 10-20 kV to an electron emitting surface. The secondary electrons thus produced were accelerated by the same potential difference and measured with a scintillation counter.) The bellows attached to the target gas chamber allow collimator K_1 and Analyzer A, to pass through a range of θ from 0° to 50°. The recoil collimator K_2 and analyzer A_2 can be rotated from 40° to 100° on their side of the incident ion beam. In 1969 Afrosimov and co-workers presented data obtained with an improved version of this apparatus (47). They replaced the magnetic analyzer A_1 and A_2 with electrostatic, parallel-plate analyzers capable of accurately measuring E_1 and E_2. The high resolution spectra in fig. 17 were obtained in this way, through the measurement of not only E_0, θ and ϕ, but E_1 and E_2 as well.

A similar apparatus, of radically different design was outlined in fig. 12. This apparatus was constructed by Kessel (43) for collision studies in the low MeV energy range (55). The construction is unusual in that the 1.3 m diameter chamber serves as the "target gas cell", within which are three separate vacuum enclosures containing collimators, analyzers and detectors. The design of this apparatus has the advantage that there are no bellows to limit the angular motion of either detecting chamber. At these higher energies the electrostatic analyzers must be considerably larger than for the lower energies, hence the need for a 1.3 m diameter outer vacuum enclosure. A cutaway view of this chamber is shown in fig. 22.

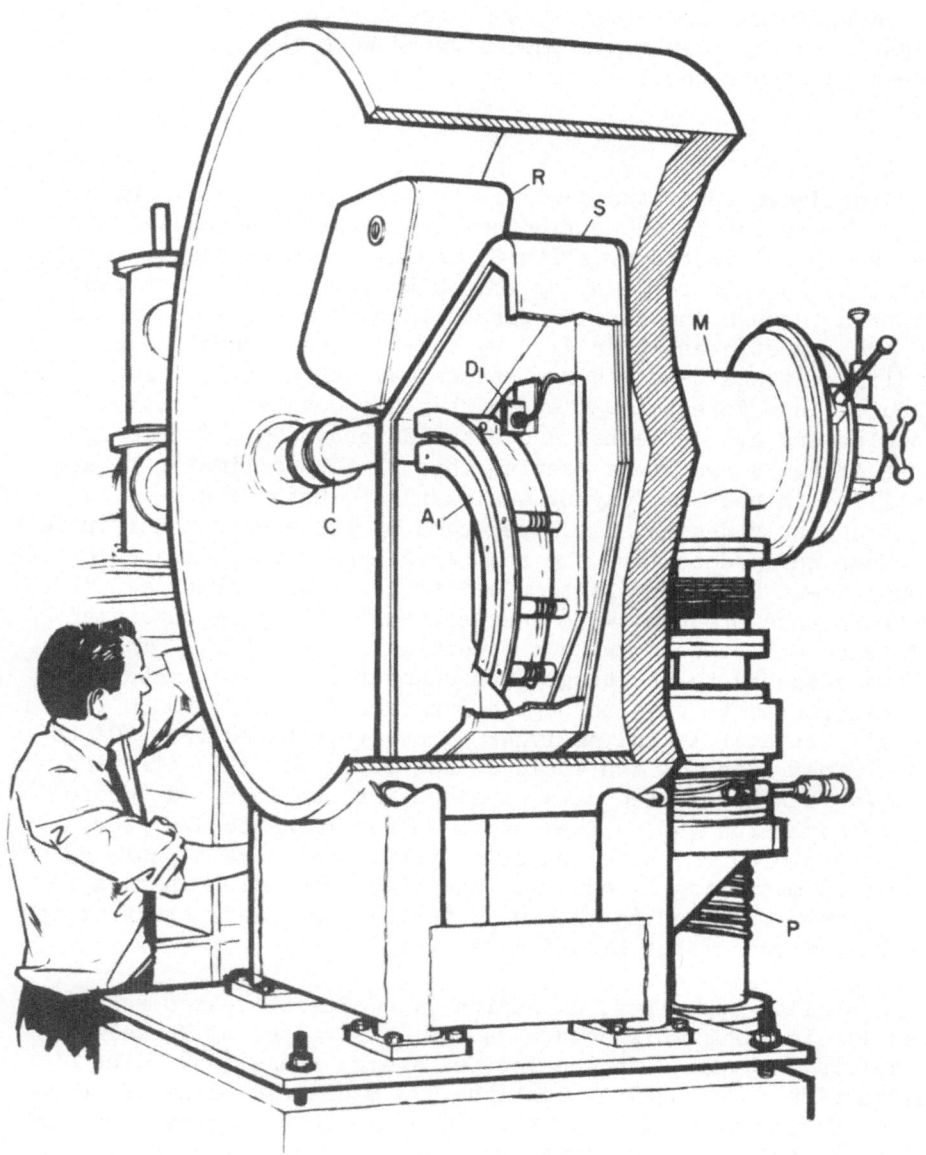

Figure 22. View of large scattering chamber designed for investigation of 0.2 - 20 MeV collisions (Kessel (43)).

3.5   Scattered-Ion--X-Ray/Electron Coincidence Apparatus

A schematic drawing of another coincidence apparatus, designed to observe Auger electrons from ion-atom collisions with a
gas target, in coincidence with scattered ions is shown in fig.
23. This apparatus was originally designed by Thomson, Laudieri
and Everhart (56) for the purpose of observing the Auger electron
spectrum from collisions with specified scattering angle (or impact parameter) as well as known scattered-ion charge state. The
apparatus was later modified by Thoe and Smith (57) to measure x
rays in coincidence with charge state selected scattered ions.
An ion beam was accelerated to an energy in the tens of keV range
and passed into a target-gas collision region. Ions scattered by
the gas target under single-collision conditions into a cone at a
fixed half angle of 20° ± 2°, pass through annular slits (con-
centric with the beam) and are refocused onto an electron multiplier for detection, passing through the electrostatic charge-
state analyzer. It was necessary to collect the whole cone of
scattered ions at all azimuthal angles in order to obtain a
reasonable coincidence rate, since the x rays and Auger electrons
are emitted in all directions and only a small fraction are
collected. Emitted x rays or Auger electrons are detected in
delayed coincidence (i.e., with compensation for the difference
between ion and x ray/electron flight times) with the scattered
ions. In the ion-electron coincidence experiment, the electrons
pass through an electrostatic energy analyzer before detection.
The soft (Ar L) x rays are detected in a Cairns-type end-window
proportional counter equipped with a stretched-polypropylene
window having a nominal thickness of one micron and a high transmission for the x rays observed. Measurements were made of the
number of L x rays found in coincidence with scattered ions for
several different ion beam energies and all observed scattered-
ion charge states. The data are discussed in section 4. These
experiments determine the impact-parameter dependence of electron
or x-ray emission cross sections associated with certain scattered
ion final charge states, but without specification of Q or of the
recoil-particle charge state. To date, no one has attempted the
"ultimate" -a triple-coincidence experiment: scattered-ion--
x ray--recoil-ion in coincidence!

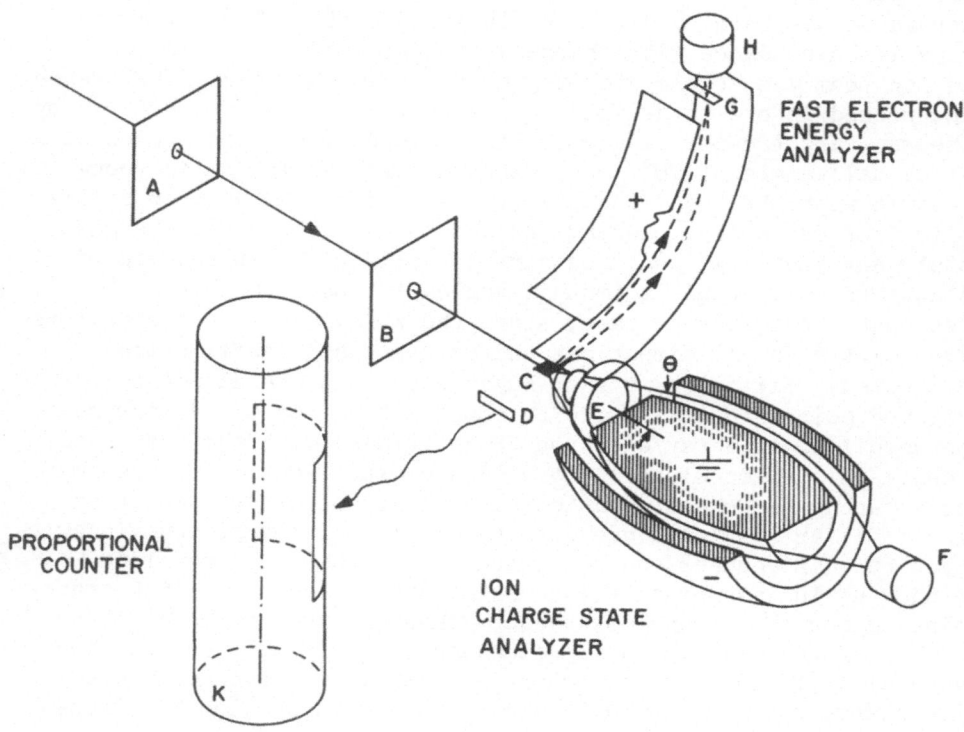

Figure 23. Schematic diagram of scattered-ion--electron/x-ray
           coincidence apparatus of Thomson, et al. (56), as
           modified by Thoe and Smith (57).

4. Discussion of Typical Data

The data obtainable from the kinds of apparatus described
fall into the following categories:

a) Final ionization states of scattered and recoil ions (with the
    impact parameter specified).
b) Inelastic energy loss in single collisions (with the impact
    parameter specified).
c) Electron emission total cross sections.
d) X-ray emission total cross sections (discussed in greater de-
    tail in the following article by Cairns and Feldman).
e) Ion-electron and ion-x-ray coincidence data (with the impact
    parameter specified).

The relative virtues of each type of data for the interpretation
of collision phenomena will be discussed in the next several
sections of this article.

4.1 Ionization States

The charge-state distributions for I-Xe collisions shown in
fig. 14 show the degree of ionization that may occur as well as
how this ionization depends on the scattering angle and the
energy. However, this tells us little about whether or not
inner-shell vacancies are being produced by the collision. An
indication of this, however, may be had by plotting the average
charge-state $(\bar{m})$ values for these data versus the scattering
angle θ and versus R_0, the distance of closest approach. This is
done in fig. 24 and it is readily seen, fig. 24a, that while $\bar{m}$
generally increases with increasing angle, no strong angular
dependence is observed. On the other hand, fig. 24b shows a
marked dependence of $\bar{m}$ on the collision's distance of closest
approach. The horizontal bracket in fig. 24b indicates the range
of R_0 corresponding to initial interpenetration of the L shells
surrounding the two nuclei. This suggests that the degree of
ionization is strongly dependent upon the shell structure of the
colliding ions. Although Kessel and co-workers had shown earlier
(55) that 3 and 6 MeV collisions did produce L x rays, the data
in fig. 24 can only be said to be consistent with the earlier
findings. They do not establish a definite relation between the
x-ray emission and the increase in ionization shown in fig. 24.
In fact multiple M-shell excitation, perhaps of the $M_{2,3}$ shell,
could account for this rise and only a coincidence measurement
between the scattered ion and the x ray can give a definite
answer to the question. (See Section 4.5).

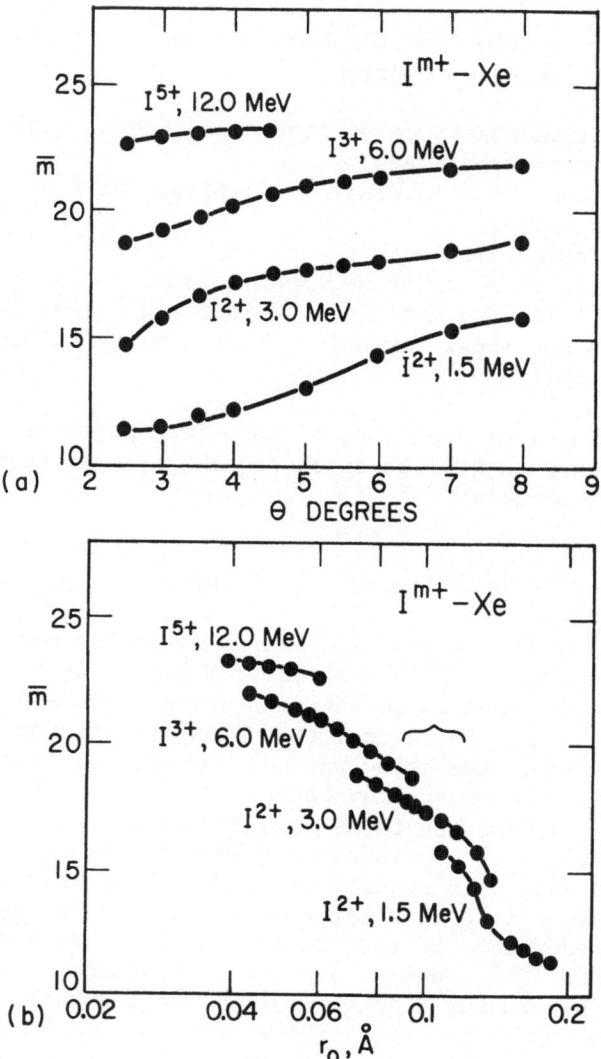

Figure 24. (a) Average charge $\bar{m}$ for the scattered iodine ions is
shown as a function of the scattering angle θ for each
of four incident-ion energies. (b) The average charge
$\bar{m}$ is plotted versus the corresponding distance of closest
approach r_0. The horizontal bracket indicates the range
of r_0 approximagely equal to the diameters of the L
shells of iodine (Kessel (44)).

4.2   Inelastic Energy Loss

In the case of Ar^+ - Ar collisions, combining charge-state measurements with inelastic energy loss measurements did allow specific conclusions to be made following a comparison of the two kinds of data. Shown in figs. 25 and 26 are the quantities $\bar{n}$ (the average charge state of both the scattered and recoil ions) versus R_o, and Q versus R_O for Ar^+ - Ar collisions. The data correspond only to the low-energy Kessel and Everhart data of fig. 26. It is observed that general features of these curves are the same. In particular, it is seen that for 25 keV collisions having a value of R_O of 0.23Å, there is an abrupt rise in the average charge state with an increase in $\bar{m}$ from about one to three. The inelastic energy loss Q for these same collisions

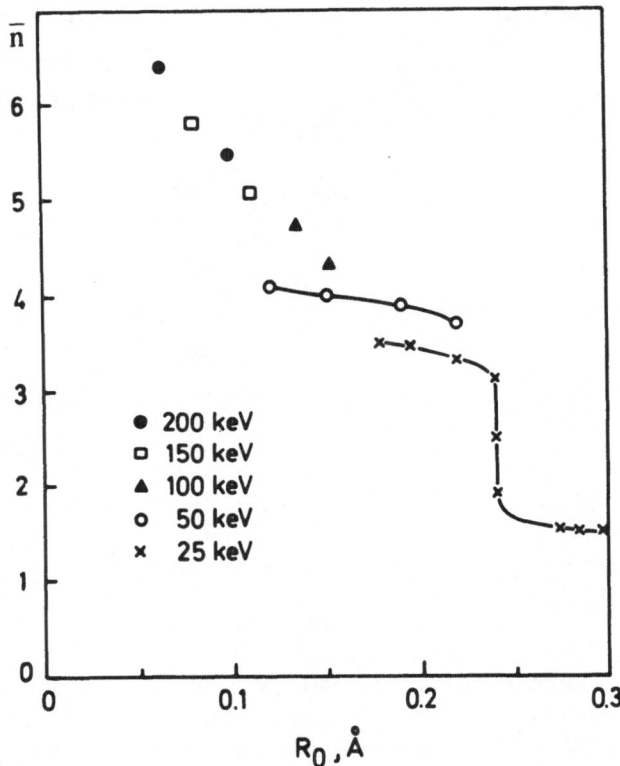

Figure 25. The average charge state $\bar{n}$, of scattered and recoil ions
           plotted versus the distance of closest approach for Ar^+-
           Ar collisions (Kessel and Fastrup (4)).

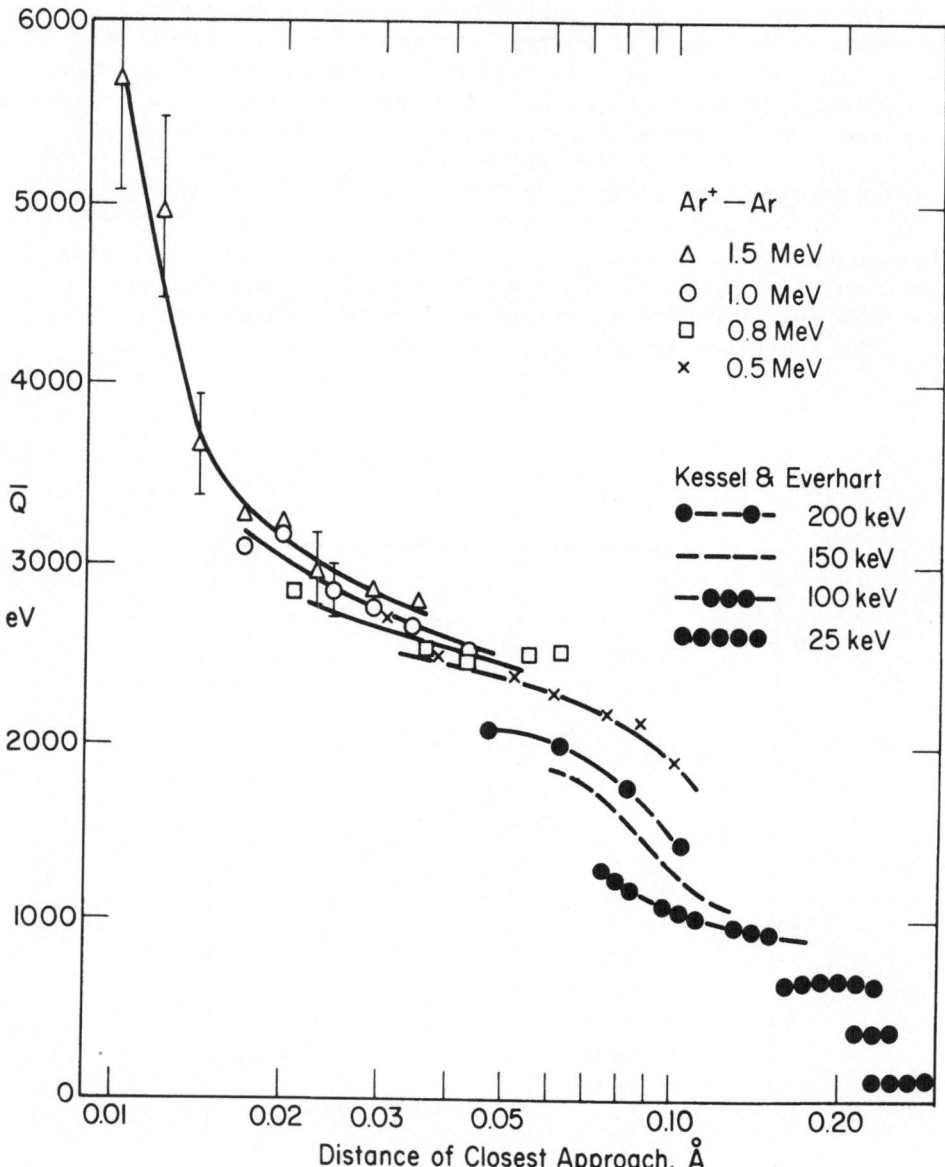

Figure 26. The average inelastic energy loss $\bar{Q}$ is plotted versus
the distance of closest approach, with the incident energy
E_O as a parameter. The data of Kessel and Everhart (28)
are indicated (Kessel et al. (55)).

increases from approximately 100 eV to 600 eV. Since 0.23Å
corresponds to a collision in which the L shells of the two ions
interpenetrate, these data imply the existence of an L-shell
excitation of some sort. This situation provides an example of
the incompleteness of differential measurements which measure
only excitation energies and not de-excitation energies. It was
clear from these data, obtained in the middle nineteen sixties,
that an important excitation was being observed, but the nature
of the excitation was not clear. It was to explain this excitation that Fano and Lichten originally put forth their molecular
orbital model suggesting that L-shell vacancies were being produced during the collision through promotions via the $4f\sigma$ MO.
They recommended looking for the de-excitation products that
would result from the filling of these vacancies. Shortly thereafter, Snoek and co-workers (58) and Rudd and co-workers (50)
(fig. 18) confirmed the emission of L-shell Auger electrons from
the collisions. Ordinarily a coincidence measurement would have
been necessary to confirm that the 0.23Å collisions were responsible for the emission of the observed electrons. However, in
this case the Auger electrons have energies of approximately 240
eV and since the Q values of collisions having R_O greater than
0.23Å are only 100 eV, it is clear that the 240 eV electrons must
have their origin in the smaller R_O collisions. Historically,
these were the measurements which demonstrated that x-ray and
Auger-electron emission phenomena could be investigated through
inelastic energy-loss measurements. It is evident that the
differential nature of the inelastic energy-loss measurements
allow them to make unique contributions toward the understanding
of the production of inner-shell vacancies.

When discrete energy losses can be resolved, as in figures
16 and 17, the resulting information is especially useful. A
good example of this is found in Ne^{++} - and Ne^+ - Ne collisions.
The Q_{II} peak in fig. 16 is due to the production of a K-shell
vacancy in one of the neon ions, while the Q_I peak is due to
those collisions in which there is no K-shell excitation. The
areas under each peak represent the relative probabilities of the
respective events. From data such as these, the probability of
producing a K vacancy may be determined as a function of collision energy and distance of closest approach. This was done by
Fastrup and co-workers (48) and their results are shown in fig.
27 where the excitation probability for a K-shell vacancy is
plotted versus both the product $E_O\theta$ and R_O. These data, and
those of McCaughey and co-workers (60) provide striking confirmation of Lichten's (9) prediction of this phenomenon. He postu-

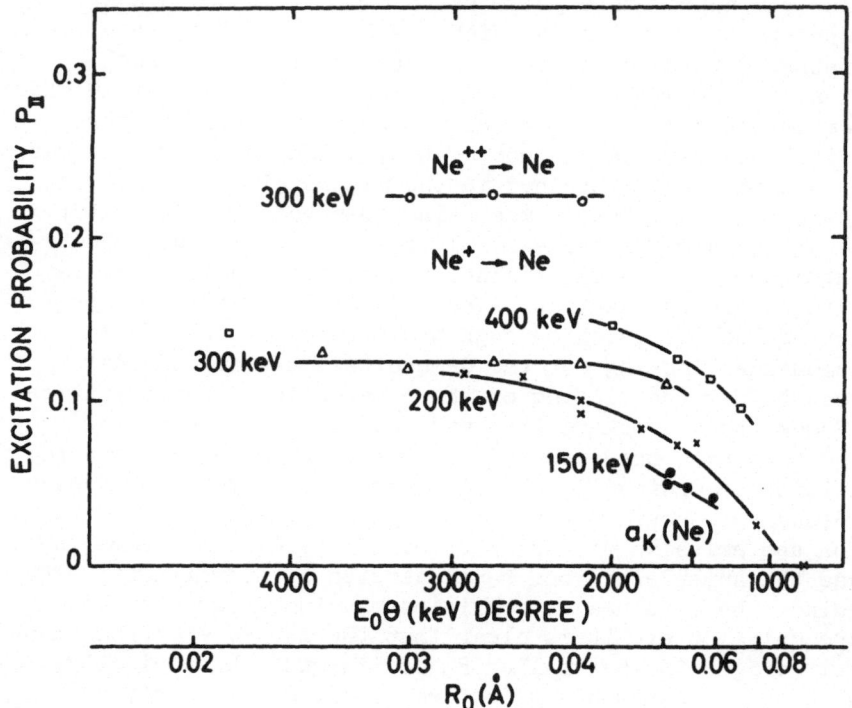

Figure 27. The probability of producing one K-shell vacancy, P_{II}, in Ne^+-Ne and Ne^{++}-Ne collisions plotted versus both the product $E_0\theta$ and R_0. The collisions for which R_0 is approximately equal to the radius of the K shell of a neon atom $a_K(Ne)$, are indicated by the arrow (Fastrup et al. (48)).

lated that the $2p\sigma$ - $2p\pi$ transition could not take place unless there was a vacancy the $2p\pi$ MO, and that the excitation probability should depend on the number of such vacancies. Specifically, Lichten predicted that the Ne^{++} data in fig. 27 should have probabilities twice those of the Ne^+ data. This concept of "open and closed channels" is very useful in discussing MO excitation phenomena.

4.3 Electron Emission Cross Sections

While the type of data shown in fig. 27 is important, in that they show both the velocity dependence encountered in a rotational MO excitation and the R_0 dependence, similar informa-

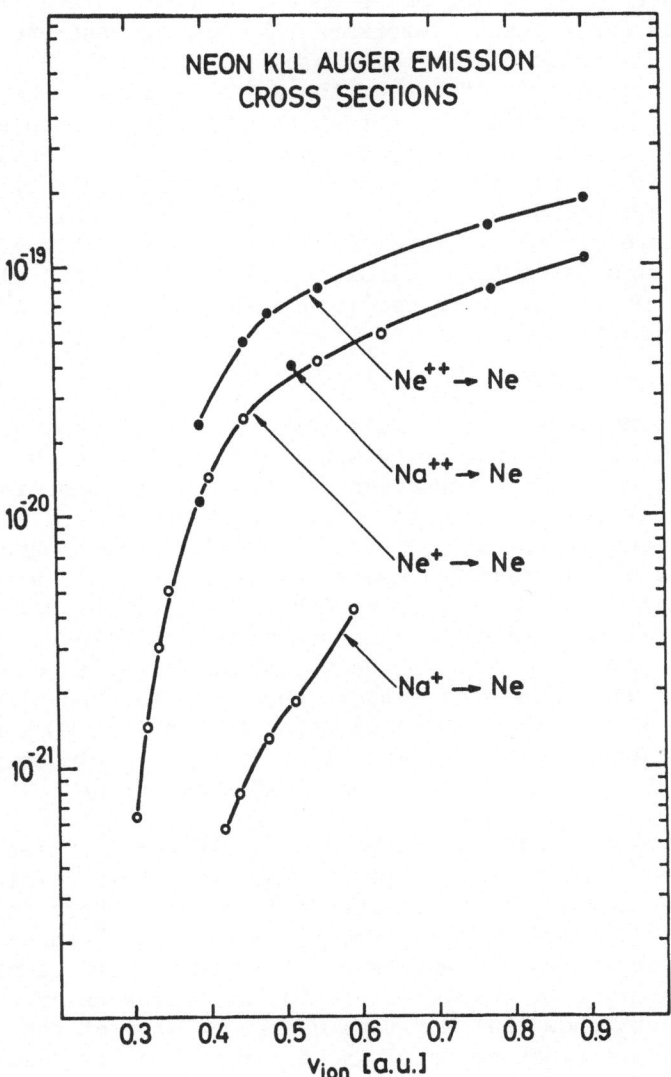

Figure 28. Total Auger electron emission cross sections plotted versus ion velocity for Ne^{++}-, Na^{++}-, Ne^{+}- and Na^{+}-Ne collisions (Fastrup and co-workers (61)).

tion may be obtained through electron spectroscopy. The number of KLL Auger electrons emitted must also be proportional to the number of vacancies in the $2p\pi$ MO. An example of this is shown in fig. 28; these total Auger emission cross sections are from the experiments Fastrup and co-workers (61) and demonstrate the open and closed channel effect quite clearly. The Ne^{++}-Ne cross sections are roughly twice those of the Ne^{+} - Ne data. The Na^{+} - Ne data provide an example of a situation where there should be no vacancies in the $2p\pi$ MO formed during the collision because the 2p levels of the separated atoms are both completely filled, and hence K-shell vacancies should not be produced. While it is clear that this excitation channel is not completely closed, K-vacancy production is certainly inhibited. For Na^{++}- Ne collisions, the number of $2p\pi$ vacancies is the same as for the Ne^{+}-Ne collisions and it is seen that the resulting cross sections are similar.

Briggs and Macek (62) have made several calculations of excitation probabilities for $2p\sigma$ - $2p\pi$ excitations through rotational coupling at small internuclear distances. This semiclassical calculation is based on earlier work for H^{+} - H collisions by Bates and Williams (35). It starts with the probability for there being one or more $2p\pi$ vacancies, and then uses a two-state approximation to calculate the likelihood for the $2p\pi$ vacancies to be filled from the $2p\pi$ MO through rotational coupling. This approach gives excellent fits to the total cross section curves of fig. 28 and reasonably good fits to the more exacting differential data of fig. 27. The conclusion is that the MO model is reasonably quantitative for K-shell excitation in symmetric collisions at relatively low energies ($\sim$ 100 keV for Ne).

Another type of coupling has been investigated by Olsen and has been used to explain data obtained from asymmetric collisions by Meyerhof and co-workers and Stolterfoht and co-workers (63). This is a type of radial coupling sometimes known as "Demkov coupling." It may allow the existence of certain excitations previously thought to be forbidden within the framework of the MO model. These data show that K x rays characteristic of the heavier ion are sometimes observed in asymmetric collisions. Reference to fig. 10 shows that there is no obvious connection between the $1s\sigma$ MO and higher order MOs to account for such an excitation. However, for large values of the internuclear distance, the $1s\sigma$ and $2p\sigma$ states lie close together when $Z_1 \approx Z_2$; and it may be that radial coupling between these states is sufficient to explain the existence of these "forbidden" K x rays.

The application of the MO promotion model to L-shell vacancy and x-ray production is less certain than for the K-shell case.

Evidence for multiple L-shell vacancy production in Ar^+ on Ar gas
collisions is found in the energy-loss data of Kessel and Everhart,
fig. 26. For relatively gentle collisions having $E_0\theta$ of about
600 keV-degrees, 2p vacancies are created in the L-shell of one
or both colliding partners as a result of the promotion (and then
"trapping" into excited or autoionizing states) of one or both of
the two $4f\sigma$ electrons in the Ar_2^+ quasi-molecule. This gives
rise to the triple energy loss peaks for relatively large values
of R_0. The triple Q values from fig. 17 correspond to the step
increases at 0.23A in fig. 26. When R_0 is inside the critical
distance of closest approach at which the steeply-rising $4f\sigma$
curve makes a large number of crossings with outer-shell MO's,
the probability is nearly unity that both of the $4f\sigma$ electrons
will be excited and perhaps autoionized after the collision.
More violent collisions produce additional 2p excitation by means
of close-in crossing of other MO curves that also correlate to
the 2p state in the separated atom limit. This gives rise to the
additional increase in the average inelastic energy loss $\bar{Q}$
(fig. 26); the <u>total</u> L-vacancy production cross section is,
however, dominated by the gentle collisions with the largest
impact parameters. In the Ar_2^+ system, the total cross section
becomes nearly constant (fig. 30) and as large as it can be
(geometric), once the L-shell of the two colliding partners begin
to interpenetrate. Such a geometric cross section is much larger
than the maximum Coulomb excitation cross sections.

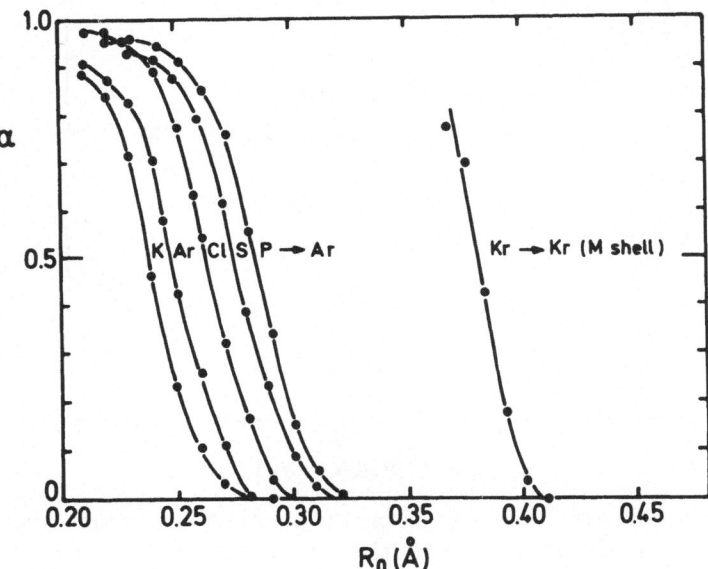

Figure 29. The one-electron promotion probability α, plotted versus
R_0 for Z^+ -Ar and Kr^+ -Kr collisions (Fastrup and co-
workers (46,64)).

Similar data are available for the promotion of $4f\sigma$ electrons in <u>asymmetric</u> $(Z_1 \neq Z_2)$ collisions of several ions with Ar
targets. The correlation diagram for near-symmetric collisions
is given in fig. 10 where it is seen that the $4f\sigma$ MO correlates
with the 2p level of the lower-Z atom in the separated atom
limit. Although when the difference between Z_1 and Z_2 is only
one, there is some mixing of levels, experiments have generally
verified this simple picture. Figure 29 shows how the $4f\sigma$ excitation
probability depends on R_o for some of these collisions. Also
included is a similar curve for the promotion of $6h\sigma$ electrons in
Kr^+ - Kr collisions.

The use of coincidence techniques to give information about
impact-parameter dependences of inner-shell vacancy production
has proven to be a very useful tool, particularly when detailed
comparisons with collision theory are to be made. Rather different types of coincidence techniques have been used in a few
cases (section 4.5 in this article) to observe the impact-para-

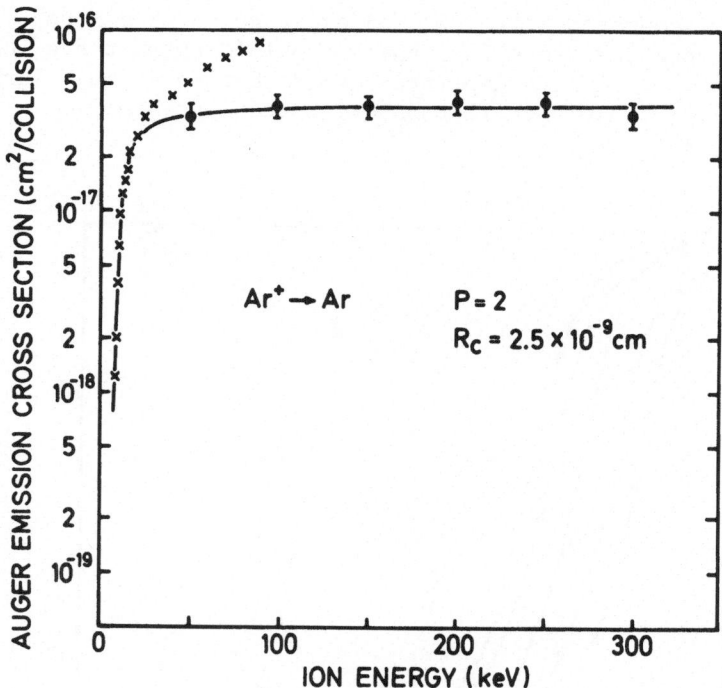

Figure 30. L-Auger electron emission cross section (solid data
          points) plotted versus the incident ion energy. The
          crosses are from the x-ray data of Saris and Onderdelinden
          (66) and have been normalized to the electron data at 20
          keV (Cacak et al. (65)).

meter dependence of x-ray and Auger-electron emission. X-ray scattered ion coincidence measurements, in particular, are important in investigations of the very strong dependence of L-shell fluorescence yields on the number of vacancies in the M-shell.

4.4 Fluorescence Yield Effects

Figure 30 compares the energy dependence of the total cross section for L-shell Auger electron and L x-ray emission in Ar^+ - Ar collisions. The Auger electron cross section rises very rapidly as the L-shells of the two colliding atoms begin to interpenetrate and then levels off to a constant value almost exactly equal to twice the geometric cross section πR_c^2. A glance at fig. 8 shows that the effective collision radius R_c is very close to the distance of closest approach at which Fano and Lichten show the $4f\sigma$ MO rising steeply and interacting with a large number of other MO's. Figure 29 also indicates that in-elastic energy-loss measurements show the single-electron pro-motion probability for Ar^+ --Ar collisions has leveled off at unity for R_0 values smaller than $\sim 0.25\text{Å}$. By contrast the nor-malized Ar L x-ray cross section in fig. 30, measured with a proportional counter by Saris and Onderdelinden (66) continues to rise after the Auger cross section has leveled off. Saris attri-buted this rise to a non-constant fluoresence yield which in-creases as the number of outer-shell vacancies produced increases. Such a result is not surprising. During heavy ion-atom colli-sions the outer shells are multiply excited and the higher the collision energy, the greater this excitation. For the case of a partially-ionized outer shell the probability of inner-shell decay by an Auger process (a two-electron process) may be af-fected.

The fluorescence yield, defined as the ratio of the x-ray emission rate to the sum of the rates for all processes that deplete a given inner shell or subshell, is much less than unity for low Z atoms, due to competition with Auger-electron emission and sometimes with radiationless Coster-Kronig processes (fig. 4). A detailed guide to the extensive literature on fluorescence yields can be found in the review by Bambynek and co-workers (7). The Z dependence is shown in fig. 31.

As has already been emphasized by Larkins (67), Bhalla and Walters (68), Rudd (69) and others, the L-shell fluorescence yield is particularly sensitive to the outer-shell configuration. Each individual Z must be looked at carefully, as one can see from fig. 32, which compares the estimated large variations of L-

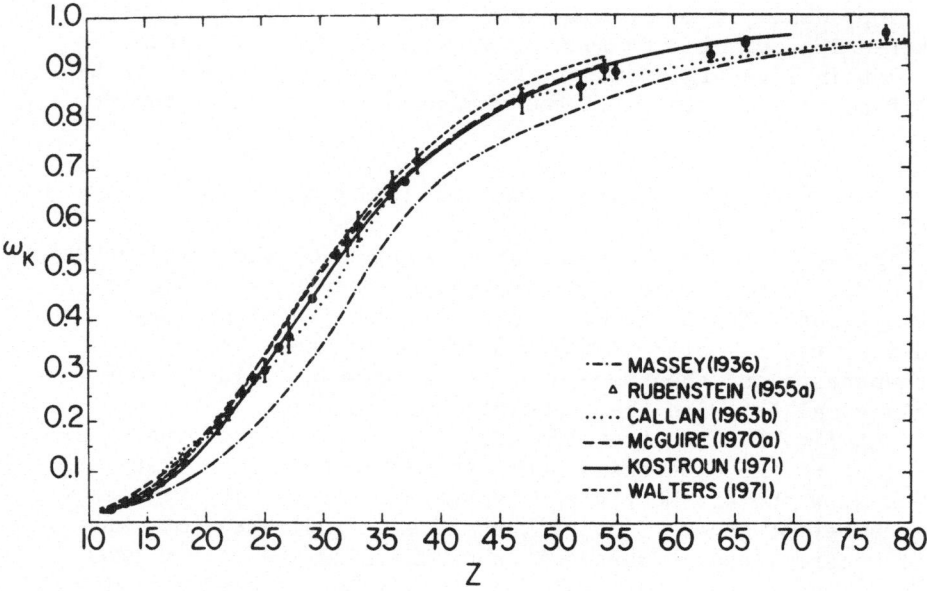

Figure 31. K-shell fluorescence yield as a function of Z, accord-
           ing to various calculations. (from Ref. 7, where refer-
           ences are given for each calculation.)

shell fluorescence yield in Ar with the rather weak variations in
Cu. The differences in the predicted fluorescence yield in the
two cases can be accounted for rather easily in qualitative terms
by scaling the Auger electron and x-ray emission rates (70). If
the x-ray emission occurs by an electron transition into the L-
shell from a partially filled M sub-shell, the x-ray rate will be
reduced relative to the rate for a <u>full</u> subshell by approximately
the fraction of a full subshell of electrons actually present
before the x-ray emission. In the case of the competing Auger-
electron emission, where for example, a single L-vacancy leads to
two M vacancies from the same subshell after electron emission,
the Auger rate is reduced relative to the full subshell rate
according to the fractional number of <u>pairs</u> of electrons in the
relevant M subshell initially. In the argon case, L-shell x rays
arise predominantly from a <u>3s--2p</u> transition. The Auger transitions in argon usually involve the <u>3p</u> electrons. Removing the
3p electrons successively in argon thus produces a rapid increase
in the fraction of $L_{2,3}$ (or 2p subshell) vacancies that are
filled by x-ray emission. The situation with copper is different

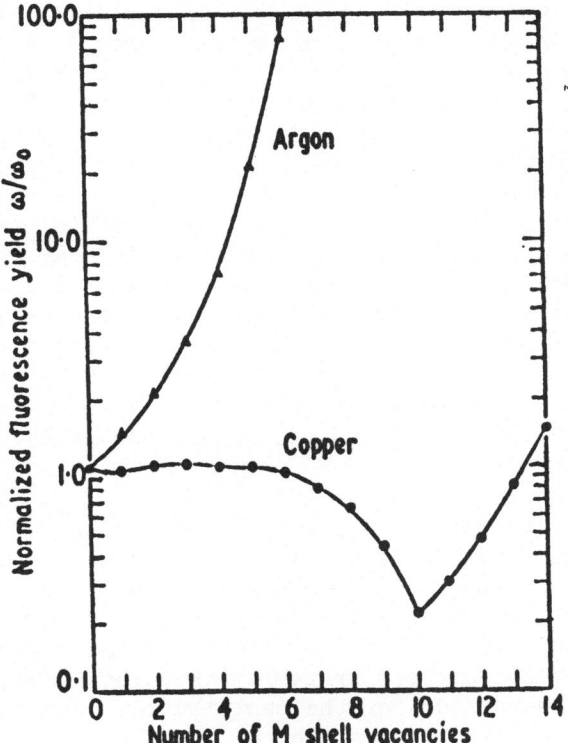

Figure 32. Normalized L-shell fluorescence yields ω/ω_o, plotted
vs. the number of M-shell vacancies for both argon and
copper (From Fortner, et al. (70)).

because the x rays filling the 2p vacancy are mostly from 3d - 2p
transitions, because of a larger transition matrix element and
because there are more 3d than 3s electrons in the neutral copper
atom. Since the competing Auger transitions in the Cu case also
involve, to a great extent, the same 3d electrons, the rate ratio
(fluorescence yield) appears to be quite insensitive to the
number of 3d electrons until most of them have been removed.
When all the 3d electrons in copper have been removed, the yield
increases rapidly, as in the argon case. The L-shell situation
in copper and heavier atoms with 3d electrons is perhaps worth
further experimental study.

4.5 X-ray--Scattered-Ion Coincidence Data

The apparatus shown in fig. 23 was fitted with a special
end-window proportional counter suitable for detecting the soft (λ
$\sim$ 55A, hν $\sim$ 200 eV) Ar L x rays from Ar$^+$ - Ar collisions in the

R_O range where $4f\sigma$ promotion occurs from the 2p shell of argon.
The purpose was to make direct x-ray measurements in coincidence
with Ar ions of known charge state and scattering angle. This
experiment, by Thoe and Smith (57), made it possible to observe
the impact-parameter dependence of the fluorescence yield effects
seen (fig. 30) in the L x-ray total emission cross section.
Comparisons were made with the fluorescence yield calculations by
Larkins and by Bhalla & Walters (67,68). The initial expectation,
somewhat naive, was that the relative number of x-ray coincidences
found with the various charge states would reflect the number of
3p vacancies and the predicted monotonically-rising $L_{2,3}$ subshell
fluorescence yield as a function of the number of such vacancies.
This belief was based in part on the assumption that the number
of inner-shell 2p vacancies would be constant at a specified
impact parameter (constant E_O and θ), regardless of the number of
3p vacancies produced.

The actual coincidence results observed for Ar^+-Ar collisions are shown in fig. 33. There is a <u>maximum</u> in the normalized
x-ray yield which shifts to <u>lower</u> charge states as the beam
energy increases. The reasons for the peak were not immediately
obvious, but became clear after the effects of the scattered-ion
charge-state distribution were properly taken into account and
after corrections were made for the contribution to the Ar L x-
ray signal from recoiling target atoms. With these corrections,
the data were satisfactorily fitted to the theoretical fluorescence yield calculations of Bhalla and Walters (68) for three
plausible initial electron configurations before x-ray emission,
this comparison is shown in fig. 34. The quantity plotted is not
the fluorescence yield itself but a related quantity defined by

$$\Omega_m = (N_x(m)/\eta N_O - R)\, p_m \qquad (5)$$

Here m is the final scattered-ion charge state, N_O is the fixed
number of scattered ions of charge state m used for normalization,
η is the net efficiency factor for the proportional counter, including window absorption and geometrical factors, R is a semi-
empirical recoil contribution and p_m is the probability of detecting
a scattered Ar^{+m} ion at a given incident energy E_O and scattering
angle θ.

The three types of initial electron configurations compared
with experiment in fig. 34 all involve one or more collision-
induced 2p vacancies in the Ar_2^+ quasimolecule. The calculated
curves display the following possibilities for the outer shell:

a) ionization of enough 3p electrons to give rise to the observed
final charge state, or b) ionization of one 3s electron and the

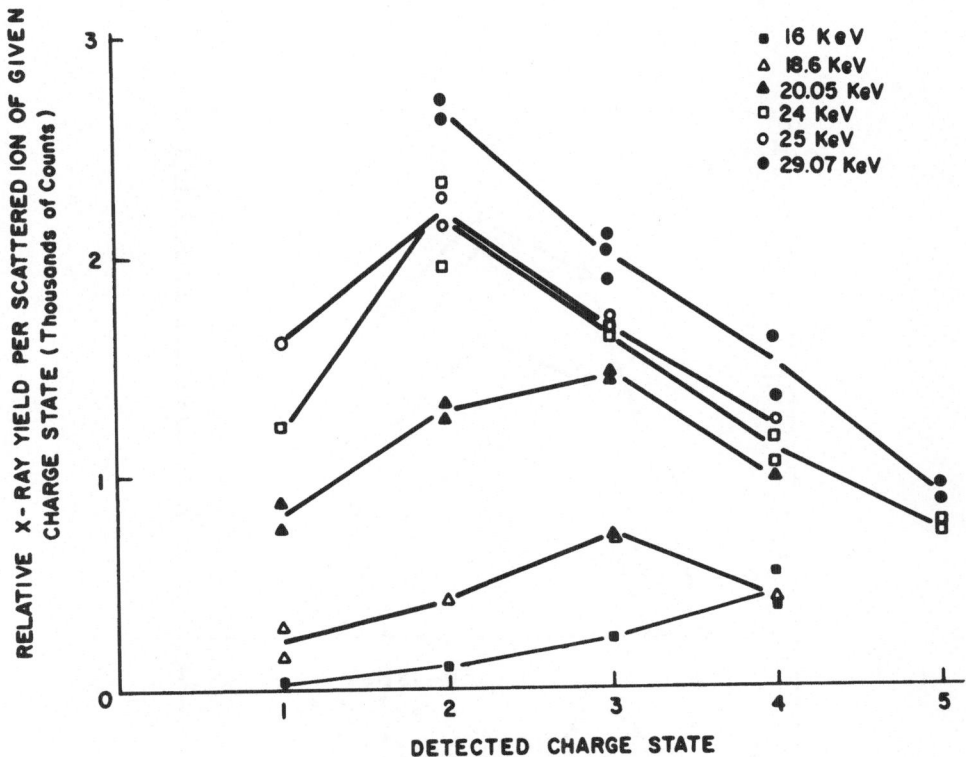

Figure 33. Relative Ar LM x-ray coincidence yields per scattered
          ion as a function of the charge state of the ion,
          at $\theta_{lab} = 20° \pm 2°$. Statistical errors are less
          than the size of the points (integration time 1-20
          hours/point) (From Thoe and Smith (57)).

balance in 3p ionizations, or c) excitation of a single 3d electron
with additional 3p ionization. Configurations of type b) reduce
the effective fluorescence yield with respect to type a), while
the fluorescence yield for configurations of type c) is enhanced
by a factor of 2-3. The data are consistent with substantial excitation of configurations of type c), at $E\theta = 400$ keV-degrees
and above, where the one-electron promotion probability of fig. 29
is close to unity. This conclusion supports an earlier suggestion
of Cunningham and co-workers (51), based on the non-coincidence
x-ray spectra of fig. 19. They attributed the 260 eV peak in the
50 keV data to 3d excitation. Thus there is a kind of complemen-
tarity between x-ray spectra taken with good enough resolution

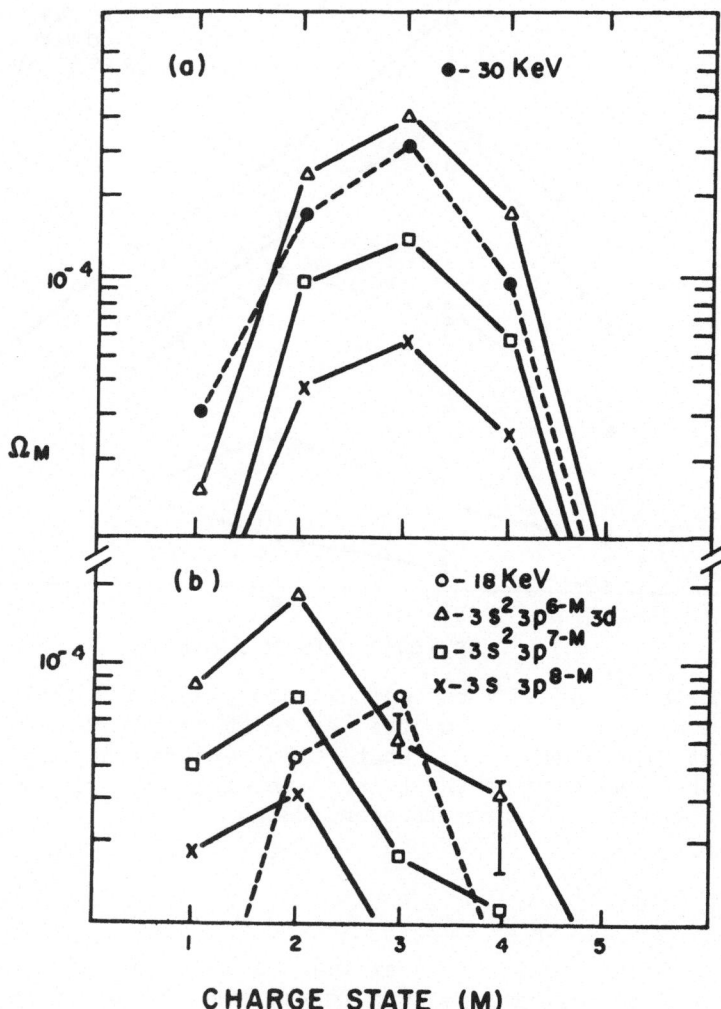

Figure 34. Comparison of experimental (dashed line connecting the points) and theoretical (solid line) values of the effective x-ray yield parameter Ω_m vs. detected ion charge state m. (From (57)).

to see specific configurations and charge-state effects but without impact parameter dependences, as in fig. 19, and the low-
resolution (57) x-ray coincidence work which we are describing
here, that gives detailed information about the impact parameters
at which the onset of a certain excitation process may occur but
without specifying precisely the initial configuration prior to
x-ray emission.

Experimental error and uncertainties in the distribution of
double 2p vacancies between projectile and recoil atom, make it
very difficult to make an unambiguous reduction of the data in
the symmetric collision case of fig. 33. The situation is much
clearer for asymmetric collisions like S^+ -- Ar, where energy
loss measurements (64) confirm the prediction of the MO model
that the 2p vacancies induced by electron promotion are predominantly in the lower -Z collision partner. In such an asymmetric collision, one can ignore the small contribution from
recoil-atom x rays when the projectile has lower Z, and set R = 0
in Eq. (3). When the collision is violent enough so that the
one-electron 4fσ promotion probability $\alpha \simeq 1$, there will be two
2p vacancies in the projectile. Then one can write a simple
theoretical expression for Ω_m to compare with the experimental
value in eq. (5)

$$\Omega_m(\text{theory}) = q_{m-1}(\omega_m + \omega'_{m-1}), \qquad (5a)$$

where q_{m-1} is the probability for a scattered ion at angle θ
to have charge state (m-1) before the two inner shell vacancies
are filled. The fluorescence yield ω_m applies if the Auger
electron is emitted first, followed by the x ray; the second term
involving ω'_{m-1} applies if the LM x ray is emitted first from an
initial configuration with two 2p vacancies, followed by Auger
emission to change the final charge state to m.

For low charge states, all fluorescence yields ω_m are <<1.
This means that Auger electron emission is by far the most
probable mode of vacancy decay, and we can ignore the possibility
of double x-ray emission. Where there are two 2p vacancies in
the same atom, the measured charge-state probability $p_m = q_{m-2}$,

i.e. q_{m-1} in equation (5a) is p_{m+1}. When Ω_m from (5) is divided

by p_m, we obtain from (5a): $(\omega_m + \omega_{m-1})\dfrac{p_{m+1}}{p_m} = \dfrac{N_{x(m)}}{\eta\, N_0}$. This gives

the explanation for the peak in the coincidence data of Figures
33 and 34: the monotonically increasing function of m, $(\omega_m + \omega_{m-1})$
is multiplied by the monotonically decreasing function of m, (p_{m+1}/p_m),

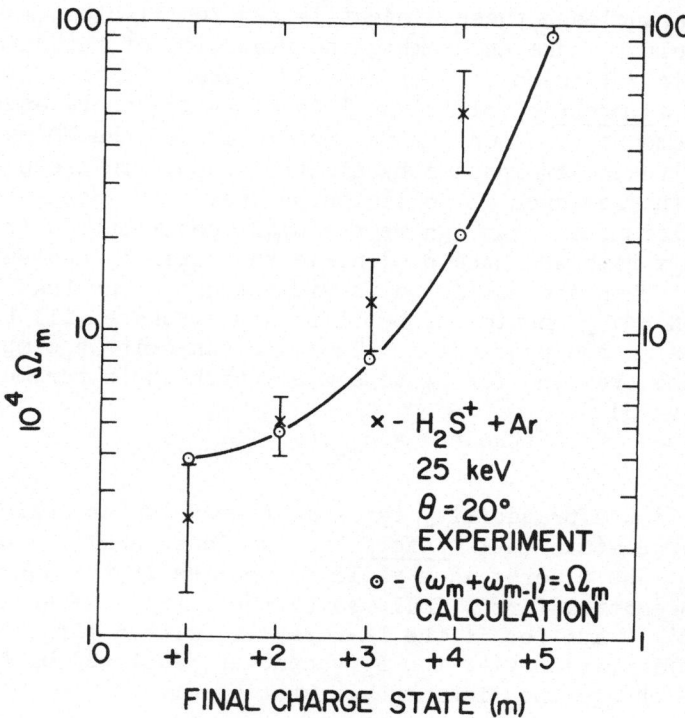

Figure 35. Comparison of effective L x-ray yield Ω_m/P_m vs. detected scattered ion charge state m, for collisions of 25 keV H_2S^+ ions with Ar (Thoe and Smith (72)).

with the ratio of the two probabilities evidently decreasing faster than the fluorescence-yield factor increases for high charge states m.

Figure 35 is a plot of Ω_m/p_m for asymmetric sulfur-argon collisions (unfortunately using a beam of H_2S^+ instead of a pure sulfur beam), showing good agreement between the fluorescence yield calculations of Bhalla (71) within experimental error. We have also observed x-ray--ion coincidences in P^+ -- Ar collisions, where preliminary data suggest an anomalously large fluorescence yield for the highest charge states (+4, +5) observed (72). Similar conclusions based on better data, but without impact parameter specification, were reached by Fortner (73) from L x-ray total cross section measurements.

When L-shell vacancies are produced in heavy particle colli- sions of low Z partners, the probability of vacancy decay by x- ray emission is quite low compared to the probability of Auger electron emission. The impact parameter dependence of the Auger electron emis ion can be measured directly by the technique of ion-electron coincidences. The apparatus of figure 23, used for the x-ray--ion coincidence work on Ar^+ -Ar collisions, was originally used to observe the impact-parameter dependence of the Auger-electron emission in coincidence with scattered ions of specified final charge state by Thomson, Laudieri and Everhart (56) and by Thomson, Smith and Russek (74). An example of data (corrected for electrons emitted from the target ion) obtained for 25 keV Ar - Ar collision at 21° scattering is presented in fig. 36. Low electron-analyzer resolution and kinematic broadening prevent detection of individual Auger lines in the spectrum. The electron energies are characteristic of $L_{2,3}MM$ transitions and the spectral-peak centroids shift systematically toward lower energies by approximately 10eV with each unit increase in final charge state. This is expected because of the reduction in screening as each successive 3p electron is removed. The low energy cut-off of the peaks in figure 36 can be used to make inferences about the relative lifetimes of outer-shell and inner- shell electronic excitations that are produced in these rather violent collisions. As an example, we note for the electrons that appear in coincidence with scattered Ar^{+3} ions, there are many electrons with energies below 157eV. This is the lowest energy ($L_{2,3}M_1M_1$) that can be expected from an ion with all outer-shell electrons in their ground states. Single-electron excitation in the outer shell can produce a depression in the fast-electron energy of up to about 10 eV. However, there are many electrons that lie below even these reduced limits. This fact strongly implies that there is a substantial probability for the presence of multiple excitation of the outer shell at the time of $L_{2,3}-M_{2,3}M_1$ and $L_{2,3}-M_1M_1$ Auger transitions (74).

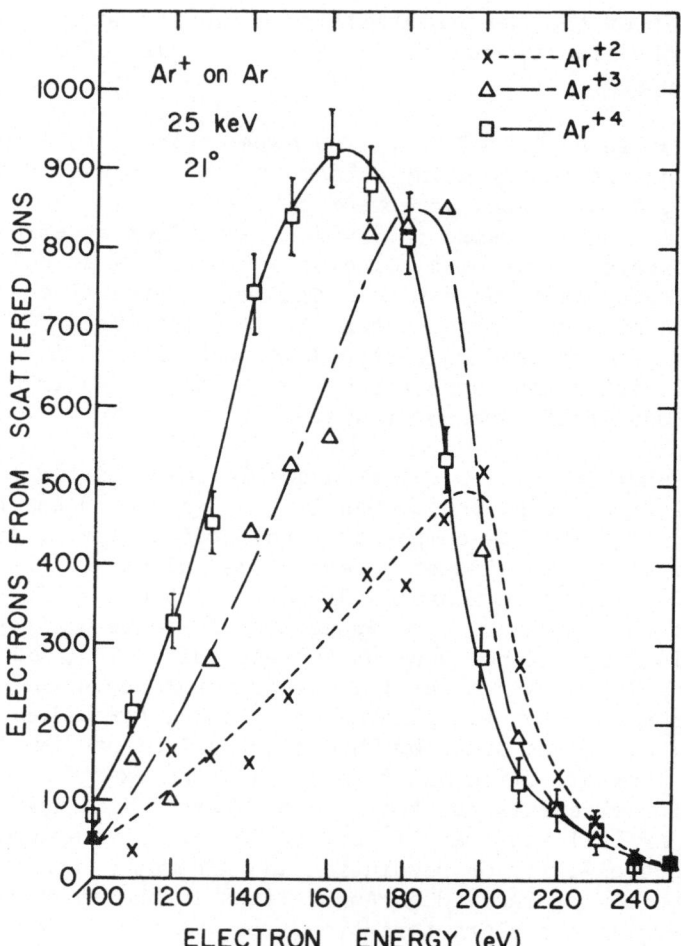

Figure 36. Fast-electron coincidence spectra associated with argon
ions whose final charge state is specified. In this
figure, the electrons associated with the undetected recoil ion and a continuous background have been subtracted,
and the spectra should apply to electrons emitted from
the scattered ion alone (Thomson, Smith and Russek (74)).

     X-ray--scattered ion coincidence data with K x rays have
been obtained by Sackmann, Lutz and Briggs (75) for Ne⁺ -Ne
collisions and appear to give a much more sensitive confirmation
of the Briggs-Macek calculation (62) than the total cross section
data. Their results are shown in fig. 37. Similar coincidence
measurements have been made in the MeV range by Stein, Lutz,
Mokler and Armbruster (76) and by Laegsgaard, Anderson and
Feldman (77).

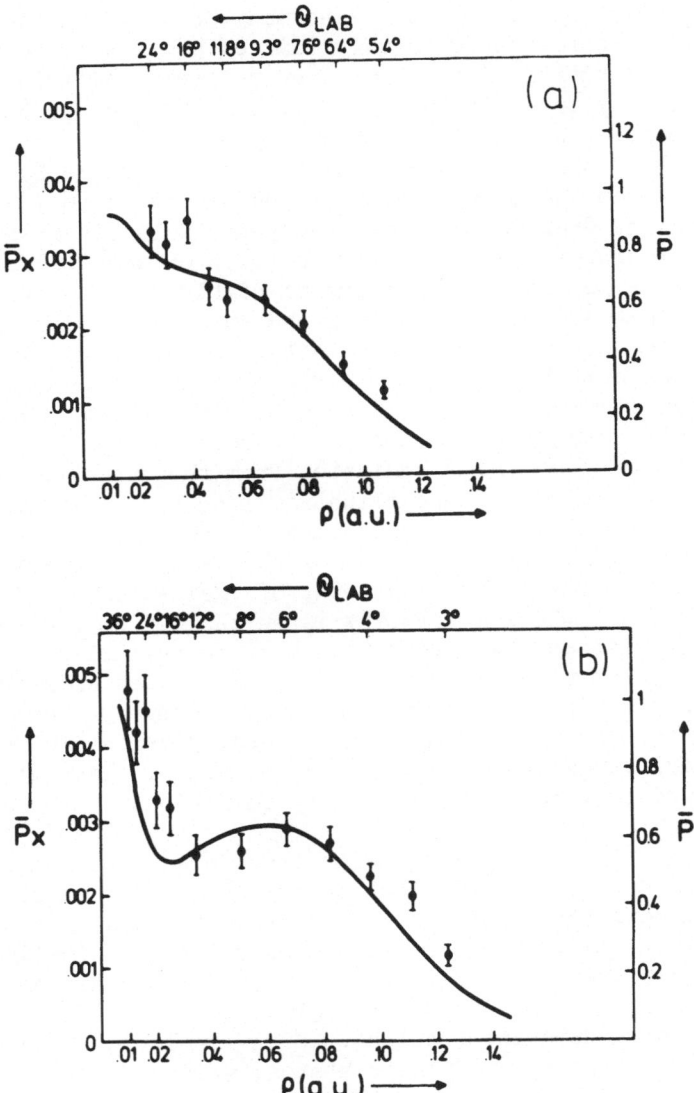

Figure 37. Experimental x-ray production probability (left-hand
scale) fitted with theoretical vacancy-production probability (solid curve, right-hand scale) in Ne$^+$ --Ne
collisions, as a function of impact parameter ρ.
a) 235 keV ion beam, b) 363 keV beam (Sackmann, Lutz and
Briggs (75)).

4.6 X Rays from Highly Stripped Fast Ion Beams

At energies from $\sim$ 100 keV to many MeV, substantial progress in atomic spectroscopy of highly ionized atoms has been made through observation of excitation by multiple collisions in thin solid foils by investigators using the techniques of beam-foil spectroscopy. This field has been reviewed in a book and several articles by Bashkin (78,79). In recent years there has been a tendency to do these experiments at MeV energies (6,79) with tandem Van de Graaffs and other heavy-particle linear accelerators. Photon emission in the x-ray region is often seen in such experiments. In some cases, with beams of highly-stripped foil-excited ions, it is found that both x-ray and autoionized electron emission can be observed from metastable electronic states of ions in the beam itself. An example of this can be found in the MeV oxygen experiments of Sellin and collaborators (80,81). Carbon-foil excitation of such ions causes appreciable emission of the 21.8 Å ($2^3P_1-1^1S_0$) radiation from the two-electron (helium-- like) O^{6+} ion. This intercombination line has an anomalously long lifetime of 1.5 nsec, which can be measured by the beam-foil technique of observing the decay of the x-ray intensity as a function of distance from the point of excitation. This is possible since the ion velocity is in the 10^9 cm/sec range. Similar transitions have also been seen in heliumlike ions such as Ar^{16+} by Marrus and Schieder (82). Forbidden emission lines from highly-stripped ions are sometimes observed. Some of these in the x-ray region are important in astrophysics, for example in the coronal lines from the sun (83) and also in laboratory plasmas (84), where they may be perturbed by collisions.

In addition to <u>x-ray</u> emission from metastable beams in the low MeV range, keV-energy <u>electron</u> emission has been observed (85) from metastable autoionizing states of highly-stripped ion beams (figure 38). The quartet states of the lithiumlike iso-electronic sequence, for example, are metastable against Coulomb autoionization because of their high spin; they decay via spin-orbit and spin-spin interactions. High-resolution electron spectra from the decay of these states, which are produced abundantly (86) by multiple collisions in foil excitation, have been obtained as a function of distance from the foil. From such data the lifetimes of many of these states may be determined (87). X-ray emission from such metastable beams is also observed, as in figure 39 (88). Attempts to observe such high-spin metastable states (for example, 1s2s2p ^4P states of the three-electron ions) in single collisions of <u>unexcited</u> ion beams with <u>gas targets</u> have not succeeded thus far, although metastable doublet states (such as $1s2s^2$ ^2S) have recently been seen in gas collisions (89).

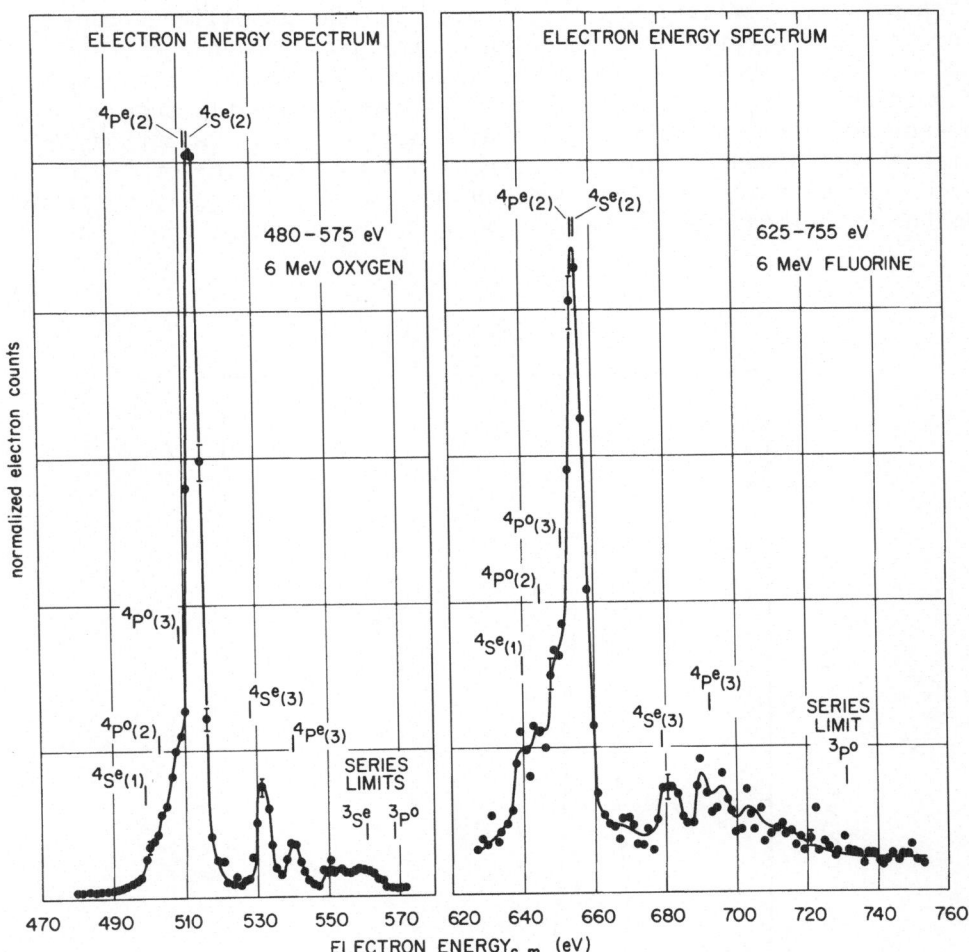

Figure 38. a) Portion of the spectrum of autoionized electrons
emitted from core-excited quartet (spin 3/2) states
of the three-electron oxygen ion, produced by foil
excitation of a 6 MeV beam. The two lowest-energy
peaks, designated $^4P^o(1)$ and $^4P^e(1)$, of odd and even
parity respectively, occur at energies of 417 and 429
eV in the ionic rest frame and are not shown here
details in (87)).

b) Similar spectra from quartet metastable autoionizing
states of fluorine (Sellin and co-workers (85)).

High-resolution projectile x-ray spectra produced by MeV
beams of heavy ions on a solid target have been observed for some
time now. Of particular interest has been the satellite structure in such spectra as a function of incident ion charge state
(90). These experiments have the drawback that one cannot obtain
a pure charge state in a thick solid target. Recently such
measurements have been made with both high and low resolution for
single collisions of highly-stripped ions with gas targets. In
such gas collisions the incident charge state is uniquely specified,
so the experiments are "clean." As might be expected from our

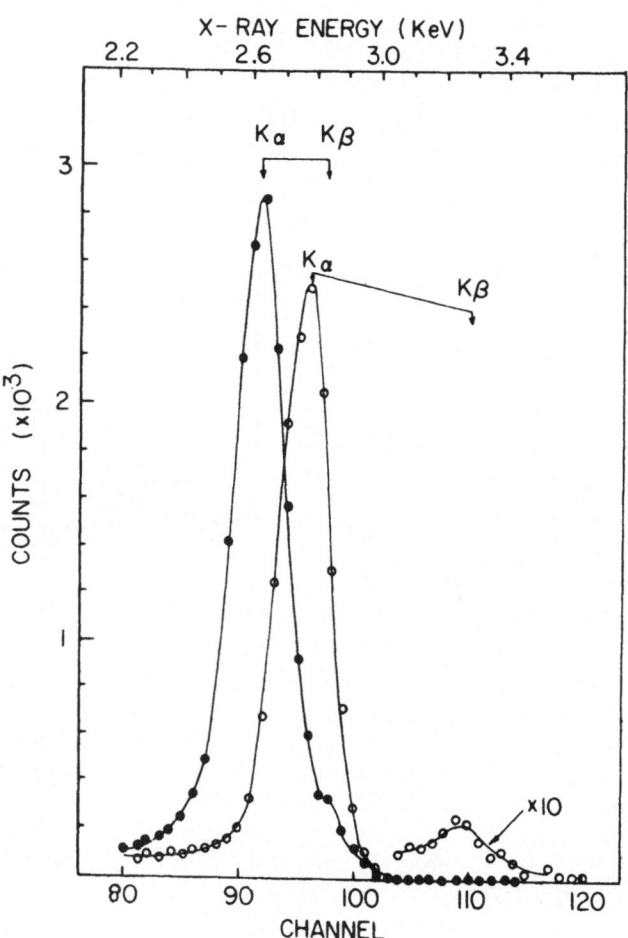

Figure 39. X-ray energy spectrum from a 45-MeV chlorine beam viewed
          2 cm downstream from the exciting foil (open circles).
          An x-ray spectrum displaying x-ray energies characteristic
          of removal of a single 1s electron from neutral chlorine
          produced by proton bombardment of NaCl is also shown
          (closed circles) (Cocke et al. (87)).

previous discussion of the variability of fluorescence yields
with ionic electron configuration, strong dependences of x-ray
yield on projectile charge are observed in these experiments.
Mowat and co-workers (91) made low resolution spectral and yield
measurements of the projectile-charge dependence of Ar K and Ne K
x rays using 80 MeV highly-stripped Ar ion beams on Ne gas targets.
The Ne K x-ray yield increased by more than a factor 1000 when
the argon ion charge was varied from +6 to +17. The Ne K yield
was an exponential function of Ar charge state and the Ar K yield
had an increasing slope as a function of charge state on a semi-

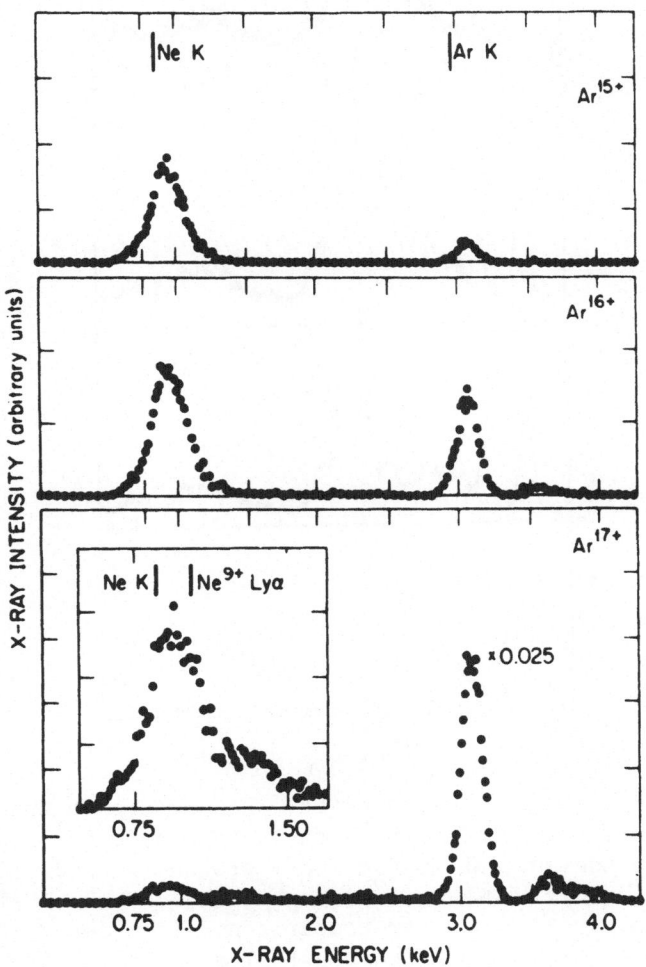

Figure 40. X-ray spectra resulting from collisions of 80-MeV Ar
ions with Ne. The Si(Li) detector window and dead
layers attenuate the neon peak by a factor of ∿ 10
compared with the argon peak (Mowat et al. (91)).

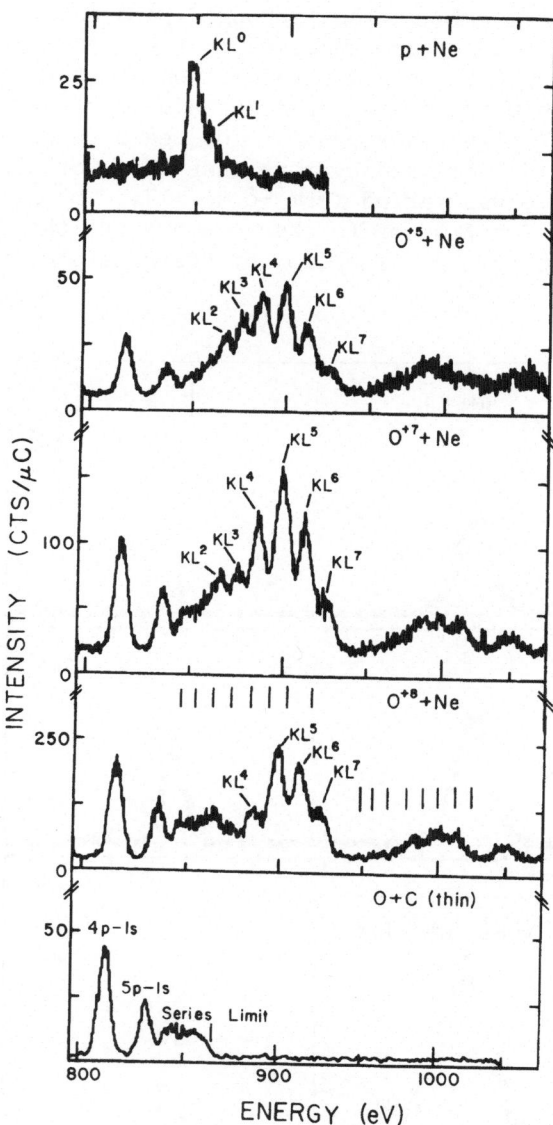

Figure 41. Resolved Kα spectra of Ne. KL^n refers to transitions
from states with one K-shell vacancy and n L-shell
vacancies. Shown are 2-MeV proton-induced, 30-MeV O^{+5}
-induced, 30-MeV O^{+7}-induced, and 30-MeV O^{+8}-induced
spectra of Ne. For comparison H-like decays of oxygen
excited by a thin carbon foil are shown. The ordinate
is given in counts/μC of beam (after passing through the
gas cell). Two groups of eight vertical lines indicate
calculated positions of Kα transitions with single and
double K-shell vacancies (Kauffman et al. (2)).

log plot (faster than exponential). These effects can be qualitatively understood in terms of a variable fluorescence yield,
though the interpretation is not completely clear. An interesting
feature of the spectra is that the Ne target sometimes loses
almost all of its 10 electrons in a collision with Ar^{17+}, giving
rise to the possibility that several electrons transfer to the
highly-stripped argon ion producing strong Ar K x ray emission.
The spectra are shown in figure 40. Kauffman et al. (2) have
made high-resolution measurements of the Ne K satellite structure
in collisions of 30 MeV oxygen ions of specified charge state
with Ne gas, using a curved crystal x-ray spectrometer. These
data are shown in fig. 41. Variations in the fluorescence yields
and evidence of charge exchange between the highly-stripped
projectile and the target Ne atom are also indicated by these
data (92).

5. Summary

     This review has led us to a number of broad, somewhat impression-
istic conclusions concerning the excitation of inner-shell vacancies and the resultant x-ray emission in gas collisions:

     1. A variety of experimental techniques: measurements
of inelastic energy loss, electron spectra, x-ray spectra, total
cross sections, angular distributions, scattered ion x-ray/electron
coincidences and fluorescence yields, all provide complementary
methods for studying inner-shell excitation. Each method has its
peculiar advantages and limitations.

     2. The MO (or Fano-Lichten) electron-promotion model
provides a heuristic guide and often quantitative predictions
of vacancy production cross sections for many low-velocity collision systems. It does so in regions where the various Born and
semiclassical calculations break down, but there is no satisfactory
general theory for collisional x-ray production in many-electron
atoms at all velocities.

     3. Experiments which are differential with respect to
the angle of scattering and which determine what excited states
are produced are particularly useful in testing the MO promotion
model. This is because these experiments establish the impact-
parameter dependence of the excitation and therefore allow direct
comparison of the data with predictions of the MO model.

     4. Fluorescence yields for K shells are relatively in-
sensitive to multiple vacancies in outer (M and higher) shells;
however, for shells just inside the outermost shell, x-ray yields

can be quite sensitive to outer-shell configuration and therefore to chemical binding effects.

5. At high collision energies multiple outer-shell vacancies are the rule, not the exception, with heavy ion excitation.

6. Very large (geometric) x-ray emission cross sections are possible with heavy ion excitation; this is true even for low-Z atoms that normally have a small fluorescence yield, if they are in high states of ionization.

7. Many of the generalizations found with gas collisions can be expected to carry over into collisions with solid targets.

REFERENCES

1.   A. R. Knudson, D. J. Nagel, P. G. Burkhalter and
      K. L. Dunning, Phys. Rev. Letters 26, 1149 (1971).

2.   R. L. Kauffman, F. F. Hopkins, C. W. Woods, and P.
      Richard, Phys. Rev. Letters 31, 621 (1973).

3.   D. J. Volz and M. E. Rudd, Phys. Rev. A 2, 1395 (1970);
      M. E. Rudd in R. W. Fink, S. T. Manson, J. M. Palms,
      P. Venugopala Rao (eds.), Proceedings of the Intnl. Conf.
      on Inner-Shell Ionization Phenomena and Future Applica-
      tions, Atlanta 1972, U.S. Atomic Energy Commission,
      Technical Information Center, Oak Ridge, Tennessee,
      CONF-720404; p. 1489.

4.   Q. C. Kessel and B. Fastrup, in M.R.C. McDowell and
      E. W. McDaniel (eds.), Case Studies in Atomic Physics,
      Vol. 3, North-Holland Publ. Co., Amsterdam, 1973, p. 137.

5.   J. D. Garcia, R. J. Fortner and T. M. Kavanagh, Rev. Mod.
      Phys. 45, 111 (1973).

6.   P. Richard, in B. Crasemann (ed.), Atomic Inner-Shell
      Processes, to be published by Academic Press, N.Y. (1975).

7.   W. Bambynek, B. Crasemann, R. W. Fink, H.-U. Freund, H.
      Mark, C. D. Swift, R. E. Price, P. Venugopala Rao, Rev.
      Mod. Phys. 44, 716 (1972).

8.   N. A. Dyson, X-rays in Atomic and Nuclear Physics, Longman
      (London), 1971; p. 129.

9. U. Fano and W. Lichten, Phys. Rev. Letters 14, 627 (1965); W. Lichten, Phys. Rev. 164, 131 (1967).

10. M. Barat and W. Lichten, Phys. Rev. A 6, 211 (1972).

11. D. H. Madison and E. Merzbacher, in B. Crasemann (ed.), Atomic Inner-Shell Processes, op. cit.

12. E. Merzbacher and H. W. Lewis, in S. Flugge (ed.), Encyclopedia of Physics, Vol. 34, Springer, Berlin.

13. M. Inokuti, Rev. Mod. Phys. 43, 297 (1971).

14. J. D. Garcia, Phys. Rev. A 1, 280 (1970); Phys. Rev. A 4, 955 (1971).

15. J. H. McGuire and P. Richard, Phys. Rev. A 8, 1374 (1973).

16. D. L. Walters and C. P. Bhalla, Phys. Rev. A 3, 1919 (1971); C. P. Bhalla in B. C. Cobic and M. V. Kurepa (Eds.), Electronic and Atomic Collisions, Abstracts of Papers, VIII ICPEAC, Institute of Physics, Belgrade, (1973); p. 739.

17. G. Basbas, W. Brandt, R. Laubert, A. Ratkowski, and A. Schwarzschild, Phys. Rev. Letters 27, 171 (1971).

18. J. Bang and J. M. Hansteen, Mat. Fys. Medd. Dan. Vis. Selsk. 31, No. 13 (1959); J. M. Hansteen and O. P. Mosebekk, Z. Physik 234, 281 (1970).

19. W. Brandt, R. Laubert, and I. A. Sellin, Phys. Rev. 151, 56 (1966).

20. E. Merzbacher, in R. W. Fink, S. T. Manson, J. M. Palms, P. Venugopala Rao (Eds.), Proceedings of the Intnl. Conf. on Inner-Shell Ionization Phenomena and Future Applica- cations, Atlanta 1972, U.S. Atomic Energy Commission, Technical Information Center, Oak Ridge, Tennessee, CONF-720404; p. 928.

21. S. Datz, J. L. Duggan, L. C. Feldman, E. Laegsgaard and J. U. Andersen, Phys. Rev. A 9, 192 (1974).

22. B. Fricke, M. Rashid, P. Bertoncini and A. C. Wahl, Phys. Rev. Letters 34, 243 (1975).

23. W. Weizel and O. Beeck, Z. Physik 76, 250 (1932).

24.   L. Landau, Physik. Z. Sowjetunion 2, 46 (1932); E. C. G. Steuckelberg, Helv, Phys. Acta 5, 369 (1932); C. Zener, Proc. Roy. Soc. (London) A137, 696 (1932); see also D. R. Bates, ibid. A257, 22 (1960).

25.   Predissociation is discussed in the context of the Landau-Zener model in Landau and Lifshitz, Quantum Mechanics (Nonrelativistic theory), Addison-Wesley, Reading, Mass., 1965; pp. 322-330.

26.   Mott and Massey, The Theory of Atomic Collisions, 2nd edition (Oxford Univ. Press, 1952), p. 152.

27.   V. V. Afrosimov, Yu. S. Gordeev, M. N. Panov, and N. V. Fedorenko, Zh. Tech. Fiz. 34, 1613 (1964).

28.   Much of this work is reviewed in Q. C. Kessel, "Coincidence Measurements", in E. W. McDaniel and M. R. C. McDowell (eds.), Case Studies in Atomic Collision Physics I (North-Holland Publishing Co., Amsterdam, 1969).

29.   F. P. Ziemba and E. Everhart, Phys. Rev. Letters 2, 299 (1959).

30.   J. C. Slater, Quantum Theory of Molecules and Solids, Vol. 1, McGraw-Hill, New York, 1963.

31.   D. R. Bates, H. S. W. Massey, and A. L. Stewart, Proc. Roy. Soc. (London) A216, 437 (1953); F. P. Ziemba and A. Russek, Phys. Rev. 115, 922 (1959); W. Lichten, Phys. Rev. 131, 229 (1963).

32.   See translation of ref. 27 in Soviet Phys.--Tech. Phys. 9, 1248 (1965); ibid, p. 1256-1265; Q. C. Kessel, A. Russek, and E. Everhart, Phys. Rev. Letters 14, 484 (1965).

33.   H. Gabriel and K. Taulbjerg, Phys. Rev. A 10, 741 (1974).

34.   J. N. Bardsley, Phys. Rev. A3, 1317 (1971).

35.   D. R. Bates and D. A. Williams, Proc. Phys. Soc. (London) 83, 425 (1964).

36.   A. Russek, Phys. Rev. A 4, 1918 (1971).

37.   "Carambole" is a term for a double-collision in billiards.

38.   F. W. Saris, C. Foster, A. Langenberg and J. Van Eck, J. Phys. B: Atom. Molec. Phys. 7, 1494 (1974).

39.   F. W. Saris, W. F. van der Weg, H. Tawara and R. Laubert,
       Phys. Rev. Letters 28 717 (1972); P. H. Mokler, H. J.
       Stein, and P. Armbruster, Phys. Rev. Letters 29, 827
       (1972); J. R. MacDonald, M. D. Brown and T. Chiao,
       Phys. Rev. Letters 30, 471 (1973); W. E. Meyerhof et
       al., Phys. Rev. Letters 30, 1279 (1973); H. Kubo, VIII
       ICPEAC, p. 718 (1973); L. C. Feldman, Phys. Rev.
       Letters 32, 502 (1974); G. Bissinger and L. C. Feldman,
       Phys. Rev. Letters 33, 1 (1974); J. S. Greenberg, C. K.
       Davis, and P. Vincent, Phys. Rev. Letters 33, 473
       (1974); R. S. Thoe, I. A. Sellin, et. al., Phys. Rev.
       Letters 34, 64 (1975); and several other references.

40.   W. Lichten, Phys. Rev. A 9, 1458 (1974).

41.   E. Everhart, G. Stone and R. J. Carbone Phys. Rev. 99,
       1287 (1955); F. W. Bingham, J. Chem. Phys. 46, 2003,
       (1967). Extensive tables of R_o and other classical
       parameters calculated by Felton Bingham are available
       as Document No. SC-RR-66-506 (unpublished) from Clearing-
       house for Federal Scientific and Technical Information,
       National Bureau of Standards, U.S. Department of Com-
       merce, Springfield, Va., U.S.A.

42.   M. T. Robinson, Tables of Classical Scattering Integrals
       for the Bohr, Born-Mayer, and Thomas-Fermi Potentials,
       Oak Ridge National Laboratory (ORNL-3493, UC-34-Physics,
       TID-4500, 21st ed.; ORNL-4556, UC-34-Physics (1970)).

43.   Q. C. Kessel, Rev. Sci. Inst. 40, 68 (1969).

44.   Q. C. Kessel, Phys. Rev. A2, 1881 (1970).

45.   G. H. Morgan and E. Everhart, Phys. Rev. 128, 667
       (1962).

46.   B. Fastrup, G. Hermann and K. J. Smith, Phys. Rev. A3,
       1591 (1971).

47.   V. V. Afrosimov, Yu. S. Gordeev, A. M. Polyanskii and
       A. P. Shergin, Line Widths of Discrete Energy Losses in
       Violent Collisions of Atomic Particles. In: Sixth
       International Conference on the Physics of Electronic
       and Atomic Collisions - Abstracts of Papers (MIT Press,
       Cambridge,1969) pp. 744 - 747; V. V. Afrosimov, Yu. S.
       Gordeev, A.M. Polyanskii and A. P. Shergin, Zh. Eksp.
       Teor. Fiz. 57, 808 (1969) (English transl.: Soviet
       Phys. JETP 30, 441 (1970)).

48.   B. Fastrup, G. Hermann, Q. Kessel and A. Crone,
        Phys. Rev. A9, 2518 (1974): B. Fastrup, G. Hermann
        and Q. Kessel, Phys. Rev. Letters 27, 771, 1102 (1971).

49.   M. E. Rudd and J. H. Macek, Case Studies in Atomic Physics
        3, 47 (1972).

50.   M. E. Rudd, T. Jorgensen and D. J. Volz, Phys. Rev. 151,
        28 (1966).

51.   M. E. Cunningham, R. C. Der, R. J. Fortner, T. M. Kavanagh,
        J. M. Khan, C. B. Layne, E. J. Zaharis and J. D. Garcia,
        Phys. Rev. Letters 24, 931 (1970).

52.   F. P. Larkins, J. Phys. B: Atom. Molec. Phys. 4, 1
        (1971).

53.   R. J. Carbone, E. N. Fuls and E. Everhart, Phys. Rev. 102,
        1524 (1956).

54.   V. V. Afrosimov, Yu. S. Gordeev, M. N. Panov and N. V.
        Fedorenko, Zh. Tekh . Fiz. 34, 1613, 1624, 1637 (1964)
        (English Transl.: Soviet Phys.-Tech. Phys. 9, 1248, 1256
        (1965)).

55.   Q. Kessel, P. Rose and L. Grodzins, Phys. Rev. Letters
        22, 1031 (1969).

56.   G. Thomson, P. Laudieri, and E. Everhart, Phys. Rev.
        1439 (1970).

57.   R. S. Thoe and W. W. Smith, Phys. Rev. Letters 30, 525
        (1973).

58.   C. Snoek, R. Geballe, W. F. van de Weg. P. K. Rol and
        D. J. Bierman, Physica 31, 1553 (1965).

59.   D. R. Bates and D. A. Williams, Proc. Phys. Soc. (London)
        83, 425 (1964).

60.   M. P. McCaughey, E. J. Knystautas, H. C. Hayden and E.
        Everhart, Phys. Rev. Letters 21, 65 (1968).

61.   B. Fastrup, E. Boving, G. Larson and P. Dahl (to be pub-
        lished).

62.   J. S. Briggs and J. H. Macek, J. Phys. B 6, 982 (1973);
        see also: J. S. Briggs and J. H. Macek, J. Phys. B 5,
        579 (1972) and J. H. Macek and J. S. Briggs, J. Phys. B 6,
        841 (1973).

63. R. E. Olsen, Phys. Rev. A6, 1822 (1972); Yu. N. Demkov, Sov. Phys. JETP 18, 138 (1964); W. E. Meyerhof, S. M. Lazarus, W. A. Little, T. K. Saylor, B. B. Triplet and L. F. Chase, Bull. Am. Phys. Soc. 18, 559 (1973); W. E. Meyerhof, Phys. Rev. Letters 31, 1341 (1973); and N. Stolterfoht, P. Ziem and D. Ridder, J. Phys. B: Atom. Molec. Phys. 7, L409 (1974).

64. B. Fastrup and G. Hermann, Phys. Rev. Letters 23, 157 (1969).

65. R. K. Cacak, Q. C. Kessel and M. E. Rudd, Phys. Rev. A 2, 1327 (1970).

66. Saris and Onderdelinden, Physica 49, 441 (1970).

67. F. Larkins, J. Phys. B 4, L29 (1971).

68. C. P. Bhalla and D. L. Walters, in R. W. Fink et al. (eds.), Proceedings of the Intnl. Conf. on Inner-Shell Ionization Phenomena, Atlanta, Ga., 1972, (U.S. Atomic Energy Commission, Oak Ridge, Tenn., 1972), p. 1572.

69. M. E. Rudd, in R. W. Fink et al. (eds.), ibid., p. 1489.

70. R. J. Fortner, R. C. Der, T. M. Kavanagh and J. D. Garcia, J. Phys. B 5, L73 (1972).

71. C. P. Bhalla; private communication.

72. R. S. Thoe and W. W. Smith, Bull. Am. Phys. Soc. 18, 1507 (1973); R. S. Thoe, Ph.D. Thesis, University of Connecticut, unpublished (1973).

73. R. J. Fortner, J. Phys. B 7, L240 (1974).

74. G. M. Thomson, W. W. Smith and A. Russek, Phys. Rev. A 7, 168 (1973).

75. S. Sackmann, H. Lutz and J. Briggs, Phys. Rev. Letters 32, 805 (1974).

76. J. H. Stein, H. Lutz, P. Mokler, K. Sistemich and P. Armbruster, Phys. Rev. Letters 24, 701 (1970); Phys. Rev. A 2(2575) 1970; and H. J. Stein, H. O. Lutz, P. H. Mokler and P. Armbruster, Phys. Rev. A 5, 2126 (1972).

77. E. Laegsgaard, J. U. Andersen and L. C. Feldman, Phys. Rev. Letters 29, 1206 (1973).

78.   S. Bashkin, (ed.), Beam Foil Spectroscopy, Gordon and
        Breach, New York, 1968; S. Bashkin in P. G. Sandars
        (ed.), Atomic Physics 2, Plenum Press, N.Y. 1971, pp.
        43-63.

79.   S. Bashkin, in E. Wolf, ed., Progress in Optics,
        North-Holland, p. 289ff., (1974).

80.   I. A. Sellin, Bailey Donnally and C.Y. Fan, Phys. Rev.
        Letters 21, 717 (1968).

81.   I. A. Sellin, M. Brown, W. W. Smith and B. Donnally,
        Phys. Rev. A2, 1189 (1970).

82.   R. W. Schmieder and R. Marrus, Phys. Rev. Letters 25,
        1245, 1689 (1970); R. Marrus and R. W. Schmieder, Phys.
        Rev. A5, 1160 (1972).

83.   R. L. Balke, T. A. Chubb, H. Friedman and A. E. Unziker,
        Astrophysical Journal 142, 1 (1965).

84.   B. Edlen, Physics 13, 545 (1947); G. A. Sawyer, A. J.
        Bearden, I. Henins, F. C. Jahoda, and F. L. Ribe, Phys.
        Rev. 131, 1891 (1963); B. C. Fawcett, A. H. Gabriel,
        W. G. Griffin, B. B. Jones, and R. Wilson, Nature 200,
        1304 (1963); R. C. Elton and W. W. Koppendorfer, Phys.
        Rev. 160, 194 (1967); and H. J. Kunze, A. H. Gabriel,
        and H. R. Griem, Phys. Rev. 165, 267 (1968).

85.   B. Donnally, W. W. Smith, D. J. Pegg, M. Brown and
        I. A. Sellin, Phys. Rev. A 4, 122 (1971); I. A. Sellin,
        D. J. Pegg, M. Brown, W. W. Smith and B. Donnally, Phys.
        Rev. Letters 27, 1108 (1971).

86.   W. W. Smith, B. Donnally, D. J. Pegg, M. Brown and
        I. A. Sellin, Phys. Rev. A 7, 487 (1973).

87.   D. J. Pegg, P. M. Griffin, I. A. Sellin, W. W. Smith
        and B. Donnally, in Atomic Physics, Vol. 3, S. J. Smith
        and G. K. Walters, editors, Plenum Publishing Corp., N.Y.
        p. 327 ff, (1973); D. J. Pegg, I. A. Sellin, R. Peterson,
        J. R. Mowat, W. W. Smith, M. D. Brown and J. R. MacDonald,
        Phys. Rev. A 8, 1350 (1973).

88.   C. L. Cocke, B. Curnutte and J. R. Macdonald, Phys. Rev.
        Letters 28, 1233 (1972).

89.   D. J. Pegg, H. H. Haselton, M. D. Brown, R. S. Thoe, P. M. Griffin, I. A. Sellin and W. W. Smith, Bull. Ama. Phys. Soc. II 19, 1184 (1974).

90.   A. R. Knudson, D. J. Nagel, P. G. Burkhalter and K. T. Dunning, Phys. Rev. Lett. 26, 1149 (1971); D. Burch, P. Richard and R. L. Blake, Phys. Rev. Lett. 26, 1355 (1971).

91.   J. R. Mowat, I. A. Sellin, D. J. Pegg, R. S. Peterson, M. D. Brown and J. R. Macdonald, Phys. Rev. Letters 30, 1289 (1973).

92.   N. Stolterfoht, D. Schneider, P. Richard and R. L. Kauffman, Phys. Rev. Letters 33, 1418 (1974).

CHAPTER 8

ION-INDUCED X-RAYS IN SOLIDS

J. A. Cairns

AERE Harwell, Didcot, Berkshire, England

L. C. Feldman

Bell Laboratories, Murray Hill, New Jersey 07974 U.S.A.

1.   INTRODUCTION

Over the past few years, there has been a growing interest in the observation of X-rays resulting from the bombardment of solid targets by energetic protons and heavy ions. The reasons for this interest are both fundamental and applied. Thus, for example, the X-ray emission can provide basic information about the atomic collision processes which take place as an energetic projectile penetrates a solid; in addition, the X-ray production can be used as a versatile analytical tool.

The purpose of the present chapter is to provide a rather general insight into the subject. Thus, after a review of the basic equipment required to conduct studies in the field, we subdivide the projectiles into two main classes, viz. (a) protons and helium ions and (b) ions of larger mass, and consider both their relevant fundamental properties and some of their current applications. Naturally, it is impossible to deal exhaustively with the various topics to be mentioned here. Rather it is hoped that the reader will be inspired to wider reading, and thereby gain an understanding of the workings of the equipment and an appreciation of the fascinating diversity associated with this field of study.

At the time of writing, there are two excellent review articles on the subject.[1,2]

2.   ACCELERATORS AND TARGET CHAMBERS

Since this chapter is concerned with the bombardment of solid targets by energetic ions, it seems appropriate to begin by describing the basic characteristics of the ion accelerator machine and associated equipment. We may consider this as consisting of a series of discrete components, viz. a high voltage generator; an ion source; a magnet for analyzing the beam; a beam line containing both a focusing system and a series of deflectors and slits; and finally a target chamber. Figure 1 summarizes this: it shows these components on the Harwell 500 kV Cockcroft-Walton facility. Most of the studies which concern us in the present chapter have been conducted in the energy regime of 20 keV - 5 MeV; the higher energy beams coming from Van de Graaff accelerators.

Two of the components mentioned above which merit particular attention are the ion source and the target chamber.

2.1 Ion Sources

The subject of ion sources has become so varied and complex that it merits a special field of study by itself.[3] For the present purpose, ion sources may be subdivided into two types. First, there are those which provide ions from gases or gaseous compounds, and are indeed perfectly adequate for many analytical purposes, as well as for conducting many basic studies. On the other hand, if it is necessary to use beams of ions such as gold, platinum, iridium etc., which have no convenient gaseous compounds, then a second, more versatile source must be used. Figure 2 summarizes the mode of action of one type of such a source, the Hill-Nelson sputtering ion source.[4]

2.2 Target Chambers

In designing a target chamber, there are many features to be incorporated. Among the most important ones are the following: (i) Provision should be made for speedy insertion and removal of targets. (ii) There should be a variety of ports available to accommodate various X-ray detectors, vacuum gauges, target manipulator, etc., and spare ports should be provided for possible future requirements. (iii) The beam current measured should be independent of the target material or its temperature; this demands careful attention to secondary electron supression. (iv) Care must be taken to prevent hydrocarbon buildup on the target. The basic elements of a target chamber system are illustrated in Fig. 3, whilst Fig. 4 shows an alternative arrangement, in rather more detail.

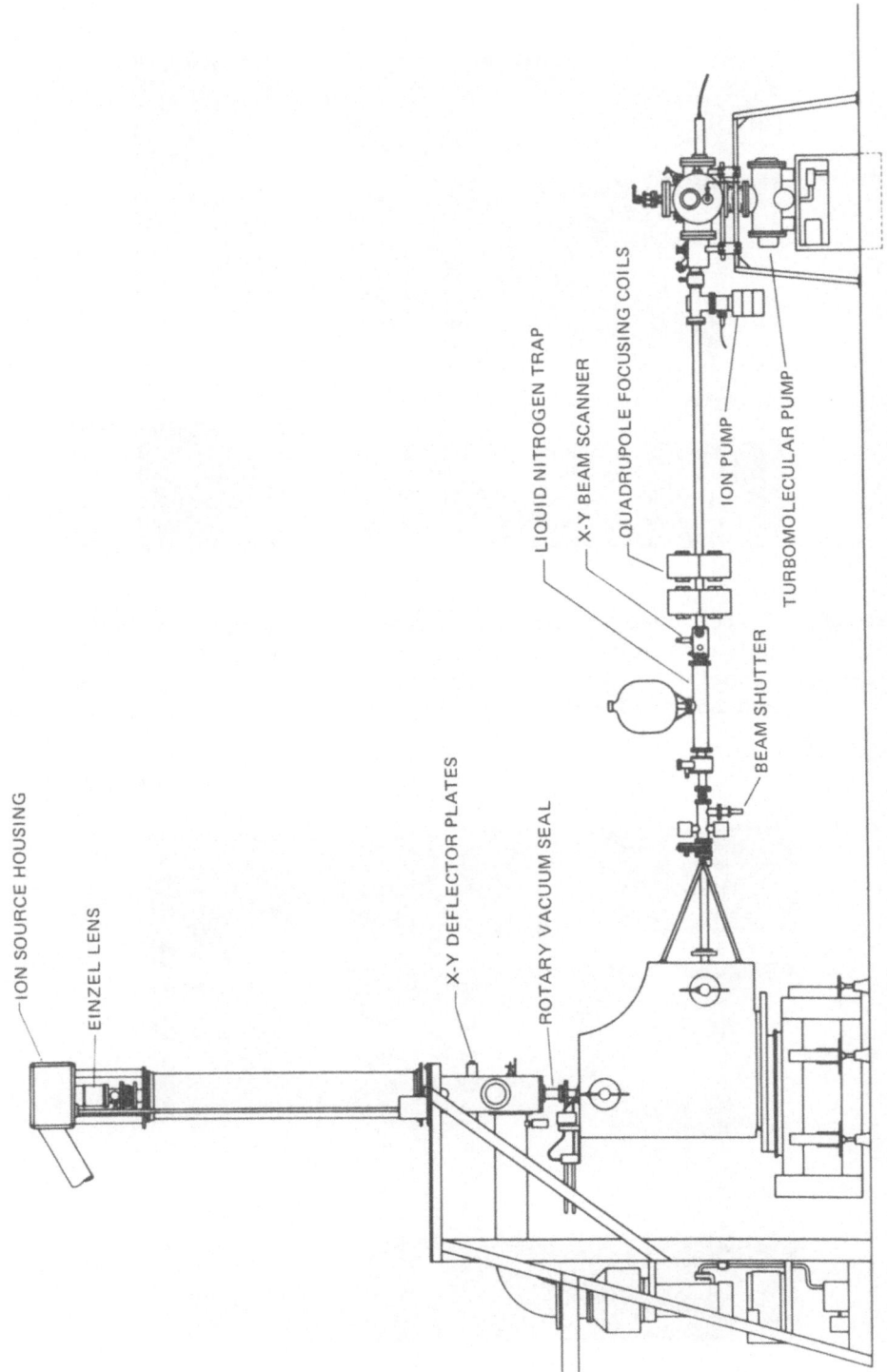

Fig. 1 500 kV Cockcroft-Walton beam line facility.

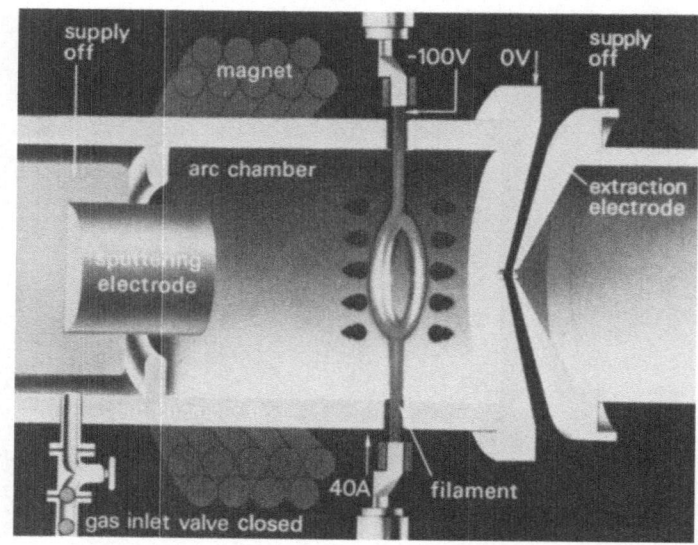

Figure 2/1   System evacuated to $\sim 10^{-6}$ Torr. Filament heated to
             emission level. Coil energised to produce axial
             magnetic field. Arc potential applied between filament
             and body. Electrons emitted from filament to body.

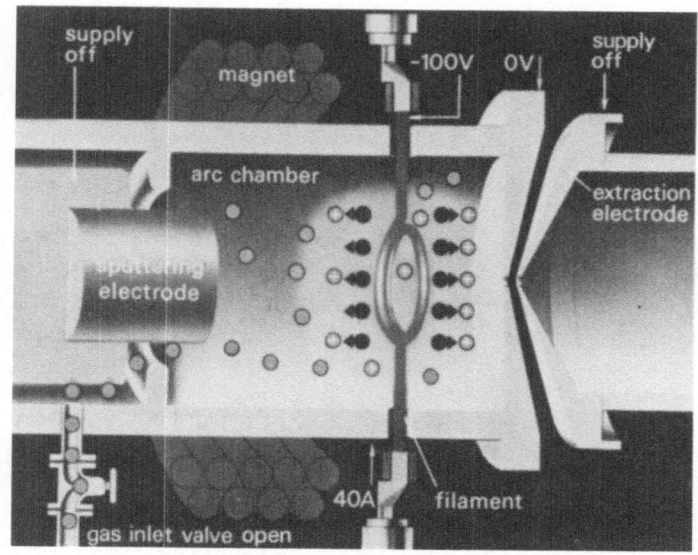

Figure 2/2   Gas valve opened.
             Gas ionised in vicinity of filament.

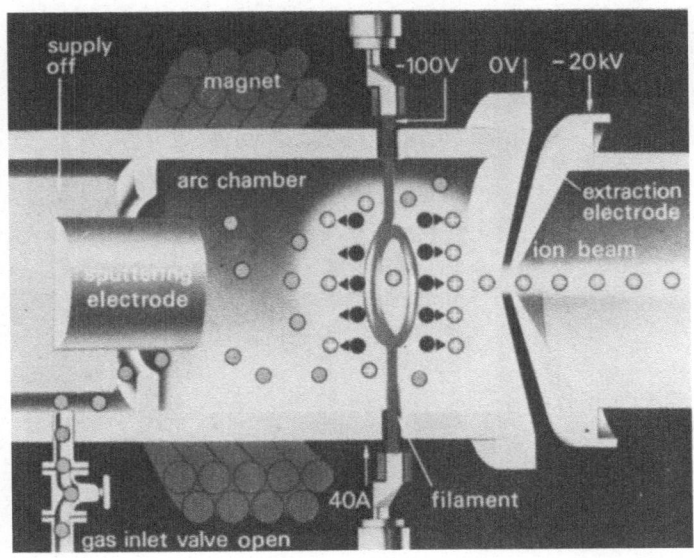

Figure 2/3   Extraction potential applied. Gas ions extracted and
accelerated. System running as gas ion source.

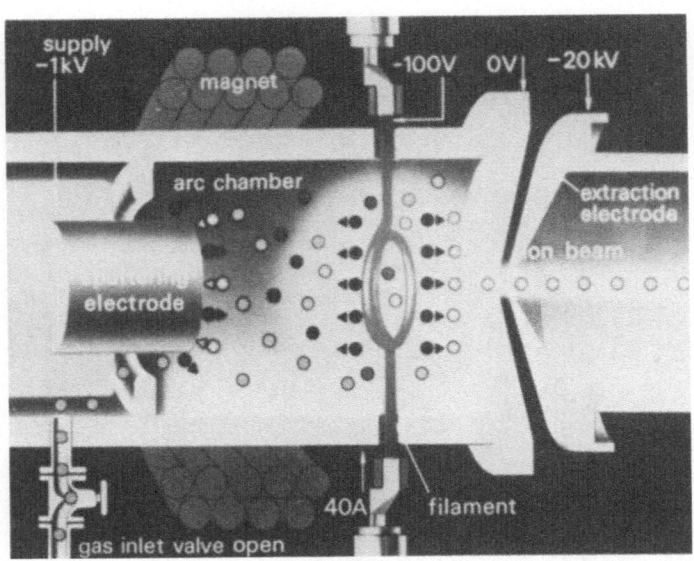

Figure 2/4   Potential of -1kV applied to sputtering electrode.
Positive gas ions accelerated towards sputtering
electrode releasing neutral atoms of metal.

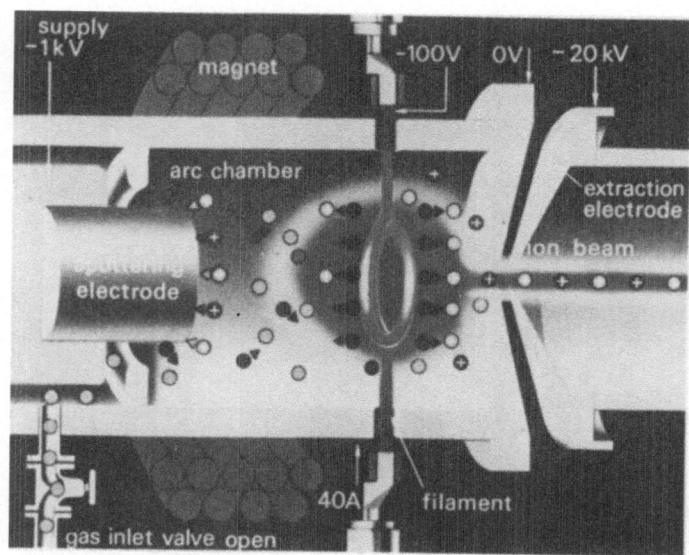

Figure 2/5   Metal atoms ionised. Mixture of positive metal ions
             and gas ions extracted. Mixture of positive metal ions
             and gas ions accelerated to sputtering electrode.
             Further release of neutral atoms of metal.

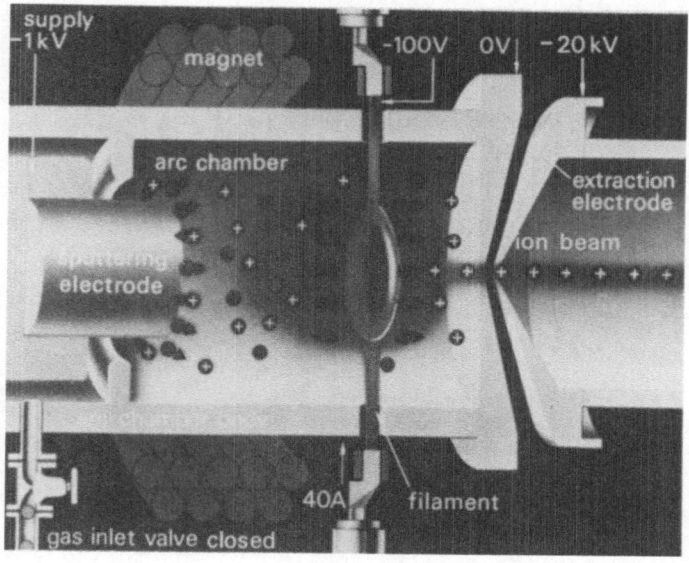

Figure 2/6   Gas valve closed. Arc maintained with metal ions.
             Metal ions extracted and accelerated.

3.   THE DETECTION AND ANALYSIS OF X-RAYS

In this section we consider the instruments used most commonly to detect X-radiation and to analyze its spectrum. The simplest of these is the gas flow proportional counter, which is relatively inexpensive and useful for soft X-rays, but lacks sufficient energy resolution to separate X-rays from elements of similar atomic number. On the other hand, the solid state detector has a much better inherent energy resolution, but often presents special difficulties when used to detect X-rays of energy less than ~500 eV, although modifications can be made to extend its energy range to include such soft X-rays. Finally, the crystal or grating wavelength spectrometer is the instrument best suited to provide the ultimate in X-ray resolution, particularly in the soft X-ray region, but has a much lower X-ray collection efficiency and does not provide an instantaneous spectrum of all the generated X-rays.

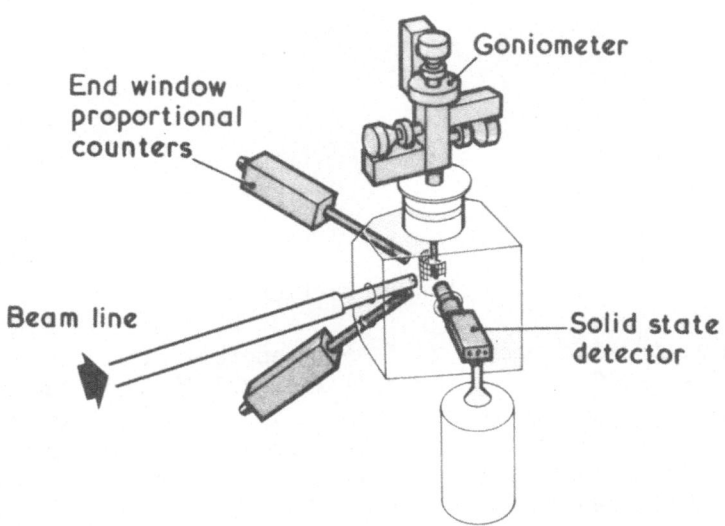

Fig. 3 Schematic illustration of the basic elements of a target chamber system.

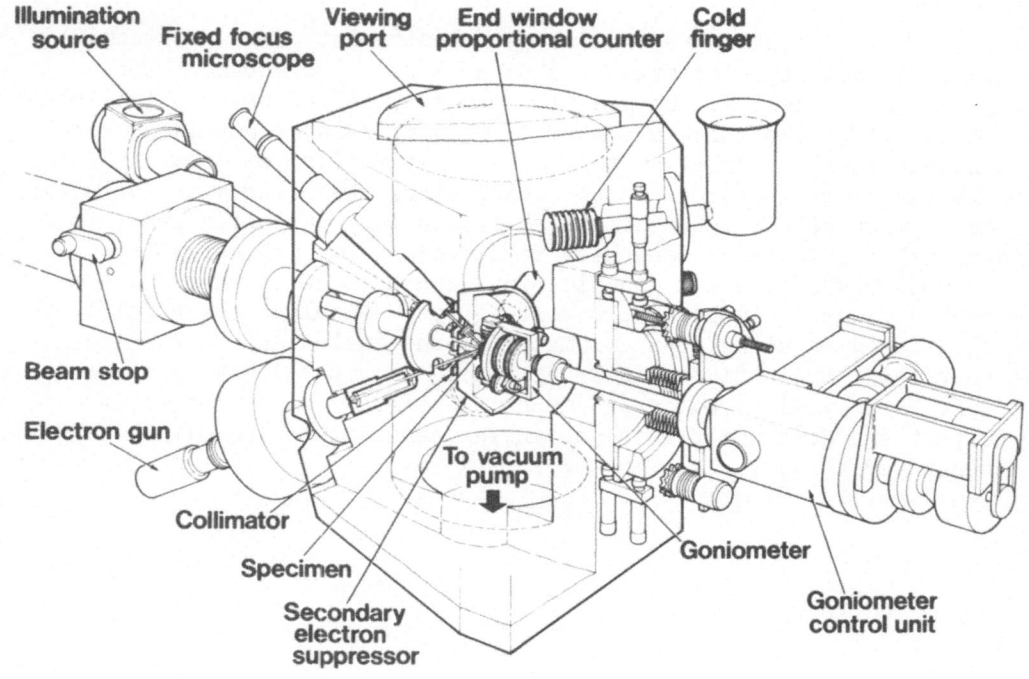

Fig. 4   Target chamber, designed for X-ray studies, incorporating
target goniometer system.

3.1   The Gas Flow Proportional Counter

This is the instrument used most generally for the detection
of soft (i.e. ∿100 eV to a few keV) X-rays. It works as follows:
incident X-rays pass through a thin window and cause ionization
of a gas (usually P10: 90% argon, 10% methane). The electrons
thereby produced then drift towards an anode wire, where they come
under the influence of an intense field, and produce further
ionization. The avalanche of electrons is collected on the wire,
whilst the remaining positive ion sheath drifts more slowly to the
earthed walls of the detector. The resultant signal is magnified
by a head amplifier, ideally mounted close to the anode in order

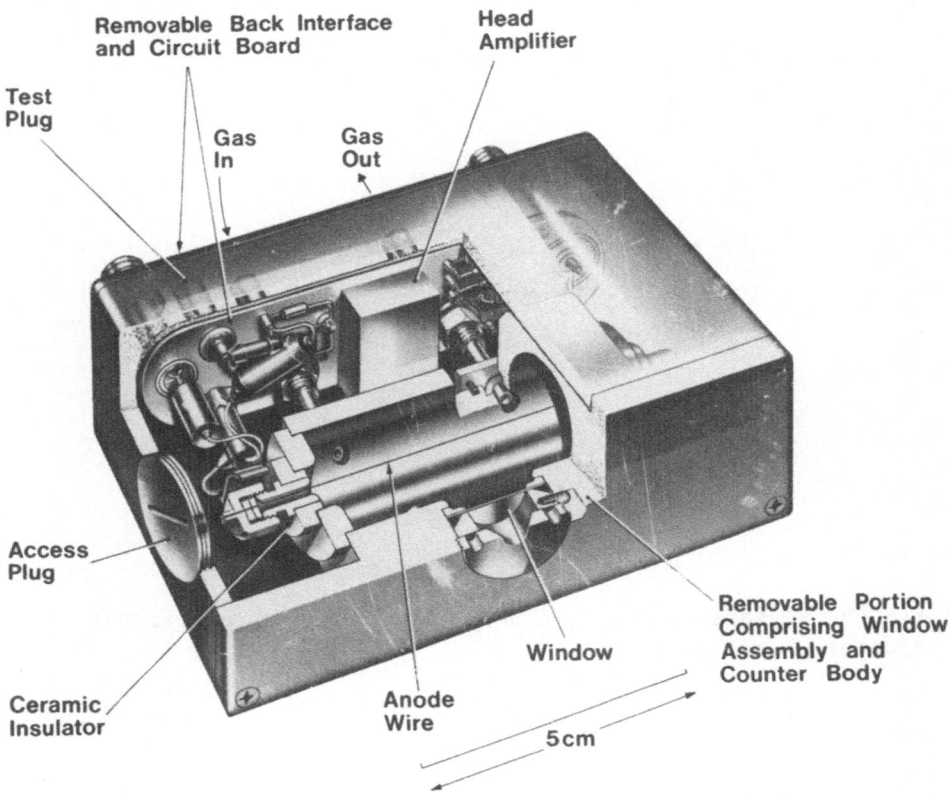

Fig. 5   Miniature side-window proportional counter (Ref. 5).

to minimize capacitance, and is then passed to a main amplifier
for further amplification and shaping. The detector is usually
constructed from a steel or brass cylinder having polished
internal walls, with an anode wire mounted coaxially. A window
is situated on the side of the cylinder, and usually has a con-
ductive coating to ensure uniformity of the internal electric
field. An example of a modern instrument is shown in Fig. 5.[5]
Since it is essential to minimize absorption of X-rays as they
enter the detector, the window is usually very thin and therefore
may not be completely gas tight over a long period of time. For
this reason, it is necessary to arrange for gas to flow through

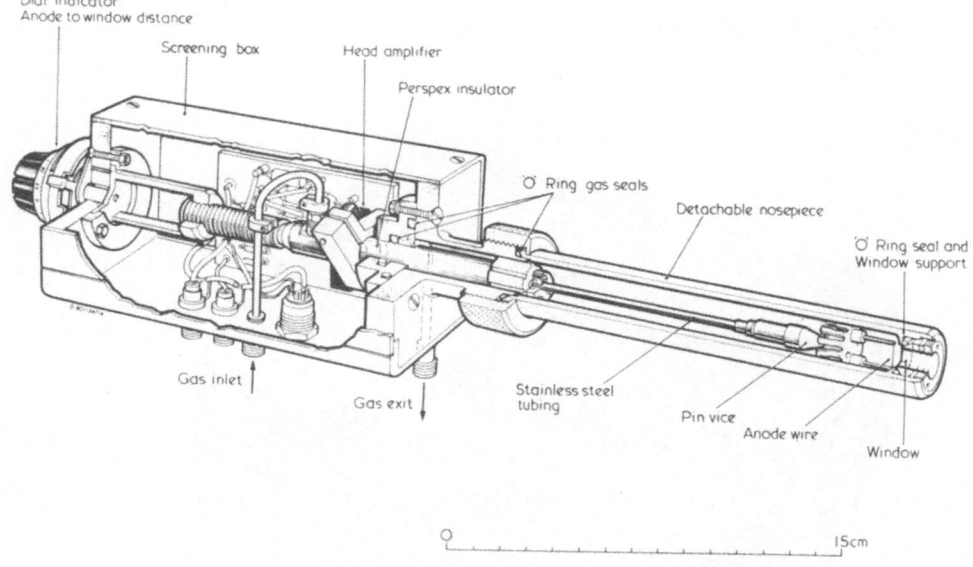

Fig. 6 Variable geometry end window gas flow proportional
counter (Ref. 9).

the detector constantly--hence its name. Typical window materials
used are 1 μm polypropylene[6] and parylene.[7] When measuring soft
X-ray yields quantitatively it is necessary to take into account
the extent to which they are absorbed by the window. This is
particularly serious if the X-ray energy is near to the absorption edge of the window. Tables of mass absorption coefficients
of the various popular window materials and flow gases are
available.[8] Gas flow proportional counters are sometimes used
non-dispersively--i.e. facing the source of X-rays directly,
without the aid of a diffraction system to improve the resolution.
For example, it may be necessary to collect as high a fraction of

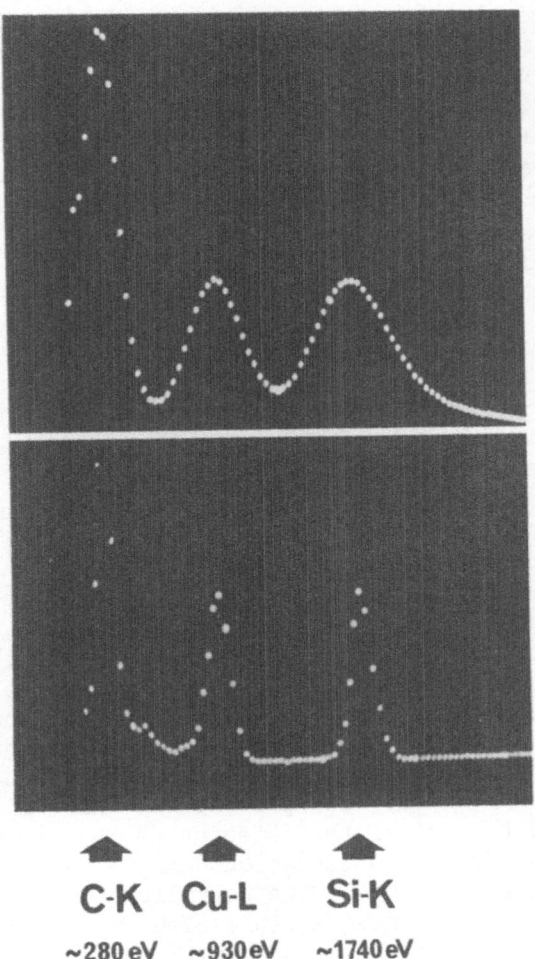

C-K   Cu-L   Si-K

~280 eV   ~930 eV   ~1740 eV

Figure 7   Contrast between the energy resolution exhibited by a gas
flow proportional counter (upper) and a Si(Li) detector,
the latter having a specially fitted 1µm polypropylene
window.

the X-rays as possible (e.g. to minimize radiation damage to the target). One way of doing this is to use an end-window proportional counter. An example of this type of instrument is shown in Fig. 6.

However, the biggest problem associated with the use of a gas flow proportional counter in a non-dispersive mode is its inherent lack of energy resolution: this is due to the relatively large energy (~25 eV) required to produce an ion pair in argon. Thus, Fig. 7[10] shows the contrast between the resolution exhibited by a gas flow proportional counter and a Si(Li) solid state detector, modified specially to detect soft X-rays as described below (3.2). As an alternative means of obtaining improved resolution, the gas flow detector may be combined with a reflecting element, such as a crystal or grating, to produce an X-ray crystal or grating spectrometer (3.3).

3.2   The Si(Li) Detector

The Si(Li) detector is a solid state device, consisting of of p-type single-crystalline silicon, carefully compensated with lithium as a donor, and capable of detecting a wide range of X-rays simultaneously, with good resolution. The improved resolution, as compared to the gas flow proportional counter described above, arises mainly because of the modest energy (~3.5 eV) required to produce an electron-hole pair in silicon. An example of the resolution obtained with a modern instrument is shown in Fig. 8. One complication is that the detector must be maintained at near 77°K, in order to prevent lithium redistribution, and to reduce background. Thus the detector must be maintained in an evacuated, gas-tight tube (to prevent water vapor condensing on to it), sealed with a thin window (usually beryllium). The presence of this window usually restricts the use of the detector to X-rays of energy greater than ~500 eV, but modifications can be made to insert a thin plastic window in place of the beryllium, or even to use it in a windowless form when mounted in a clean, evacuated target chamber. The resolutions shown in Figs. 7 and 8 were obtained with a Si(Li) detector having a 1 μm polypropylene window mounted in place of the beryllium window.

3.3   The X-Ray Crystal or Grating Spectrometer

This instrument separates X-rays according to their wavelengths by making use of Bragg's rule $n\lambda = 2d \sin \theta$. X-rays of wavelength λ are directed at an angle θ towards a reflector (which may be a crystal or a grating), having a distance d between two successive reflecting planes. The reflected X-rays are

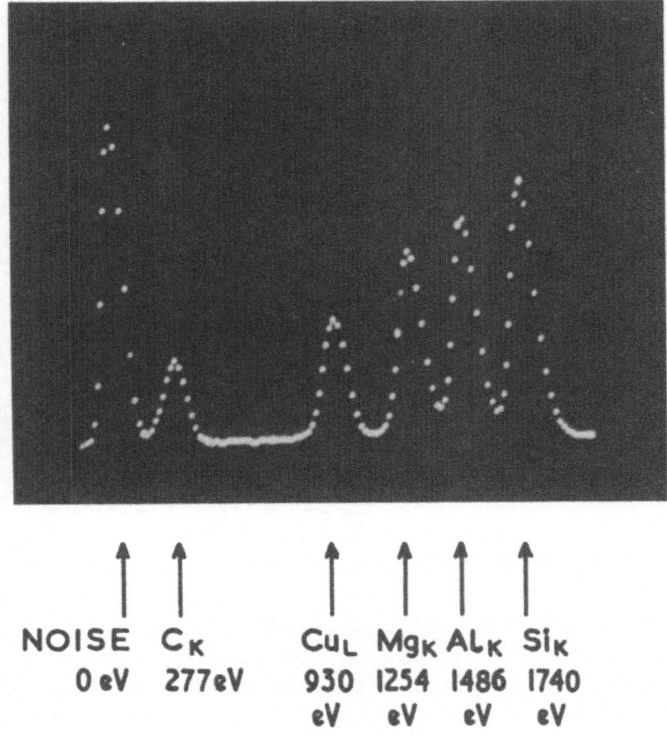

NOISE C_K   Cu_L Mg_K AL_K Si_K
0 eV   277eV   930   1254 1486 1740
                    eV   eV   eV   eV

Fig. 8 Further example of resolution obtained with a Si(Li) detector having a 1 μm polypropylene window.

detected (usually by a proportional counter) at a position corresponding to their particular wavelength. Thus although any reflector can cover a series of X-rays, by varying the angle θ, it is apparent from the Bragg equation that a reflector cannot deal with wavelengths greater than 2d. Hence it is necessary to use a series of X-ray reflectors. The most difficult region of the spectrum is the soft X-ray (long wavelength one, covering the range 10 - 100 Å). However, in recent years the situation has been improved by the development of multilayered soap films, such as lead stearate.

Some of the most popular X-ray reflectors are listed below, together with their 2d spacings.

Bragg Reflector	Full Name	2d Spacing ($\overset{\circ}{A}$)
PbSt	Lead Stearate	100
PbMy	Lead Myristate	80.5
OHM	Octadecyl Hydrogen Maleate	63.5
KAP	Potassium Acid (Hydrogen) Phosphate	26.63
Gypsum	Calcium Sulphate Dihydrate	15.19
ADP	Ammonium Dihydrogen Phosphate	10.65
EDDT	Ethylenediamine Ditartrate	8.808
PET	Pentaerythritol	8.742
NaCl	Sodium Chloride	5.614
LiF(200)	Lithium Fluoride (200)	4.028
LiF(220)	Lithium Fluoride (220)	2.848
Topaz	Aluminum Fluosilicate (hydrated)	2.712

X-ray reflectors are used generally in two forms: flat and curved. Flat Bragg reflectors yield good resolution, but they require a parallel X-ray beam; thus the X-ray emerging from the target must be collimated carefully, with the result that only a very small fraction actually enters the spectrometer. In order to improve intensity, a curved reflector may be used. This reflector can handle divergent beams of X-rays, and focus them down to a line, which is arranged to coincide with the position of the detector. One of the most commonly used geometrical conditions is the Johansson arrangement, shown in Fig. 9.

An example of the resolving power of a flat crystal spectrometer is shown in Fig. 10.[11]

4. THE USE OF PROTONS AND HELIUM IONS TO GENERATE X-RAYS FROM SOLID TARGETS

A fast charged particle moving by an atom creates a time dependent electric field which can excite or ionize electrons of the atom. The mathematical formalisms for this process has already been discussed in the Chapter "Ion Induced X-Rays from Gas Collisions". Three theoretical treatments that have been particularly successful in describing this process are: 1) the plane wave born approximation described by Merzbacher and Lewis;[12] 2) the semi-classical time dependent perturbation theory of Bang, Hansteen and Moseþekk;[13] and 3) the binary encounter approximation (BEA) of Garcia.[14] The range of validity of these treatments as well as their ability to successfully predict the observed phenomena has been discussed previously.[2] However, it is worth

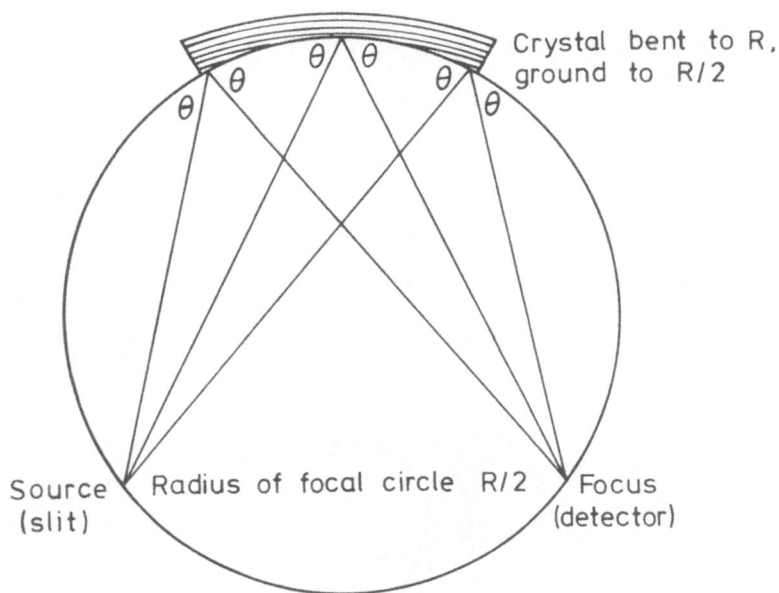

Fig. 9 Johansson focusing arrangement.

pointing out that each treatment provides different approaches and insights to the phenomena involved and they should not be considered redundant.

Because the quantitative understanding of the fundamental process is relatively well understood much of the recent work with proton and helium ion generated X-rays has been with specific applications in mind. In this section we shall first make some remarks concerning proton and helium excitation with respect to some fundamental areas that are currently receiving attention. Then we shall consider a number of specific applications. In light of the preceding chapter on collisions in gases, we should note that the solid state of the target makes little difference in most of the inelastic processes concerning protons or helium ions; this will not be the case in the following section on heavy ion induced X-rays.

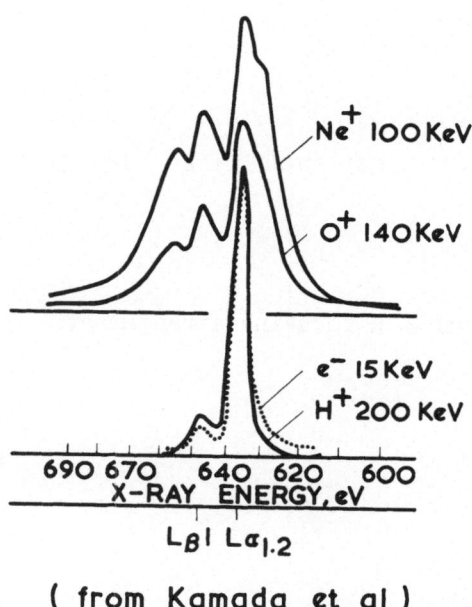

(from Kamada et al)

Fig. 10 Manganese L X-rays, generated by bombardment of a
manganese target by electrons, protons, oxygen ions and neon ions.

4.1 Current Areas of Fundamental Interest

There are a number of areas in the fundamental interaction
mechanism of protons and helium ions which is currently of considerable interest. Although most of these questions can be
explored with either gas or solid (thin) targets the solid target
is often more convenient. In some cases, the comparison of the
gas and solid target is vital in order to separate the effects of
the recoil environment. Finally, some of the following cases are
interesting in order to acquire data in particular regions where
theory is unreliable. As we shall see from the next section, it
is just this region that is often the most useful in solid state
applications.

a) <u>Low-Energy Cross Sections</u>. In a large number of problems
in solids, one is interested in exploring the surface region or
near surface region of a solid. In order to maximize the yield of
X-rays from the surface region relative to the bulk, it becomes
necessary to reduce the energy of the projectile consistent with
the stopping power and yield requirements. In the case of He^+, the
region around 1.0 MeV is particularly useful since the stopping
power is a maximum. In this region, the Coulomb deflection effects
discussed by Bang and Hansteen[13] and the binding effect discussed by
Brandt[15] become particularly important in the modification of the
simpler theoretical predictions. Furthermore, in this region
reliable experimental data are particularly scarce. For example,
the tables of Rutledge and Watson[16] do not show any data below
.25 MeV/AMU for elements with $Z_2 > 13$ in their ^{4}He-K-shell ionization tabulations. Thus fundamental work in this area is still
necessary.

b) <u>Coincidence Measurements</u>. In recent years a number of
particle-K-X-ray coincidence measurements have been made to better
test the theories of K-shell excitation.[17] Such measurements are
almost always done on solid targets for obvious experimental
reasons. In general, the theoretical predictions are confirmed.
However, recently Chemin[18] et al. have shown discrepancies with
theory of up to a factor of two for very small impact parameter
events. Since events at small impact parameters do not make a
large contribution to the total cross section, this effect would
not be readily noticed in such cross section measurements. However,
these detailed measurements do provide stringent tests of theory.
There are not any published results for the impact parameter
dependence of inner-shell ionization by He ions.

c) <u>L-Subshell Ionization</u>. Ionization of the L-shell and the
subsequent X-ray emission provide new tests and lead to further
insights of the ionization mechanism. For example, Datz[19] et al.
have explored the L-shell ionization of Au by protons and He ions.

The energy dependence of the cross sections for the different
subshells shows structure which can be related to the spatial
properties of the 2p and 2s wave functions. An interesting and
not yet understood result of this study is the difference in the
ratio of the $2p_{3/2}$ to $2p_{1/2}$ ionization cross sections for protons
and helium ions at the same velocity. Coincidence measurements,
as described above, should yield further information on this
interesting subject.

d) <u>Satellite Structure</u>. Satellites corresponding to additional L-vacancies at the time of K-shell emission are frequently
observed in high resolution X-ray spectra. Such spectra have been
discussed in the previous chapter. The intensity of these satellites has been treated within the Coulomb ionization theories by
taking the K-shell and L-shell ionization independently, with reasonable success for protons and helium ions. With respect to
solids it is worth noting that those satellites are strongly
enhanced for He excitation and structure within a satellite may be
indicative of chemical binding effects. Burkhalter et al.[20] have
demonstrated this for the case of the aluminum Kα spectrum in
aluminum metal and aluminum oxide.

e) <u>Chemical Effects</u>. Chemical effects have been observed in
high resolution X-ray spectra for a long time. For the case of
protons these effects are very similar to those observed for
electronic excitation, effects which have been well documented.
However, for He[+] ions the X-ray spectra appear quite different and
are sensitive to the chemical environment in a more complex way.[20]
Such investigations are ripe for detailed measurements and theoretical understanding.

4.2 Applications

As we have already indicated our understanding of proton and
helium induced X-ray excitation is relatively good. The most
explicit evidence being the agreement between measurement and
theory for total cross sections. This has been discussed in the
previous chapter. As so often happens when phenomena become
understood, the physicist seeks applications. In the next sections
we shall describe some applications of proton and helium induced
X-rays to the study of solid targets.

The classical method of creating inner shell vacancies and
thence subsequently producing X-rays from solid targets is to
subject the targets to bombardment by energetic electrons, as
was done by Roentgen when he observed X-rays for the first time
in 1895. This is still by far the most common method of X-ray
production. We may then ask: why should an alternative method
be sought? One of the motivations for this is that X-rays pro-

duced by electron bombardment are not 'clean'; they are accom-
panied by the simultaneous production of continuous, 'white'
electromagnetic radiation, known as bremsstrahlung (or braking
radiation), i.e. radiation emitted as the electron loses energy
within the target. The presence of such radiation is particularly
troublesome when attempting to generate characteristic X-radiation
from light elements, as their radiation, being inherently of low
energy, tends to become swamped by the background. Here arises one
of the attractions of using protons: the tendency to produce
bremsstrahlung radiation is greatly reduced. Indeed Folkman
et al.[21], having taken into account the background radiation
which <u>can</u> arise during ion bombardment, both from proton and
secondary electron bremsstrahlung, conclude that this method of
X-ray production is about three orders of magnitude cleaner than
can be obtained by electron bombardment.

    a) <u>The Proton Microprobe</u>. This application seeks to take
advantage of the improvement in sensitivity which results from
the use of protons for X-ray production, as compared to electrons.
Poole[22] suggested that this could form the basis for an instrument which would incorporate the additional feature of having the
proton beam focused to a fine spot, thereby combining analytical
sensitivity and spatial resolution. An indication of the capabilities of this instrument may be appreciated from the work of
Cookson and Pilling.[23]

    b) <u>Environmental Analysis</u>. The spectacular extension of ion
induced X-rays to environmental analysis by Cahill and his co-
workers[24] illustrates well the inherent attractive features of the
technique. That is, it is non-destructive; can be applied to a
broad range of elements with relatively uniform minimum detectable
limits; and is applicable to very small sample amounts with good
sensitivity. In addition, assuming the availability of a suitable
accelerator, the experimental equipment comprises simply a target
holder/changer system and a Si(Li) detector. Cahill has demonstrated that such a facility can be engineered to ensure reproducible specimen analysis, with rapid throughput. A complete
discussion of this work is described in another chapter in this
book.

    c) <u>The Combination of Helium Backscattering and Helium
Induced X-Rays</u>. The use of helium backscattering as a quantitative,
depth-sensitive technique for near-surface analysis is now well
established.[25] (See the appropriate chapter in this book.) In
such work the energy of a backscattered particle is indicative of
the mass of the element from which it scattered and the depth into
the solid from which it scattered. Thus there is an intrinsic ambiguity. This ambiguity is serious for materials containing ele-

ments of similar mass or when detecting light impurities in a heavy
host. However, by using helium induced X-rays, it is possible to
overcome these shortcomings. A conventional Si(Li) detector can
distinguish X-rays from any two adjacent elements in the
periodic chart.

This is illustrated by the work of Feldman et al.[26] who
studied anodically grown oxide films on GaAs. In this case it
was desired to understand the stoichiometry of the surface region
as a function of temperature, a problem that could not be
resolved by backscattering because of the close values of the Ga
and As masses. Using the grazing incidence geometry shown in
Fig. 11, it was possible to confine the X-ray excitation to a
well-defined region of the near surface and vary that depth by
adjusting the angle.

The direct X-ray spectra easily showed that the surface
region became Ga rich upon heat treatment as shown in Fig. 12.
By a continuous variation of the incident angle, it is possible
to get a profile of the film.

d) The Detection of Trace Amounts of Elements on Solid
Surfaces. It will be apparent from the preceding remarks that
the use of hydrogen and helium ion induced X-rays is well suited
to the detection of elements on surfaces, particularly when in few
monolayer to sub-monolayer amounts. Under these conditions, the
X-ray yields can be related quantitatively to the surface element
thickness. Thus Musket and Bauer have studied the surface
chromium depletion of stainless steel on vacuum annealing,[27] and
determined quantitatively single phase oxide thicknesses on
metals.[28] The technique is discussed in some detail[29] and compared to the other analytical techniques which arise also from
proton bombardment, viz., Rutherford backscattering and Auger
electron emission.

e) Lattice Location. As is described in another chapter in
this book, the technique of lattice location using channeling is
an important particle-solid application to solid-state physics.
Such studies are often done in the backscattering mode; however,
for impurities lighter than the host, other techniques are
required. In a number of cases the characteristic X-rays have
been used with considerable success.[30] Corrections have to be
made for the modification of the energy loss in channeling directions and various absorption effects. Development of such X-ray
techniques can be useful for a variety of elements in a given
host and thus allow tests of the various parameters which determine lattice location in a region of Z_1 and Z_2 not very much
explored. For example, MacDonald et al.[31] have recently deter-

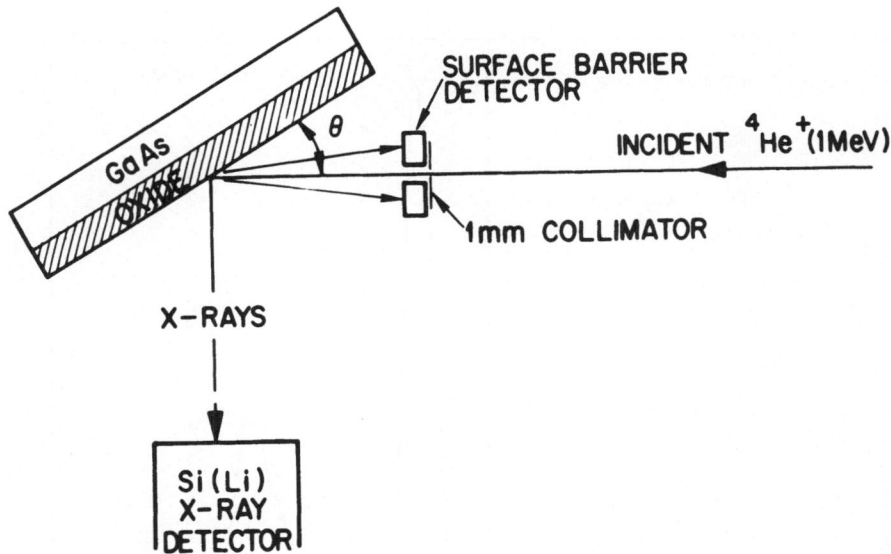

Fig. 11 Schematic diagram of the apparatus for combined back-scattering and X-ray excitation. The Si(Li) X-ray detector was mounted 5 cm from the target at 90° to the beam.

mined the location of K, Ca, and Ar implanted into Fe. Their interest was motivated by considerations of internal magnetic fields observed through the hyperfine interactions of nuclear excited state dipole moments with the internal magnetic field of Fe. These internal fields are sensitive to the final site of the implanted species and proper interpretation requires knowledge of this site.

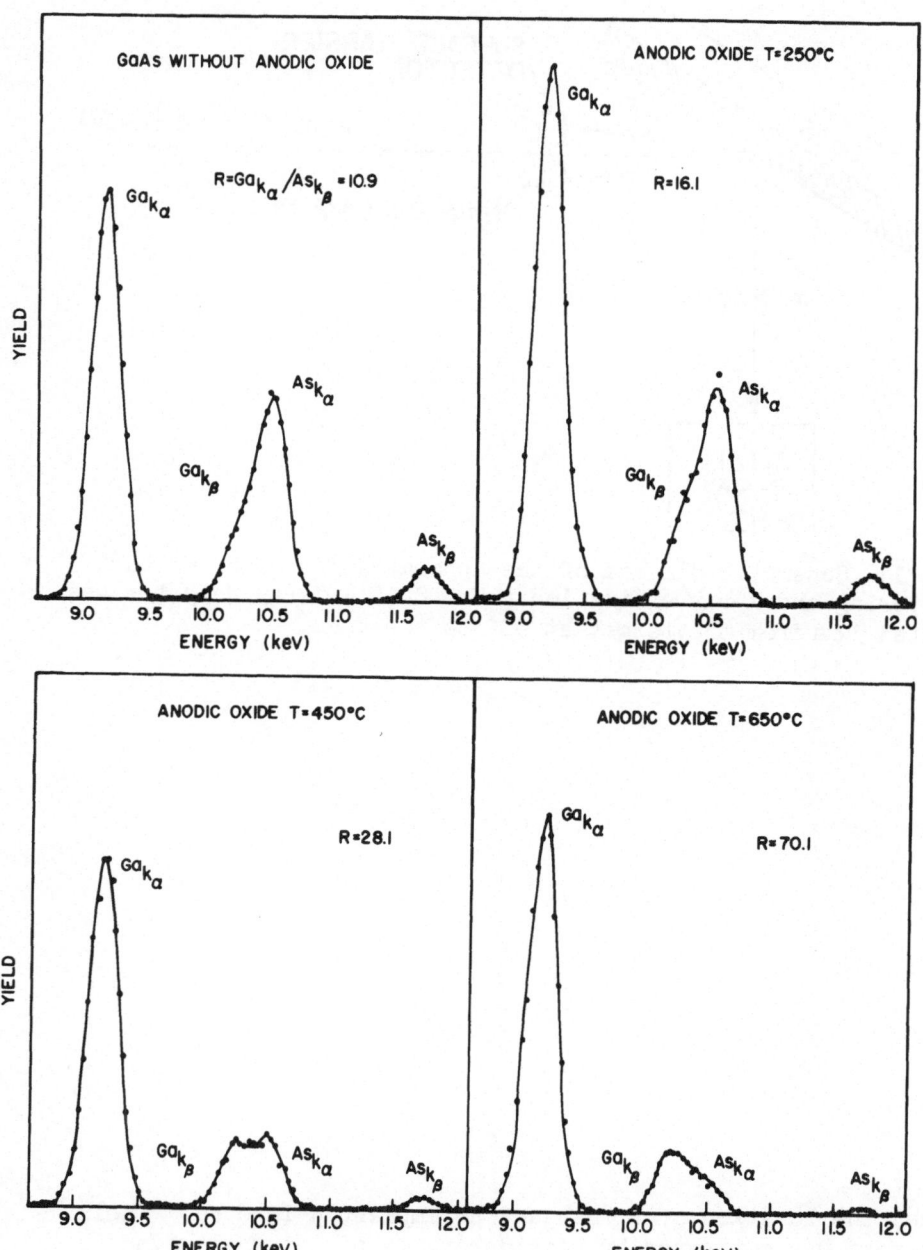

Fig. 12 X-ray spectra for GaAs without an oxide film and for anodic oxides as a function of annealing temperature. The samples were mounted at 20° grazing angle with an incident He⁺ beam energy of 1.0 MeV.

5. THE USE OF HEAVY IONS TO GENERATE X-RAYS FROM SOLID TARGETS

5.1   General Background

One of the initial motivations for the interest in using
heavy ion bombardment for X-ray generation was the observation
that when inert gas ions, ranging in energy from ~20 keV to a few
hundred keV, were used to bombard a series of targets, the
characteristic X-ray yields were much higher than would have been
expected for electron ejection by direct Coulomb interaction.
This led to the widespread acceptance of the idea that since the
projectile velocity was much slower than the electronic orbital
velocity, the electronic systems of target and projectile could
adjust during the collision and so their interaction could be
considered in terms of the creation of a pseudomolecule.[32]
Electrons from the target and projectile could then be assigned
into particular energy levels, according to fairly well defined
rules.[33] However, since the interacting nuclei were in a state
of relative motion, the molecular orbitals were non-adiabatic, or
diabatic. This means that the non-crossing rules are no longer
valid, and so it is possible for an electron to be promoted, at a
minimum distance of closest approach, from an inner shell to an
outer one, thus creating an inner shell vacancy, which persists
when the target and projectile separate after the collision. This
accounts for another observation: X-rays from target or projectile
often arise rather suddenly at a certain minimum projectile energy,
and thereafter the X-ray yield increases very steeply as a function
of projectile energy.[34] These processes are described in detail
in the preceding section of this book.

We may now begin to appreciate why X-ray production from
heavy ion/solid interactions is so much more complicated than
when light ions are used. For example,

(i) We must first construct the pseudomolecular diagram
(usually called the correlation diagram) for each target/projectile
system, in order to anticipate which X-rays are likely to arise.

(ii) We cannot predict quantitative X-ray yields very easily
because the theory is not so well developed.

On the other hand, a whole new series of possibilities is
opened up, both for studying the physics of atomic collisions and
for analytical purposes. Some examples of each of these types may
now be considered.

5.2 Physical Processes

Although much of the phenomena described above can be studied with targets in the gas phase, a number of new and interesting effects are observed in the X-ray spectrum generated by heavy ion bombardment of solids. In some cases these effects arise because the cross section for ionization is so high and the lifetime of inner-shell vacancies long enough that the incident beam traveling through a solid is in a highly ionized state. It interacts with the atoms of the solid in configurations not available in the gas phase. Other heavy-ion effects of current interest center about problems concerning the charge states of ions traveling through solids and the radiative electron capture (R.E.C.)[35] effect which samples the momentum distribution of electrons in a solid.

a) <u>Double-Scattering Mechanisms--Characteristic Lines</u>. In a number of cases in solids, characteristic X-ray lines are observed which are not predicted on the basis of the promotion mechanism and molecular orbitals.[36] For example, in the case of the $\sim$300 keV bombardment of solid Si with Ar^+, a strong yield of Si-K is observed, although apparently forbidden on the basis of the Fano-Lichten promotion scheme since the 1s level does not correlate with a vacant level. A number of authors have shown that this results from a double scattering in which an Ar 2p vacancy is created in one collision and lives long enough to provide a vacant level to the 1s-orbital of silicon in a second event. The understanding of such phenomena provides information on the atomic state of a projectile traveling through a solid. It also provides a means of measuring the lifetime of a vacancy on a moving ion as the Si-K yield is directly dependent on this parameter. It should be pointed out that an additional contribution to the yield in the Ar-Si case comes from Si-Si recoil events. The importance of this recoil effect has been demonstrated by Taulbjerg et al.[37] for the case of Ar on Al.

b) <u>Swapping</u>. As a heavy ion travels through a solid, a number of outer electrons may be stripped off. This, of course, decreases the screening, increasing the binding energy of inner shell states. The appropriate correlation diagrams are not that generated from the isolated atom but the one appropriate for the new configuration. In some cases, this change in binding energy can cause an important swapping of levels, opening excitation channels which did not previously exist. Examples of the swapping phenomena are discussed in Ref. 37. Obviously such effects shed light on the atomic state question as well.

c) <u>Non-Characteristic or Molecular X-Rays</u>. Sometimes during heavy ion/solid collisions, X-rays are observed which cannot be identified as belonging to either the projectile or the target. A particularly striking example of this arises during the

bombardment of silicon by Ar^+ ions of a few hundred keV. The
explanation for the effect, first given by Saris et al.,[38] was
that these X-rays were produced during the lifetime of the pseudomolecule. The process can be envisaged as follows: when an
argon projectile strikes the target, a vacancy is formed in the
projectile 2p shell. This vacancy exists long enough to be
carried into a subsequent collision (either with a silicon atom
or another implanted argon atom) and it may then be filled by an
electron from an outer level, with consequent X-ray emission
during the existence of the pseudomolecule. Naturally the X-ray
thereby produced is rather broad because it is emitted during the
time the two nuclei are in relative motion to each other and so
the binding energy of the orbital containing the vacancy is grad-
ually changing. However, its end-point energy gives an indication
of this binding energy at the distance of closest approach of the
nuclei. Thus studies have been made[39] of this end point for
various projectiles and as a function of energy, in order to make
some attempt to 'map out' the appropriate molecular orbital
energy levels.

Figures 13a and 13b show an example of this phenomena for
different projectiles. The end-point energy of the broad NCR
(non-characteristic radiation) moves to higher energy with
increasing Z_1 corresponding to the higher binding energy of the
pseudomolecule, Z_1+Z_2. The end-point energy as a function of Z_1
is shown in Fig. 13b. These end-point energies correspond closely
to that expected on the basis of the correlation diagrams.

d) <u>Atomic States in Solids</u>. As we have already stated the
production of X-rays in heavy-ion solid interactions yields important information on the atomic state of ions in solids. One of
the most complete studies at low energies is that of Fortner
et al.[40] By means of X-ray spectral measurements of S, Cl and Ar
moving through solid carbon targets these authors extract the
equilibrium distribution of vacancies in the valence and L-shells
of the projectiles. They conclude that the state of excitation of
these low energy ions ($\sim$90 keV) is extremely high while traveling
in the solid, and the ion leaves the foil essentially electrically
neutral. The so-called emergent charge state is formed just outside of the foil by Auger and autoionization processes.

At higher energies (2.5 MeV/AMU) the Oak Ridge group[41] has
studied the charge state dependence of Si-K X-ray production by
O^{+6}, O^{+7} and O^{+8} in gaseous SiH_4 and solid silicon targets. In
the gas case, the X-ray cross section shows a large charge state
dependence. Using this as a calibration, they "determine" the
charge state of the ion in the solid. From this comparison they
conclude that the effective charge state for X-ray production in
solid Si is in agreement with that anticipated on the basis of
emergent charge state distributions.

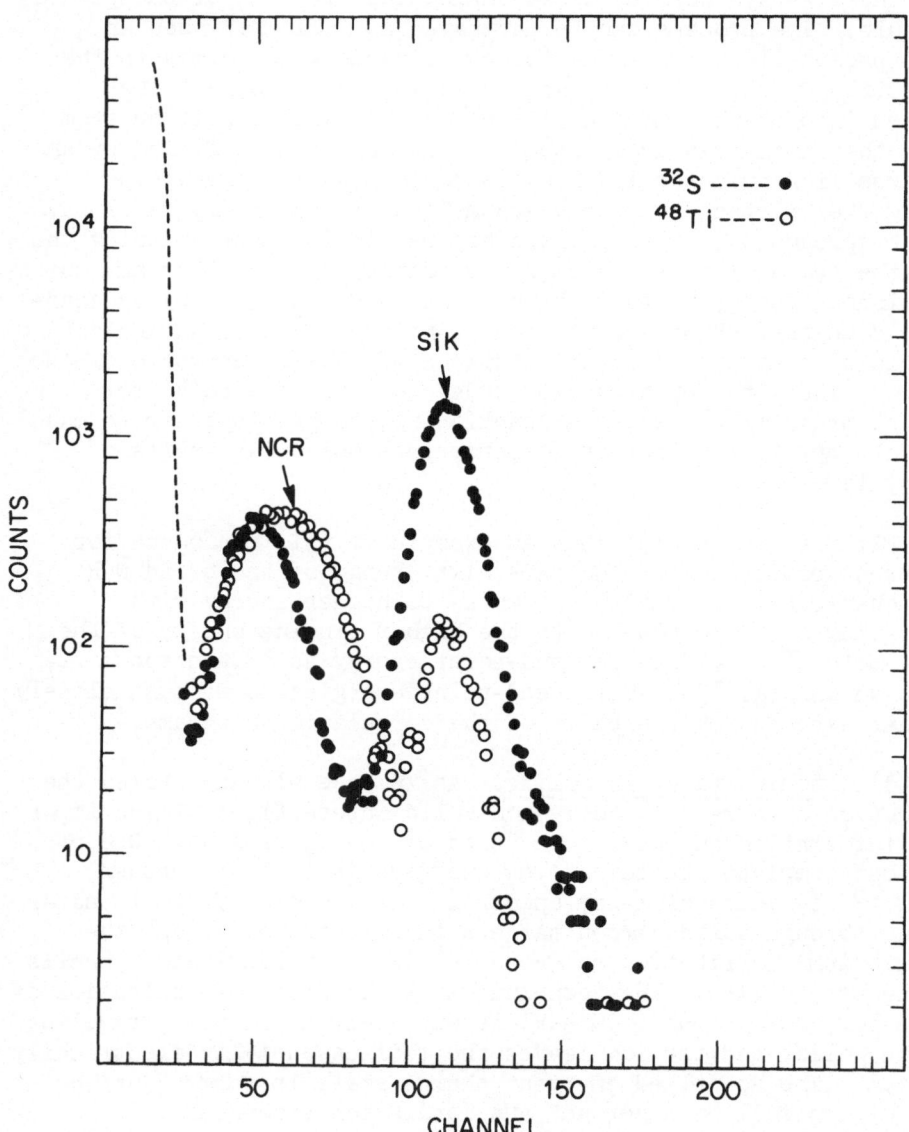

Fig. 13a X-ray spectrum from the bombardment of thick Si targets with 200 keV^{32}S and ^{48}Ti.

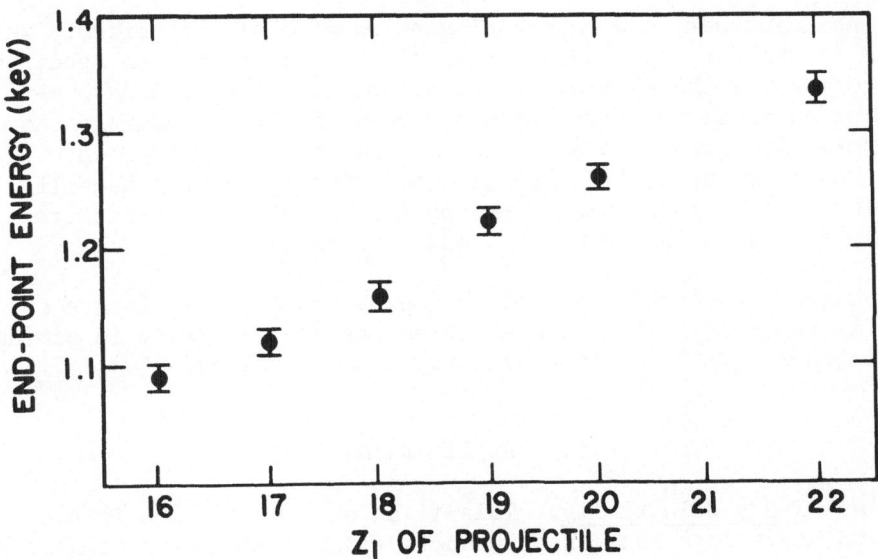

Fig. 13b   End-point energies of the non-characteristic radiation for 200 keV S, Cl, Ar, K, Ca and Ti bombardment of Si.

e) High Resolution X-Ray Measurements: Satellites.
Referring back to Fig. 10, we see an example of high resolution
X-ray spectral measurements, taken with a crystal spectrometer,
of Mn-L X-rays, generated by bombarding a manganese target with a
variety of projectiles.

The striking feature about this is that whereas the spectra
obtained from proton and electron bombardments are virtually
identical, those obtained from heavy ion bombardments are much
more complicated. Such spectra yield information concerning the
degree of electronic excitation which occurs as a result of the
collisions.

As a further, and rather elegant example of this effect, we
refer to the work of Knudson et al.[42] who used a crystal spectrometer to measure the aluminum K_α X-ray spectrum obtained by bombarding an aluminum target with 5 MeV Ne^+ ions. In addition to
the normally expected peak, arising from the transfer of an
electron from the filled L shell into a vacancy in the K-shell,
the found five satellites, corresponding to electron transfers
from L shells having 1, 2, 3, 4 and 5 vacancies.

Since the charge state of an ion has a strong influence on
its fluorescence yield, such measurements are important in aiming
for improved quantification of X-ray cross sections.

5.3 Applications

a) X-Ray Selectivity. This is an attribute which is peculiar to heavy ions and can be used to enhance analytical sensitivity
for certain elements of interest. It can be applied in at least
two ways:

(i) By choosing a projectile which is capable of exciting a
suitable X-ray from an element of interest whilst having a low
cross section for production of a potentially interfering X-ray
from another element in the sample. As an example of this,
Fig. 14[43] shows how Kr^+ ions at ~100 keV may be used to excite
SbM X-rays, even in the presence of an abundance of silicon,
since SiK X-rays are not excited at this energy.

(ii) By making use of correlation diagrams, and again choosing
the best projectile for the task in hand. Thus we see from the
correlation diagrams shown in Fig. 15 that during argon/boron
collisions, boron K X-rays should be excited, since the boron
1s electron may be promoted readily. On the other hand, during
argon/carbon collisions the carbon 1s orbital does not correlate
with higher orbitals and so carbon 1s electron promotion and thus
carbon K X-ray emission, is less likely. Indeed Fig. 16[10] proves

this deduction: it shows that whereas 100 keV H^+ bombardment of a boron/carbon target produces both boron K and carbon K X-rays, 100 keV Ar^+ bombardment of the same target produces boron K X-rays (and argon L X-rays, including a satellite), but no carbon K X-rays.

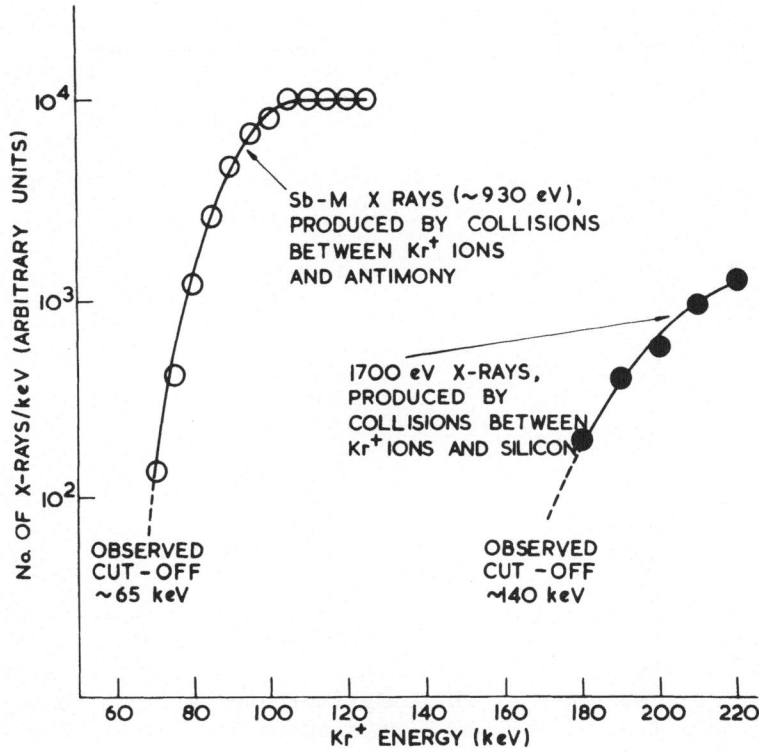

Fig. 14 X-ray selectivity of low energy (65 - 120 kV) Kr^+ ions for SbM X-ray production.

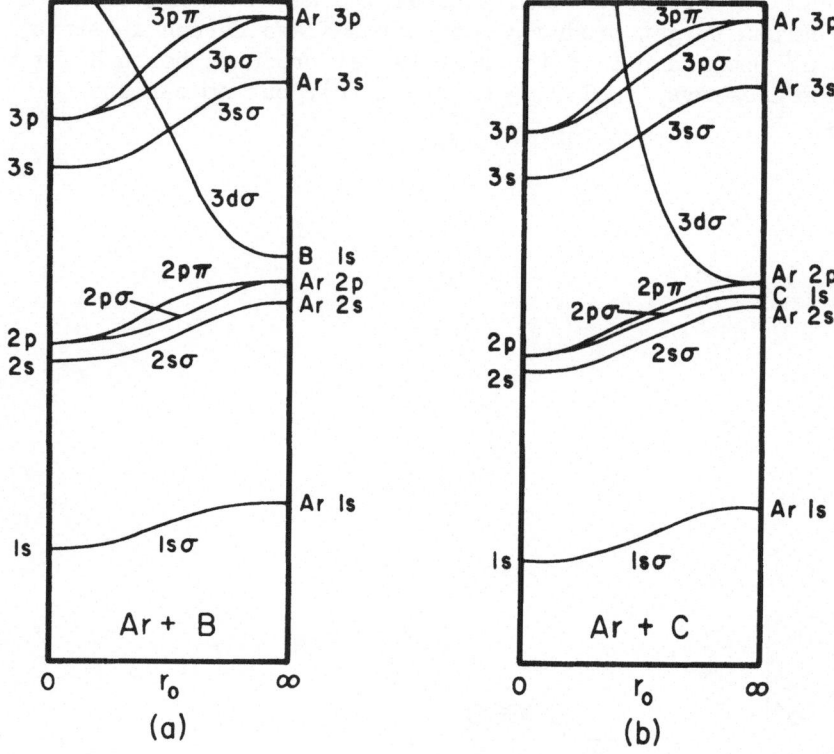

Fig. 15 Argon + boron and argon + carbon correlation diagrams.

b) <u>Surface Sensitivity</u>. If we refer again to Fig. 14, it is
apparent that the cross section for SbM X-ray production by Kr^+
falls dramatically as the Kr^+ loses energy. In practice, it was
found that for a specimen of antimony-implanted silicon bombarded
by 100 keV Kr^+, most of the SbM X-rays arose from within 100 Å of
surface. This surface sensitivity was enhanced further by the
absorption of the SbM X-rays within the silicon. Thus we see that
heavy ions can be used as a surface-sensitive probe, particularly
if used to generate soft X-ray components (for which the cross

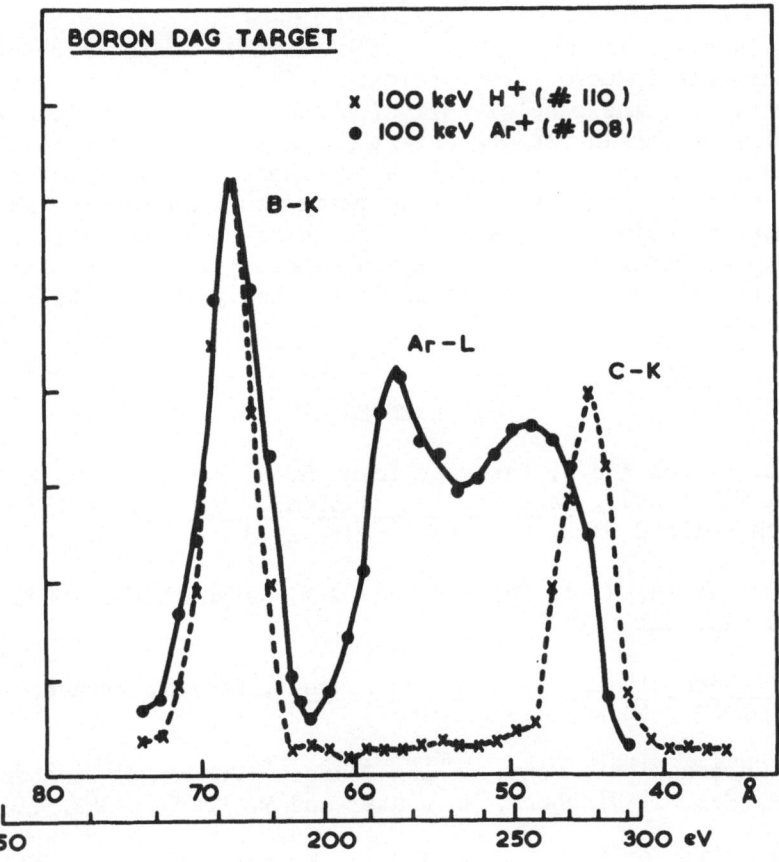

Fig. 16 X-ray production during bombardment of a boron/carbon target by 100 keV protons and 100 keV Ar$^+$ ions.

sections are often very high). On the other hand, it must be bourne in mind that the use of heavy ions involves the penalty of potential target removal by sputtering and also radiation damage, so it is essential to keep the doses low by maximizing X-ray collection efficiency.

6.   CONCLUSIONS

It is clear from a survey of the current physics literature that particle induced X-ray excitation is a subject of great interest. For the case of solids, protons and helium ions will be applied to an increasing extent for analytical purposes. For the heavy ion case, X-ray spectra provide a wealth of information on the collision processes taking place as an energetic ion traverses a solid. As the heavy-ion physics becomes better understood, applications will again be sought, perhaps making use of the selectivity available in the heavy-ion excitation mechanism.

REFERENCES

1. Q. C. Kessel and B. Fastrup, in M. R. C. McDowell and E. W. McDaniel (Eds.), Case Studies in Atomic Physics, 3, 137, North Holland Publ. Co., Amsterdam, 1973.

2. J. D. Garcia, R. J. Fortner and T. M. Kavanagh, Rev. Mod. Phys. 45, 111 (1973).

3. 1st International Conf. on Ion Sources, Saclay, France, June 1969 (INSTN Saclay, France).

   2nd International Conf. on Ion Sources, Vienna, September 1972, Eds. F. Viehbock, H. Winter and M. Bruck (SGAE, Vienna).

4. K. J. Hill, R. S. Nelson and R. J. Francis, AERE (Harwell) Report No. R6343 (1970).

5. J. A. Cairns, D. F. Holloway and G. F. Snelling, Nucl. Instr. and Meth., 111, 419 (1973).

6. A. J. Caruso and H. H. Kim, Rev. Sci. Instrum., 39, 1059 (1968).

7. M. A. Spivack, Rev. Sci. Instrum., 41, 1614 (1970).

8. B. L. Henke, R. L. Elgin, R. E. Lent and R. B. Ledingham, Norelco Reporter, 25, 112 (1967).

9. J. A. Cairns, C. L. Desborough and D. F. Holloway, Nucl. Instr. and Meth., $\underline{88}$, 239 (1970).

10. J. A. Cairns, A. D. Marwick and I. V. Mitchell, Thin Solid Films, $\underline{19}$, 91 (1973).

11. H. Kamada, T. Tamura and M. Terasawa in G. Shinoda, K. Kohra and T. Ichinokawa (Eds.), Sixth International Conference on X-Ray Optics and Microanalysis, University of Tokyo Press, 1972, p. 541.

12. E. Merzbacher and H. W. Lewis, in S. Flugge (Ed.), Encyclopedia of Physics, $\underline{34}$, 166, Springer, Berlin.

13. J. Bang and J. M. Hansteen, Mat. Fys. Medd. Dan. Vid. Selsk., $\underline{31}$, No. 13 (1959).

14. J. D. Garcia, Phys. Rev. A$\underline{1}$, 280 (1970) and Phys. Rev. A$\underline{1}$, 1402 (1970).

15. W. Brandt, R. Laubert and I. Sellin, Phys. Rev. $\underline{151}$, 56 (1966).

    W. Brandt and Grezgorz Lapicki, Phys. Rev. A$\underline{10}$, 474 (1974).

16. C. H. Rutledge and R. L. Watson, Atomic Data $\underline{12}$, 195 (1973).

17. E. Laegsgaard, J. U. Andersen and L. C. Feldman, Phys. Rev. Lett. $\underline{29}$, 1206 (1972).

    W. Brandt, K. W. Jones and H. V. Kramer, Phys. Rev. Lett. $\underline{30}$, 351 (1973).

18. J. F. Chemin, J. Roturier, B. Saboya and Q. T. Dien, Phys. Rev. A$\underline{11}$, 549 (1975).

19. S. Datz, J. L. Duggan, L. C. Feldman, E. Laegsgaard and J. U. Andersen, Phys. Rev. A$\underline{9}$, 192 (1974).

20. P. G. Burkhalter, A. R. Knudson, D. J. Nagel and K. L. Dunning, Phys. Rev. A$\underline{6}$, 2093 (1972).

    F. Abrath and T. J. Gray, Phys. Rev. A$\underline{9}$, 682 (1974).

21. F. Folkmann, C. Gaarde, T. Huus and K. Kemp, Nucl. Instr. and Meth., $\underline{116}$, 487 (1974).

22. D. M. Poole and J. L. Shaw, in 5th Intern. Congr. on X-Ray Optics and Microanalysis, Springer, New York, 1969, p. 319.

23. J. A. Cookson and F. D. Pilling, Thin Solid Films 19, 381 (1973).

24. T. A. Cahill, this book.

25. W. K. Chu, J. W. Mayer, M. A. Nicolet, T. M. Buck, G. Amsel and F. Eisen, Proceedings of International Conference on Ion Beam Surface Layer Analysis? J. W. Mayer and J. F. Ziegler (Eds.), Elsevier Sequoia S.A., Lausanne, 1973, p. 423.

26. L. C. Feldman, J. M. Poate, F. Ermanis and B. Schwartz, Thin Solid Films, 19, 81 (1973).

27. R. G. Musket and W. Bauer, Appl. Phys. Letters, 20, 411 (1972).

28. R. G. Musket and W. Bauer, J. Appl. Phys. 43, 4786 (1972).

29. R. G. Musket and W. Bauer, Thin Solid Films 19, 69 (1973).

30. J. A. Cairns and R. S. Nelson, Physics Letters 27A, 14 (1968).

    J. F. Chemin, I. V. Mitchell and F. W. Saris, J. Appl. Phys. 45, 537 (1974).

31. J. R. MacDonald, R. A. Boie, W. Darcey and R. Hensler (Phys. Rev. to be published).

32. U. Fano and W. Lichten, Phys. Rev. Letters, 14, 627 (1965).

33. M. Barat and W. Lichten, Phys. Rev. A6, 211 (1972).

34. J. A. Cairns, D. F. Holloway and R. S. Nelson, in D. W. Palmer, M. W. Thompson, P. D. Townsend (Eds.), Atomic Collision Phenomena in Solids, North Holland Publ. Co., Amsterdam, 1970, p. 541.

35. H. W. Schnopper and J. P. Delvaille in Atomic Collisions in Solids, S. Datz, B. R. Appleton and C. D. Moak (Eds.), Plenum Press, New York, 1975, p. 481.

36. J. Macek, J. A. Cairns, J. S. Briggs, Phys. Rev. Lett. 25, 1298 (1972).

37. K. Taulbjerg, B. Fastrup and E. Laegsgaard, Phys. Rev. A8, 1814 (1973).

38. F. W. Saris, W. F. van der Weg, H. Tawara and R. Laubert, Phys. Rev. Lett. 28, 717 (1972).

39. G. Bissinger and L. C. Feldman, Phys. Rev. Lett. <u>33</u>, 1 (1974).

     J. R. MacDonald, M. D. Brown and T. Chiao, Phys. Rev. Lett. <u>30</u>, 471 (1973).

40. R. J. Fortner and J. D. Garcia, in <u>Atomic Collisions in Solids</u>, S. Datz, B. R. Appleton and C. D. Moak (Eds.), Plenum Press, New York, 1975, p. 469.

41. J. R. Mowat, B. R. Appleton, J. A. Biggerstaff, S. Datz, C. D. Moak, and I. A. Sellin in <u>Atomic Collisions in Solids</u>, S. Datz, B. R. Appleton and C. D. Moak (Eds.), Plenum Press, New York, 1975, p. 461.

42. A. R. Knudson, D. J. Nagel, P. G. Burkhalter and K. L. Dunning, Phys. Rev. Lett. <u>26</u>, 1149 (1971).

43. J. A. Cairns, D. F. Holloway and R. S. Nelson, Radiation Effects, <u>7</u>, 167 (1971).

SUBJECT INDEX